EX Libris

SAF Kakar

SECOND EDITION

# Earth History and Plate Tectonics

## An Introduction to Historical Geology

**Carl K. Seyfert**
STATE UNIVERSITY OF NEW YORK
COLLEGE AT BUFFALO

**Leslie A. Sirkin**
ADELPHI UNIVERSITY

HARPER & ROW, PUBLISHERS

New York    Hagerstown    Philadelphia    San Francisco    London

Sponsoring Editor: Dale Tharp
Project Editor: Robert Ginsberg
Designer: Helen Iranyi
Production Manager: Marion A. Palen
Compositor: York Graphic Services, Inc.
Printer and Binder: Halliday Lithograph Corp.
Art Studio: J&R Services

**EARTH HISTORY AND PLATE TECTON-
ICS: An Introduction to Historical Geology,
*Second Edition***

Library of Congress Cataloging in Publication Data

Seyfert, Carl K 1938–
    Earth history and plate tectonics.

    Includes bibliographical references and index.
    1. Historical geology.   2. Plate tectonics.   I. Sir-
kin, Leslie A   joint author.   II. Title.
QE28.3.S47   1979       551.7       78-24612
ISBN 0-06-045921-2

# Contents

# Preface

The theory of plate tectonics, proposed in 1968 by W. Jason Morgan, states that the earth's surface is divided into a relatively small number of rigid plates which move relative to each other. Folding, faulting, volcanism, and mountain building take place principally at the margins of these plates. This theory is largely an outgrowth of Alfred Wegener's theory of continental drift. In 1915, Wegener proposed that during the Paleozoic all of the continents were part of a single, large landmass, which he called Pangaea. When this landmass broke apart during the Mesozoic, the resulting continents moved (drifted) to their present positions. Wegener also proposed that the geographic poles have moved (wandered) relative to the continents. He based his conclusions largely on the study of ancient climates as revealed by fossils and rocks from the continents. Wegener's ideas were not widely accepted at the time, and in fact were met with considerable resistance by a scientific community entrenched in the belief of permanence of the continents and ocean basins.

During the 1960s, study of the ancient magnetic field of the earth and of the magnetics of the ocean floor stimulated renewed interest in Wegener's ideas. Data dealing with the ancient magnetic field of the earth, that is, *paleomagnetism,* provided strong support for Wegener's concepts of continental drift and polar wandering. Marine magnetics revealed impressive evidence that large blocks of seafloor have moved relative to each other and that new seafloor is continuously being added at the crest of mid-ocean ridges.

When the first edition of this text was being written, the ideas of continental drift were still rather controversial. Today, however, most geologists accept continental drift and polar wandering, along with the newer concepts of plate tectonics and seafloor spreading. They disagree somewhat on how, when, and why these movements have taken place.

Since the first edition of this book appeared, the concept of the Wilson cycle has been proposed and has gained wide acceptance. The Wilson cycle describes the opening and closing of ocean basins throughout geologic time. Thus, evidence may now be presented that ocean basins have opened and closed at least five and perhaps as many as seven times in the last 3.8 billion years. Times of ocean opening are generally associated with widespread intrusion of basaltic dikes and extrusion of basaltic flows, and times of ocean closing are followed by continental collision and mountain building.

Acceptance of the concepts of plate tectonics and seafloor spreading has required a great many revisions in the account of the physical history of the earth. We have undertaken the task of incorporating the idea of large-scale crustal movements into the content of historical geology. We have attempted to trace the history of continental movements and mountain building from the time of the oldest known rocks (of Precambrian age) to the present. We also explore the effects of plate tectonics on long-range climatic changes and on the evolution of life. Today, a substantial body of literature has evolved which traces the role of continental drift, seafloor spreading, polar wandering, and plate tectonics through geological time. In addition to discussing these ideas, we have presented some new ideas as "food for thought" on such topics as Precambrian continental configurations and causes of plate movements. Some geologists may object to incorporating unproven ideas into a historical geology text. However, such hypothetical ideas are clearly labeled, and we feel that exposing readers to new ideas, provided they are not presented as fact, can heighten their appreciation of the subject matter.

We have chosen to take a worldwide view of the history of the earth, rather than to discuss the historical geology of a single continent. It is not possible to discuss a continent as an isolated body because its history has been affected by other continents. Whatever continent is studied, it is found that sources of sediment, glaciers, and mountain-building forces are traced beyond the limits of that continent. By taking a global approach to earth history, and by applying the concepts of plate tectonics, many previously unsolved problems, such as the origin of mountains, causes of glaciation, and origin of ocean basins, are readily explained. The science of historical geology is, however, in a constant state of rapid change as a result of the reevaluation of geologic data in the light of plate tectonics. We have come to realize that geology is truly a dynamic and evolving science that is constantly being refined by new insights, and it is not merely a body of rigid facts that are unsusceptible to change.

It is a major purpose of this text to integrate modern and classical concepts of geology and astronomy into a workable earth history. We have seen the earth from space, with its land, water, and air masses clearly delineated. We have seen close up the crater-strewn surfaces of the moon, Mars, and Mercury, and with radar observations of Venus we know of the presence of craters there as well. We are compelled to wonder to what degree the history of the earth has been influenced by meteorite impacts. There is little question that its early history was dominated by impacts, but we will also examine the possibility that the impacts of large meteorites (small asteroids) may have had an important effect on later earth history as well.

In order to make this text as useful as possible for both the science major and the general student, it has been divided into two parts. The first seven chapters introduce the reader to the fundamental concepts of historical geology. The remaining chapters are devoted to a discussion of earth history. Included in the second part are the history of plate movements, location and shape of ancient landmasses, ocean basins and mountains, and the evolution of plants and animals.

We wish to thank the many people who helped in the preparation of this text. Sections of the first and/or second edition have been read critically by Parker E. Calkin, G. Gordon Connally, Rhodes W. Fairbridge, Robert M. Finks, Richard J. Foster, Dennis S. Hodge, William M. Kaula, John E. Mack, Franklin C. MacKnight, Donald D. Mitchell, James R. Orgren, Allison R. Palmer, Paul H. Reitan, Harold F. Roellig, William D. Romey, Holmes A. Semken, James W. Skehan, James R. Spotila, Robert C. Stein, Irving H. Tesmer, Fred R. West, Grant M. Young, and E-an Zen.

Rotations of coastlines and paleomagnetic poles were performed using the HYPERMAP program of Robert L. Parker modified by Paul Orgren. Karen M. Seyfert, the wife of Carl Seyfert, provided much valuable assistance editing the manuscript and drafting many of the illustrations.

CARL K. SEYFERT
LESLIE A. SIRKIN

# Introduction 1

I N THE 1960s the geological sciences experienced a revolution comparable to the revolution in astronomy that resulted from the Copernican concept of a heliocentric universe, the revolution in biology following Darwin's work on evolution, and the revolution in physics caused by Einstein's theory of relativity. The revolution in geology has resulted from the acceptance of the concept of *plate tectonics*. The central idea in this concept is that the earth's crust is divided into a number of comparatively rigid plates and that these plates move relative to each other. Polar wandering, continental drift, and seafloor spreading are all aspects of this movement. These ideas have radically changed our views on the stability of the earth's crust. For years geologists believed the earth to be a rigid body in which continents and ocean basins were permanent, but the concept of plate tectonics has provided a dynamic rather than a static model for the evolution of earth's continents and ocean basins.

The German meteorologist Alfred Wegener was the first to carefully compile geologic and paleoclimatic evidence that supported the concepts of polar wandering and continental drift. Wegener proposed a sequence of reconstructions of the continents that showed their relative positions at various times in the past, including reconstructions that showed the breaking apart of a supercontinent, Pangaea. Consequently, the revolution in geology has been attributed to the work of Wegener. He first published his ideas in 1915, but it took almost 55 years for the concepts of continental drift and polar wandering to gain general acceptance among geologists. This did not occur until oceanographic and seismological studies of the ocean basins brought forth impressive evidence documenting the spreading of the seafloor.

One of the important factors in the recent advances in the geological sciences is the interdisciplinary nature of many of the investigations. Studies by scientists in such fields as geophysics, geochemistry, geobiology, astrogeology, and astrophysics have contributed to our understanding of the origin and development of the earth. Paleomagnetic, paleontologic, paleoclimatic, seismological, and geochronological studies have provided additional support for Wegener's model of the classical episode of continental drift. Interdisciplinary research has also benefited investigations into the history of crustal plate movements prior to the break up of Pangaea. As plates join together, mountain belts are created along converging margins and as they are rifted apart, new ocean basins form. Impressive evidence has been accumulated in recent years indicating that the plates have experienced a number of episodes of joining together and rifting apart. Interdisciplinary interpretation of data gathered by recent space probes and manned lunar landings has helped in the reconstruction of the early history of the earth, about which we had

**FIGURE 1.1**  Fault scarp produced during a violent earthquake near West Yellowstone, Montana, in August, 1959. The uplift of the block in the background occurred in less than a minute. (Photo by J. B. Hadley, courtesy of the U.S. Geological Survey.)

little information due to the absence of rocks older than about 3.8 billion years in the earth's crust. These studies have confirmed the theory of impact origin for most lunar craters and have shown the presence of large craters of probable impact origin on Mars, Venus, and Mercury. Many scientists now consider that meteorite and comet collisions played an important part in earth history.

## HIGHLIGHTS IN EARLY GEOLOGIC STUDIES

Prior to the seventeenth century, investigations of the earth were confined primarily to descriptions of small-scale geologic features, with minimal speculation as to their origin. There was little scientific value in these studies, since a framework did not exist into which new ideas and observations might be placed. It was not until the seventeenth and eighteenth centuries that the fundamental geological principles were established. These principles provided the basis

for assembling a reliable account of the formation and development of the earth and the life on it.

The ancient Greek and Roman philosophers believed that the earth was at the mercy of capricious gods, a supposition that all but discouraged the study of geologic features. However, Xenophanes in the fifth century B.C. studied the fossil fish and seashells of Italy and concluded that the sea had once covered the areas where these fossils were found.

The importance of water in shaping the land, both by erosion and deposition, was perceived in the fourth century B.C. by Aristotle who pointed out that the Nile River Delta had been built up by the slow deposition of sediments from the river. He also suggested that deposition of these sediments displaced seawater and thus caused a rise in sea level. Strabo, a Greek who lived in Asia Minor during the first century A.D., stated that land was often uplifted following earthquakes (Fig. 1.1), and that this type of uplift had been important in shaping the

**FIGURE 1.2** Raised marine terraces on Middleton Island, Alaska Gulf Region, Alaska. Such terraces provide evidence of slow uplift of the land relative to the sea. (Photo by S. R. Capps, courtesy of U.S. Geological Survey.)

earth. He also recognized that land could be uplifted quite slowly without perceptible disturbance (Fig. 1.2), and that active volcanoes (such as Mt. Etna) have structures similar to ancient volcanoes.

With the fall of the Roman Empire, the people of Europe were more concerned with defense against foreign invasions than with the cause of science. Although the center of learning shifted to the Middle East for several hundred years, little scientific progress was made beyond that of the Greeks. During the eleventh century, the philosopher Avicenna noted that soft mud hardened into stone with time and that water dripping from the roof of a cave also turned into stone. Avicenna concluded that the cause of the petrification was a mystic "congealing virtue." His ideas concerning erosional processes were more sound. For example, he stated that mountain peaks were shaped by wind and streams over a long period of time and that softer rocks were eroded to form valleys between ridges of more resistant rocks.

Four hundred years after Avicenna, Leonardo da Vinci (1452–1519) observed and speculated on the origin of a number of geologic

features. He recognized the organic nature of fossils and postulated that marine forms which were found in rocks well above sea level indicated either that the land had been uplifted or that sea level had been lowered.

## STENO'S PRINCIPLES

It was not until the seventeenth century that detailed investigations were made of rocks and the fossils they contain. In 1669 Nicolas Steno published a study of the mountainous region around Tuscany, Italy, and proposed in this work two very basic geologic principles. The first is that beds of sedimentary rock are initially deposited in horizontal layers (Fig. 1.3). This principle, known as the *principle of original horizontality,* is of great importance in establishing whether or not rocks have been deformed. In general, if rock layering is tilted or arched upward into an anticline or bowed downward into a syncline it can be assumed that the rocks were not deposited that way but were folded by compressive stresses. There are, however, exceptions to this rule. Sedimentary layering may have a gentle initial dip if beds

**FIGURE 1.3** The horizontal strata of the Grand Canyon record 300 million years of geologic history.

were deposited on a slope with a gentle initial dip. Furthermore, beds, especially sandstones deposited in a desert environment, may have laminations (cross-stratification) with initial dips of up to 30° or so, but these are generally easy to distinguish from normal bedding planes (see Chapter 4).

Steno's second important principle is the *principle of superposition,* which states that younger beds are deposited on top of older beds. This principle, like the first, seems fairly obvious, but it is very important in the determination of relative ages of rock layers. It also helps establish whether or not folding has caused beds to be overturned (turned upside down) and if they have been displaced by a fault. During intense folding, rock layers may be overturned so that older beds overlie younger beds. Application of the principle of superposition leads to the conclusion that overturned beds were not deposited that way, but rather that they were deformed by folding.

Reverse faulting also commonly results in displacement of older beds over younger beds.

There are other cases in which older beds overlie younger beds, but these are not exceptions to the principle of superposition since they were not originally deposited that way. For example, younger sediments may be deposited beneath older beds in the roof of a cave, and younger beds may be deposited beneath older beds below an overhang, such as might occur along the banks of a stream (Fig. 1.4).

Steno was far ahead of his time however, and most of his ideas were rejected by his contemporaries. It was not until the nineteenth century that these principles finally received general acceptance.

## CATASTROPHISM

Until the nineteenth century, clergymen and scientists alike believed that the earth had been created only a few thousand years ago. With

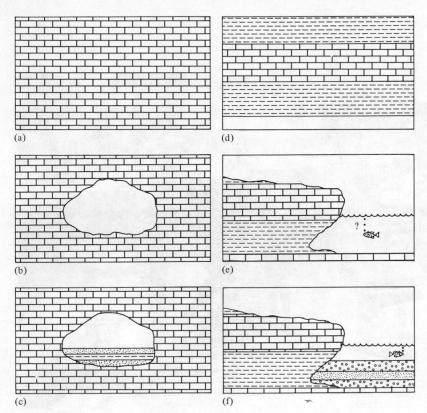

(a)

(b)

(c)

(d)

(e)

(f)

**FIGURE** 1.4 Diagrams showing two cases in which younger sediments are deposited below older sediments. These are two exceptions to the principle of superposition of beds.

so little time available for the formation and development of the earth, it is not surprising that many scientists believed that sudden, violent forces (catastrophes) had shaped the earth's surface. This philosophy has become known as "catastrophism." Some catastrophists believed that canyons, such as the Grand Canyon, were simply giant cracks in the ground formed during a series of violent earthquakes and that tidal waves (tsunamis) and floods were caused by huge meteorites or comets that struck the earth.

Other catastrophists postulated that the thick sequences of sedimentary rocks in the earth's crust had been deposited during the worldwide flood depicted in the Bible, that is, Noah's flood. They thought that fossils found in these rocks were the remains of organisms killed by the deluge. Still others realized the difficulty of accounting for the apparent changes in fossils

from older to younger beds with only one flood. William Buckland, a theologian who became the popular professor of geology at Oxford University during the early nineteenth century, accommodated this discrepancy by adding two more worldwide floods to the original deluge. This led to the belief that practically all life had been successively wiped out by such floods and replaced by new forms, which resulted from separate, divine creations.

Georges Cuvier (1769–1832), a leading vertebrate paleontologist of his day, proposed that all the organisms that had existed from time to time in Europe had perished in a series of such floods. Cuvier postulated that these organisms had subsequently been replaced by migrations from other localities. He believed that the floods were caused by "instantaneous rather than gradual" episodes of crustal subsidence,

**FIGURE 1.5** Anchorage, Alaska, following the "Good Friday" earthquake of March 27, 1964. Such displacements are produced by the slumping of unconsolidated sediments as a result of the passage of seismic waves. (Courtesy of the U.S. Geological Survey.)

which alternated with sudden crustal uplifts. Cuvier was an influential man whose catastrophist teachings were widely accepted in both Europe and North America.

Although most of the catastrophists' proposals were related to observation of natural events, they were generally distorted by gross exaggerations in scale. For example, uplift and subsidence of the crust do occur, but generally at a very slow rate. Earthquakes do cause cracks in the earth, but the width of such cracks is generally not more than a few feet (Fig. 1.5). Floods are common, but always limited in extent, never engulfing an entire continent, much less the entire earth.

## NEPTUNISTS

The neptunist school was established in the late eighteenth century by Abraham Werner, a professor at the Freiberg Mining Academy. He adopted a theory, originally proposed by Johann Lehmann, which stated that the majority of the earth's rocks had been chemically precipitated in a universal sea. Although he had observed only the rocks in the vicinity of Freiberg, Werner assumed that the rock formations were the same everywhere. A persuasive and popular teacher, Werner presented his theories with authority and tolerated no criticism. Students came from all over Europe to study under this man who inspired so much confidence in his ideas.

Werner called the oldest formation of the earth the "Primitive" or "Primary." It consisted of granite and presumably had been precipitated out of a murky, universal sea onto the nuclei of the continents. It was assumed to have become subsequently mixed with gneiss, schist, porphyry, basalt and marble. The overlying "Transitional" strata also consisted mainly of precipitates, but included some sediments weathered from the primary islands that were exposed as the sea retreated. Rocks belonging to the Transitional strata included sparsely fossiliferous slate, graywacke, and limestone.

According to the neptunists, sandstone, conglomerate, limestone, chalk, and coal became more abundant as the seas retreated further. This unit, the *"Flötz,"* was characterized by steeply inclined bedding planes, due presumably to precipitation on the slopes of undersea mountains and to slumping of soft sediments on inclined slopes. The youngest deposits, the "alluvial," were deposited by streams following the retreat of the sea from the continents. This category included clay, peat, sand, gravel, and some volcanic materials. Werner believed that volcanic eruptions were of minor importance and were caused by the burning of coal beds in the *"Flötz."* In spite of his apparent lack of acceptance of magmatic processes, Werner stands as an early geochemist in his understanding of the process of chemical precipitation.

## UNIFORMITARIANISM

In opposition to the geological hypotheses of the catastrophists and the neptunists, a school of geologists emerged in England that stressed the uniformity of nature. They reasoned that the development of the earth's crust could be best understood by observing geologic processes at work today. In 1788 James Hutton, a Scottish geologist, published a paper entitled "Theory of the Earth"; this paper laid the groundwork for the modern science of geology and led to the eventual abandonment of neptunism and catastrophism.

The formation of ancient rocks, according to Hutton, could be explained without invoking any processes other than those which could be directly observed. Hutton believed that most geological phenomena could be understood through careful observations of modern processes, and that all geologic processes that operated in the past also operate at the present time. Faith in the constancy of these processes became the essence of the *principle of uniformitarianism.* The history of the earth could then be interpreted without positing unpredictable catastrophes, divine intervention, or processes that could not be observed or tested. The phrase "the present is the key to the past" sums up the concept of uniformitarianism.

When Hutton published his ideas, Werner had been teaching at Freiberg for 20 years, and the neptunist philosophy was widely accepted. In opposing neptunism, Hutton insisted that granites had crystallized from hot molten rock reservoirs or "magmas" situated deep in the earth and that basalts had flowed from volcanoes whose superficial parts had, in many cases, long since been obliterated by erosion (Fig. 1.6). He further observed that sand and mud were transported to the sea by streams and rivers and that these sediments were deposited in the ocean. Ultimately, sandstone and shale were formed by the hardening or "lithification" of these sediments, whereas the cementation of tiny fragments of calcareous shells produced limestone. Hutton's ideas seemed impossibly overdramatic to the neptunists, who persisted in the belief that all important rock types were chemical precipitates from cold water.

Even though Hutton's ideas were backed by careful field observations, his paper was written in such a difficult style that it was not widely read. An associate, John Playfair, realized this fault and in 1802 published what has become a

**FIGURE 1.6** Vertical granitic dike cutting Triassic sandstones and diabase. Observations such as this led Hutton to conclude that granites were intruded in the molten state rather than precipitated on the sea floor. (NERC Copyright. Reproduced by permission of the Director, Institute of Geological Sciences, London.)

scientific classic, *Illustrations of the Huttonian Theory of the Earth* **(1)** in which Hutton's ideas were clarified.

Uniformitarianism received increasing support from a group of geologists who were known as the "plutonists" because of their belief in the importance of heat as a factor in the formation of certain rocks. (Pluto was king of the underworld in Roman mythology.) The debate which ensued between the neptunists and plutonists was one of the most bitter and drawn out clashes in the discipline of geology. The neptunists persisted in claiming that basalts were chemically precipitated from the sea, and the plutonists believed them to be crystallized from magma.

The work of Nicholas Desmarest on the basalts of Auvergne, France, in the latter part of the eighteenth century, greatly strengthened the plutonists' case. Desmarest traced basalt to ex-tinct volcanic cones, which were identical in form to active cones such as Mt. Vesuvius, and he concluded that the basalt had crystallized from lava. Such empirical observations, combined with Werner's eventual retirement as a teacher, led to the ultimate abandonment of neptunism.

Gradually catastrophism also lost considerable support in favor of uniformitarianism. Charles Lyell used the principle of uniformitarianism as the unifying theme in his highly successful textbook, *Principles of Geology,* the first volume of which appeared in 1830. Lyell became interested in geology when, as a law student at Oxford, he took a course in geology from William Buckland. Vivid descriptions by the Reverend Buckland of the conflict between the neptunists and the plutonists sparked Lyell's interest.

Lyell used one of his vacations to study the problem in the field and became convinced of the value of direct observation. Becoming a staunch supporter of uniformitarianism, he postulated that geologic processes occurred at about the same rate in the past as they do at present. Uniformity in the rate of change became known as *gradualism,* a concept which is often confused with uniformitarianism and which is now thought to be an oversimplification. There is much in the geologic record to indicate that geologic processes do *not* always operate at the same rate today as they did in the past. Volcanic activity, glaciation, and sedimentation, among other processes, have varied considerably throughout geologic history.

## GEOLOGIC MAPS

Before the eighteenth century, the geologic literature consisted of descriptions of the rocks at different localities. No attempt was made to relate the rocks of one locality to those of another. In the latter part of the eighteenth century, William Smith, an English surveyor, observed that definite relationships existed between rock units in different localities. While supervising the building of a canal in the west of England, Smith traced certain rock units over hundreds of square miles and always found them in the same distinctive sequence.

In his travels throughout the country, Smith also observed that each stratigraphic level had its own distinctive fossil content. These observations gave rise to the *principle of faunal and floral succession*. Fauna and flora constitute all of the animal and plant species living in a given place at a given time. The principle of faunal and floral succession states that in areas not disturbed by faulting or folding, fossil faunas and floras succeed one another in a definite, orderly, and predictable progression with the more complex, highly evolved species found in rocks overlying those containing simpler, less advanced species.

Using lithologic character and fossil content as a means of tracing strata, William "Strata" Smith, as he came to be known, published in 1815 a hand-colored geologic map of England, Wales, and part of Scotland. The map showed the boundaries between different rock units (Fig. 1.7). The accompanying detailed descriptions of the rock units, which included thicknesses, rock types, and fossils, were the forerunners of modern *columnar sections* (Fig. 1.8). These vertical sections graphically depicted the sequence and resistance to weathering of rock units from the oldest on the bottom to the youngest on the top of the column. Smith also made *geological cross sections* showing the rock strata exposed in hypothetical vertical slices in the earth's crust. Through the use of geologic maps, columnar sections, and cross-sections, Smith illustrated the stratigraphy and outcrop pattern of the rock units of England and provided the science of geology with some of its most valuable tools.

## BUILDING THE DATA BANK

The nineteenth century and early twentieth century may be characterized as a time of extensive geological discoveries brought about by field mapping in previously unexplored parts of North America, South America, and elsewhere. The science of geology began to diversify and geologists began to make detailed reconstructions of earth history. The Swiss scientist Louis Agassiz learned about Alpine glaciation and the deposits of former glaciers from the German naturalist Charpentier. This experience led him to develop a "theory of the ice ages" in which he proposed that glaciers covered much of Europe and North America during a time when worldwide climate was much cooler than today.

Many European geologists, such as Charles Lyell and Louis Agassiz, visited North America and influenced the growing science of geology through their active field investigations and their teaching. William McClure traveled widely and wrote of the geology of North America. Eventually he went on to found the American Geological Society (now called the Geological Society of America). As geological mapping extended westward past the Appalachian Mountains, David Owen, a Scot who was associated with McClure, mapped the Paleozoic strata and coal beds in the midcontinent of North America. William Logan mapped the ancient rocks of the Canadian Shield and produced the first geological map of Canada. Another geologist of Scottish descent, Hugh Miller, discovered and described many forms of ancient life including an early form of fish, the ostracoderm.

Many American geologists were influenced by the developments in the science overseas, and many contributed significant new concepts to the evolving science. Benjamin Silliman, a professor at Yale University, lectured to audiences across the United States on geological phenomena, and founded the American Journal of Science, a journal devoted to geology. Similarly inspired to teach as well as study geology was Amos Eaton, who mapped the geology of New York State along the Erie Canal. He also founded a school to train geologists and engineers, the Rensselaer Institute (now RPI). One

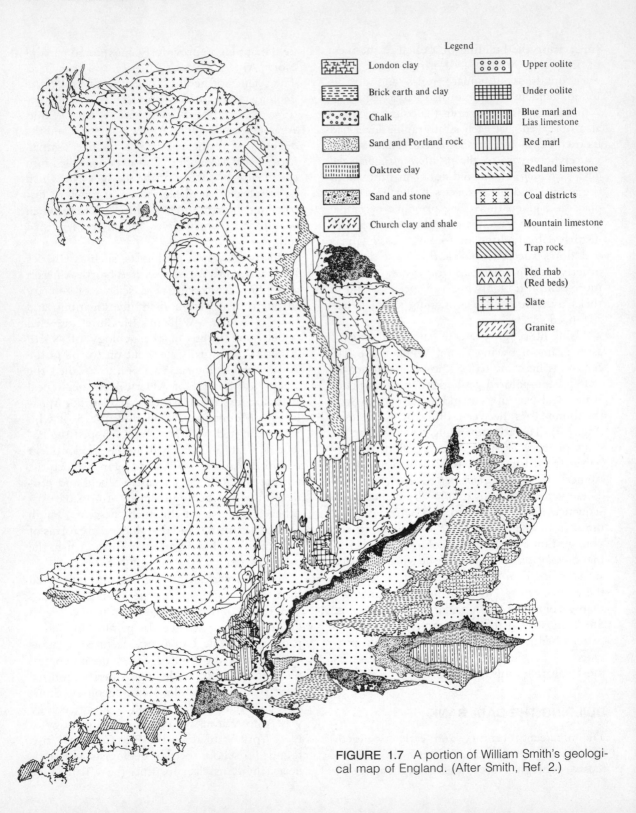

Legend

| | | | |
|---|---|---|---|
| London clay | | Upper oolite | |
| Brick earth and clay | | Under oolite | |
| Chalk | | Blue marl and Lias limestone | |
| Sand and Portland rock | | Red marl | |
| Oaktree clay | | Redland limestone | |
| Sand and stone | | Coal districts | |
| Church clay and shale | | Mountain limestone | |
| | | Trap rock | |
| | | Red rhab (Red beds) | |
| | | Slate | |
| | | Granite | |

FIGURE 1.7 A portion of William Smith's geological map of England. (After Smith, Ref. 2.)

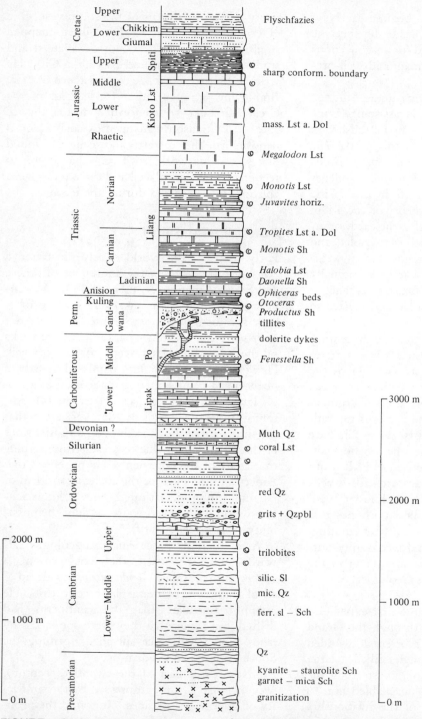

**FIGURE 1.8** An example of a columnar section. This one is of deposits in the Spiti region in the western Himalayas. (After A. Gansser, Ref. 3.)

of his protégés, James Hall, became the New York State paleontologist, and he became known throughout the world for his collections and descriptions of the abundant and well-preserved fossils of New York.

James Dana studied geology under Silliman and while he was Silliman's assistant, Dana wrote the *System of Mineralogy* first published in 1837. A condensed version of this work, *The Manual of Mineralogy,* after numerous revisions is still a widely used textbook in many colleges today. Dana studied coral reefs in the Pacific ocean and developed a new theory on the origin of mountains. He believed that the earth originated in a molten state, and as it cooled and contracted it formed mountains due to wrinkling of the crust where land and sea originally met. This idea gained widespread acceptance at the time, but has now been largely rejected in the light of plate tectonics.

The second half of the nineteenth century saw the opening of the western United States with the railroad pushing its way across the plains and mountains to the Pacific. Geologists led or accompanied railroad and government sponsored surveys along the route of the railroad. Among the geologists were Clarence King, a student of Dana, and Ferdinand Hayden, who was an assistant of James Hall. Both of these men discovered many rich fossil beds and sent tons of previously unknown fossil specimens, including many dinosaur bones, east to universities and museums for later curating. Hayden led the Geology and Geography Survey of 1869, and King became director of the newly formed United States Geological Survey in 1877.

John Wesley Powell led the first expedition down the Colorado River through the Grand Canyon in 1868. Traveling in small wooden dories, Powell mapped and described a large section of the American Southwest along the Colorado River. His mapping enabled him to devise a workable theory of the relationship between mountain building and erosion, and he also correctly ascertained the fault origin for many of the mountains of the southwest. One of Powell's assistants, G. K. Gilbert, mapped significant parts of the Colorado Plateau and studied the region around Salt Lake City. He was the first to recognize the existence of Lake Bonneville, a much larger predecessor of the Great Salt Lake which formed during the ice ages. R. D. Salisbury mapped extensive glacial deposits in the eastern and central United States, and his work led T. C. Chamberlin to recognize that four glacial and three interglacial episodes had occurred during the ice ages.

## CONTINENTAL DRIFT

Early maps of the world clearly illustrated a "jigsaw puzzle fit" of the shorelines of Africa and South America. As early as 1801, Alexander von Humboldt among several other prominent scientists of his day pointed out that not only were the coastlines parallel, but the rocks of the opposite coasts were similar as well. However, von Humboldt postulated that these phenomena had resulted from erosion of a single large continent by marine currents (4). Antonio Snider-Pellegrini in 1858 suggested that the similarity of the shorelines on opposite sides of the Atlantic was due to a catastrophic splitting apart of the continents. He proposed that America might even be the ancient continent of Atlantis (3). At that time, however, the scientific world did not seriously consider the possibility of continental drift.

As early as 1885, the Austrian geologist Edward Suess proposed the existence of a supercontinent called *Gondwanaland* made up of all of the southern hemisphere continents. He did not postulate that the continents had drifted, but rather that they were connected by lands which had since subsided, forming the South Atlantic and Indian oceans.

In the early part of the twentieth century, Wegener also became impressed by the similarity of opposing coastlines. He found the problem intriguing and for several years gathered data on ancient climates, paleontology, and the

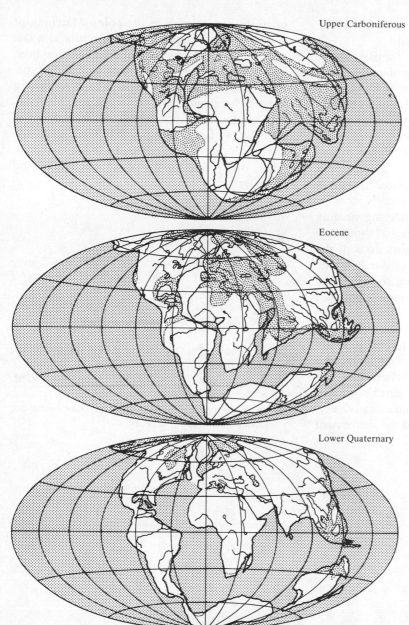

Upper Carboniferous

Eocene

Lower Quaternary

**FIGURE 1.9** Wegener's reconstruction of the continents during three different intervals of time. (After Wegener, Ref. 5, printed by permission of Dover Publications, Inc., New York.)

structural history of the continents. In 1915, Wegener put forth his observations and interpretations in a book entitled *Die Enstehung der Kontinente und Ozeane* (*The Origin of Continents and Oceans*). He proposed that the present continents once comprised one large landmass, which he named Pangaea (Fig. 1.9).

According to Wegener, Pangaea began to break apart and the individual continents started to move toward their present positions during the Mesozoic era. Wegener's work also provided evidence of movement of the earth's rotational poles relative to Pangaea. It is now believed that there is both a shift of the crust of the earth relative to the poles as well as displacement of individual continents.

Wegener's book, which was translated into English in 1924, created a storm of controversy among geologists and geophysicists. Most of them rejected Wegener's conclusions and the evidence supporting continental drift because the author had failed to offer a really convincing mechanism, and because of some minor inconsistencies and errors.

With the work of numerous later investigators (see Chapters 6 and 7), Wegener's ideas on continental drift and polar wandering became widely accepted. What is remarkable about Wegener's work is that his reconstructions of the continents, which were based on far less data than there is available today, are so much like reconstructions of the continents published within the last few years.

Since the time of Wegener, the science of geology has matured and proliferated in many areas as a result of the works of numerous geologists, geophysicists, geochemists, and others. Measurements of the earth's ancient magnetic field have provided quantitative evidence of seafloor spreading and the movement of continents relative to the poles. Matching of continental outlines in reconstructing the last major episode of continental drift has now been supplemented by the correlation of rock type, radiometric age, and structure in determining the relative position of crustal plates. The concept of plate tectonics, which was introduced in 1968 by W. Jason Morgan, has been substantiated by many new discoveries about the earth. These discoveries are discussed in the following chapters.

## REFERENCES CITED

1. J. Playfair, 1956, *Illustrations of the Huttonian Theory of the Earth:* Facsimile reprint by University of Illinois Press, Urbana.
2. W. Smith, 1815, *A Delineation of the Strata of England and Wales with Part of Scotland:* S. Gosnell, London.
3. A. Gausser, 1964, *Geology of the Himalayas:* Wiley, New York.
4. A. V. Carozzi, 1970, *Geo. Soc. Amer. Bull.,* v. 81, p. 283.
5. A. Wegener, 1966, *The Origin of Continents and Oceans:* Dover, New York.

# Evolution and the Fossil Record

**2**

Fossils were once defined as any strange object dug out of the earth. This definition covered inorganic structures, such as crystals and concretions, as well as the remains of pre-existing life. In modern usage, the term *fossil* denotes any evidence of former life. *Paleontology* is the study of ancient life based on fossil evidence. In the development of the geological sciences, paleontology has provided an indispensable means of determining relative age of rock units. Fossils are also an aid in geologic mapping and in the reconstruction of ancient geography, environments, and climates. They provide basic evidence in support of the concepts of evolution and continental drift.

## THE NATURE OF FOSSILS

Early naturalists speculated about the nature of fossils, but few pursued any kind of systematic study or understood the organic nature of such specimens. Some had remarkably bizarre ideas. One sixteenth century writer thought that fossils were the remains of organisms that grew within a rock, after having been carried there by germ-laden air. Medieval scholars thought that fossils were the remains of poor, unfortunate individuals who had been trapped in Noah's flood. These men were not entirely wrong, because some mode of burial is necessary for the preservation of fossils. However, such early explanations of the origins of fossils were largely the result of attempts to fit the history of life into a 6000-year period, in accordance with contemporary notions of the age of the earth.

As late as the eighteenth century, Johann Beringer, a professor at the University of Wurzburg, maintained that fossils need not have an organic origin. His evidence consisted of replicas of the sun, moon, planets, and Hebrew letters, which he found along with true fossils. Beringer was so fascinated by the "fossils" he found that he published descriptions and illustrations of the specimens (Fig. 2.1). Some time later, Beringer found a "fossil" in his own likeness with his name on it. Only then did he realize that the specimens had been "planted" where he could find them. Eventually, the culprits turned out to be two envious colleagues (1).

The paleontological sciences have often been subjected to this type of hoax. Two other infamous frauds were the Piltdown man, a faked "missing link," and the Cardiff giant, a solid stone replica of a "giant man fossil."

## METHODS OF FOSSILIZATION

Fortunately, fossils are quite common in spite of the vast amount of organic debris that is destroyed and eventually reenters the organic cycle. The most important factors in fossilization are rapid burial and the presence of hard parts (resistant structures) in the original organism. Rapid burial prevents the remains of an organism from being broken and scattered or

**15**

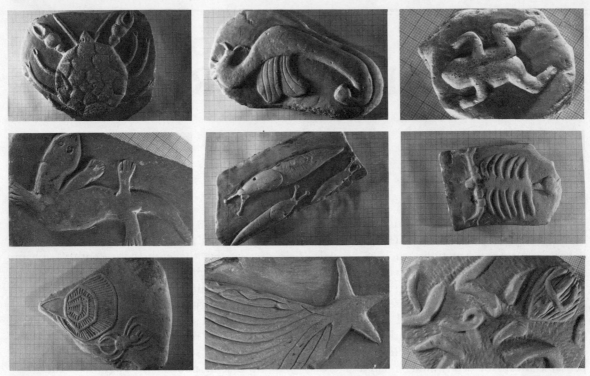

**FIGURE 2.1** Johann Beringer's "fossils." (From Jahn, Ref. 1. Originally published by University of California Press; reprinted by permission of the Regents of the University of California.)

destroyed by weathering. Even after burial, the fleshy soft parts of an organism are decomposed by bacteria, so that generally only the hard parts are preserved. In rare instances, however, entire organisms have been preserved.

### Original Remains

Essentially unaltered calcareous shells of clams and snails are common in relatively young sedimentary deposits (Fig. 2.2a). Such shells may retain the original irridescent mother-of-pearl luster, the leathery connective tissues, and the chitinous outer layers. In older deposits, the original shell may have been recrystallized, but it generally has the same composition as the original shell.

Occasionally decay has been retarded to the extent that the soft parts of an organism retain their original form, although greatly reduced in volume as a result of dehydration. This type of preservation may occur in dry caves, bogs, and tar deposits. Freezing has preserved the entire remains of large Pleistocene mammals in the permafrost of the Arctic region. Some of these animals, such as woolly mammoths, appear to have been trapped in bogs or swamps. The entire animal, including hair, muscles, blood, and even stomach contents, has been preserved.

The cell walls of certain plants and valves (shells) of certain marine animals are composed of nearly indestructible compounds of carbon, oxygen, hydrogen, and nitrogen. Cellulose and chitin are examples of such materials. Pollen and spores are plant cells which are extremely

FIGURE 2.2 Methods of fossilization: (a) Original remains, pelecypod shells in a coquina. (b) Original remains, pollen (1000X). (c) Carbonized root. (d) X-ray showing soft part of trilobite, approximately natural size (courtesy of Stuermer, Ref. 2, copyright © 1970 by the American Association for the Advancement of Science). (e) Caddis fly in amber (courtesy of F. M. Carpenter, Ref. 3, copyright © by American Association for the Advancement of Science). (f) Internal mold of pelecypod shell (courtesy of the Smithsonian Institute). (g) Brachiopod shell replaced by silica (photo by G. A. Cooper, courtesy of the Smithsonian Institution). (h) Worm trail. (i) Sedimentary rock showing evidence of bioturbation. (j) Coprolites on the floor of Rampart Cave (courtesy of Paul S. Martin).

resistant and are often preserved as original remains. They provide good indications of ancient vegetation and climate. (Fig. 2.2b)

## Carbonization

Soft organic material may be preserved by carbonization or distillation, a process in which nitrogen, oxygen, and hydrogen are lost and a carbon replica of the organism is produced. Jellyfish, fish, and parts of trees have been fossilized in this manner (Fig. 2.2c). Such fossils are generally found in black shales that were deposited under reducing conditions. X-rays of black shales have revealed carbonized films of the soft parts of organisms which would not otherwise be visible (Fig. 2.2d). Insects in amber are distilled remains of animals that were trapped by and then engulfed in a pine tree's sticky sap which hardens to produce resin. Amber is fossil resin (Fig. 2.2e).

## Casts and Molds

The original shell of an organism will leave an impression or *mold* of itself in the enclosing sediments. If the original shell is later dissolved, the mold may be filled with sediment or mineral deposits. The characteristics of the original shell, which have been preserved in the mold, are impressed on the sedimentary filling to form a *cast* of the original (Fig. 2.2f).

## Replacement

The original remains of an organism may be partially or wholly replaced by other minerals, such as quartz (silica), pyrite, or galena, that are carried by aqueous solutions within the confining sediments (Fig. 2.2g). In the related process of *permineralization,* minerals are precipitated inside the cell walls of wood or bone. Subsequently, the wall material may be removed and additional minerals precipitated in the open spaces. Thus a hard, compact petrification may

be formed in which the original cell walls, growth rings, or other structures are preserved in detail. This kind of preservation occurs in Petrified Forest National Monument, Arizona (Fig. 12.19). The petrified trees are believed to have been buried in stream deposits where replacement occurred. Permineralized dinosaur bone is common in the Colorado Plateau of eastern Utah and western Colorado. One of the most notable occurrences is in Dinosaur National Monument near Vernal, Utah (Fig. 12.20).

## Tracks and Trails

Bottom-dwelling organisms may leave tracks or trails in wet sediments. Such markings are called *trace fossils*. *Tracks* are isolated impressions left by the feet of an organism. *Trails* are continuous grooves or furrows left by an animal which plowed through the sediment with all or parts of its body. The shape of some trails indicates that they were probably made by worms (Fig. 2.2h), but not all are readily identified. Many small organisms, such as worms, churn up the upper few centimeters (inch or so) of lake or marine sediment as they feed. The flat-lying muds become contorted by this activity, known as *bioturbation* (Fig. 2.2i). Trails and bioturbation may be the only record preserved in rock of soft-bodied organisms.

## Coprolites

Coprolites, which are fossilized fecal matter, are also valuable indications of the presence of certain animals. Not only can coprolites often be keyed to a certain species, but they may contain evidence of the type of food eaten. The pollen and spores which the coprolites contain may give an indication of the plants found in the area where the animal lived. In the southwestern United States, large quantities of dung from the now extinct giant ground sloth have been found in caves where the sloths lived during the last glaciation (Fig. 2.2j).

# CLASSIFICATION OF FOSSILS

By the middle of the eighteenth century, the biologic origin of fossils had been accepted. Fossils could then be classified according to the systems used by biologists for living organisms. This system was described by the Swedish botanist Carolus Linnaeus (Carl von Linné, 1707–1778). In his text, *Systema Naturae,* Linnaeus formulated the general principles of biological classifications, which include the ordering and naming of plant and animal groups according to structural similarities.

As it is used today, the Linnaean system includes a hierarchy of seven basic categories:

Example: Humans (*Homo sapiens*)

| | |
|---|---|
| Kingdom | Animalia |
| Phylum (pl. phyla) | Chordata |
| Class | Mammalia |
| Order | Primates |
| Family | Hominidae |
| Genus (pl. genera) | *Homo* |
| Species | *sapiens* |

Additional groupings have been added by using the prefixes super- and sub-, to form such categories as superfamilies and subclasses. Linnaeus proposed that each organism be designated by two descriptive Latin names, in keeping with the custom of scholars of that day who Latinized their own given and family names. In this scheme, known as the *system of binomial nomenclature,* the following rules are universally applied:

1. The first letter of the generic name (genus) is capitalized.
2. The first letter of the specific name (species) is in lower case.
3. Both genus and species are italicized or underlined.
4. Generic names may be used only once, but specific names may be used a number of times with different genera.
5. The genus and species should be Greek or Latin words or have Greek or Latin endings.
6. The name should be derived from a characteristic of the organism, the geographic area where the species is found, a person who did some work on the species, or the geologic horizon in which the fossil was found.

The species is the basic unit of classification. A living species is comprised of an interbreeding population of individuals, which is capable of producing fertile offspring. In paleontology, particularly when there are no living representatives, species designation depends on comparative morphology of fossilized structures and, when available, physiological and ecological characteristics. Modern paleontologists establish the relationship of similar fossils by precisely measuring various dimensions, such as shell length, width, thickness, shape, and number and type of surface ornaments, of large numbers of fossil specimens. Computer analysis of these statistics provides even more accurate species determinations. Closely related species are grouped into a single genus, related genera are incorporated into a single family, and so on.

Some biologists and paleontologists use a classification system in which organisms are assigned to two kingdoms; Plantae (plants) and Animalia (animals). However, some one-celled organisms exhibit characteristics of both plants and animals. For example, some are mobile like animals and yet manufacture their own food like plants. Consequently, a third kingdom, Protista, has been established to include all one-celled organisms (protists). In a third classification system, an additional kingdom is added, Monera, consisting of all organisms such as bacteria and blue-green algae that do not have a separate nucleus.

A brief description of the major fossil groups is given in Table 2.1 and Appendix B. These sections contain supplemental information for those who do not have a background in biology.

TABLE 2.1  **Phyla That Are Important in the Fossil Record**

| KINGDOM | PHYLUM | CLASS | BRIEF DESCRIPTION |
|---|---|---|---|
| Protista | | | Most are one-celled organisms |
| | Cyanophyta | | Blue-green algae—form stromatolites |
| | Chrysophyta | | Diatoms, coccoliths |
| | Pyrrophyta | | Dinoflagellates, etc. |
| | Protozoa | | One-celled animals with calcareous shells (Foraminifera) or siliceous shells (Radiolaria) |
| Animalia | | | Multicellular animals |
| | Porifera | | Sponges |
| | Coelenterata | | Coelenterates |
| | | Scyphozoa | Jellyfish |
| | | Hydrozoa | Stromatoporoids |
| | | Anthozoa | Corals—most have calcareous skeleton |
| | Bryozoa | | Bryozoans—"moss-animals" |
| | Brachiopoda | | Brachipods |
| | Mollusca | | Mollusks |
| | | Pelecypoda | Pelecypods—clams, oysters, etc. |
| | | Gastropoda | Gastropods—snails |
| | | Cephalopoda | Cephalopods—nautiloids, ammonoids, belemnities, squids |
| | Arthropoda | | Jointed-legged organisms |
| | | Trilobita | Trilobites |
| | | Crustacea | Crustaceans—crabs, lobsters |
| | | Insecta | Insects |
| | Echinodermata | | Echinoderms |
| | | Cystoidea | Cystoids |
| | | Blastoidea | Blastoids |
| | | Crinoidea | Crinoids |
| | | Echinoidea | Echnoids—sea urchins, sand dollars |
| | | Asteroidea | Starfishes |
| | Chordata | | Chordates—Subphyla Hemichordata (graptolites, etc.) and Vertebrata (vertebrates) |
| | | Graptolithina | Graptolites |
| | | Agnatha | Primitive jawless fish |
| | | Placodermi | Placoderms—primitive fish with jaws |
| | | Chondrichthyes | Sharks |
| | | Osteichthyes | Bony fish |
| | | Amphibia | Amphibians |
| | | Reptilia | Reptiles |
| | | Aves | Birds |
| | | Mammalia | Mammals |
| Plantae | | | Plants |
| | Bryophyta | | Mosses, etc. |
| | Psilopsida | | Primitive land plants |
| | Lycopsida | | Club mosses and scale trees |
| | Sphenopsida | | Horsetail rushes, etc. |
| | Pteropsida | Filicinae | Ferns and seed plants |
| | | Gymnospermae | Gymnosperms—seed ferns, cycads, ginkgoes, conifers |
| | | Angiospermae | Angiosperms—flowering plants |

# EVOLUTION

The eighteenth century conflict between the catastrophists and those who took a uniformitarian approach carried over into the study of fossils. Georges Cuvier, for example, attempted to show that the succession of different fossils in the strata of the Paris Basin resulted from a series of extinctions brought about by catastrophic floods. Each flood was followed by the creation or immigration of new life forms. Although the catastrophic doctrine was ultimately rejected, Cuvier and other catastrophists, such as Alexandre Brongniart, made significant contributions to descriptive paleontology.

According to the concept of evolution, new life forms develop over a long period of time from more primitive organisms. Although this idea had been suggested by some of the early Greek philosophers, the French biologist Jean de Lamarck developed the first fully consistent theory of evolution. In 1809 Lamarck proposed that characteristics acquired during the life of an organism could be inherited by its offspring. Accordingly, a giraffe's long neck would have resulted from the stretching of the neck by the giraffe's browsing ancestors. Lamarck had many followers including the Soviet biologist Trofim Lysenko, who, until recently, had considerable influence in biology in the Soviet Union, in part because his ideas seemed to fit in with Communist ideology. Modern genetic research has firmly established that acquired characteristics cannot be inherited, and Lamarck's and Lysenko's ideas on evolution can no longer be accepted.

The modern version of the theory of evolution stems from the observations of Charles Darwin (4). Darwin collected considerable biological data during his five-year voyage aboard the H.M.S. *Beagle* in the southern hemisphere (1831–1836). His observations led him to conclude that evolution occurred by a process of natural selection among organisms in the struggle for food and space. Darwin proposed that the characteristics which enabled an organism to compete successfully were often passed on to their offspring, so that eventually a new species would develop. He also believed that the offspring were slightly different from the parents, thus enhancing gradual change. In 1859 Darwin published his theories in *The Origin of Species,* in which he documented evidence in support of evolution. This work was a radical departure from nineteenth-century thought and has become the basis for modern concepts of evolution. His later book, *Descent of Man* (1871), created new controversy over the process of evolution when Darwin proposed that both human beings and the apes could be traced to a common ancestor.

Darwin adopted Charles Lyell's concept of gradualism and believed that evolution occurred gradually. While the fossil record cannot be used to support the concept of gradual evolution, neither does it support the alternate idea of rapid evolutionary bursts.

Many scientists believe that new species develop in small populations that are under some sort of stress and that have been genetically isolated from the main gene pool. Successful populations are likely to maintain themselves unchanged.

The paleontologists Stephen Jay Gould and Niles Eldredge (5) have proposed that the history of evolution is not one of uniform, gradual change, but rather is punctuated occasionally by the rapid evolutionary development of new species in an isolated population. By this process, termed *punctuated equilibrium,* the new species may eventually challenge the static, ancestral population, causing change within that group as well.

Darwin's theory of evolution profoundly altered our perception of the world by providing a dynamic rather than static view of the history and development of life. In this sense, the theory of evolution complements the Copernican view of a dynamic earth in space.

## Evidence for Evolution

Virtually all scientists accept the idea that species develop through evolution from simpler, less advanced forms. However, the evolution of a new species has not been observed in the laboratory; therefore, the evolutionary process can *not* be described as fact, but rather as an interpretation which is supported by overwhelming evidence.

EVOLUTION AND EARTH HISTORY. The fossil record provides extensive documentation of the evolutionary development of many groups of terrestrial and marine organisms. This evidence will be discussed in detail in Chapters 9 through 14. It is evident that a definite pattern exists in the appearance of major groups of organisms. Only primitive organisms are found in very ancient rocks, while more advanced forms occur only in younger sequences. Such a pattern is the natural consequence of the evolution of the advanced organisms from simple organisms.

SIMILARITY IN BODY CHEMISTRY. The chemical composition of the blood of many animal groups is strikingly similar. Furthermore, many of the ions present in seawater are also present in blood in approximately the same abundances. These similarities have led to the suggestion that all animals had a common origin within the early oceans.

Another similarity in body chemistry is seen in the nucleic acids. The chromosomes in the nucleus of each living cell contain deoxyribonucleic acid and ribonucleic acid, two nucleic acids commonly known by their abbreviations, DNA and RNA. The molecules of DNA are enormously long and are shaped like two intertwined coiled springs, a double helix. Each of these molecules is incredibly complex and contains thousands of subunits lined up in a precise order, like numbers on a tape. The pattern forms a code of instructions which dictates what each cell of an organism will be, how it will grow, what substances will be produced, and what raw materials will be required.

The basic building materials of all living organisms are proteins, which are composed of various combinations of about 20 amino acids. The DNA code specifies how the amino acids are to be joined in order to form the proteins. Studies of the DNA molecule have shown that, with few exceptions, a given code "word" in the DNA molecule always specifies the joining of the same amino acids. What is startling about this discovery is that this relationship holds true for *all* organisms, from the protists through the vertebrates. The similarity in the DNA code provides additional evidence for the idea that all life on earth had a common origin.

SIMILARITY IN BODY STRUCTURES. The skeletal structures of terrestrial and marine vertebrates are remarkably similar. The bones in the human arm closely correspond in number and distribution to those in the appendages of reptiles, birds, and other mammals, including marine forms (Fig. 2.3). The teeth of many vertebrates are also quite similar.

PRESENCE OF VESTIGIAL STRUCTURES. Vestigial structures are small, imperfectly developed parts of organs which were more fully developed in earlier generations. The human appendix is of no use now, but it corresponds to a digestive organ in some herbivorous mammals. The presence of the appendix serves as an indication that humans evolved from a herbivorous ancestor. The coccyx, a small triangular bone forming the lower extremity of the human spinal column, may be part of a vestigial tail.

STAGES IN THE GROWTH OF AN ORGANISM. In its development from the embryonic to the mature stage, an organism may undergo changes similar to the evolutionary changes that have occurred during the development of the species. This process is technically referred to as the *biogenetic law* which states that

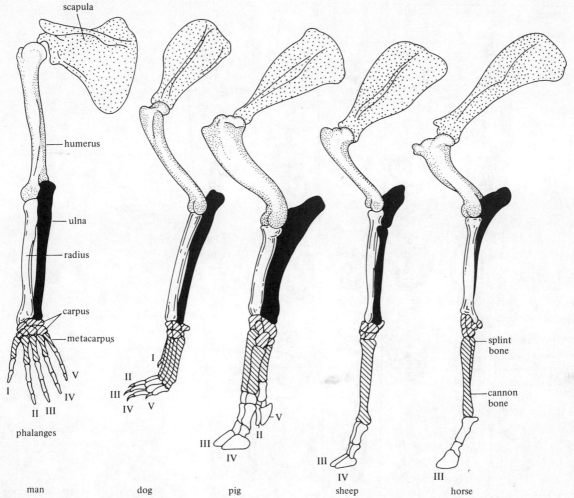

scapula

humerus

ulna

radius

carpus

metacarpus

V

I

IV

II III

phalanges

man

I

II

III

IV V

dog

I

II

III

V

IV

pig

III

IV

sheep

splint bone

cannon bone

III

horse

**FIGURE 2.3** Comparison of the bones of various vertebrates. (From Moody, Ref. 6, after LeConte.)

"ontogeny recapitulates phylogeny." For example, the human embryo passes through stages in which it resembles, in turn, the embryos of fish, amphibians, and reptiles (Fig. 2.4). Thus the stages in the development of the human embryo suggest that human beings evolved from fish, through amphibian and reptile, to mammal.

The feet of the embryo of the modern horse develop in stages similar to those through which the ancestors of the horse evolved, as indicated by fossil evidence. At an early stage of

development, the embryo has three toes, as did horses in early Tertiary time (Fig. 2.5). The mature modern horse has one toe, with only vestiges of the other two.

## Divergent and Convergent Evolution

When a species is dispersed from its point of origin into new areas, new species may develop. Eventually, ecologically similar but taxonomically different organisms may be

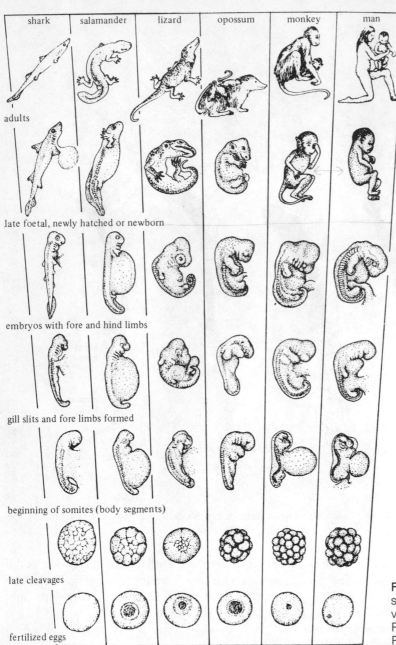

FIGURE 2.4 Comparison of the stages of development of some vertebrate embryos. (From Moody, Ref. 6, after Gregory and Roigneau, Ref. 7.)

found in comparable habitats. This process is known as *divergence (divergent evolution)* or *adaptive radiation.*

Excellent examples of this were found by Charles Darwin in the Galapagos Islands, 600 miles west of Ecuador. Darwin observed that the fauna of these islands was in many ways significantly different from its apparent ancestral stock on the mainland of South America. Adaptation coupled with complete isolation

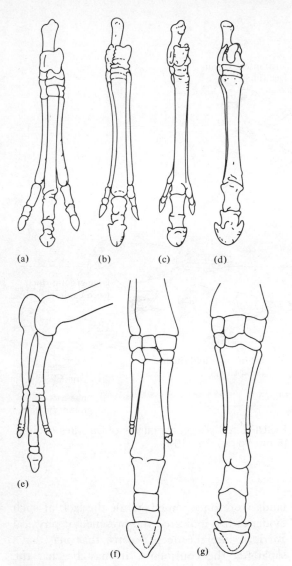

(a)          (b)          (c)          (d)

(e)

(f)          (g)

**FIGURE 2.5** Comparison of the feet of Cenozoic horses with the feet of the embryo of a modern horse: **(a)** *Hyracotherium,* a horse which lived during the Eocene (from Moody, Ref. 6, after Cope). **(b)** *Miohippus,* a horse which lived during the Oligocene. **(c)** *Merychippus,* a horse which lived during the Miocene [**(b)** and **(c)** from Moody, Ref. 6, after Romer, Ref. 8]. **(d)** *Equus,* the modern horse (from Moody, Ref. 6). **(e)** Limb of embryo of *Equus* at six weeks, showing three toes. **(f)** The same at eight weeks, showing the side toes much reduced relative to the middle toe. **(g)** The same at five months [**(e)**, **(f)** and **(g)** from Wells *et al.,* Ref. 9, after Prof. Cossar Ewart; reprinted by permission of Collins-Knowlton-Wing, Inc.].

from the continent gave rise to new species probably over a period of several million years. Apparently the species which initially occupied the islands diversified and occupied many of the available habitats.

The finches on the islands have had little competition and have produced 14 distinct species (Fig. 2.6). They developed a number of modifications that allowed them to better exploit the various food supplies in the islands. The most notable changes are in the beaks, which have become variously adapted for woodpecking and eating insects or cactus (**11**). One group, whose beaks are too short for probing rotten wood for grubs, developed the technique of poking small twigs or thorns into holes to retrieve the grubs (Fig. 2.7). This is one of the few examples of the use of a tool by one of the lower animals.

The fossil record also contains numerous examples of *convergent evolution,* resulting from the development of organisms with a similar external morphology by quite different evolutionary paths. In this process, animals or plants of different taxonomic groups may acquire similar forms or characteristics as a result of living in similar habitats. For example, the shark, (a fish), the ichthyosaur (a Mesozoic marine reptile), and the dolphin (an aquatic mammal) are all similar in appearance as a result of similar selective forces in the marine environment (Fig. 2.8). However, these animals evolved from different basic stocks. Similarities produced as a result of parallel evolution are also evident in birds, the pterosaur (a Mesozoic flying reptile), and the bat (a mammal). These all developed wings and other structures which permitted them to adapt to an aerial environment.

## Paleontology and Plate Tectonics

Correlation of similar fossil floras and faunas between certain modern landmasses has provided some of the evidence showing that these

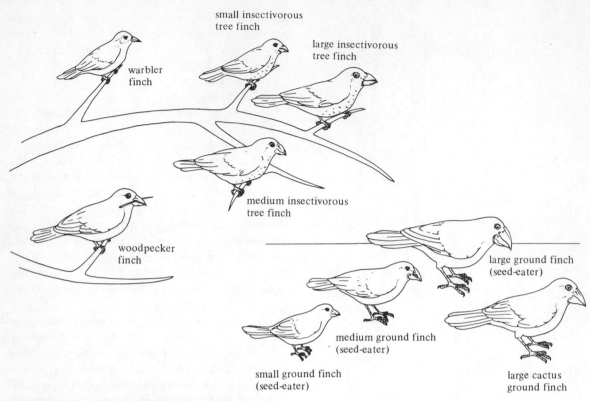

small insectivorous
tree finch

large insectivorous
tree finch

warbler
finch

medium insectivorous
tree finch

woodpecker
finch

large ground finch
(seed-eater)

medium ground finch
(seed-eater)

small ground finch
(seed-eater)

large cactus
ground finch

**FIGURE 2.6** Representatives of Darwin's finches. (From Volpe, Ref. 10.)

**FIGURE 2.7** Galapagos Islands finch using a cactus spine to extract insects from a log. (Photograph by Irenaus Eibesfeldt, Max-Planck-Institut.)

lands were once joined; while the lack of such evidence has indicated the presence of physical barriers, like the deep oceans, that may have separated the continents. It may be that the disappearance or extinction of some fossil groups may be related to the destruction of certain environments during continental collisions. Alternatively, the appearance of new forms may be related to the development of new environments following plate movements.

## REFERENCES CITED

**1.** M. E. Jahn and D. J. Woolf, 1963, *The Lying Stones of Dr. Johann Bartholomew Adam Beringer, Being His Lithographiae Wirceburgensis:* University of California Press, Berkeley.

**FIGURE 2.8** Comparison of the external morphology of **(a)** the shark, **(b)** the ichthyosaur, and **(c)** the dolphin.

**2.** W. Stuermer, 1970, *Science,* v. 170, p. 1300, Dec. 18.

**3.** E. O. Wilson, F. M. Carpenter, and W. L. Brown, Jr., 1967, *Science,* v. 157, p. 1038, Sept. 1.

**4.** C. Darwin, 1964, *The Origin of Species,* a facsimile of the 1st edition, with introduction by E. Mayr: Harvard University Press, Cambridge, Mass.

**5.** S. J. Gould and N. Eldredge, 1977, *Paleobiology,* v. 3, p. 115.

**6.** P. A. Moody, 1970, *Introduction to Evolution,* 3rd ed.: Harper & Row, New York.

**7.** W. K. Gregory and M. Roigneau, 1934, *Introduction to Human Anatomy:* American Museum of Natural History, New York.

**8.** A. S. Romer, 1966, *Vertebrate Paleontology,* 3rd ed.: University of Chicago Press.

**9.** H. G. Wells, J. S. Huxley, and G. P. Wells, 1929, *The Science of Life:* Doubleday, Garden City, N.Y.

**10.** E. Peter Volpe, 1970, *Understanding Evolution:* Brown, Dubuque, Iowa.

**11.** D. Lack, 1947, *Darwin's Finches:* Cambridge University Press, New York.

# Geologic Time **3**

IN RECONSTRUCTING THE HISTORY of the earth, geologists have developed various methods of relative and absolute dating of rocks and structures so that geologic events may be arranged in a meaningful, chronologic sequence. The relative age of a rock unit is determined by its position with respect to the adjacent rock units. Absolute age is given in terms of years before the present and is determined through the use of radioactive elements in a rock sample.

The development of the geological time scale has provided a standardized chronology with worldwide applicability. Thus the histories may be compared not only between regions on a given continent, but between continents as well. Absolute dating has enabled geologists to relate the history of the earth to that of the moon and the other planets of the solar system. This has yielded valuable data for the formulation and testing of theories on the origin of the earth–moon system and planetary systems in general.

## RELATIVE AGES

Relative ages are established through the use of the sequence of sedimentary beds, primary structures, unconformities, cross-cutting structures, and craters.

### Sequence of Beds

As was discussed in Chapter 1, younger beds are deposited on top of older beds. Therefore the relative ages of beds may be determined by their position in a sequence of beds if they have not been disturbed by folding or faulting.

### Primary Structures

Primary structures are produced at the time of deposition of a layer of rock, and some may be used to determine relative ages. These include cross-stratification, graded bedding, ripple marks, fossils, and pillow lavas. *Cross-stratification* may be formed by wind or subaqueous currents. Cross-stratified layers are inclined at angles up to 40° to the horizontal. Erosion may truncate the tops of these layers at a moderate angle, but the bottoms of cross-stratified layers are generally parallel to the bottom of the bed (Fig. 3.1a). Thus, the relative position of the truncated and tangent portions of the cross-strata may be used to establish the relative ages of units within a sedimentary sequence. Furthermore, the concave side of cross-strata generally faces in the direction of the younger beds (tops) (Fig. 3.1a).

In *graded beds,* the particle size varies from coarse at the bottom of a bed to fine at the top of that bed (Fig. 3.1b). Graded bedding is formed by intermittent submarine currents, most commonly turbidity currents. *Turbidity currents* are dense masses of water charged with sediment. Such currents flow rapidly down a slope, and as the current slows, coarser particles settle out first, followed by successively finer particles. For this reason, particle size decreases

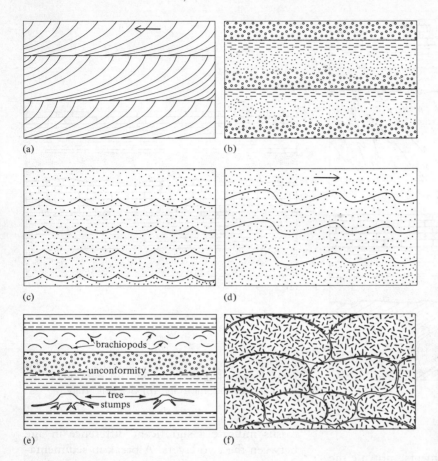

**FIGURE 3.1** Diagrammatic sketches of primary structures which may be used to determine the relative age of beds: **(a)** cross-laminations, **(b)** graded bedding, **(c)** oscillatory ripple marks, **(d)** current ripple marks, **(e)** fossils, **(f)** pillow lavas.

in the direction of the top of a bed. A *turbidite* is a rock formed from a turbidity current deposit.

*Ripple marks* are small, wave-shaped structures that are produced by both wind and water currents. Oscillating currents of water (that is, those that move back and forth) produce nearly symmetrical ripples with pointed crests and rounded troughs (Fig. 3.1c). The crests point toward the younger beds. Translational currents (that is, those that move in one direction only) produce ripples with a steep face in the direction of current flow (Fig. 3.1d). The crests of current ripples are generally more pointed than the troughs, and therefore their shape may be used to determine relative ages.

The orientation of fossils is a commonly used method for determination of relative ages. For example, the valves (shells) of clams or brachiopods when they have been disarticulated (broken apart) generally lie with their convex side facing up in the direction of the younger beds (Fig. 3.1e). This orientation is the most stable position in submarine currents; valves lying concave upward may be flipped over, while those facing convex up are less likely to do so. The orientation of fossil tree trunks may also be used to determine relative ages of beds (Fig. 3.1e). Tree stumps will often be in their original growth position with roots pointing generally downward.

Pillow lavas are masses of volcanic rock that were probably produced by submarine volcanic eruptions. Somewhat rounded in cross section, they have convex upper and concave basal surfaces. Projections on the bottoms of some pillows point toward the older beds (Fig. 3.1f).

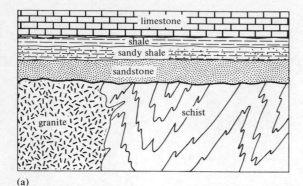

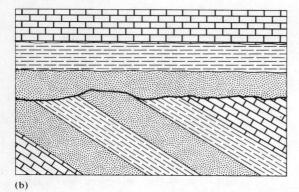

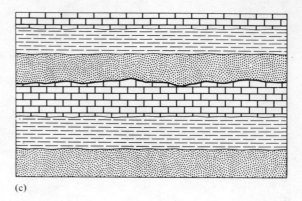

FIGURE 3.2   Diagrammatic sketches illustrating three types of unconformities: **(a)** nonconformity, **(b)** angular unconformity, **(c)** disconformity.

## Unconformities

In many sequences of sediments, not all the layers that were originally deposited are preserved. Uplift may have resulted in erosion surfaces which may have been subsequently covered by younger sediment. Buried erosion surfaces are termed *unconformities* and they may be used to determine the relative age of beds. Some major types of unconformities are nonconformities, angular unconformities, disconformities, and diastems.

At a *nonconformity,* sedimentary rocks overlie older metamorphic or plutonic igneous rocks (Fig. 3.2a). An *angular unconformity* separates tilted or folded sedimentary rocks from younger, less deformed sedimentary rocks (Fig. 3.2b). A *disconformity* is a buried erosion surface between beds that are essentially parallel to each other (Fig. 3.2c). A break in sedimentation between two layers may be recognized if fossils indicate a significant difference in age between the two layers. A break in sedimentation, representing a short time interval, is termed a *diastem* or *depositional hiatus*.

## Cross-Cutting Structures

Cross-cutting igneous intrusions, such as *dikes, stocks,* and *batholiths,* are always younger than the youngest beds which they intrude and older than the oldest bed which unconformably overlies the intrusion (Fig. 3.3a). The age of one igneous intrusion relative to another may also be determined by cross-cutting structures. If two intrusions are in contact, dikes of the younger intrusion commonly intrude the older intrusion (Fig. 3.3b).

It is also possible to determine the relative time during which movement along a fault occurred. The final movement along a fault must have occurred after the deposition of the

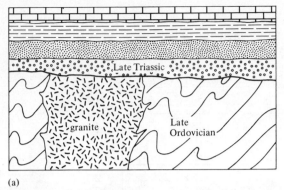 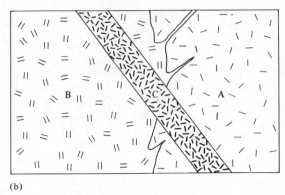

(a) (b)

**FIGURE 3.3** Relative age determination of igneous rocks: **(a)** Granite pictured was emplaced between the Late Ordovician and the Late Triassic. **(b)** Multiple episodes of igneous activity—intrusion A was emplaced first, followed by B and then C.

youngest beds that are displaced by the fault and before the deposition of the oldest bed that unconformably overlies the fault (Fig. 3.4). Similarly, the age of folding of a sequence of rocks is always younger than the youngest beds that are folded and older than the oldest overlying beds that are not folded (Fig. 3.4). The age of metamorphism of a sequence is always younger than the youngest bed that is metamorphosed and older than the oldest overlying bed that is not metamorphosed (Fig. 3.4).

## THE GEOLOGICAL TIME SCALE

One of the first geological time scales was developed in 1756 by Johann Lehmann (and later Werner), who recognized three ages of rocks:

Primitive: all crystalline rocks, such as granite and gneiss
Secondary: consolidated sedimentary rocks containing fossils
Alluvial: soils and gravels

A subsequent time scale, proposed in 1760 by Giovanni Arduino, an Italian mining geologist, listed four ages:

Primitive: crystalline rocks in the cores of mountains

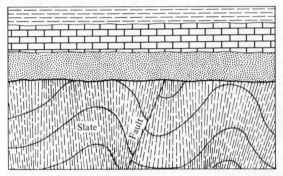

**FIGURE 3.4** Fault and slatey cleavage terminating at an unconformity.

Secondary: sedimentary rocks
Tertiary: unconsolidated sediments
Volcanics: extrusive igneous rocks

Of these terms, "Tertiary" alone has survived in the modern system of chronologic nomenclature.

### The Modern Time Scale

The modern time scale was developed during the nineteenth century (Table 3.1). Following the lead of William Smith, geologic periods were established on the basis of relative ages and the fossil content of sequences of rock. Periods were grouped into eras, based on the fossils that they contained. The periods of the

Table 3.1 **The Geologic Time Scale**

| | Era | Period | Age of beginning, millions of years ago | | |
| | | | (*Kulp* [5]) | (*Holmes* [6]) | (*Harland et al.* [7]) |
|---|---|---|---|---|---|
| Phanerozoic | Cenozoic | Quaternary | 1 | 1 | 1.5–2 |
| | | Tertiary | 63 | 70 | 65 |
| | Mesozoic | Cretaceous | 135 | 135 | 136 |
| | | Jurassic | 180 | 180 | 190–195 |
| | | Triassic | 230 | 225 | 225 |
| | Paleozoic | Permian | 280 | 270 | 280 |
| | | Pennsylvanian | 310 | | 325 |
| | | Mississippian | 355 | 350 | 345 |
| | | Devonian | 405 | 400 | 395 |
| | | Silurian | 425 | 440 | 430–440 |
| | | Ordovician | 500 | 500 | 500 |
| | | Cambrian | 600 | 600 | 570 |
| Precambrian | Proterozoic | Late | 1000 | | |
| | | Middle | 1700 | | |
| | | Early | 2500 | | |
| | Archean | | 4600 | | |

Paleozoic era (from the Greek: ancient life) were established through the work of several English geologists. In 1831 Adam Sedgwick began a detailed study of a sequence of deformed sedimentary and volcanic rocks in northern Wales. After several years, he published a description of the rocks in this area and proposed that the sequence be named the "Cambrian system" after Cambria, the Roman name for Wales. At about the same time, Roderick Murchison published a description of the rock sequence in southern Wales, which he named the "Silurian system," after an ancient Welsh tribe, the Silures.

A study of the fossils in Sedgwick's Upper Cambrian System showed that they were almost identical to those of Murchison's Lower Silurian System. Evidently, these two rock units were overlapping, and a bitter feud developed between the two men over where to place the disputed rocks. The conflict was not re-

solved until 1879, when Charles Lapworth recognized that this intermediate sequence was in fact distinctive and proposed that it be placed in a new system, the "Ordovician system," named after another ancient Welsh tribe, the Ordovices.

The "Carboniferous system" was named in 1822 by two English geologists, William Coneybeare and William Phillips. The name was chosen because large quantities of coal occur in many sequences of this age. North American geologists felt that the Carboniferous should be subdivided into two systems, the Mississippian and the Pennsylvanian. The Mississippian system, defined by Alexander Winchell in 1870, corresponds approximately to the Early Carboniferous. The Mississippian system was named by Henry Shaler Williams in 1891, and it takes its name from the Mississippi River, along which strata of this age are exposed. The Upper Carboniferous is approximately equivalent to the Pennsylvanian system. Rocks of this age are widely exposed in Pennsylvania.

Sedgwick and Murchison together studied a thick marine sequence in the County of Devon in southwestern England. At first they thought that this sequence was either Cambrian or what is now called Ordovician, based on the lithology of the beds. However, the paleontologist William Lonsdale studied the fossils in this sequence and concluded that they were intermediate in development between those of the Silurian and Carboniferous systems. Ultimately, Sedgwick and Murchison accepted Lonsdale's conclusion and assigned these rocks to the "Devonian system."

Murchison is also credited with defining the "Permian System." On invitation from the Czar of Russia, Murchison went to the European part of Russia to study the strata of the region. Just west of the Ural Mountains, he found a very thick sequence of rocks which contained fossils younger in appearance than those of the Carboniferous system and yet older than those of a unit known as the "Triassic." He named this sequence after the Russian province of Perm.

The Mesozoic era (from the Greek: middle life) consists of three periods, the Triassic, Jurassic, and Cretaceous. The type region for the "Triassic system" is in central Germany where it consists of two continental rock sequences separated by a marine sequence. The Triassic was named by Von Alberti in 1834. The name Triassic refers to the tripartite nature of the unit.

The "Jurassic system" is named for the Jura Mountains which are located along the border between France and Switzerland. Jurassic rocks, first studied by the German geologist von Humboldt in 1799, are noted for their wealth of well-preserved fossils.

The "Cretaceous system" includes the chalk units exposed in the Chalk Cliffs of Dover and in other parts of the London and Paris basins. The name is derived from the Latin word for chalk, *creta*. Cretaceous deposits were studied extensively by the Belgian geologist d'Halloy who in 1822 subdivided the system into seven stages based on their characteristic fossils.

The Cenozoic era (from the Greek: recent life) is comprised of two periods, the Tertiary and Quaternary. The Tertiary is derived from the time scale of Arduino and the Quaternary, or fourth unit, is a holdover from that early style of classification. The deposits of these two periods are largely unconsolidated sediments that contain a wide variety of modern-looking fossils. The Tertiary contains several epochs, the Paleocene, Eocene, Oligocene, Miocene, and Pliocene, which were named by Charles Lyell in 1883 based on the ratio of extinct to modern forms. The epochs of the Quaternary are the Pleistocene and Recent (or Holocene). The former was first used by Desnoyers in 1829 for the unconsolidated glacial deposits in France; the latter has achieved status through use, but generally signifies a post-Pleistocene age. The term Holocene dates from 1869 when it was proposed by Gervais for generally postglacial deposits. Some geologists prefer the use of the terms Paleogene for the lower Tertiary (Paleocene, Eocene, and Oligocene epochs) and Neogene for the upper Tertiary (Miocene and Pliocene) because there is a natural paleontological break between the Paleogene and Neogene.

The term *Precambrian* has been used to identify all rocks older than Cambrian. While the origin of the term is not associated with any one geologist, it is widely used in the literature. However, some geologists today prefer the terms *Prephanerozoic* or *Prepaleozoic,* or the eras, *Proterozoic* (former life) and *Archean* or *Archeozoic* (most ancient).

## Dating Using Fossils

A rock that contains fossils is generally easier and less expensive to date by paleontologic techniques than by radiometric dating. Furthermore, most sedimentary rocks do not contain

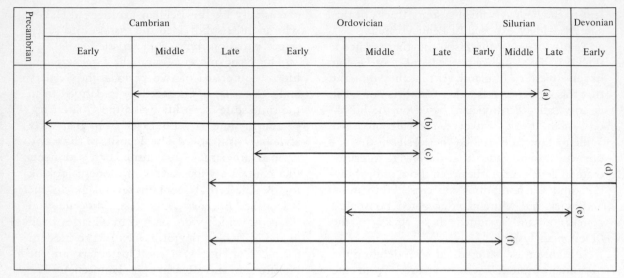

**FIGURE 3.5**  An assemblage of six fossil species was found in a single rock unit. The diagram shows the total time ranges during which each of the species lived. The only time period when all of these organisms could have been living together is the Middle Ordovician. This is the only time during which all of the ranges overlap.

sufficient quantities of radioactive materials to be dated radiometrically. The procedure for paleontologic dating is basically:

1. Collection of a suite of fossils from the rock unit to be dated
2. Identification of the fossils by comparison with specimens in museum collections, in photographs in books and journals, or by consultation with a paleontologist
3. Determination of the range of time that each of the fossils collected is known to have lived

References like *Index Fossils of North America* (**2**) and the *Treatise of Paleontology* (**3**) illustrate the important fossils that serve as guide or index fossils. If one or more index fossils are present in the collection, they readily indicate the age of the unit. A good index fossil should:

1. Have a short time range
2. Have a wide geographic distribution
3. Be abundant
4. Be easily identifiable

The trilobite *Olenellus* is a good guide fossil because it is restricted to Early Cambrian deposits, found in many parts of North America and Europe, relatively abundant, and has a distinctive morphology.

A group of several different fossil species which are consistently found together comprise a *fossil assemblage*. If an assemblage is restricted in geological range, it may be used in determining the age of the rock unit. A comparison of the time ranges of each of the fossils within an assemblage serves to determine when they all could have lived together (Fig. 3.5).

## EARLY ESTIMATES OF THE DURATION OF GEOLOGIC TIME

In 1658 Archbishop James Ussher, Primate of Ireland, set the year of the earth's creation as 4004 B.C. This date, based on biblical chronology, was not acceptable to the geologists of the late eighteenth and nineteenth centuries because most geologists, even neptunists such as Werner and Buffon, felt that at the least tens of thousands of years were required for the deposition of the great thicknesses of sedimentary rocks in the earth's crust.

## Rate of Deposition

As early as the fifth century B.C. the Nile River Delta had been studied by Herodotus, who reasoned that the delta must be thousands of years old based on the y arly rate of deposition of sediment. Similar attempts have been made to determine the age of the earth based on estimates of the maximum thickness of sedimentary rocks and their rate of deposition. A number of attempts to date the earth were made using this method in the latter part of the nineteenth century and early part of the twentieth century. It was estimated that between 33,000 and 100,000 m (100,000 and 335,000 ft) of sedimentary rock have been deposited since the formation of the earth. However, estimates of the rate of sedimentation vary considerably, from 51 to 3000 m (170 to 10,000 ft) per million years. Consequently, estimates of the age of the earth varied from 17 to 1584 million years (**1**).

## Salt in the Oceans

Another early attempt to determine the age of the earth involved calculating the total amount of salt in the ocean compared with the amount of salt added yearly. In this scheme, it was assumed that the oceans originated early in the earth's history as bodies of fresh water. It was estimated that the total amount of salt in the ocean was 16 quadrillion ($16 \times 10^{12}$) tons and that 160 million tons of salt was added yearly, carried there in solution by rivers and derived from the weathering of rocks and soil. Using this line of reasoning, the age of the earth was placed at approximately 100 million years. However, these calculations ignored the vast quantities of salt which are present in sedimentary sequences in many parts of the world.

## Rate of Cooling

Lord Kelvin, the nineteenth-century English physicist, set the age of the earth at 70 million years based on the earth's present cooling rate. Kelvin assumed that the earth originated as a molten ball which had been pulled out of the sun and that the original temperature of the earth was that of the melting point of an average igneous rock. The fallacy in these calculations is that Kelvin also assumed that no internal source of heat existed within the earth. At that time, it was not known that radioactive decay releases a significant amount of heat.

## NUCLEAR CLOCKS

In 1896 Antoine Henri Becquerel observed that uranium-bearing minerals, such as uraninite, caused a covered photographic plate to darken —an effect previously associated only with X-rays (Fig. 3.6). Subsequently, it was demonstrated that uranium decays spontaneously and

**FIGURE 3.6** Autophotograph of the uranium-bearing mineral uraninite from Grafton Center, New Hampshire. The photograph was made from a negative that had been enclosed in black paper while the specimen lay upon it. (Courtesy of Ward's Natural Science Establishment.)

gives off energy in the form of particles and electromagnetic radiation (i.e., radioactivity). The particles radiated are helium nuclei (alpha rays) and electrons (beta rays), and the electromagnetic radiation is in the form of gamma rays, which are similar to X-rays but have shorter wavelengths. For the geologist, radioactive decay has become important as a means of determining the age of the earth and its rock formations.

Although the atoms of a given element always have the same number of protons, many elements have atoms that differ in the number of neutrons they contain. These are known as *isotopes* of the element. For example, uranium has two naturally occurring isotopes, uranium 235 and uranium 238. The numbers 235 and 238 represent the *mass numbers* of the two isotopes of uranium and are the sums of the protons and neutrons in the atoms. Both isotopes of uranium have 92 protons, but uranium 235 has 143 neutrons, whereas uranium 238 has 146 neutrons.

Some isotopes are unstable and with time will decay into one or more isotopes with a different mass number, a different number of protons, or both. The process of radioactive decay proceeds at a rate that is unique for each isotope. The time required for one half of

the original amount of an isotope to decay is termed the half-life of that isotope.

The ratio of the remaining amount of the original isotope to the accumulated decay products is used in determining the age of rocks containing radioactive minerals. The utilization of "nuclear clocks" for the determination of geologic age relies on the basic assumptions that the decay occurs at a constant rate and that the radioactive minerals have suffered no loss or addition of decay products. Laboratory studies of short-lived radioactive isotopes have confirmed the first assumption.

Radiometric dating has enabled geologists to assign absolute ages to episodes of metamorphism, igneous intrusion, and deposition of sediments. Radiometric dating is often the only reliable method of dating metamorphic rock units. It provides a means of assigning absolute ages to the geological time scale.

## Uranium–Lead (U–Pb) Dating

Both uranium 235 and uranium 238 decay spontaneously with the loss of an alpha particle. The half-life of uranium 238 is 4.5 billion years. This means that starting with 1 gm of uranium 238, 0.5 gm will remain after 4.5 billion years, 0.25 gm after 9 billion years, and so on

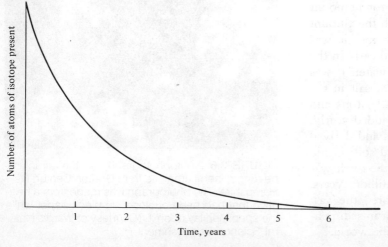

FIGURE 3.7 Decay curve of a hypothetical radioactive isotope with a half-life of one year.

(Fig. 3.7). Uranium 238 breaks down to form uranium 234, which in turn decays rather rapidly through a series of daughter products to form the isotope lead 206. Similarly, uranium 235 decays to lead 207, and thorium 232 decays to lead 208. As a result of radioactive decay, uranium-bearing minerals continuously accumulate lead.

The procedure for determining the age of a rock using the uranium-lead method of dating is as follows:

1. A uranium-bearing mineral such as zircon is separated from the rock which is to be dated. First the rock is crushed into particles a fraction of a millimeter in size. The fragments are then placed in a heavy liquid (a liquid of high density), in which the heavy minerals including zircons will sink to the bottom. The remaining light minerals float and they may be decanted off with the liquid. The zircons are then recovered by separating them from the other heavy minerals.
2. The amount of uranium and lead in the zircons is determined by careful chemical analysis.
3. The relative amount of each isotope of lead (*isotopic ratio*) is measured with an instrument called a mass spectrometer. The lead is vaporized and the lead ions are passed through a strong magnetic field (Figs. 3.8 and 3.9).

**FIGURE 3.9** An Avco mass spectrometer. The ion source is on the left, and the collector is on the right. (Courtesy of Avco Electronics Division.)

The amount of deflection of the lead ions in this field depends on the mass and the charge of the ions. Because ions of lead 204 are lighter than ions of lead 206, the lead 204 will be deflected more than the lead 206 ions. The isotopic ratio of the lead is measured either photographically, (by measuring the degree of darkness of lines on a negative) or with a photoelectric cell.

4. The apparent age of a rock is calculated or computed from the ratios of uranium 238 to lead 206, uranium 235 to lead 207, thorium 232 to lead 208, or lead 206 to lead 207. Each of these ratios will provide an age for the rock. If the ages agree reasonably well, they are *concordant*. If they disagree significantly, they are *discordant*. Discordant ages are often due to a later heating or metamorphism of a rock.
5. If more than one uranium–lead measurement is made for a suite of rocks from a single unit, the ratios of uranium 238 to lead 206 and uranium 235 to lead 207 may be plotted on a concordia diagram (Fig. 3.10). In this diagram, concordant ages will plot on the concordia curve, while discordant ages will plot on a straight line (chord). The intersec-

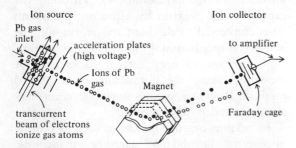

**FIGURE 3.8** Diagrammatic sketch of a mass spectrometer. An ion of mass $M_1$ and electric charge $e$ will be separated from another ion of mass $M_2$ and the same charge as it passes through a magnetic field. (After Hurley, Ref. 5. Copyright © 1959 by Educational Services, Incorporated.)

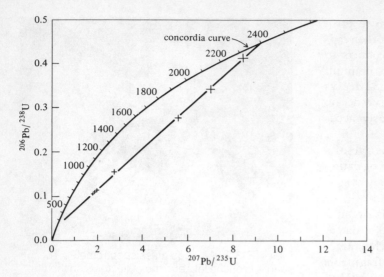

FIGURE 3.10 Concordia plot used in uranium-lead dating of 7 samples from the Baggot Rocks Granite, Medicine Bow Mountains, Wyoming. The crosses are data points which lie along a line intersecting the concordia curve at 2400 million years, the probable age of crystallization of the granite. (From Hills *et al.*, Ref. 8.)

tion of the chord with the concordia curve at its upper end gives the original age of metamorphism of a metamorphic rock or the original age of crystallization of an igneous rock. The lower intercept of the chord may give a time of later heating or metamorphism of the rock. This method of dating, known as the uranium–lead concordia method, is thought to give the most accurate age of intrusion or original metamorphism. However, even this method may at times give incorrect results. For example, radiogenic lead may be present in zircons in a sedimentary rock before its metamorphism, or in zircons that were incorporated into a magma by melting or assimilation of an older rock into which an intrusion has been emplaced. Such zircons will tend to give an age older than the original metamorphism or intrusion of a rock.

## Thorium-Protactinium ($^{230}$Th-$^{231}$Pa) Dating

Age determinations made by thorium 230-protactinium 231 dating utilize decay products of uranium 238 and uranium 235. The relatively short half–lives of thorium 230 (75,000 years) and protactinium 231 (32,480 years) permit dating of sediments and corals less than 250,000 years old. These isotopes are useful in dating marine cores and other marine deposits in studying changes in sea level.

## Potassium-Argon (K-Ar) Dating

Naturally occurring potassium has three isotopes. The stable isotopes, potassium 39 and potassium 41, are far more abundant than the unstable isotope potassium 40. Through the capture of an electron by a proton, potassium 40 is converted to the inert gas argon 40. The half-life of potassium 40 is 11.85 billion years. Thus the age of very ancient rocks may be measured using the potassium-argon (K-Ar) method. Rocks as young as 40,000 years may also be dated by this method.

Potassium-bearing minerals, such as biotite, muscovite, hornblende, sanidine, glauconite, and glaucophane, as well as whole rocks, may be dated by the potassium-argon method. This method of dating is widely used owing to the abundance of potassium-bearing minerals in most rocks. The potassium content of a mineral or rock is determined by chemical analysis, and the percentage of argon is measured indirectly by a mass spectrometer. Since argon is a gas, it may be lost from the mineral or rock during metamorphism. For this reason, the potassium-argon method dates the time of the last heating of a granitic rock, time of the last metamorphism of a metamorphic rock, or the time of last important uplift and erosion of a region. Potassium-argon dates are generally regarded as minimum ages for a rock because of the possibility of argon loss. There are, however, cases in which a potassium-argon date may be too old. Argon 40 may be inherited from preexisting minerals in a sedimentary rock, and if it is not completely lost during metamorphism, a potassium-argon date on a metamorphic rock may be older than the time of metamorphism. This may be a problem in dating less intensely metamorphosed rocks such as a slate. Furthermore, some minerals may incorporate argon which was not formed within the mineral into their structure (**9**). A potassium-argon measurement on such a mineral would give a date which would appear to be too old.

## Argon 40–Argon 39 ($^{40}$Ar-$^{39}$Ar) Dating

In the recently developed argon 40–argon 39 method of dating, the ratio of argon 40 to argon 39 is carefully measured with a mass spectrometer. A potassium-bearing rock, which produces argon by radioactive decay, gives off argon when heated (**10**). If subsequent heating has not been too intense, argon released at high temperatures will generally have an isotopic ratio characteristic of the time of initial metamorphism of a metamorphic rock or intrusion of an igneous rock, whereas argon released at lower temperatures will have an isotopic ratio characteristic of the time of last metamorphism, uplift, or heating of a rock. Thus, using the argon 40–argon 39 method, it may be possible to "see" beyond the time of last heating of a rock, something which it is not possible to do with the potassium-argon method. The time of first metamorphism of a metamorphic rock or initial intrusion of an igneous rock may be shown as a "plateau" in the argon release diagram (Fig. 3.11).

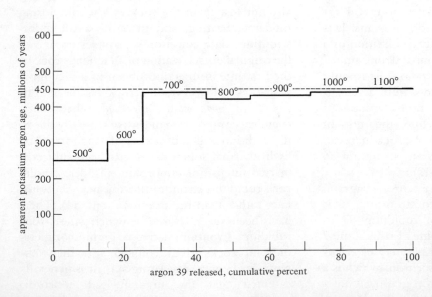

FIGURE 3.11 Argon 40-argon 39 age spectra for biotite from a hypothetical igneous rock which was intruded 450 million years ago and then was subjected to a mild heating about 250 million years ago. The plateau in the argon release spectrum is at 450 million years.

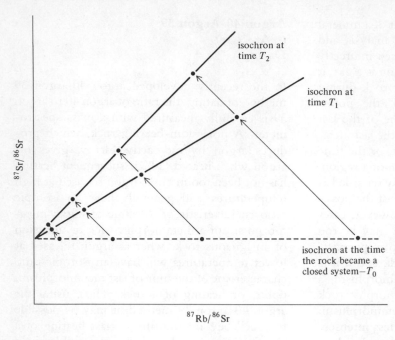

isochron at
time $T_2$

isochron at
time $T_1$

$^{87}Sr/^{86}Sr$

isochron at the time
the rock became a
closed system—$T_0$

$^{87}Rb/^{86}Sr$

**FIGURE 3.12** Change with time of the slope of the isochrons in a whole-rock rubidium-strontium isochron plot. Since both the strontium 87-strontium 86 and rubidium 87-strontium 86 ratios change with time, the four data points (shown as black dots) move diagonally.

## Rubidium–Strontium (Rb–Sr) Dating

Most rock-forming minerals contain small quantities of rubidium. The unstable isotope rubidium 87 decays with the emission of an electron to strontium 87. The 50 billion year half-life of strontium 87 makes it useful for dating Paleozoic and Precambrian events. It is assumed that at the time of crystallization of a rock, all its minerals contained strontium 87 and strontium 86 in the same ratio. The amount of strontium 87 accumulated within a rock is directly proportional to the original amount of rubidium 87 present. Therefore, the ratio of strontium 87 to strontium 86 of a rock increases and the ratio of rubidium 87 to strontium 86 decreases with time. (Fig. 3.12).

In a rubidium-strontium isochron diagram, the ratio of strontium 87 to strontium 86 is plotted against the ratio of rubidium 87 to strontium 86. Since the half-life of rubidium 87 is known, the age of the sample may be calculated from the slope of the isochron, which is a line connecting points on the graph. Whole-rock rubidium–strontium dates are determined through the use of a rubidium–strontium isochron diagram (Fig. 3.13). Since biotite and muscovite may lose strontium when metamorphosed, rubidium-strontium dating of these minerals provides the age of the last heating of the rock. On the other hand, strontium is usually not lost from the rock as a whole during moderate heating, and therefore a rubidium-strontium date commonly provides an age of the original crystallization of an igneous rock or the first intense metamorphism of a metamorphic rock. However, if a rock has been subjected to *metasomatic alteration* (changes in chemical composition of the rock usually due to the addition and subtraction of ions carried by hydrothermal solutions) or has been subjected to two intense metamorphic episodes, whole-rock rubidium strontium dating may show the later rather than the original event (**11**). There have been several cases in which whole-rock rubidium-strontium dating of granitic or metamorphic rocks gives younger ages than uranium-lead concordia dating (**11**). It is generally assumed that the uranium-lead concordia

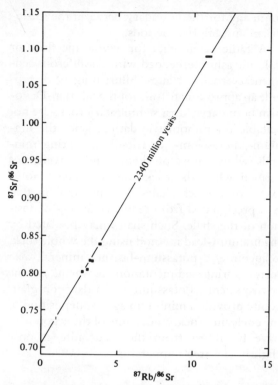

FIGURE 3.13 Rubidium-strontium whole-rock iso-chron diagram for the Baggot Rocks Granite. Eight samples (shown as rectangles or squares) fall along a straight line whose slope indicates an age of approximately 2340 million years. Notice that this age is close to but not exactly equal to the uranium-lead concordia age determined from the same granite body. (From Hills *et al.*, Ref. 8.)

method gives a more accurate date for the original metamorphism or intrusion of a rock. However, in most cases, the whole rock rubidium-strontium method is quite reliable when it is applied to rocks of felsic (granitic) composition. It is less reliable when applied to mafic igneous rocks such as basalts or diabases or to sedimentary rocks.

## Fission Track Dating

The spontaneous fission of uranium 238 and uranium 235 within a crystal or glass produces a *fission track* which is a linear zone in the crystal

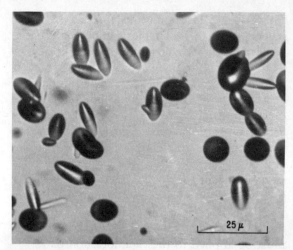

FIGURE 3.14 Fission tracks in glass. (Courtesy of R. L. Fleisher.)

or glass about .0075 mm (.0003 in) long and .000001 mm (.0000004 in) wide (**12**). Fission tracks are due to a defect in the crystal lattice or the breaking of chemical bonds in glass (Fig. 3.14). It is thought that these tracks are caused by fission fragments moving at high speed. The number of tracks per unit area is a function of the uranium concentration and the age of the material. Once the uranium content is estimated, the age of the material may be determined by counting the number of tracks in a given area. Because the fission tracks have such small diameters, they are enlarged by etching in an acid solution. An important advantage of this method is that very small samples may be dated. A disadvantage is that heating tends to destroy fission tracks, so that this method gives only minimum ages.

## Radiocarbon Dating

The isotope carbon 14 ($^{14}C$) is produced in the upper atmosphere through collisions of neutrons with nitrogen 14 ($^{14}N$). Entering the carbon cycle as carbon dioxide, carbon 14 is fixed in the tissues of plants and animals as well as in shells and bones during the life of an organism.

This isotope of carbon is unstable and decays in several steps to the stable isotope, carbon 12. Willard Libby realized that if the ratio of carbon 14 to carbon 12 remains constant and if carbon 14 is produced at a fixed rate, then the ratio of carbon 14 to carbon 12 could be used to date carbon-bearing materials. Recent radiocarbon dating of very old trees whose age is known from counts of tree rings has shown that the rate of production of carbon 14 has been relatively constant for the last 2500 years, but was somewhat lower between 2500 and 4500 years ago. This has tended to yield ages too great in this time range (13), and a correction factor must be applied to these radiocarbon dates.

The relatively short half-life of 5730 years limits this method to materials less than about 40,000 years old, although ages in the 60,000 year range may be obtained if large samples are used and shielding is used to eliminate background radiation. A recently devised technique of radiocarbon dating using a cyclotron also shows promise of extending the time range of carbon 14 dating (14).

Dating by the carbon 14 method has greatly advanced our understanding of human prehistory, the late Pleistocene ice age, and sea level changes. The major drawbacks to radiocarbon dating are its time span limitation, the effects of atmospheric testing of nuclear bombs on carbon 14 production, and the increase in carbon 12 in the atmosphere owing to the increasing use of fossil fuels.

## RADIOMETRIC DATING OF THE GEOLOGICAL TIME SCALE

Radiometric dating of suitable igneous and sedimentary rocks has provided ages for the boundaries between geologic periods and epochs (Table 3.1 and Appendix A). For example, a date on a granite that intrudes Late Ordovician sedimentary rocks and is unconformably overlain by Early Silurian beds gives a maximum age for the boundary between the Ordovician and Silurian periods.

A radiometric date for a volcanic flow or volcanic ash interbedded with fossiliferous sedimentary rocks of latest Silurian age will provide an approximate time for the Silurian-Devonian boundary. Most sedimentary rocks are not suitable for radiometric dating, since the uranium-, potassium-, or rubidium-bearing minerals which most of them contain were not formed when the rock was deposited. However, some black shales contain uranium that was precipitated from seawater during deposition of the shale. Such shales may be dated by the uranium-lead method using the whole rock. Glauconite, a potassium-bearing mineral, also forms during sedimentation in warm, marine environments. Potassium-argon dates on glauconite provide a minimum age for deposition of the enclosing rock, since some of the argon may have been lost from the glauconite at only slightly elevated temperatures.

## DENDROCHRONOLOGY

The study of the growth pattern and age of trees (*dendrochronology*) can provide an absolute age for a piece of wood. Generally, trees in temperate climates add one growth ring per year. Normal or *complacent* tree rings vary in thickness depending primarily on climatic conditions such as temperature and rainfall. The pattern of spacing of growth rings provides a date for a tree if it is compared against a standard model for tree ring spacing which has been calibrated by radiocarbon dating.

## AMINO ACID DATING

The character of amino acids in plant or animal tissues may also be used in dating. Amino acids are present in all organisms, and after the organism dies, the amino acids change the orientation of their optical axes. Determinations have been made of the rate of this change, and this has led to a technique for dating very small organic samples. This method of dating may be used for samples up to about 200,000 years old, but has proven to be temperature sensitive, that is, the rate is slowed down by extended periods of cold such as occur during glaciations.

## PALEOMAGNETIC DATING

The study of the fossil magnetism of rock, *paleomagnetism,* has provided a method of dating rocks. Paleomagnetic studies (discussed in detail in Chapter 7) provide the orientation of the magnetic field of a rock. Such studies have shown that the direction of magnetization of some rocks is exactly the opposite of that in adjacent strata (Fig. 3.15). In relatively young rock sequences, the direction of magnetization is such that the north-seeking pole of the magnetic axis of the rock is pointed in the direction of the magnetic North Pole. These rocks have a *normal polarity*. When the north-seeking pole points in the direction of the magnetic South

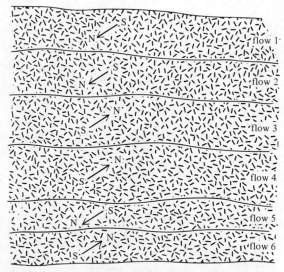

**FIGURE 3.15** Directions of magnetization of a sequence of volcanic flows. Flows 1, 2, and 5 are reversely magnetized and flows 3, 4, and 6 are normally magnetized.

Pole the rocks are said to have a *reversed polarity*. It is now generally accepted that these differences in polarity are due to the earth's magnetic field having reversed itself in the past. Rocks that have a normal polarity were formed when the earth's polarity was normal—in other words, about the same as it is today. Rocks that have a reversed polarity were formed when the earth's polarity was reversed. The principal evidence for reversals in the earth's magnetic field is that all samples of rock which are younger than 690,000 years old have a normal polarity and samples between 690,000 years and 890,000 years old have a reversed polarity. Recognition and dating of the reversals in the earth's magnetic field has led to the construction of the geomagnetic time scale (**15**) (Fig. 3.16). This time scale identifies times of normal and reversed polarity based on measurements of the K-Ar age and polarity of various samples of volcanic rock. The results were then plotted and the time scale constructed from this data (Fig. 3.16, columns 1 and 2).

Studies of the strength of magnetization of rocks formed during a reversal in the earth's magnetic field indicate that the field decreases in intensity by a factor of 10 during a reversal and that the position of the pole migrates slowly during the reversal. The switching of the poles takes between 1000 and 10,000 years to be completed (**16**).

The geomagnetic time scale has been especially useful in dating sediments recovered from the ocean basins. In an undisturbed sequence, the uppermost sediments will be of normal polarity, and the boundary between normal and reversely magnetized sediments is dated at 690,000 years. Older ages may be obtained by comparing the pattern of reversals of polarity of the sediments with the geomagnetic time scale.

The age of a rock may also be determined by the position of a paleomagnetic pole from that rock on a radiometrically dated path of apparent polar wandering. This method is especially useful if it is not possible to date a rock either

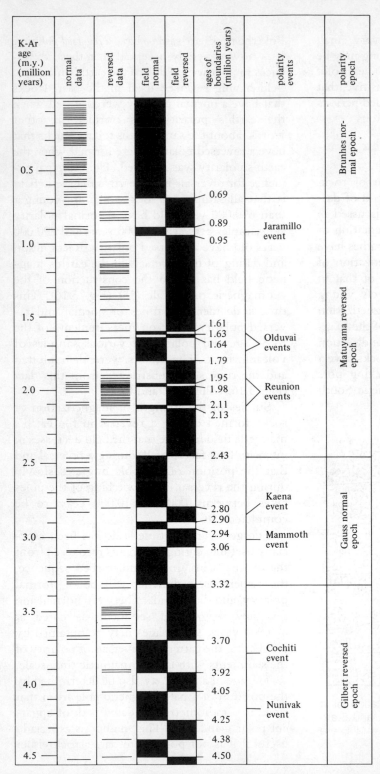

FIGURE 3.16 Geomagnetic time scale for the past 4.5 million years. Each short horizontal line on the second and third columns shows the age and polarity (samples with normal polarity in column 2 and samples with reversed polarity in column 3) of volcanic flows collected from various parts of the world. Periods during which the earth's magnetic field was dominantly of one polarity are called *epochs* (such as the Brunhes normal epoch) and short periods of opposite polarity within these epochs are termed *events*. (From Cox, Ref. 15; copyright © 1968 by American Association for the Advancement of Science.)

by paleontologic or radiometric methods. These concepts will be discussed in detail in Chapter 7.

## VARVES

Varve counts may assist in the determination of the ages of events during the retreat of the last ice sheet. *Varves* form on the bottom of lakes and consist of alternating light and dark layers of silt or clay (Fig. 3.17). The light-colored layers are generally coarser than the dark layers. They form during the spring, summer, and fall when streams carry coarse sediment along with dissolved oxygen into the lake. During the winter when the lakes are frozen, only fine silt and clay are deposited. The organic material in these sediments is not oxidized on the stagnant lake bottom. This material gives a dark color to the layers deposited during the winter. A varve set consists of a light-colored, relatively coarse-grained layer and a darker fine-grained layer. Each varve set represents one year of deposition. Varved sediments are typically formed in lakes near glacial margins. By radiocarbon dating of one of the varve sets, an absolute chronology of the lake history may be reconstructed by counting the total number of varve sets.

## IMPACT CRATERS AND DATING

Craters formed by meteorite impact may be used to date planetary surfaces, and may themselves be dated by various methods. Relative ages of the surfaces of the moon, Mars, and Mercury may be obtained by measuring the density of craters on them (Fig. 3.18). This is

**FIGURE 3.17** A number of varve sets from glacial Lake Hitchcock in Willimansett, Massachusetts. Each set consists of one dark layer and one light layer. (Photo by K. L. Verosub, University of California, Davis.)

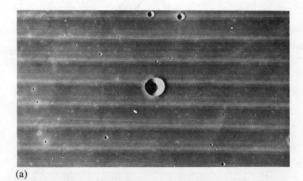

(a)

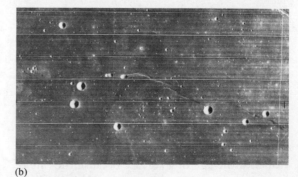

(b)

**FIGURE 3.18** **(a)** Crater density on Mare Imbrium as compared to **(b)** Mare Tranquillitatis (both on the moon). (From G. E. McGill, ref. 17.)

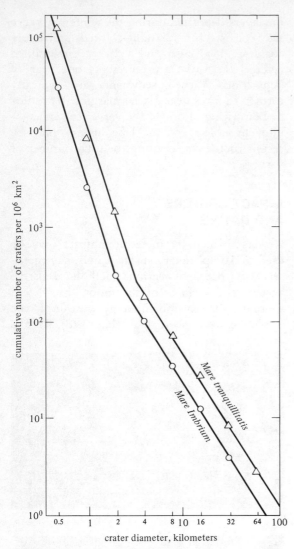

cumulative number of craters per 10⁶ km²

*Mare tranquillitatis*

*Mare Imbrium*

crater diameter, kilometers

**FIGURE 3.19** Log-log diameter versus frequency plots for Mare Imbrium and Mare Tranquillitatis for craters more than 0.5 km (0.3 mi) in diameter. (From G. E. McGill, ref. 17.)

usually done by measuring the size of craters in a given area and making a plot of the numbers of craters versus their relative age (Fig. 3.19). The density of craters is not directly proportional to the age of a surface because the rate of cratering in the solar system has declined sharply since its formation some 4.6 billion years ago. Radiometric dating has established

the age of some of the surfaces on the moon and the density of craters on these surfaces may be used to help determine the absolute age of other lunar surfaces. Furthermore, if the rate of cratering at any one time has been more or less constant throughout the solar system, crater density on lunar surfaces of known age may be used to determine the age of surfaces on other planets. This technique will not work for the earth because the rate of cratering is too low to use on young surfaces, and the rate of erosion is too great to use on old surfaces (many older craters are destroyed by erosion).

The relative ages of craters may be determined by the degree to which craters are smoothed by erosion. In a relatively young crater, the crater rim will be sharp, but with time the rim will be rounded and flattened and the crater may be filled in by material from the rim. (Fig. 3.20). On the moon, young craters have rays radiating from them (Fig. 3.21). Another technique in determining the relative age of craters is to study the *ejecta blanket,* which is the material blasted out of the floor of a crater. A crater which cuts the ejecta blanket is younger than the crater from which the ejecta blanket is derived, and conversely a crater which is partly filled by an ejecta blanket is older than the crater which produced the ejecta blanket.

In craters 10 km (6 mi) or more in size, the heat generated by the impact is sufficient to melt some of the rocks in the floor of the crater and this melt cools quickly to glass. A potassium - argon or fission track date on this glass will give the approximate age of impact, provided the glass was not heated up at a later date. Another way of determining the time of impact is by paleontologic dating of the materials filling the crater and in its wall. The age of impact will always be younger than the youngest disturbed rocks in the wall of a crater and older than the oldest rocks filling the crater. An absolute age range during which the impact occurred can be determined by consulting a radiometrically dated, geological time scale. For example, suppose that a crater cuts Middle Or-

FIGURE 3.20 Lunar craters more than 45 km (27 mi) in diameter showing the effects of erosional processes. Craters are arranged in order of increasing degradation and thus in order of increasing age. (From G. E. McGill, ref. 17.)

*(a) 64 km*  *(b) 100 km*  *(c) 87 km*

*(d) 91 km*  *(e) 64 km*  *(f) 105 km*

*(g) 98 km*  *(h) 96 km*  *(i) 110 × 126 km*

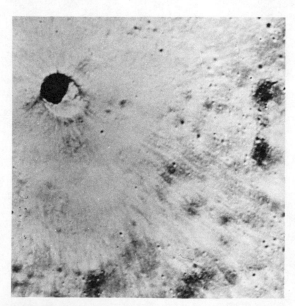

FIGURE 3.21 Lunar crater showing rays. (Courtesy of NASA.)

dovician, Trentonian sedimentary rocks and it was filled with Upper Ordovician, Richmondian sedimentary rocks. Appendix B shows that the Trentonian began 444 million years ago and the Richmondian ended 395 million years ago. Therefore, the impact must have occurred between 444 and 395 million years ago.

## REFERENCES CITED

1. L. R. Wager, 1964, *Quart. J. Geol. Soc. London,* v. 120s, p. 13.
2. H. W. Shimer and R. R. Shrock, 1944, *Index Fossils of North America:* Wiley, New York.
3. R. C. Moore, ed., *Treatise on Invertebrate Paleontology:* Geological Society of America.
4. J. L. Kulp, 1961, *Science,* v. 133, p. 1105.
5. A. Holmes, 1959, *Trans. Edinburgh Geol. Soc.,* v. 17, p. 183.

**6.** W. B. Harland, A. G. Smith, and B. Wilcock, eds., 1964, *The Phanerozoic Time-Scale: A Symposium:* Geological Society of London, v. 120s.

**7.** P. M. Hurley, 1959, *How Old Is the Earth?:* Doubleday, Garden City, N.Y.

**8.** F. A. Hills, P. W. Gast, R. S. Houston, and I. G. Swainback, 1968, *Geol. Soc. Amer. Bull.,* v. 79, p. 1757.

**9.** F. R. Hart and R. T. Dodd, Jr., 1962, *J. Geophys. Res.,* v. 67, p. 2998.

**10.** M. A. Lanphere and G. B. Dalrymple, 1976, *Earth Planet. Sci. Lett.,* v. 32, p. 141.

**11.** W. R. Van Schmus, E. M. Thurman, and Z. E. Peterman, 1975, *Geol. Soc. Amer. Bull.,* v. 86, p. 1255.

**12.** J. D. Macdougall, 1977, *Sci. Amer.,* v. 235, p. 114. R. L. Fleischer and P. B. Price, 1964, *Geochim. Cosmochim. Acta,* v. 28, p. 755.

**13.** E. K. Ralph and H. N. Michael, 1974, *Amer. Scientist,* v. 62, p. 553.

**14.** R. A. Muller, 1977, *Science,* v. 198, p. 489.

**15.** A. Cox, 1969, *Science,* v. 163, p. 239.

**16.** J. R. Dunn, M. Fuller, H. Ito, and V. A. Schmidt, 1971, *Science,* v. 172, p. 840. N. D. Opdyke, D. V. Kent, and W. Lowrie, 1973, *Earth Planet. Sci. Lett.,* v. 20, p. 315. K. L. Verosub, 1975, *Science,* v. 190, p. 48.

**17.** G. E. McGill, 1977, *Geol. Soc. Amer. Bull.,* v. 88, p. 1102.

# Reconstructing Ancient Environments 4

*Paleoecologists* can determine the environment in which a sedimentary rock was deposited from its color, rock type, grain size, degree of sorting, and fossils. Such environmental conditions as chemistry, depth, and clarity of water, as well as geographic setting and climate may be inferred by a comparison of the properties of a rock with sediments being deposited at the present time in various environments. A comparison of ancient and modern climates may provide evidence for changes in the latitude of a continent which results from the movements of lithospheric plates. Moreover, paleogeographic reconstructions have been very useful in oil and mineral exploration.

## RELATIONSHIP OF ORGANISMS TO THE ENVIRONMENT

Under natural conditions, organisms become adapted to their environment through the process of evolutionary selection. They may be classified according to the ecological niche which they occupy. *Benthonic organisms* live on the surface of the land, on the floor of the oceans, or on the bottom of streams, rivers, or lakes. They are either *vagrant benthonic,* meaning that they move about, or *sedentary (sessile) benthonic,* that is, permanently attached to the surface. Examples of vagrant benthonic organisms are horses, dogs, and lizards on land and snails, horseshoe crabs, and starfish in the

oceans. Sessile benthonic organisms include trees on the land, and corals, sponges, and bryozoans in the oceans. *Pelagic organisms* live above the land surface or above the bottom of the oceans, streams, rivers, or lakes. They are either *planktonic,* that is, floating organisms, or *nektonic,* which means that they swim or fly. Examples of planktonic organisms are many types of algae, foraminifera, and radiolaria. Birds, bats, fish, porpoises, and cephalopods are nektonic organisms.

## SEDIMENTARY ENVIRONMENTS

Marine environments consist of those below mean sea level, and continental environments are those above mean sea level. Since sea level exhibits considerable short-term variations, mainly because of tides, there is a complex transitional zone between marine and continental environments. Within each of these environments, the nature of the sediments encountered depends on such factors as agents of transportation, energy of currents, source of the sediment, kinds of organisms present, water depth, and the chemistry of the water.

### Marine Environments

Sampling of sediments from ocean basins by coring of the subbottom sediments has provided modern equivalents of the depositional

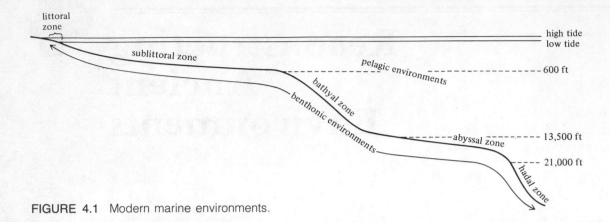

FIGURE 4.1 Modern marine environments.

environments of ancient marine sediments and sedimentary rocks. Such studies have shown that the most reliable criteria for the recognition of marine deposits are the presence of fossils that belong to taxa (A *taxon* is a group of organisms such as a species, genus, family, etc.) found only in marine environments and the absence of fossils that belong to taxa found only in freshwater or brackish water today. Modern corals, for example, live only in marine environments, and it is very probable that ancient corals were similarly restricted. The presence of deposits of salt or gypsum does not necessarily indicate deposition in a marine environment because some evaporite deposits form in enclosed terrestrial basins like the Great Salt Lake basin.

Organisms interact with the overall marine environment and produce changes in the chemistry of the seawater and changes in the character of the sediment deposited. The lithology of a sedimentary rock and the fossils and sedimentary structures which it contains may be used to infer such conditions as water depth, oxygen content, salinity, and clarity of the seawater.

WATER DEPTH. The sea floor below low tide level is the *benthic environment,* which may be divided into the sublittoral, bathyal, abyssal, and hadal zones (Fig. 4.1). The *sublittoral zone* is located between low tide and a depth of 200 m

(600 ft). The lower limit of sunlight penetration, the *photic zone,* is about 50 m (150 ft). Life is abundant above this level, but is less so below it. Under normal conditions, the shallow parts of the sublittoral zone are affected by wave action down to wave base. The term *wave base* refers to the depth to which average waves will provide the energy to move sediment. During times of storms, waves of large amplitude (height) will scour sediment below wave base. Because marine organisms are affected by wave action, protective shells tend to be heavier in the sublittoral environment than elsewhere.

Where the influx of clastic sediment is not too great, organisms of the sublittoral zone may build moundlike *bioherms* composed of a great variety of life such as corals, sponges, foraminifera, bryozoa, sea urchins, sea lilies (crinoids), brachiopods, clams, and oysters (**1**). The tops of bioherms are typically just below low tide level. *Reefs* are a type of bioherm which is more than 1.66 km (1 mi) in length. Many reef organisms do not grow at depths greater than 20 m (65 ft) (**1**), and the maximum depth at which corals are active in building reefs is 46 m (150 ft) (**2**).

Another characteristic feature of the sublittoral environment, especially in warm climates, is *algal mats*. These are dense, sheetlike masses of intertwined algal filaments 0.1 to 5.0 mm (1/250 to 1/3 in.) thick (**3**). Algal mats containing blue-green algae commonly form calcareous (containing calcite or aragonite) structures known as *stromatolites* (Fig. 9.41). Because algae must have sunlight to live, stromatolites can form only in the shallow waters of the photic zone. Some ancient stromatolites formed at depths of about 45 m (140 ft) (**4**) near the bottom of the photic zone.

Algae are important agents in the formation of several other types of calcareous deposits, including many types of limestones. Some algae contain tiny needles of calcium carbonate in the form of aragonite in their cell walls. Aragonite has the same chemical composition as calcite, but it has a different crystal structure and therefore is a different mineral. These needles of aragonite are released when the algae die. Other types of algae aid in the precipitation of calcium carbonate by removing carbon dioxide from the seawater. The amount of calcium carbonate contained in solution is dependent on the carbon dioxide in the seawater. When carbon dioxide is removed, calcium carbonate is precipitated. Some types of algae trap tiny particles of calcium carbonate on their sticky filaments. The presence of thick limestones suggests that deposition occurred within the photic zone in waters less than 50 m (150 ft) deep.

*Ooliths* (also called oolites and ooids) are small, concentrically layered grains generally composed of calcium carbonate (calcite or aragonite). They form in the shallow waters of the sublittoral zone, and their presence in a limestone indicates deposition in a shallow water environment (Fig. 4.2). A gentle wave action is necessary to maintain the rounded shape of the oolith. Oscillatory ripple marks are associated with wave action and therefore are produced only in shallow waters.

*Deltas* extend from continental environments into the sublittoral zone. Deltaic deposits of the sublittoral zone include primarily clastic sediments such as sands, silts, and muds. *Clastic sediments* consist of fragments of rocks or organic structures which have been transported from their place of origin.

The *bathyl zone* extends from 200 to 4100 m

**FIGURE 4.2** Hand sample of an oolithic limestone. (Photo by W. T. Huang.)

(600 to 13,300 ft) in depth, and part of it is approximately coincident with the continental slope and the upper part of the continental rise at the base of the continental slope. Seismic reflection profiling indicates that little sediment accumulates on the relatively steep slopes in the upper bathyal region and that thick wedges of sediment are deposited at the base of the continental slope in the lower bathyal and upper abyssal zones. Core samples from the base of the slope are mostly poorly sorted turbidites. These deposits were laid down by rapidly moving turbidity currents that flowed down the continental slope. The bathyal zone also includes the crest and flanks of oceanic ridges. Here the sediments tend to be pelagic deposits. Calcareous deposits are abundant here because calcium carbonate is dissolved in the deeper parts of the ocean. The calcareous *tests* (skeletons) of planktonic foraminifera sink toward the sea bottom when the organism dies. In areas far from terrestrial sources of sediment, these tests accumulate as a lime mud or ooze at depths down to about 3000 m (10,000 ft) (**5**). However, at the great pressures below that depth, calcareous tests begin to dissolve. Therefore, foraminiferal oozes are not generally found below a depth of 3000 m (10,000 ft) un-

less the rate of sedimentation is unusually high. The depth below which calcium carbonate is rapidly dissolved is called the *carbonate compensation level*.

The *abyssal zone* extends from 4100 m (13,500 ft) to 6180 m (21,000 ft) and covers most of the deep ocean basins. Abyssal sediments include turbidites, clay, and biogenic oozes. The turbidites are found as much as 1300 km (780 mi) from the nearest land (**6**) which demonstrates that turbidity currents can travel great distances. The clay tends to be reddish or brownish in color and forms from sediment suspended in ocean water. The oozes are composed largely of the remains of siliceous plankton such as radiolaria. The lithified equivalents of these sediments are graywacke, claystone, and chert. Sedimentary rocks formed in the abyssal zone are often associated with *ophiolite sequences*. These consist of serpentinites and/or peridotites at the base, overlain consecutively by gabbros, diabase dikes, pillow lavas, and pelagic sediments (often radiolarian cherts). Ophiolites are thought to be sections of the ocean crust and possibly also uppermost mantle. They are incorporated into the continental crust during collision of plates.

The *hadal zone* extends below a depth of 6180 m (21,000 ft) and is confined to oceanic trenches. Sediments in this area are primarily turbidites derived from continental areas or island arcs bordering deep ocean trenches (see Chapter 6).

Although sunlight does not penetrate the bathyal, abyssal, or hadal zones, both nektonic and benthonic organisms are found there. Photographs have revealed the presence of brittle stars as well as the tracks and burrows of other organisms at pressures greater than 60,000 gm/sq cm (2000 lbs/sq in.). But life is not abundant in the bathyal zone, and it is rare in the abyssal and hadal zones. For this reason, fossils are for the most part uncommon in deep water deposits. Deep water deposits may however contain fossils of planktonic and nektonic forms that lived in shallower water and drifted into deeper water when the organisms died or that

were carried from shallower water by turbidity currents.

OXYGEN CONTENT. The amount of free oxygen dissolved in seawater affects the color and fossil content of the sediment deposited in that environment. Rocks deposited under *reducing conditions* (no free oxygen present in the water) are generally black in color due to the presence of abundant unoxidized carbon. Under *oxidizing conditions* (free oxygen present) carbon is oxidized and the rock is much lighter in color. Only a few types of organisms, such as worms and certain brachiopods, can live under reducing conditions. Fossils are generally rare, except in those cases in which organisms were carried into the reducing environment after death.

SALINITY. Typical marine waters have salinites of about 34 parts per thousand of dissolved salts, such as halite, gypsum, and sylvite, while freshwater has less than 10 parts per thousand of dissolved salts. Brackish water would be between 34 and 10 parts per thousand dissolved salts. Salinity of seawater in which a rock was deposited may be determined by comparison of the fossils found in the rock with modern organisms whose salinity preference is known. Certain species prefer brackish water, while others prefer normal marine waters.

pH. The acidity or alkalinity (pH) of seawater in which a rock was deposited may be determined by comparison with modern organisms. The composition of the rock may also indicate something about the pH. Limestone is not deposited under very acidic conditions and chert is not deposited under very alkaline conditions.

CLARITY OF THE SEAWATER. The amount of suspended matter affects the clarity of the water, and it may be inferred from the type of rock present. Shales and poorly-sorted sandstones, such as graywackes, are generally deposited from muddy waters while limestones

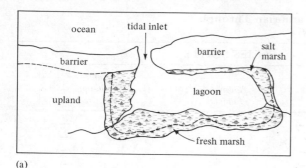

(a)

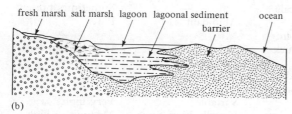

(b)

FIGURE 4.3  Transitional environments: **(a)** map view, **(b)** cross section.

and well-sorted sandstones and conglomerates are generally deposited in clear waters. Furthermore, certain organisms prefer clear water while others prefer muddy water; therefore, fossil content may be an additional guide to water clarity.

## Transitional Environments

The transitional environments lie within the littoral zone between average high and average low tide levels, and include beaches, tidal flats, lagoons, estuaries, and coastal marshes (Fig. 4.3). The littoral zone may vary from a few feet to a few miles in width, since the extent of tidal flooding depends on the slope of the beach or tidal flat and the height of the tides.

BEACHES.  In the high-energy environment of a beach, sediments may range from medium-grained sand to gravel. Sandstones deposited in such an environment may retain a variety of characteristic sedimentary structures such as ripple and swash marks (Fig. 4.4). Relatively few varieties of organisms are found in the beach zone, primarily because wave action severely restricts the inhabitable niches. Organisms that have adapted with varying success to this environment include burrowing crustaceans, heavy-shelled pelecypods and gastropods, plants rooted to holdfasts, and nektonic forms such as fish. The fossil remains of many organisms are destroyed or broken and abraded by wave action, and shell and bone fragments may be found in thin beds along with coarse sand and small pebbles in this zone.

FIGURE 4.4  Swash marks on a sandy beach on the island of Eleuthera in the Bahamas. Swash marks are thin, wavy lines of sand, shell fragments, bits of seaweed, or other debis left at the point where each wave begins to recede from the beach.

TABLE 4.1   **Facies Characteristics as Recorded in Cambrian Through Devonian Carbonates of the Central Appalachians**[a]

| Facies Characteristics | Tidal Flat | Shallow, Subtidal | Deep, Subtidal | Organic Build-ups (e.g., reefs) |
|---|---|---|---|---|
| | | FACIES SUITES | | |
| Mud cracks | Typical | — | — | — |
| Scour and fill structures | Typical | — | — | — |
| Pebble conglomerates | Typical | — | — | — |
| Laminations | Typical | — | — | — |
| Early dolomite | Typical | — | — | — |
| Cross-bedding | Small scale | Medium scale | — | Sometimes present |
| Burrow-mottling | Rare | Common | Abundant | Rare |
| Oolites | — | Often present | — | — |
| Bedding | Thin–medium | Medium–thick | Thick–massive | Unbedded–massive |
| Algal structures | Present | Present | Absent | Present |
| Burrows | Vertical | Vertical and horizontal | Horizontal | Rare |
| Fossil abundance | Low | Very high | Variable | Very high |
| Fossil diversity | Low | Medium | Usually high | Medium to high |
| Major taxa | Trilobites and/or ostracods | Calcareous algae, pelmatazoa, brachiopods | Brachiopods, bryozoa, trilobites, and pelmatozoa | Tabulate, rugose and stromatoporoid corals |
| Vertical facies variations | Sharp and frequent | Transitional and common | Very gradual and infrequent | Complex |
| Areal facies variations | Outcrop scale | Relatively persistent | Basinal scale | Outcrop scale to several miles |
| Facies strike | Variable | Parallel to basin margin | Parallel to basin axis | Variable |

[a] From L. F. Laporte, personal communication.

Tidal pools common to rocky coasts may contain a great variety of organisms. The rocks of the tide pool are often coated with attached invertebrates and algae. Mobile benthonic forms such as crabs and starfish live among the rocks. A tidal pool would yield abundant fossils after burial.

TIDAL FLATS.   Tidal flats generally occur in low energy littoral environments that have developed a low profile through deposition of fine-grained sediments such as silt, clay, and lime mud (Table 4.1). Clastic sediments accumulate in areas adjacent to highlands, whereas lime muds, which are the precursors of limestone and dolostone, are deposited in warm climates in the absence of nearby sources of clastics. Intermittent subaerial exposure (exposure to the air) of the tidal flats may produce mud cracks that will be filled in by sediments; the resulting sedimentary rock will preserve these structures (Fig. 4.5).

LAGOONS.   Barrier islands and coral reefs may isolate quiet, shallow bodies of water from the sea. These bodies of water are called *lagoons*, and their deposits generally consist of terrigenous (land derived) clastic sediment and calcareous muds.

The nutrient level is often high in lagoons and this results in a large number and a great variety of organisms. Restricted circulation in

(a)  (b)

FIGURE 4.5 Comparison of modern mud cracks with those in an ancient sedimentary rock: **(a)** Mud cracks produced by the drying of a small pond (photo by David Leveson). **(b)** Mud cracks in argillaceous siltstone of Permian age, Kansas (photo by Ada Swineford).

an arid climate may result in the deposition of salt (halite), gypsum, or anhydrite in the lagoon.

ESTUARIES. Mixing of marine and fresh water is also characteristic of estuaries. Estuaries are open bodies of brackish water that form at the mouths of rivers because of tidal changes and mixing of fresh and seawater. Estuarine sediments are generally terrigenous clastics and commonly contain current-produced, depositional structures such as ripple marks, cross strata, and channels (Figs. 4.6 and 4.7).

Estuaries have distinct faunal and floral assemblages that are adapted to the brackish water environment. Recent studies of estuarine fauna have revealed distinct salinity preferences on the part of certain species. Some species are able to withstand a wide range of salinities and can live in both fresh and seawater, while other species are restricted to fairly narrow salinity tolerances.

COASTAL MARSHES AND SWAMPS. A marsh is an area of low, wet land that contains various species of grasses and herbs but few, if any, trees and shrubs. Coastal marshes lie within the tidal zone and grade into freshwater marshes above sea level. The sediments in marshes are similar to those found in tidal flats. The major difference between coastal marshes and tidal flats is that water movement in the marshes is sufficiently restricted that marsh plants (usually salt-tolerant species of grass) may become established there. Such plants act as filters trapping fine sediments such as silt and clay. Under suitable conditions, marsh plants may form peat.

Swamps differ from marshes in that they contain trees and a few shrubs. The tropical and subtropical mangrove swamps occur in both salt and brackish water environments of the littoral zone. Brackish-water swamps of the transitional zone may grade into freshwater swamps, such as the cypress swamps of the

(a)
(b)

**FIGURE 4.6** Comparison of modern and ancient ripple marks: **(a)** Ripple marks on the shore of Plum Beach, New York (photo by David Leveson). **(b)** Current ripple marks on sandstone, northern Scotland (NERC Copyright. Reproduced by permission of the Director, Institute of Geological Sciences, London).

**FIGURE 4.7** Stream channel filled with sediments. Following deposition of the relatively flat layers a stream channel was cut into them. Later the stream was filled with its own sediments. (Courtesy of Richard V. Fisher.)

southeastern United States. Sediments deposited in swamps include silt, mud, clay, and peat.

Coastal marshes and swamps provide quiet environments for smaller invertebrates and a breeding ground for many marine vertebrates. These environments, which are distributed throughout the world, provide models of coal-forming processes.

## Continental Environments

Most continental deposition occurs in the channels and flood plains of rivers and streams, in lake beds, swamps, and deserts, and at the margins of glaciers.

STREAMS AND RIVERS.   Streams and rivers transport and deposit a variety of sediments ranging from boulders and pebbles to sand and mud. Deposition is greatest on flood plains or in deltas, where the transporting currents are sufficiently diminished for sedimentation to occur. Stream and river (fluvial) deposits are characterized by a variety of depositional structures, such as channel fillings and assymmetrical ripples (Fig. 4.7).

Fluvial deposits commonly consist of medium- to coarse-grained sandstone and conglomerate. The conglomerates are rather well-sorted and their pebbles are generally well-rounded. Such conglomerates consist largely of white quartz pebbles. While fluvial conglomerates may resemble beach deposits, they tend to occur in linear bands. The orientation of the pebbles indicates that the current flowed parallel to the bands.

The color of fluvial deposits may be useful in establishing their environment of deposition. Many fluvial deposits in arid regions are red in color due to the oxidation of iron-bearing minerals. In humid regions fluvial sediments acquire lighter colors because water seepage carries the oxides away. Marine clastic deposits, on the other hand, are normally gray, tan or black, since ocean water is generally not strongly oxidizing. However, the presence of marine fossils

in some red beds indicates that red coloring alone is not conclusive evidence of continental deposition. The red coloring in such deposits may have developed subaerially prior to redeposition of the sediments in the marine environment.

Coalescing fluvial deposits in desert areas form extensive *bajadas*. The deposits in bajadas are generally coarse and when lithified form coarse-grained, moderately well-sorted sandstones and conglomerates. Such deposits commonly grade into finer-grained playa deposits which contain fine-grained sandstones, siltstones, and shales. *Playas* are intermittent lakes formed on the flat floor of the desert valley.

Fluvial environments are less likely to preserve fossils than marine environments, since swift currents and turbulence discourage colonization and often destroy fossils. Though uncommon in fluvial deposits, fossils present may include clams, snails, fish, amphibians, reptiles, and mammals.

SAND DUNES.   Sandstones derived from wind-laid deposits are very common in the geological record. Dune sands are well-sorted and commonly have grains that have been frosted as a result of repeated high-speed impacts with other grains. The microtexture of dune sands may also be distinctive (Fig. 4.8). Dune sandstones characteristically contain large-scale cross stratification which is oriented parallel to the slip face of the sand dunes (Fig. 4.9).

LAKES.   Lakes are formed by a variety of erosional, depositional, and tectonic processes, such as scouring of a depression in bedrock by a glacier, damming of meltwater between the ice and an end moraine, or damming of river valleys by lava, fault blocks, and landslides. *Lacustrine* (lake) deposits are generally composed of fine-grained clastics, and their floral and faunal associations are similar to those of fluvial environments. Lake muds may contain a variety of microscopic plants and ani-

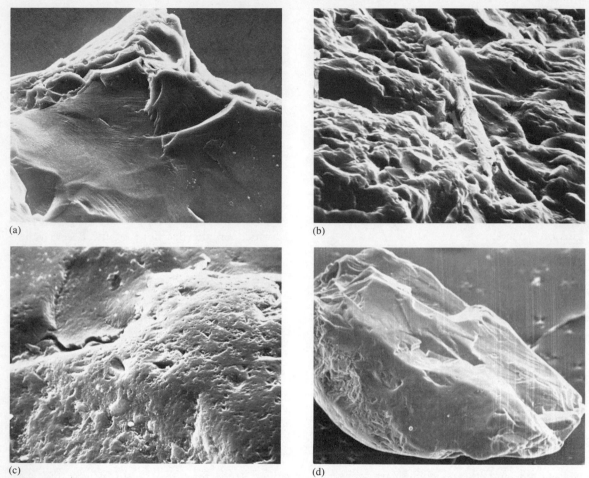

**FIGURE 4.8** Scanning electron photomicrographs of sand grains: **(a)** Edge of quartz sand grain showing irregular fractures probably caused by grinding in glacial ice; this grain was taken from a core of Miocene age taken from the North Pacific. Magnification 900X. **(b)** Quartz grain from a dune sand collected near Cairo, Egypt; upturned impact plates probably resulted from impact and crack propagation across crystallographic planes. Magnification 1300X. **(c)** Quartz sand grain from a high-energy beach near La Jolla, California; this grain shows V-shaped patterns superimposed on worn breakage blocks. Magnification 1500X. **(d)** Quartz sand grain from Late Pleistocene deposits of Nassau County, Long Island. The irregular fractures on the upper surface and the angular outline suggest glacial action; the lower left-hand portion of the grain is rounded as a result of later dune action. Magnification 75X. (Photos courtesy of David Krinsley and Stanley Margolis.)

mals, as well as clams, snails, and fish. The relatively short-lived nature of most lakes makes their sediments poor repositories of long geological records, but their rates of deposition are beneficial in preserving detailed evidence of events within a geologic epoch.

The color of lake sediments is an indication of their environment of deposition. Black lake shales point to deposition in a reducing environment on a stagnant lake bottom, while a red color indicates deposition in a strongly oxidizing environment such as in a playa. Under nor-

(a)

(b)

**FIGURE 4.9** Comparison of modern sand dunes with ancient dune sandstone: **(a)** Sand dune in the Calanscio Sand Sea, Libya (courtesy of British Petroleum Co.). **(b)** Cross-bedding in dune sandstone of the Triassic new red sandstone (NERC Copyright; reproduced by permission of the Director, Institute of Geological Sciences, London).

mal conditions, lacustrine deposits are grey to brown.

SWAMPS, MARSHES, AND BOGS. Poorly-drained lands favor the development of freshwater swamps, marshes, and bogs. These environments include the muskeg swamps of the Arctic, peat bogs of the midlatitudes, and tropical swamps. A bog is a tract of wet, spongy ground underlain by an accumulation of peat. In contrast to swamps and marshes, the

vegetation is primarily mosses, sedges, and/or heaths which grow actively at the surface but decay slowly beneath the surface. The peaty deposits are generally dark in color due to reducing conditions.

Swamps, marshes, and bogs furnish a vast breeding ground for insects, fish, and aquatic birds. These environments retard decomposition and provide rapid burial of organic remains and are thereby capable of producing extensive coal deposits.

GLACIERS.  Glacial deposits vary considerably in texture and composition and range from stratified to unstratified deposits. Stratified drift is deposited by the runoff from melting glaciers and is, therefore, a *fluvioglacial deposit*. Such deposits closely resemble other fluvial deposits, but their origin is evident if they are associated with other types of glacial deposits.

*Till* is deposited directly by a glacier, and it is largely unsorted and unstratified with particles ranging from clay-size (less than $1/256$ mm) to boulders as big as a house. Clay-sized particles may consist of mineral fragments pulverized by the ice without significant chemical weathering. In other types of sediments, clay-sized material is composed mainly of clay minerals formed by the weathering of other minerals. The sand-sized mineral grains in till may bear distinctive surface markings such as microstriations and roughly broken surfaces formed by glacial abrasion (Fig. 4.8).

The presence of pebbles, boulders, or cobbles with striated and/or faceted surfaces within a deposit is a strong indication (but not conclusive proof) of a glacial origin (Fig. 4.10). If such

FIGURE 4.11  Striated bedrock in front of the Athabasca Glacier in Jasper National Park, Canada.

a deposit rests on a grooved, striated, and polished *basement* (underlying bedrock), it is virtually certain to be of glacial origin (Fig. 4.11). The lithified counterpart of till is called *tillite, diamictite,* or *mixtite*. The term tillite should only be used to describe a lithified till, ie., a rock of glacial origin. Diamictite and mixtite are descriptive terms used to designate unsorted and unstratified deposits in general.

Glacial deposits generally contain few fossils due to the scarcity of fauna and flora near a glacier and to the extremely rapid rate of glacial deposition. Wood, a common fossil in till, was incorporated in the sediment when the glacier overrode a forest.

## PALEOCLIMATOLOGY

Climate plays an important role in determining the nature and composition of sediment in a given area. Generally, glacial deposits are associated with cold climates, and coral reefs and

FIGURE 4.10  Striated and faceted cobble from the Rapitan Group in the Mackenzie Mountains, Canada. (Courtesy of Grant Young.)

FIGURE 4.12 Roche moutonnée in the Sierras, California. Ice moved from right to left in this photograph.

thick limestones are associated with warm climates. Dune sands, evaporites, arkosic red beds, and dolostones are commonly deposited in arid or semi-arid climates. Ideally, the *paleoclimate* (ancient climate) of an area is reconstructed on the basis of not just one, but several criteria, termed *paleoclimatic indicators*. Comparison of paleoclimate with modern climatic belts may indicate the *paleolatitude* (ancient latitude) of a region at the time of deposition. Comparison of present latitude with paleolatitude is a way of determining large scale movements of plates.

### Evidence of Glacial Climates

Glacial climates are commonly indicated by the presence of such deposits as tills, tillites, varved clays, and ice-rafted detritus. Of these, tills and tillites are the most easily recognized and consequently the most commonly reported.

Bedrock beneath till or tillite may be smoothed and shaped as *roches moutonnées* (Fig. 4.12). These features are elongated parallel to the direction of ice flow and have a gently inclined *stoss* side in the direction from which

the glacier flowed and a steeply inclined *lee* side in the direction of ice flow. Striations on bedrock are also parallel to the direction of glacial flow.

### Indications of Arid Climate

Arid climates are generally found at low to middle latitudes, approximately 10° to 40° from the equator. However, arid climates also may be found in the rain shadow of a mountain range and in the center of a large continent. Sand dunes, evaporites, and red beds are characteristic of such environments.

The average dip (inclination) of cross stratification in dune sandstones is in the direction of the prevailing wind at the time of deposition.

Extensive dune sandstones generally indicate deposition in an arid or desert climate. However, in some moist regions, such as northeastern North America and southern Alaska, the combination of an excessive supply of sand and high winds gives rise to dune fields in regions of moderate or high rainfall. Thus the presence of dune sandstones alone may not be conclusive evidence in determining ancient dry climates, but may supplement other indicators of aridity, such as red beds and evaporites.

RED BEDS.   The origin of the red coloring in red beds has been the subject of considerable debate among geologists (**7,8**). Red, lateritic soils may be produced in moist tropical and subtropical regions owing to deep weathering. Most silicates are chemically weathered into red clays rich in aluminum and iron. Sediments eroded from a lateritic soil and deposited in an oxidizing environment would presumably retain the red color. However, there are no known areas in which red beds of this type are being formed at the present time. Furthermore, the absence of feldspar in lateritic soils indicates that red arkoses cannot have been formed by erosion of such soils.

Red sediments are common in desert regions such as the Kalahari Desert of southern Africa, northern Columbia, and northwestern Mexico (**8**). In these regions rain is infrequent and much of what there is comes down in cloudbursts. For this reason, feldspar and iron-bearing minerals are eroded before they can be completely weathered chemically. The first step in the formation of the red coloring in the sediment is the weathering in place of the iron-bearing minerals, such as hornblende and biotite, to limonite (hydrated iron oxide) within the sediment. In a hot, arid climate the limonite is slowly dehydrated to form hematite. Because feldspar is relatively stable in a dry climate, it is likely that the red arkosic sandstones are produced in this manner. In moist temperate or polar climates, red beds generally will not form because iron is usually carried away in solution. Red arkoses, therefore, are a useful rock type as an indicator of hot, arid conditions.

EVAPORITES.   *Evaporites* are deposits of salt (halite), anhydrite, gypsum, sylvite, and other salts formed by the evaporation of sea water. Primary requirements for the deposition of large evaporite bodies are low rainfall, low humidity, and high temperatures. For many years, it was assumed that all evaporites were formed in enclosed basins, but today many investigators believe that most evaporites form in *sabkhas,* flat regions bordering an ocean just above mean high tide (**9**). It is more likely that evaporites commonly occur in both regions. Many major evaporite deposits apparently developed in large basins that were bordered either by land or coral reefs (**10**) (Fig. 4.13). Such basins allowed little exchange of water with the open ocean, and thus significant concentrations of dissolved salts were formed. The Mediterranean Sea is an enclosed basin in an area of relatively low rainfall, but no extensive evaporites are accumulating in this area because the inflow of normal marine water is balanced by an outflow of more saline water through the Strait of Gibraltar. There are, however, indications that the Mediterranean Sea was the site of extensive evaporite deposition during the Tertiary period (see Chapter 13).

While most evaporites occur in shallow water, deep water evaporite deposition has also been demonstrated. Coring by the drilling ship *Glomar Challenger* has shown that there are extensive evaporites in the Gulf of Mexico; and seismic reflection profiling has suggested the presence of salt domes in several parts of the Atlantic ocean. In both of these areas the water depth is greater than 4 km (2.5 mi). These evaporites must have been formed when the continents were closer together and the basins of deposition were more restricted than they are today.

For precipitation of large quantities of dissolved salts to occur, the rate of evaporation must exceed the inflow of freshwater from rainfall, streams, and rivers. Such conditions are found today in arid climates of low to middle latitudes, leading climatologists to believe that large evaporite deposits are indicative of arid climates. The deposition of potassium-bearing salts (e.g., sylvite) requires almost complete evaporation of seawater. Therefore, the presence of such salts may indicate deposition in a

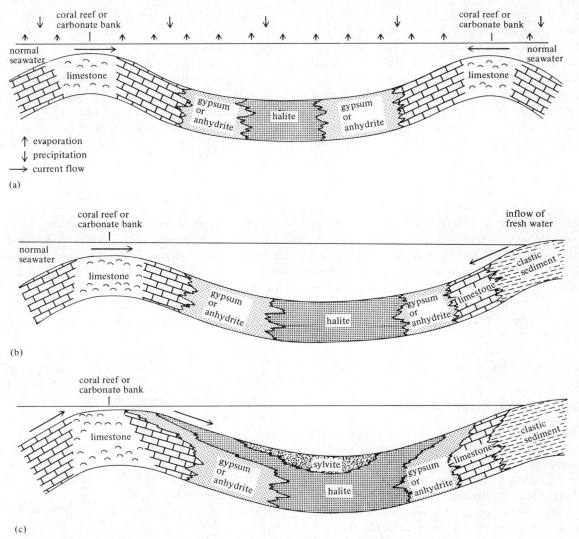

coral reef or
carbonate bank

coral reef or
carbonate bank

normal
seawater

limestone

gypsum
or
anhydrite

halite

gypsum
or
anhydrite

limestone

normal
seawater

↑ evaporation
↓ precipitation
⟶ current flow

(a)

coral reef or
carbonate bank

inflow of
fresh water

normal
seawater

limestone

gypsum
or
anhydrite

halite

gypsum
or
anhydrite

limestone

clastic
sediment

(b)

coral reef or
carbonate bank

limestone

gypsum
or
anhydrite

sylvite

halite

gypsum
or
anhydrite

limestone

clastic
sediment

(c)

**FIGURE 4.13** Formation of evaporites in an enclosed basin: **(a)** Basin bordered by coral reefs or carbonate banks. **(b)** Basin bordered partly by coral reefs and partly by land. **(c)** Basin with very restricted inflow due to regression of the sea or up-building of carbonate bank.

very arid climate. Thinner salt deposits, including salt crusts and halite crystal casts on bedding planes, may form in semi-arid climates.

Large, enclosed marine basins do not presently exist on the continents and consequently there are no extensive marine evaporites forming today. However, minor amounts of marine evaporites are being deposited in the Persian Gulf, on the east and west coasts of Baja, California, and on the northern border of South America (**11**). These areas are located between 10° and 30° N latitude and have arid climates.

DOLOSTONES. The origin of dolostones is a widely debated topic among geologists. Experimental and observational data indicate that dolomite, which is the principle mineral of dolostone, is seldom if ever precipitated directly from seawater. At the present time, however, dolomite has been observed replacing calcite and aragonite in the sediments above mean high tide in areas of high aridity (**12**). Evaporation of the water trapped in sediments in the supratidal zone results in the formation of a very saline *brine*. This evaporation process also results in an increase in the magnesium content of the brine. Under such conditions, primary calcium carbonate may be dolomitized before the sediment is lithified. Dolomitization may also occur after lithification in relatively porous rocks such as reef deposits. Concentrated brine moving through the reefs results in replacement by dolomite (**13**). The climatic significance of these processes is that concentrated brine is produced only where rainfall is low and evaporation is high. Consequently, dolomitization should be widespread only in arid climates. The common association of dolostone with thick evaporites and arkosic red beds supports this inference.

## Indication of Warm Climates

Precipitation of large quantities of calcium carbonate, both by organic and inorganic processes, requires a rather warm climate. As in the case of evaporites, large volumes of seawater are required because of the low concentration of calcium carbonate in seawater. Warm temperatures are favorable for the precipitation of calcium carbonate, because the rate of evaporation increases and the solubility of calcium carbonate decreases with increasing temperature. Limestones are being formed today in the warm waters of Bermuda, the Bahamas, and the Persian Gulf.

Modern reef-building corals have skeletons composed of calcium carbonate and are always found in warm waters. The mean winter temperature where reefs grow is between 27° and 29°C (81° and 84°F) (**1**). Such waters are found today in warm-moist or warm-dry climates

within 33° or so of the equator. Modern coral reefs are generally found between 35° N latitude and 32° S latitude (**1**) (Fig. 4.14). Thus the presence of thick limestones and extensive coral reefs suggests deposition in a warm climate relatively near the equator. However, it should be pointed out that less extensive carbonate-rich sediments and small coral bioherms are forming today in relatively cool waters at latitudes as high as 71° N.

## Indication of Moist Climates

Thick coal sequences have long been associated with the lush growth of tropical swamps and marshes. However, coal-forming swamps and marshes may form in a climatic zone in which the rainfall exceeds potential evaporation plus biological transpiration. Peat is forming today in tropical, temperate, and cold climates (Fig. 4.15).

Tropical coals are important in the location of paleoequators at the time of formation of the coal. Some of the criteria for the recognition of coals formed in tropical climate based on the coal-forming vegetation (**15**) are:

1. Growth rings are generally absent in woody plants (trees).
2. Vegetation tends to be very lush, and it is often of very large scale.
3. Much of the vegetation is succulent in nature.
4. Plant tissues have large, thin-walled cells and organs have large intercellular spaces.
5. Many plants have large, drooping fronds, and leaves tend to be large.
6. Some plants have subaerial roots.
7. There are a great number of species in tropical regions. For example, in a tropical rain forest, there are often 100 species and sometimes as many as 200 species of trees in a two hectare (five acre) plot. This compares with about ten species in the same size area in a typical New England forest.

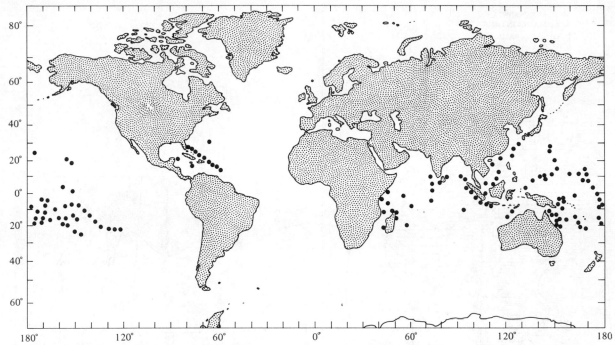

**FIGURE 4.14** Map of the world showing distribution of modern coral reefs. Reef areas are indicated by large dots. There are no important coral reefs at latitudes beyond approximately 30°. (After Shrock and Twenhofel, Ref. 14. Copyright © 1953 by McGraw-Hill, Inc. Used with permission of McGraw-Hill Book Co.)

(a)                                              (b)

**FIGURE 4.15** Modern peat-forming environments: **(a)** Peat bog in Ireland with a low escarpment where peat has been removed (photo by Eleanor Catena). **(b)** Mangrove swamp in southern Florida.

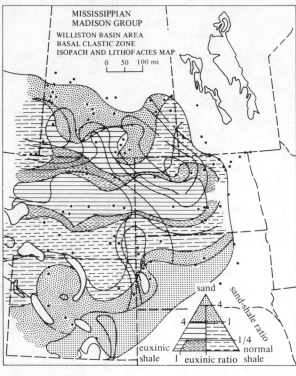

(a)

(b)

**FIGURE 4.16** Using isopach and lithofacies map in the construction of a paleogeographic map: **(a)** Inferred paleogeographic map at the time of deposition of the Madison Group. **(b)** Isopach and lithofacies map of the Mississippian Madison Group. (From H. G. Thomas *in* Sloss, *et al.,* Ref. 16. Copyright © 1960 Wiley, New York.)

In the absence of recognizable plant macro-fossils, the composition of the fossil floras may be indicated by the types of pollen present, which in turn may be used to indicate the climate during deposition of the coals.

An environment unfavorable for coal formation is a hot, dry climate because peat-forming vegetation requires moisture. It is interesting to note that coals are not generally found in sequences containing abundant evaporites, dune sandstones, and red beds (Figs. 10.3, 11.1, and 12.1).

## PALEOGEOGRAPHIC MAPS

Paleogeographic maps show depositional environments at some point in time (Fig. 4.16a). These maps may show the locations of such geographic features as mountains, rivers, shallow seas, deep seas, and islands. Mountains may be located by examining changes in thickness and grain size of a sedimentary unit. In general, sediments get thicker and coarser in the direction of their source, and if the sediments are very thick and very coarse, it is likely that they were deposited close to the mountain range. River mouths may be located where a tongue of coarser sediment extends into finer sediment (Fig. 4.16b). Shallow and deep seas are indicated by sediment type as discussed earlier. Shorelines are located where sedimentary deposits indicate a change from continental to marine environments. Moreover, shorelines tend to parallel the boundaries between rock types within a unit (Fig. 4.16b).

A major problem arising in the preparation of paleogeographic maps for continental areas is the lack of subsurface data for regions in which relatively young deposits cover older sequences. This problem is especially acute near the continental margins, where sedimentary units thicken greatly. Many paleogeographers have assumed that an area was above sea level at a given time if direct evidence of marine deposits of that age have not been found. However, unless marine deposits of that age are known to be absent, this assumption can be erroneous. Areas in which marine deposits of a given age are known to be absent may have had such deposits removed by erosion.

## RECOGNITION OF IMPACT CRATERS

Craters are produced by volcanic activity and by the impact of meteorites. Young impact craters have rims which have a relatively constant ratio of rim height to crater depth for a given size crater. On the other hand, some volcanic craters have no rim, while others are perched on top of volcanoes which are large in proportion to the size of the crater. Furthermore, relatively young impact craters commonly contain meteorite fragments in their ejecta blanket.

*Fossil craters,* which are impact craters considerably modified by erosion, are more difficult to recognize. The intense shock of a meteorite impact commonly produces *shatter cones,* which are conical structures with grooves extending down their sides (Fig. 4.17a); *breccias,* which are rocks containing angular fragments (Fig. 4.17b); planar features in quartz (Fig. 4.17c); high pressure phases of silica such as the minerals coesite and stishovite; and a glass formed by shock disordering of feldspar called maskelenite. The high temperatures produced by meteorite impact in craters more than 10 km (6 mi) in diameter results in a melting of some of the rocks in the floor of the crater. This melt cools quickly to a glass. *Tektites* are glass objects several centimeters (an inch or so) in size which are ejected from a crater and travel through the earth's atmosphere considerable distances (several tens of km or more) from the site of the impact (Fig. 4.17d). Impacts commonly produce a mixture of glass and angular rock fragments called *suevite* (Fig. 4.17e). Impact craters larger than 10 km (6 mi) commonly have a *central uplift* produced by rebound of the rocks in the floor of a crater following impact (Fig. 4.17f). Circular folds and faults are commonly associated with impact structures, and even deeply eroded impact craters commonly preserve a circular shape (Fig. 4.17g).

## ANCIENT ENVIRONMENTS AND PLATE MOVEMENTS

The reconstruction of ancient environments has been a major factor in providing evidence in favor of the large-scale movements of plates. The seemingly anomalous presence of fossils of tropical vegetation in regions which are now near the poles and the presence of ancient glacial deposits on or very near the present equator was at one time thought to indicate world-wide climatic changes. However, it has been established that cold climates occurred in regions that are hot at the present time *at the same time* as hot climates occurred in polar regions. This can

FIGURE 4.17 Features produced by meteorite impact: **(a)** Shatter cone from Manicouagan, Quebec. **(b)** Impact breccia from Manicouagan. **(c)** Planar features in quartz. **(d)** Sketch of tektite which was remelted on entering the earth's atmosphere (arrows show direction of air flow). **(e)** Suevite from Manicouagan. **(f)** Central uplift in a crater on Mercury (photo courtesy of NASA). **(g)** Satellite view of the Manicouagan structure, Quebec. Manicouagan is almost certainly an impact crater (astrobleme) formed about 205 million years ago (photo courtesy of NASA).

not be explained by world-wide climate changes. It is best explained in terms of large-scale movements of plates relative to the poles. This topic is discussed in more detail in Chapter 7.

## REFERENCES CITED

1. N. D. Newell, 1963, *Sci. Amer.,* v. 208, no. 2, p. 76.
2. T. W. Vaughan and J. W. Wells, 1943, *Revisions of the Suborders, Families, and Genera of the Scleractinia,* Geol. Soc. Amer. Special Paper 44.
3. C. D. Gebelein and P. Hoffman, 1973, *J. Sed. Petrol.,* v. 43, p. 603.
4. P. E. Playford and A. E. Cockbain, 1969, *Science,* v. 165, p. 1008.
5. R. Hesse and A. Butt, 1976, *J. Geol.,* v. 34, p. 505.
6. D. R. Horn, J. I. Ewing, and M. Ewing, 1972, *Sedimentology,* v. 18, p. 247.
7. P. D. Krynine, 1949, *N.Y. Acad. Sci. Trans.,* series II, v. 2, p. 60.
8. F. B. Van Houten, 1968, *Geol. Soc. Amer. Bull.,* v. 79, p. 399.
9. F. B. Van Houten, 1972, *Geol. Soc. Amer. Bull.,* v. 83, p. 2761.
9. W. E. Dean, 1975, *Amer. Assoc. Petrol. Geol. Bull.,* v. 59, p. 534.
10. L. L. Sloss, 1969, *Amer. Assoc. Petrol. Geol. Bull.,* v. 53, p. 776.
11. F. B. Phleger, 1969, *Amer. Assoc. Petrol. Geol. Bull.,* v. 53, p. 824.
12. K. S. Deffeyes, F. J. Lucia, and P. K. Weyl, 1965, *in* L. C. Pray and R. C. Murray, eds., *Dolomitization and Limestone Diagenesis:* Society of Economic Paleontologists and Mineralogists Special Publ. 13, p. 71.
13. R. L. Jodry, 1969, *Amer. Assoc. Petrol. Geol. Bull.,* v. 53, p. 957.
14. R. R. Shrock and W. H. Twenhofel, 1953, *Principles of Invertebrate Paleontology:* McGraw-Hill, New York.
15. P. W. Richards, 1973, *Sci. Amer.,* v. 229, no. 6, p. 58. J. M. Schopf, 1973, *in* D. H. Tarling and S. K. Runcorn, eds., *Implications of Continental Drift to the Earth Sciences, Vol. I:* Academic Press, New York, p. 609.
16. L. L. Sloss, E. C. Dapples, and W. C. Krumbein, 1960, *Lithofacies Maps of the United States and Southern Canada:* Wiley, New York.

# Methods of Correlation and the Facies Concept

## 5

IN MOST PARTS OF THE WORLD, exposures of rock are not continuous over great distances. This is particularly true in temperate and humid regions where soil and vegetation cover much of the land surface. For this reason, it is often not possible to observe the relationship of rocks in one area to those in another. Various methods of *correlation* are used in equating rocks in different areas.

Rock sequences are said to correlate if they are equivalent to one another. Rocks that are lithologically similar and are continuous are equivalent rock units. Rocks that were deposited during the same time interval are equivalent time-stratigraphic units. Correlation of rock units and time-stratigraphic units requires determination of many physical, chemical, and paleontological characteristics of the rocks.

The concept of *facies* implies that a rock unit having certain characteristics was deposited in a specific environment. Because a number of depositional environments may coexist at the same time in any large area, contemporaneously deposited rocks of varying facies grade into one another. This transition between facies is termed a *facies change*.

## ROCK UNITS

*Rock units* are observable stratigraphic bodies that have a distinctive lithologic (rock) type and recognizable physical boundaries. *Formations* are the fundamental rock units. They should be thick enough and distinctive enough to be shown on a geological map. A formation may consist of one rock type or several rock types. For example, one formation might contain only limestone, while another might be composed of an interbedded sequence of sandstone and shale. In the latter case, the individual beds of sandstone and shale would generally be too thin to be mapped at the usual scale of a quadrangle map. Most detailed geological mapping is done at a scale of 1 : 24,000 (7½-minute quadrangle map) or 1 : 62,500 (15-minute quadrangle map).

The boundaries or contacts between formations are generally located at breaks in sedimentation, such as changes in rock type or unconformities. A lithologically or paleontologically distinct portion of a formation may be designated as a *member* of that formation. Related formations may be combined into a *group*.

The site at which a rock unit was first described is the *type locality* of that unit, and an exposure of that unit at the type locality is the *type section* for that unit. This exposure is the standard to which similar sequences may be compared when correlations are made. The name of the rock unit generally comes from a geographical locality in the vicinity of the type section. For example, the thick shale found near Cody, Wyoming, is called the Cody Shale. The interbedded dolostone and limestone found near Lockport, New York, is named the Lockport Formation.

William Smith was one of the first to trace

and correlate rock units over large areas. His geological map of England (Fig. 1.7) shows a number of distinctive rock units around London. Smith's French contemporaries, Georges Cuvier and Alexandre Brogniart, mapped a similar sequence of rock units around Paris. Their detailed maps and cross sections provided a basis for the correlation of the rock units near London with those near Paris (Fig. 5.1).

Rock units may be correlated by a lateral continuity, rock type, and geophysical techniques.

## Correlation by Lateral Continuity

The most reliable method of correlating rock units is in establishing their physical continuity.

Where rock units are well exposed, as in arid regions where there is little vegetation or soil cover, rock units can be followed on the land surfaces by foot, traced on a topographic contour map using the pattern of contour lines, or traced on aerial photographs through differences in shading and the shape of land forms.

However, where exposures of rock are rare, rock units must be found by other techniques. Small rock fragments (float) in the soil or the color and texture of distinctive soils may serve to locate an underlying rock unit. In structurally uncomplicated regions, rock units may be traced by projecting their contacts laterally for short distances (Fig. 5.2). Rock units may also be traced laterally by following continuous land forms, such as ridges that are supported by a resistant unit and valleys cut in softer rocks.

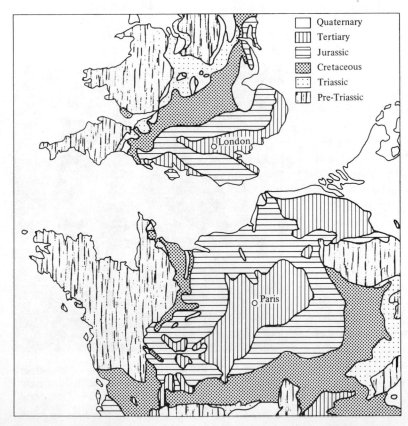

**FIGURE 5.1** Geologic map of part of Europe showing the correlation of the rocks near London with those near Paris.

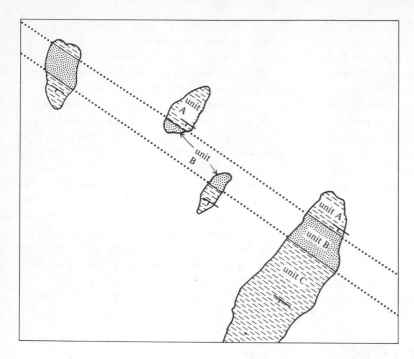

**FIGURE 5.2** Sketch illustrating correlation of units along strike in an area in which the exposures are not continuous.

## Correlation by Lithology

Correlation by rock type alone is reliable only on a local basis owing to the repetition of rock types in the geologic record. This type of correlation requires other lines of evidence, such as fossil content, mineral composition, analysis of the frequency distribution of grain sizes, and careful study of the nature of adjacent rock units.

## Correlation by Geophysical Techniques

Before the widespread use of geophysical techniques, subsurface correlations between oil wells were made primarily on the basis of descriptions of rock cuttings or rock cores from different wells. The samples recovered from the wells were inspected by a geologist in the field for composition, structures, and fossils (Fig. 5.3). Subsequently, porosity, permeability,

**FIGURE 5.3** Geologists and petroleum engineers inspect a core sample from deep within the earth. (Courtesy of Humble Oil and Refining Company.)

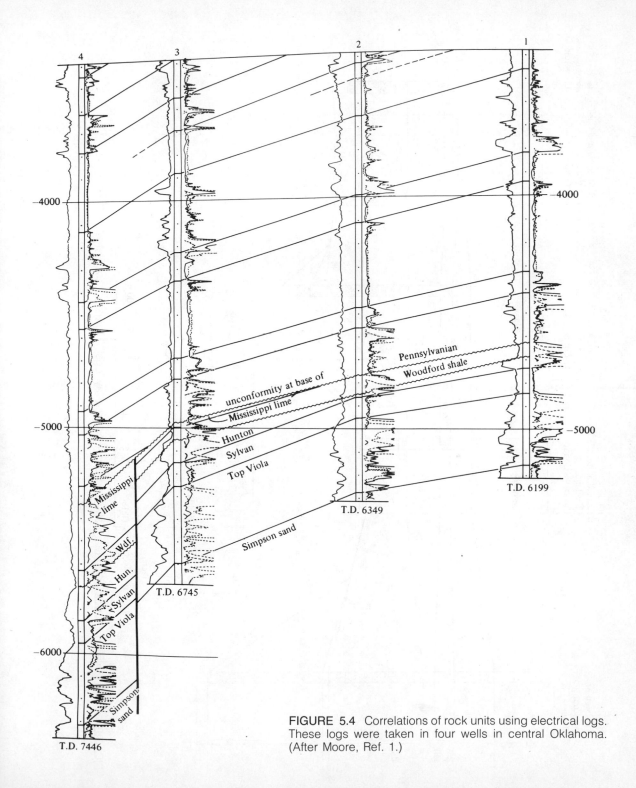

**FIGURE 5.4** Correlations of rock units using electrical logs. These logs were taken in four wells in central Oklahoma. (After Moore, Ref. 1.)

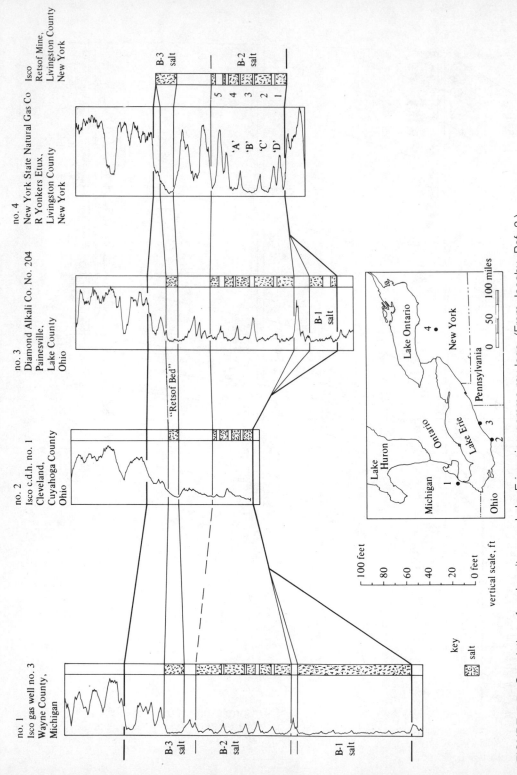

FIGURE 5.5 Correlation of rock units near Lake Erie using gamma-ray logs. (From Jacoby, Ref. 2.)

and gas content of the rocks were measured in the laboratory. Today, these techniques are supplemented by various geophysical techniques which greatly increase the accuracy of subsurface correlations.

Geologists have found that variations in electrical conductivity may be measured directly in the drill hole. As an electrical current passes through different rock units, variations in conductivity are recorded on an *electric log* (*E log*). Rock units with a high content of salt water generally have a high electrical conductivity. Correlations of rock units can be made by matching peaks of electrical conductivity (Fig. 5.4).

Correlation may also be aided by measurement of the natural radiation of rocks (Fig. 5.5). Rocks show considerable variation in uranium content. Black shales, for example, often contain significant quantities of uranium, while most limestones contain very little. The modern composite well log provides a relatively complete record of such properties, which helps geologists locate subsurface accumulations of oil and gas.

Seismic techniques provide control in correlating rock units in areas that have not been sampled by drilling. Seismic reflection measurements have been especially useful in delineating subsurface structures such as folds, faults, salt domes, and igneous intrusions. In this type of study, seismic waves are generated by explosions of small dynamite charges (Fig. 5.6), by generating intermittent bursts of electrical energy, or by dropping large weights. The resulting seismic waves are reflected from subsurface horizons to a set of receivers on the ground surface and are recorded on a set of seismograms (Fig. 5.7). Individual reflecting horizons produce a series of deflections on the seismogram. These deflections are used to trace rock strata and structures.

## TIME-STRATIGRAPHIC UNITS

A *time-stratigraphic unit* represents rocks that were deposited during a specified time interval and may include a wide variety of rock units. The Cambrian system is a time-stratigraphic unit that was deposited during the Cambrian period. The time-stratigraphic units and their matching units of geologic time are given in Table 5.1.

Time-stratigraphic correlations have been based mainly on paleontological data, but recently radiometric and paleomagnetic measurements have come into widespread use. The establishment of time-stratigraphic units has permitted intercontinental correlations and has provided a means of comparing the geologic histories of all the continents. These comparisons are a basic line of evidence in continental drift reconstructions.

**FIGURE 5.6** Typical seismograph "shot" taken during seismic profiling in west Texas. (Courtesy of Humble Oil and Refining Company.)

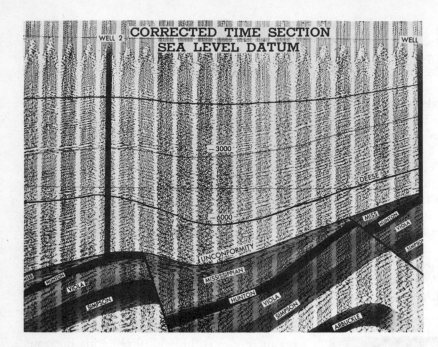

FIGURE 5.7 Correlation of rock units using seismograms. (From Dunlap, Ref. 3. Published with permission of The American Association of Petroleum Geologists.)

## Correlation by Fossils

Within Cambrian and younger sequences, most correlations of time-stratigraphic units are made through study of their fossil content. The unique succession of assemblages of fossils makes this possible. By identifying the fossils within a rock unit and determining their time ranges, the geologist may establish the age of the unit, which will in turn make it possible to compare that unit with others of the same age.

TABLE 5.1 **Relationship Between Time-Stratigraphic and Geologic Time Units.**

| TIME-STRATIGRAPHIC UNITS | GEOLOGIC TIME UNITS |
| --- | --- |
| — | Eon |
| Erathem | Era |
| System | Period |
| Series | Epoch |
| Stage | Age |

Fossils may be used to subdivide time-stratigraphic units into zones. The *zone* is the basic unit of biostratigraphy and is defined as a bed or group of beds characterized by the presence of an *index fossil* or *assemblage* of fossils. The zone takes its name from a characteristic fossil of that zone. Zones may cross the boundaries between rock units, in which case it is likely that the rocks within the zone were deposited at the same time. This is especially true if zones are not repeated. If the beds were deposited contemporaneously, the faunas are said to be *synchronous*. The Jurassic ammonite zones are probably of this type. However, some of the Devonian brachiopod zones may be repeated several times in a given undisturbed section; in this case, the zones would not have been deposited contemporaneously.

## Correlation by Radiometric Ages

Radiometric dating is important in the correlation of time-stratigraphic units, especially in igneous and metamorphic terrains. The accuracy with which units may be correlated depends on the method(s) of dating and materials dated. A potassium-argon date on mica from a metasedimentary rock provides a minimum age for the deposition of the original sediment. A uranium-lead or whole-rock, rubidium-strontium date for the basement rock on which the sediment was deposited gives a maximum age for the deposition of the sediment. Sequences that were metamorphosed at the same time and were deposited on basements of similar age are likely to be approximate time-stratigraphic equivalents.

### Correlation by Paleomagnetism

Time-stratigraphic units may be correlated by measuring the direction of magnetization (paleomagnetism) of rocks or sediments. This method has been especially useful in the correlation of cores taken in oceanic sediments (Fig. 5.8).

Correlation may involve matching the pattern of normally and reversely magnetized sediments, sedimentary rocks, and volcanic rocks.

The direction and inclination of the magnetic field of rocks varies with time as the continents move relative to the earth's poles. Therefore, correlations may also be made on the basis of matching units with similar magnetic directions and inclinations.

## RELATION BETWEEN ROCK UNITS AND TIME-STRATIGRAPHIC UNITS

In any area, it is common to find more than one depositional environment at a given time. Each environment is represented by sediments of a distinctive physical and chemical character. In coastal regions, for example, beaches may receive well-sorted sands and gravels, while at the same time clay and silt may be deposited in shallow lagoons, and carbonate may be deposited on offshore carbonate banks. The rocks resulting from the consolidation of these sediments will change laterally in lithology and in the fossils they contain. These changes make correlation of time-stratigraphic units difficult.

The recognition that time-stratigraphic units show a considerable variation in lithology and fossil content has led to the introduction of the concept of *facies*. Unfortunately, the term facies has several meanings. In North America, facies generally refers to a lateral subdivision of a stratigraphic unit. For example, the Edinburgh Formation in Virginia is of Middle Ordovician age, and is subdivided into two dissimilar facies—a fossiliferous limestone called the Lantz Mills Facies and a relatively unfossiliferous limestone and black shale named the Liberty Hall Facies. The fossiliferous limestone represents a shallow water environment, whereas the unfossiliferous limestone and associated black shale represent a deeper water environment. Thus, a facies represents a variation of a sedimentary deposit. The variation may be in rock type or fossil content as long as the rock units are contemporaneous.

In Europe the term facies is commonly used without reference to a specific stratigraphic unit. In this context, facies is the environment of deposition of a sedimentary rock at a given locality as determined by fossils and rock type. Thus, "the red beds facies" may be used to designate any sequence of red sandstones and shales that were probably deposited in a subaerial environment. Similarly, "the black shale facies" refers to sequences containing dark shales and sandstones deposited in relatively deep water.

A *lithofacies* is the rock deposited in a specific sedimentary environment. For example, the limestone lithofacies of the Onondaga Limestone in New York State represents a shallow, clear-water environment. A lithofacies map shows the distribution of rock types formed at a specific time. Such maps are useful in the construction of paleogeographic maps.

A *biofacies* is an assemblage of fossils representing a particular sedimentary environment.

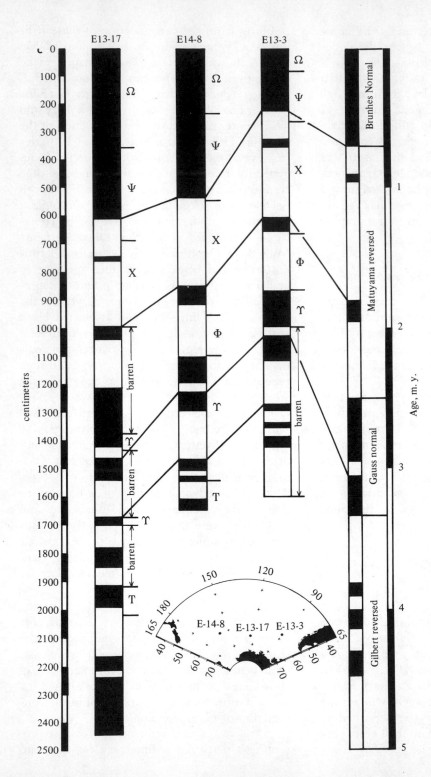

**FIGURE 5.8** Correlation of cores taken in the deep ocean near Antarctica using paleomagnetic and paleontological data. The dark strips are normally magnetized sediments, while the light strips are reversely magnetized sediments. (From Hays and Opdyke, Ref. 4. Copyright © 1967 by American Association for the Advancement of Science.)

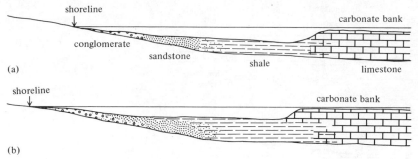

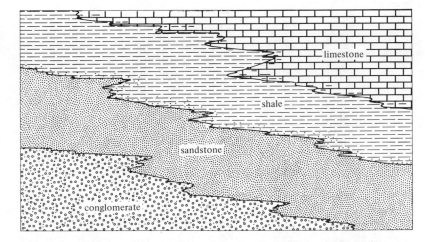

**FIGURE 5.9** Diagrammatic cross sections illustrating facies changes: **(a)** In sedimentary rocks formed during a time of stable sea level. **(b)** In sedimentary rocks formed during a time of a transgressing sea when the shoreline migrates inland.

**FIGURE 5.10** Diagrammatic cross section of sedimentary rocks deposited during a marine transgression.

For example, the Onondaga Limestone contains a coral-rich biofacies representing a shallow, clear-water marine environment.

The sizes of sedimentary particles that may be deposited in any one location are controlled by the depth of water, the distance from the source, and the energy of currents. If the water level has not changed relative to the land surface and if the sediments have been deposited in a subsiding basin, the contacts between lithofacies would be essentially vertical. However, if the water level has risen or fallen relative to the land, changes would occur in the depositional environments, and thus in the lithofacies. A rising or *transgressing* sea generally results in the migration of the shoreline inward toward the center of a continent. A transgressing sea will result in beach sands deposited over lagoonal or

freshwater deposits. Subsequently, finer marine sediments would overlap the beach sands (Figs. 5.9 and 5.10). The decrease in grain size is related to decreasing energy of the sediment-transporting currents in the deepening water.

With a lowering or *regression* of the sea the shoreline generally migrates outward away from the center of a continent. A regressing sea will result in coarse coastal sediments overlapping finer marine sediments (Fig. 5.11). During a transgression or regression of the sea, the contacts between lithofacies are at a relatively low angle to time-stratigraphic units (Figs. 5.10 and 5.11). Times of general regression of the sea are commonly interrupted by short episodes during which the sea transgresses inland, and times of general transgressions are interrupted by brief regressions of the sea. The result of

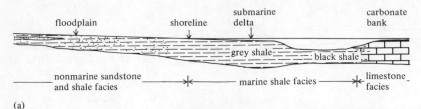

(a)

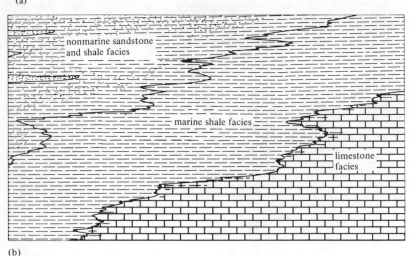

(b)

FIGURE 5.11 Diagrammatic cross sections illustrating the effect of regression on facies changes: **(a)** In marine and continental sedimentary rocks formed during a time of stable sea level. **(b)** In sedimentary rocks deposited during a marine regression.

such fluctuations in sea level is an irregular interfingering of lithofacies (Figs. 5.10 and 5.11). This interfingering is a way of showing facies changes within a time-stratigraphic unit.

The recognition of facies changes may be aided by the presence of *marker beds* or *key horizons* with broad geographic distribution (Fig. 5.10). Volcanic ash beds and their weathered equivalents, *bentonites,* provide excellent time planes because they are deposited essentially instantaneously. A change in lithofacies below the ash or bentonite would indicate a difference in depositional environment at the time the ash fall occurred, and would establish the existence of a facies change.

## BIOSTRATIGRAPHIC UNITS

*Biostratigraphic units* (or *zones*) are bodies of rock that are defined by their fossil content. There are three types of zones: assemblage zones, range zones, and concurrent range zones.

*Assemblage zones* are characterized by groups or assemblages of fossils which occur together within a body of rock. However, the ranges of some of the individual fossils within the assemblage zone may cross the zone boundaries. *Range zones* are bodies of rock defined by the vertical and horizontal limits of the occurrence of a single fossil group, usually a genus or species. Range zones may overlap each other, in contrast to assemblage zones which are mutually exclusive. *Concurrent range zones* are bodies of rock that are defined by groups of overlapping ranges of fossil groups (usually genera or species). The boundaries of this type of zone are placed at the first occurrence of one fossil group and the last occurrence of another fossil group.

## CORRELATION AND CONTINENTAL DRIFT

When Charles Lyell compared the rock units in North America to their likely counterparts in

Europe, he was struck with their similarities. While relating these rocks to similar environments, he was not aware at that time that these were the same rock units. Today we realize that these rocks are located on opposite sides of the Atlantic as a result of continental drift. The geologic history resulting from plate movements will be developed in more detail in subsequent chapters.

## REFERENCES CITED

1. C. A. Moore, 1963, *Handbook of Subsurface Geology:* Harper and Row, New York.
2. C. H. Jacoby, 1969, *Amer. Assoc. Petrol. Geol. Bull.,* v. 53, p. 136.
3. R. C. Dunlap, 1956, *Amer. Assoc. Petrol. Geol. Bull.,* v. 40, p. 1462.
4. J. D. Hays and N. D. Opdyke, 1967, *Science,* v. 158, p. 1001.

# Dynamics of the Earth's Interior

**6**

O<small>N</small> N<small>OVEMBER</small> 14, 1963, fishermen observed black clouds coming from the sea south of Iceland. Upon investigation, they found that a new island had been built up from the sea bottom by volcanic eruptions. The birth of the island was accompanied by a plume of steam four miles high, formed as cold seawater came in contact with hot lava. The volcano hurled millions of tons of rock onto the surface of the island so that after 16 days of activity, the island was 1000 m (3300 ft) long and 250 m (750 ft) above sea level at its highest point (Fig. 6.1).

The island, which was christened Surtsey after a fire demon of Norse mythology, is located on a vast and largely submarine range of mountains known as the Mid-Atlantic Ridge. Earthquakes are common along the ridge, which runs the length of the Atlantic Ocean. Submarine depth profiling has located numerous volcanic seamounts covering the crest and flanks of the ridge (**1**). The Mid-Atlantic Ridge is part of a worldwide system of oceanic ridges more than 60,000 km (36,000 mi) long.

Volcanic eruptions and earthquakes are also very common in the "ring of fire," a zone around the Pacific Ocean basin in which many of the world's great mountain belts are located. The Alpine-Himalayan region comprises another major zone of instability.

Observations such as these have led scientists to conclude that the interior of the earth is very active. Geophysical investigations of the earth's interior have provided insights into processes related to this activity. These studies have indicated that there are large-scale convection currents within the earth's mantle and that it is these currents which cause the movements of plates. Moreover, geological data suggest that plate movements have played an important role in the development of oceanic ridges, island arcs, and mountain belts.

## EARTH STRUCTURE AND COMPOSITION

The solid earth is divided into three layers: the crust, mantle, and core (Fig. 6.2). The crust is between 5 and 70 km (3 and 42 mi) thick, and it makes up less than one percent of the earth by both weight and volume. The four most common minerals in the earth's crust are feldspar, quartz, pyroxene, and hornblende. Continental crust underlies the continents and shallow oceans and is between 25 and 70 km (15 and 42 mi) thick. It is composed of a variety of igneous, sedimentary, and metamorphic rocks. Common plutonic (intrusive) igneous rocks are granite, diorite, and gabbro. Common volcanic (extrusive) igneous rocks are rhyolite, andesite, and basalt. Sedimentary rocks in continental crust include shale, sandstone, limestone, dolostone, and conglomerate. Common metamorphic rocks are gneiss, schist, marble, quartzite, greenstone, amphibolite, and granulite. The upper crust is less dense than the lower crust and is probably richer in granites and felsic

FIGURE 6.1  The violent birth of a new island, Surtsey, in November, 1963. (Courtesy of Solarfilma, Reykjavik.)

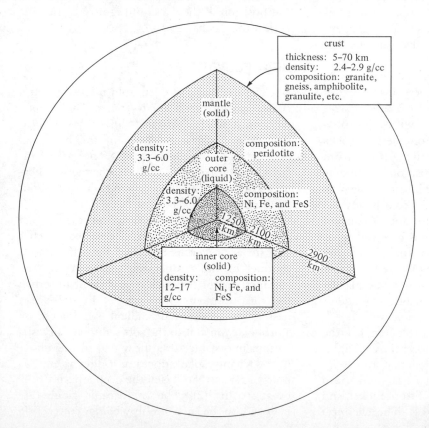

FIGURE 6.2  Diagrammatic cross section through the earth.

TABLE 6.1 **Chemical Composition of the Earth in Weight Percent of the Earth's Crust, Mantle, and Core.**

| Oxide | Composition of the Earth as a Whole | | | | |
|---|---|---|---|---|---|
| | *Continental Crust* | *Ocean Crust* | *Average Crust* | *Mantle* | *Core* |
| SiO₂ (silicon oxide) | 59.6 | 50.5 | 57.8 | 44.5 | 0.0 |
| TiO₂ (titanium oxide) | 0.7 | 1.5 | .9 | 0.5 | 0.0 |
| Al₂O₃ (aluminum oxide) | 17.3 | 13.7 | 16.6 | 3.0 | 0.0 |
| Fe₂O₃ (ferric oxide) | 3.3 | 3.2 | 3.3 | 1.5 | 0.0 |
| FeO (ferrous oxide) | 3.3 | 10.0 | 4.6 | 8.5 | 0.0 |
| MgO (magnesium oxide) | 2.7 | 6.0 | 3.4 | 37.5 | 0.0 |
| CaO (calcium oxide) | 5.8 | 8.7 | 6.4 | 3.0 | 0.0 |
| Na₂O (sodium oxide) | 3.5 | 2.8 | 3.4 | 0.5 | 0.0 |
| K₂O (potassium oxide) | 2.0 | 1.0 | 1.8 | 0.1 | 0.0 |
| H₂O (water) | 1.2 | 2.0 | 1.4 | 1.0 | 0.0 |
| FeS (troilite–iron sulfide) | 0.0 | 0.0 | 0.0 | 0.0 | 20.0 |
| Fe (uncombined iron) | 0.0 | 0.0 | 0.0 | 0.0 | 73.5 |
| Ni (uncombined nickel) | 0.0 | 0.0 | 0.0 | 0.0 | 6.5 |

gneiss than the lower crust, which is probably rich in mafic granulites and amphibolite. The average composition of continental crust is approximately that of anorthosite (see Appendix C for definitions of rocks) or andesite (Table 6.1).

Oceanic crust underlies the deep oceans and averages 5 km (3 mi) in thickness. It consists of 3 layers: layer 1, sediments; layer 2, pillow lavas and diabase dikes, and layer 3, which is probably gabbro, serpentinite, and olivine-rich rocks that have formed by crystal settling from gabbros. The average composition of oceanic crust is approximately that of basalt (Table 6.1).

The mantle constitutes approximately 83 percent of the earth by volume and 66 percent by weight. The upper mantle is believed to be almost entirely peridotite, a rock made up largely of olivine and enstatite with minor amounts of diopside, spinel, garnet, and plagioclase (Table 6.1). The lower mantle is made of high pressure phases of these minerals.

The earth's core comprises 16 percent of the earth by volume and 33 percent by weight. Seismic evidence has indicated that the outer core is liquid and the inner core is solid. Both are thought to be composed of iron, nickel, and troilite (iron sulfide) (Table 6.1). The chemical composition of the earth as a whole (Table 6.2) is estimated from the percentage, density, and composition of the rocks thought to be present within the earth.

## HEAT FLOW

Rock in the walls of deep gold mines in South Africa is almost too hot to touch. The miners, who must work shortened shifts to avoid heat prostration, are well aware of the heat that escapes from within the earth. The rate of increase in temperature with depth, the geothermal gradient, has an average value of about 30°C per km (80°F per mi).

The observable heat flow represents that portion of the earth's internal heat that reaches the surface. Mathematically, heat flow is a product of the temperature gradient and the conductivity of the rock or sediment in which the measurements were made. In the ocean basins, the geothermal gradient is measured by a probe dropped into the soft sediment. Investigations in oceanic areas show that the heat flow in the vicinity of midoceanic ridges is significantly greater than it is in the deeper parts of the ocean basins (**2**). The midoceanic ridges are believed to be sites of upwelling convection

TABLE 6.2  **Chemical Composition of the Earth. Mantle compositions are assumed to be approximately that of peridotite and the composition of the core is assumed to be similar to the composition of nickel–iron meteorites or the metallic and sulfide phases of ordinary chondrites.**

| OXIDE | COMPOSITION OF THE EARTH AS A WHOLE PERCENT BY WEIGHT | ELEMENT | PERCENT BY WEIGHT | ATOM PERCENTAGE |
|---|---|---|---|---|
| $SiO_2$ | 29.7 | O | 29.9 | 48.88 |
| $TiO_2$ | 0.3 | Si | 13.9 | 14.00 |
| $Al_2O_3$ | 2.0 | Ti | 0.2 | 0.028 |
| $Fe_2O_3$ | 1.0 | Al | 1.1 | 1.30 |
| FeO | 5.7 | Fe | 33.0 | 18.87 |
| MgO | 25.0 | Mg | 15.1 | 12.50 |
| CaO | 2.0 | Ca | 1.4 | 0.46 |
| $Na_2O$ | 0.3 | Na | 0.2 | 0.64 |
| $K_2O$ | 0.1 | K | 0.1 | 0.056 |
| $H_2O$ | 1.0 | H | 0.1 | 0.12 |
| FeS | 8.3 | Ni | 2.0 | 1.40 |
| Fe | 22.6 | S | 3.0 | 1.40 |
| Ni | 2.0 | — | — | — |

currents which carry hot mantle material from deep within the earth.

Geologists have long believed that the earth's crust attempts to maintain a balance, sinking under the load of thick sediment accumulation or rising due to unloading, as mountains erode or glaciers melt. This process, known as *isostasy,* continually adjusts the level of the crust in attempting to achieve a balance. However, other factors, such as changes in heat flow, are also involved.

## THE EARTH'S VISCOSITY

Many solids behave plastically when they are near their melting temperature. The flow of glaciers is a dramatic example of this phenomenon. The deeper levels of continental and alpine glaciers are subjected to high pressures owing to the weight of the overlying ice. At temperatures near its melting point, ice is capable of plastic flow.

The mantle of the earth is thought to be similarly capable of flowing in response to relatively small, directed stresses applied over long periods of time. This is well illustrated in the process of crustal rebound which follows the

melting of large ice sheets. The area in the vicinity of Hudson Bay near the center of ice accumulation has been uplifted more than 300 m (900 ft) since the melting of the Pleistocene ice sheet. Such uplift must have been accompanied by the flow of a large volume of mantle material into the region below Hudson Bay.

The rate at which rock flows in response to pressure is a function of its viscosity. Viscosity is a measure of the internal friction of a fluid in motion. The higher the viscosity of a fluid, the more sluggish its flow. Although rocks within the mantle are thought to behave as plastic solids rather than viscous fluids (**3**), it is possible nonetheless to consider the apparent viscosity of the mantle.

The apparent viscosity of that part of the upper mantle at a depth between 75 and 900 km (45 and 540 mi) is between $10^{21}$ and $10^{23}$* P (*Poise,* the unit of viscosity) based on studies of the velocity of seismic waves and the rate of postglacial rebound (**4**). For comparison, water has a viscosity of 0.01 P, whereas a heavy

*$10^{23}$ is 1 with 23 zeros after it.

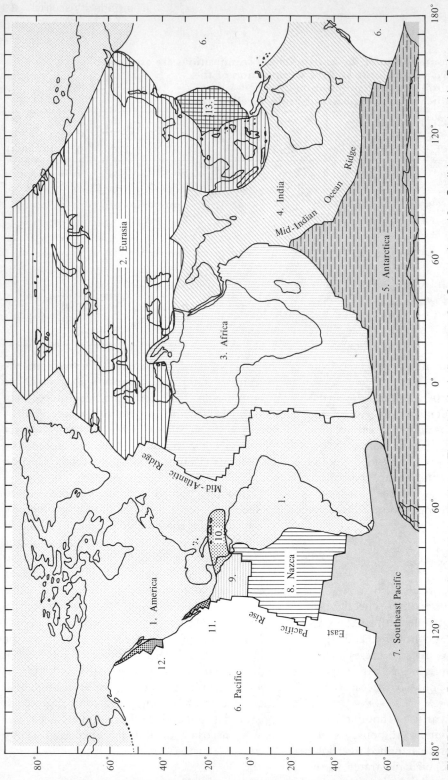

**FIGURE 6.3** Division of the earth's crust into 13 plates. The unlabeled plates are: 9—Cocos plate; 10—Caribbean plate; 11—Baja plate; 12—Juan de Fuca plate; and 13—Marianas plate. (After Morgan, Ref. 6.)

weight of motor oil has a viscosity of about 1 P. Seismic studies and studies of the acceleration of the earth's rotation indicate that the lower mantle has an apparent viscosity between $10^{22}$ and $10^{26}$ P (**4,5**).

## SEISMOLOGY AND PLATE TECTONICS

It is now generally believed that the movements of the continents and of the seafloor are both part of large-scale movements of plates. In 1968, W. Jason Morgan (**6**) introduced the concept of *plate tectonics* in which the earth's surface is considered to be divided up into a number of rigid plates bounded by midocean ridges, oceanic trenches, great faults, and active fold (mountain) belts (Fig. 6.3). At midocean ridges, like the Mid-Atlantic Ridge, the plates are moving apart or *spreading*. Oceanic trenches, like the Peru-Chile Trench, are zones of descending ocean crust and upper mantle. Oceanic trenches are bordered by *island arcs* or mountains. Island arcs are linear chains of islands made up of volcanic rocks or deformed seafloor sediments. The San Andreas Fault in California is an example of a great fault which occurs where plates move laterally past each other. The Himalayan Mountain belt is an active fold belt where the crust is being crumpled along a plate boundary as the mantle descends beneath the crust. Midocean ridges occur at *divergent plate boundaries,* while oceanic trenches, island arcs, and active fold belts occur at *convergent plate boundaries.*

Strong support for the concept of plate tectonics comes from the study of earthquakes. A plot of earthquake epicenters shows that earthquakes are most common in linear belts at or near the margins of plates, either along midocean ridges, beneath island arcs, along great faults, or under mountain ranges (Fig. 6.4). From studies of the pattern of seismic waves measured in many parts of the world, it has been found that normal faulting is the dominant type of faulting over midocean ridges (**8**). Normal faults are thought to be caused by extension of the earth's crust (tension). Those associated with midoceanic ridges are the result of separation of plates over rising convection currents.

Earthquakes also occur along faults that offset midoceanic ridges. The sense of motion along these faults is strike-slip, which means that the movement along the fault is horizontal, parallel to the fault. However, the movement along these faults is actually in the opposite direction to the apparent offset of the ridge (Fig. 6.5). If a midoceanic ridge appears to be offset in a left-lateral sense (far side of fault moves to the left), the movement along the fault would be right-lateral (far side of fault moves to the right). The movement along faults that offset midoceanic ridges is exactly what one would expect if the ridges were underlain by upwelling convection currents (Fig. 6.5). The existence of such faults, termed *transform faults,* was predicted by J. Tuzo Wilson before they were observed (**9**).

The movement of plates in the vicinity of oceanic trenches is rather complex. Here earthquakes occur along more or less planar zones which dip under a continent or island arc and extend to depths as great as 700 km (420 mi) (**10**). These are named *Benioff zones* after the seismologist Hugo Benioff (Fig. 6.6).

Studies of the attenuation (damping out) of earthquake waves indicate that the earthquakes occur within relatively cool, brittle slabs of lithosphere that have been thrust under continents and island arcs. The earthquakes are the result of both compressional and tensional stresses on the slabs (**11**). Linear belts along which the seafloor is being consumed are called *subduction zones.*

### Spreading Pole

Any movement of one plate relative to another on a sphere (such as the earth) can be expressed as a rotation about an axis. This axis is desig-

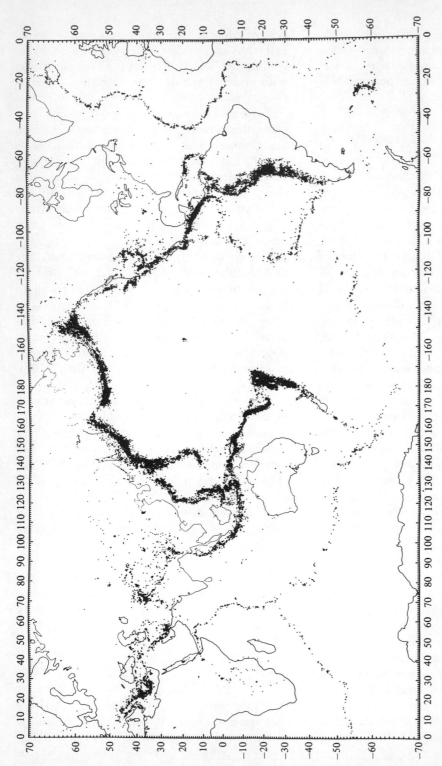

**FIGURE 6.4** Worldwide distribution of all earthquake epicenters for the period 1961–1967, as reported by the U.S. Coast and Geodetic Survey. (From Barazangi and Dorman, Ref. 7.)

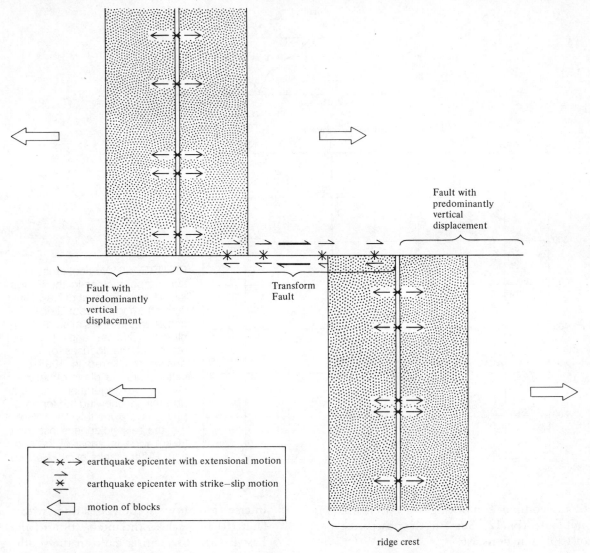

Fault with predominantly vertical displacement

Fault with predominantly vertical displacement

Transform Fault

←✳→  earthquake epicenter with extensional motion

✳  earthquake epicenter with strike–slip motion

⇦  motion of blocks

ridge crest

**FIGURE 6.5** Idealized map view of a transform fault. Notice that the ridge appears to be offset in a left-lateral sense, but the motion on the transform fault between the ridge crests is right-lateral.

nated as the *pole of spreading* (Fig. 6.7). The relative movement of two plates is parallel to transform faults along the junction of the two plates, and these transform faults are small circles which are concentric about the pole of spreading. In order to locate the pole of spreading, perpendiculars are drawn to the tangents of transform faults. The spreading pole is located at the intersection of these perpendiculars.

**Triple Junction**

*Triple junctions* occur at the margins of plates where three plates join. They may be composed

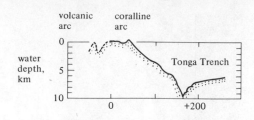

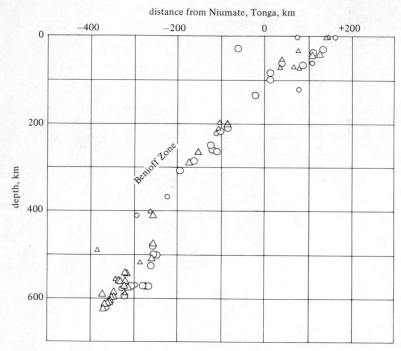

distance from Niumate, Tonga, km

FIGURE 6.6 Vertical cross section perpendicular to the Tonga Island Arc northeast of New Zealand showing the spatial distribution of earthquakes of small magnitude. Circles and triangles represent the focuses of earthquakes monitored in 1965 projected onto the plane of section. Vertical exaggeration of the upper inset showing the topography is approximately 13 : 1. Note that the seismic zone is less than 20 km wide for most of the section. (After Isacks *et al.*, Ref. 8.)

of any combination of ridges (R), trenches (T), and faults (F) (**12**). Some of the observed types of triple junctions are:

1. Joining of two sections of a ridge with a third ridge (RRR), as for example the joining of the Mid-Indian Ocean Ridge with the Atlantic-Indian Rise and the joining of the East Pacific Rise and the Galapagos Rift
2. Joining of two sections of a trench with a third trench (TTT), such as the junction of the Japan Trench with the Ryukyu Trench and the joining of the Mariana Trench with the Philippine Trench

3. Intersection of two sections of a ridge with a fault (RRF), such as the intersection of the East Pacific Rise with a fault on the southwestern border of the Nazca Plate (Fig. 6.3)
4. Intersection of two sections of a ridge with a trench (RRT), as for example, the intersection of the East Pacific Rise, the Mid-Indian Ocean Ridge and a trench extending southward from New Zealand (Fig. 6.3)

## CONVECTION CURRENTS

*Convection* is defined as the vertical transfer of heat by the circulation or movement of a gas,

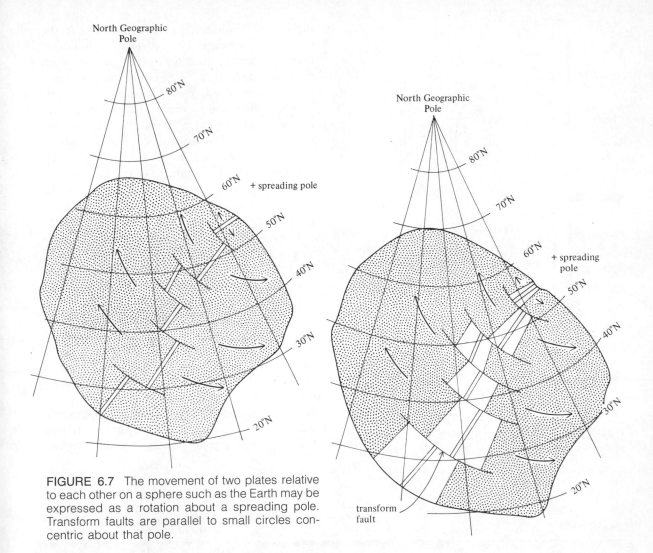

**FIGURE 6.7** The movement of two plates relative to each other on a sphere such as the Earth may be expressed as a rotation about a spreading pole. Transform faults are parallel to small circles concentric about that pole.

liquid, or plastic solid. Such a motion results from unequal heating, which produces differences in density. Rapidly moving convection currents may be observed by heating water in a glass coffee pot or beaker (Fig. 6.8). As the water in the bottom of the container is heated, it expands and becomes less dense. The warm water rises while the cooler water at the top and sides of the container sinks. This results in a heat transfer within the container. In this case,

convection takes the form of a cell in which the fluid moves in a circular fashion in cross section. However, convection currents within the earth probably do not have this pattern.

F. A. Vening Meinesz was the first to provide geophysical evidence of the existence of convection currents within the earth's mantle (**13**). His gravity measurements in the East Indies indicated that large negative gravity anomalies are associated with the oceanic trench bordering

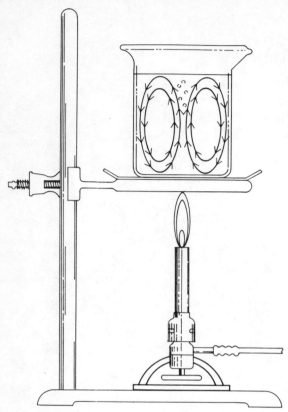

**FIGURE 6.8** Convection currents within a beaker filled with water.

Indonesia. Vening Meinesz believed that the lower-than-average force of gravity indicated that the earth's crust was being pulled down into the mantle by sinking convection currents. This, he proposed, would create a mass deficiency which would cause a negative gravity anomaly. In support of convection within the mantle, it has been calculated that viscosities equal to or less than $10^{25}$ P within the mantle will permit the development of convection currents. Thus, convection is possible within the upper mantle and may be possible within the lower mantle as well.

Convective flow within the mantle is commonly shown as convection cells with more or less circular cross sections. However, rocks behave as semiplastic solids during deformation and would not necessarily flow in the same manner as fluids. Egon Orowan (**3**) has stated that at the temperatures and pressures obtaining within the mantle, convection within a semiplastic solid would take the form of rising hot dikes (Fig. 6.9). Such convective motion would resemble the diapiric movement of solid masses of salt during the formation of salt ridges and salt domes (**14**) but would be quite different from the convective motions within a liquid.

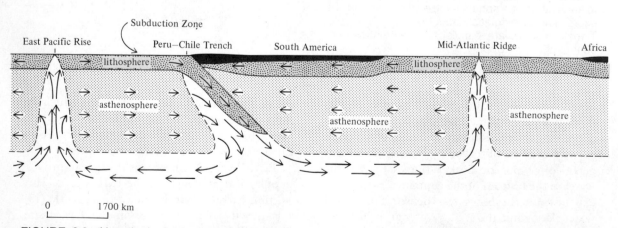

**FIGURE 6.9** Hypothetical cross section through the crust and upper mantle of the earth showing how diapiric convective motions within the mantle may produce horizontal movements of large blocks.

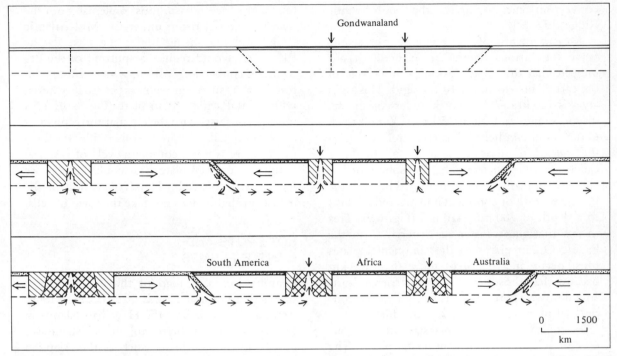

**FIGURE 6.10** Hypothetical cross section showing accretion of mantle onto blocks as diapiric instrusions of mantle produce horizontal movements of blocks.

Diapiric intrusions of mantle material would cause lateral separation of rigid plates as the current rises. These plates are referred to as *continental plates* if they include a continent and *oceanic plates* if they do not include continental crust.

During injection of dike material from the mantle into a plate, the outer, more rapidly cooled parts of the intrusion are accreted onto the sides of the plate. Any subsequent intrusions are in the center of earlier intrusions, because this zone is warmer and less rigid than the walls of the intrusion (Fig. 6.10). Therefore, if a dike of semiplastic mantle were to split a continent, the center of the dike would remain midway between the resulting continents. This may explain why the Mid-Atlantic Ridge and the Mid-Indian Ocean Ridge are approximately centered between the adjacent continents.

For most continental plates, little if any material is lost as new material is added beneath midocean ridges. In fact, the Pacific margins of continental plates appear to be growing by accretion of oceanic sediments and volcanic rocks. Therefore, it appears that the continental plates are increasing in area and volume. Since the earth does not appear to be expanding or contracting significantly (**15**), the plates beneath the Pacific Ocean must be decreasing in area and volume at approximately the same rate as the continental plates are increasing. A transfer of mantle material from the Pacific Ocean plates to the continental plates would be required. Transfer may occur through downward movement of material in the vicinity of oceanic trenches. The mantle material would then move horizontally under the continental plates, replenishing the mantle material of the rising convection current under the Mid-Atlantic Ridge (Fig. 6.10).

Measurements of the velocity of seismic waves at various depths under ocean basins indicate that the upper 75-km (45-mi) layer of the earth's interior is relatively rigid. This rigid layer is 120 km (72 mi) thick under continental interiors and as little as 50 km (30 mi) thick under mountain belts. This layer, referred to as the *lithosphere,* overlies a more plastic layer called the *asthenosphere*. The lithosphere of Pacific Ocean plates bends downward beneath the Pacific margin of continental plates and extends to a depth of 700 km (420 mi) (Fig. 6.9). This means that the transfer of material from the Pacific Ocean plates to the continental plates must occur at a depth of more than 700 km because the lithosphere acts as a barrier to horizontal flow. Therefore, it is likely that the continental plates are at least 700 km thick.

Heat flow measurements support the concept of relatively thick continental plates. The average value of heat flow for the continental plates is almost exactly the same as the average heat flow for the ocean basins (**16**). This is surprising, since the continental crust contains greater amounts of radioactive heat-producing isotopes, such as uranium 235, uranium 238, thorium 232, and potassium 40, than the oceanic crust. Apparently the mantle beneath the continents is depleted of radioactive elements relative to the mantle beneath the ocean basins. In order to maintain the similarity in heat flow during continental drift, the continents and the underlying depleted mantle must move as a single unit (**17**). This would be the case if the continental plates were more than 700 km thick. The Pacific Ocean plates may be of the same thickness as the continental plates, although there is little direct evidence for this supposition. A layer of low seismic velocity at a depth of 800 km (480 mi) (**18**) may be located at the bottom of both continental and Pacific Ocean plates.

The material descending at the margins of the Pacific Ocean basin must replenish both the mantle material rising under the Mid-Atlantic Ridge and that rising under the East Pacific Rise. This would require a splitting of the descending convection current, which would then result in a spiral movement of the material within the Pacific Ocean plate (Fig. 6.9). This motion contrasts with the circular movement of convection cells within viscous fluids. Furthermore, there is not a complete cycle of material within continental plates (as would be the case in a convection cell). Thus convection in the mantle probably does not take the form of cells.

## HOT SPOTS AND MANTLE PLUMES

Examination of a map of the Pacific Ocean shows that volcanic islands often occur in long, straight chains such as the Hawaiian Islands in the central part of the ocean, and in island arcs bordering the continents such as the Aleutian Islands (Fig. 6.11). The linear chains of volcanic islands are often associated with *seamounts* (rounded, submarine volcanoes) and *guyots* (flat-topped submarine volcanoes). The age of volcanic activity generally changes systematically along the chains with a progression from active volcanoes at one end of the chain to extinct volcanoes at the other end. Relative ages of volcanic activity are indicated by the degree of erosion of the islands. As age increases, there is a progression from an active volcano with little erosion to an inactive volcano showing a moderate amount of erosion, to a deeply eroded volcano, to an *atoll* (coral reef topping a submerged volcano), and then to a guyot or seamount. Radiometric dating of volcanic rocks on the islands, seamounts, or guyots along the chain provides an absolute chronology for its development.

The Hawaiian Islands show a progressive increase in the age of volcanic activity in a west-northwest direction. Mauna Kea on the big island of Hawaii at the east end of the chain

Legend:
≡ Midocean ridge crest
-·-·- Island chain
ᐱᐱᐱ Seamount chain

**FIGURE 6.11**  Island chains in the Pacific Ocean.

is active, while deeply eroded volcanoes occur on the west end, and a linear chain of seamounts, atolls, and guyots extends west-north-west from these islands. At the Milwaukee Seamount Group, the chain takes a 60° bend to the north-northwest along the Emperor Seamount Chain. To the south, there are also several other linear chains of volcanoes which parallel the Hawaiian chain (Fig. 6.11).

In the Atlantic Ocean, there are similar features, but they are mostly submarine and appear to emanate from the Mid-Atlantic Ridge (Fig. 6.12). The best-developed linear island chains are the Rio Grande Ridge and the Walvis Ridge in the South Atlantic. Unlike spreading

ridges, these have no earthquake activity and are called *aseismic ridges*. Both ridges end at the Mid-Atlantic Ridge at an active volcanic island, Tristan da Cunha. Iceland, which has many active volcanoes, has linear volcanic ridges extending both eastward and westward from it.

The volcanically active areas at the end of linear chains are termed *hot spots*. W. Jason Morgan (**19**) has proposed that hot spots are underlain by *mantle plumes* which are upwellings of hot, plastic (but mostly solid) mantle material. Plumes have more or less circular cross sections and are thought to originate deep in the mantle. As a plate moves over a mantle plume, a linear chain of volcanoes would be

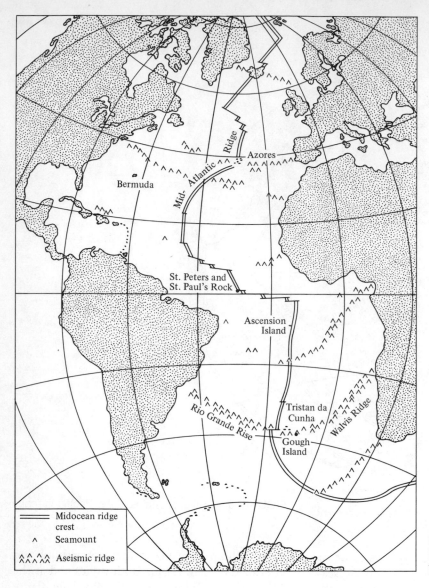

FIGURE 6.12 Volcanic islands and aseismic ridges in the Atlantic Ocean.

produced (Fig. 6.13). The age of the volcanic rock increases in the direction of movement of the plate. Hot spots under stationary plates, like the African plate, do not produce linear island chains.

## WHY PLATES BREAK APART

The mechanism that causes plates to break apart has not been identified to the satisfaction of many geologists. Most feel that convection plays an important role in this process, and

Morgan has suggested that plumes are involved (20). In relating plumes to upwellings in the mantle, he seems to imply that they originate as a result of convection in the mantle, but why a plume is initiated in a particular area has been something of a mystery. One idea is that thinning of the crust due to erosion at the base localizes heat flow and creates a plume.

It is possible that mantle plumes and the resulting hot spots are the result of meteorite impacts (21). An impact of sufficient size might initiate a mantle plume by removing a great

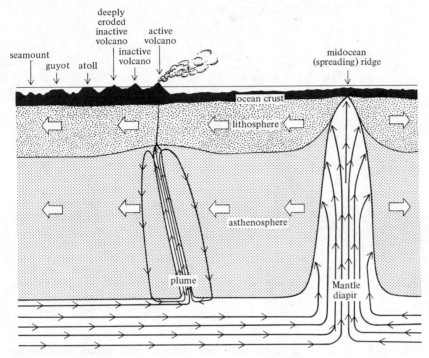

FIGURE 6.13 Diagrammatic sketch of formation of a linear chain of volcanic islands or aseismic ridge as a result of movement of a plate over a mantle plume.

thickness of crustal and, in some cases, mantle rock during the formation of a crater. Isostatic adjustments would cause upwelling of mantle material beneath the crater. Elastic rebound following the impact would cause further upwelling, and this might serve to trigger a continuing upwelling which would be a mantle plume. If more than one closely spaced impact occurred, plumes might become linked by impact-generated fractures Mantle upwelling along such fractures might produce midocean (spreading) ridges. In support of this hypothesis, there seems to be a strong correlation between the age of major impacts and times of major plate movements (**21**). This relationship will be discussed in detail later in the book.

## GEOSYNCLINES

James Hall, State Geologist of New York in the 1850s, observed that the folded sedimentary rocks in the Appalachian Mountains are much thicker than the equivalent strata that had been deposited in the midwestern part of the United States. Hall noted that the Paleozoic deposits thickened from 1500 m (4500 ft) in the Mississippi Valley region to 12,500 m (40,000 ft) in the Appalachians. In 1857 Hall proposed that the sediments of the Appalachian Mountains had been deposited in an elongated, slowly subsiding trough. Later, James Dana proposed the name *geosyncline* for this type of structure.

Both Hall and Dana believed that there was a connection between the geosynclines and the mountains that subsequently developed from them. Hall postulated that when the trough became filled with sediments, the rocks were folded by collapsing into the center of the trough. Dana, on the other hand, believed that the earth was cooling and contracting, and that the geosyncline was folded as a result of a horizontal compressive force produced by this contraction.

Some geologists prefer to restrict the use of

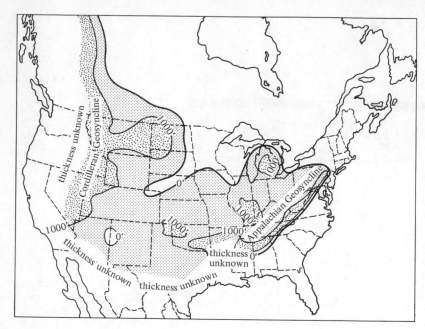

**FIGURE 6.14** Thickness of Mississippian strata in North America. Contours are in feet. Area is where the rate of sedimentation was greater than 13 m (40 ft) per million years. The Appalachian and Cordilleran geosynclines are shown as elongated zones with this sedimentation rate. (Modified after Sloss, Dapples, and Krumbine, Ref. 22.)

the term geosyncline to an elongated depositional trough in which clastic sediments and volcanic rocks are deposited in moderately deep water. However, this definition is quite different from Hall's definition of a geosyncline. Hall used the Appalachian Miogeosyncline to illustrate the character of a geosyncline, and this geosyncline contains a great abundance of shallow-water carbonates and clastics with no associated volcanics. The term geosyncline should denote an elongate zone in which a thick sequence of sedimentary and/or volcanic deposits has accumulated, regardless of the depth of water in which they were laid down.

The following criteria are suggested for the identification of both modern and ancient geosynclines:

1. The length of the geosyncline should be more than twice its width.
2. The rate of accumulation of the deposits within the geosyncline should have been more than 13 m (40 ft) per million years.
3. The basement on which the geosynclinal

deposits were laid down should have been downwarped into a broad syncline during accumulation of these deposits.

One method of locating a geosyncline is to plot and contour the thickness of a time-stratigraphic unit on a map. The result is an *isopach map*. A plot of the thickness of Mississippian sedimentary rocks in North America shows several areas in which the rate of sedimentation was more than 13 m (40 ft) per million years (Fig. 6.14). Four of the areas are somewhat circular in outline and are more appropriately termed basins, rather than geosynclines. However, two of the areas, the Appalachian and Cordilleran geosynclines, are distinctly elongated. Thus, it is likely that the geosynclinal deposits were downwarped into a broad syncline during deposition. Paleogeographic reconstructions indicate that the areas of thick sedimentary deposits were bounded by *geanticlines* (mountain ranges that are adjacent to geosynclines and that furnish much of their sediment).

## Miogeosyncline

Miogeosynclinal deposits include limestone, shale, quartz sandstone, and conglomerate. The presence of such primary structures as mud cracks, ooliths, and algal stromatolites indicates deposition of these rocks in relatively shallow water. Igneous intrusions and volcanics rarely occur within a miogeosyncline.

Miogeosynclines are located either near the margin of a continent or adjacent to a mountain range within a continent. Those near a continental margin, such as the Cordilleran Geosyncline, are generally adjacent to eugeosynclines and are (or were) near subducting plate margins (Fig. 6.14). Miogeosynclines adjacent to mountain ranges, such as the Appalachian Geosyncline during the Late Paleozoic, form in areas which are near a former plate margin. The presence of a miogeosyncline in the center of a continent may be used to infer that the miogeosyncline was formerly at the margin of a continent that was subsequently joined to another continent.

## Eugeosyncline

Eugeosynclines are usually associated with island arcs on subducting plate margins. The total thickness of sedimentary and volcanic rocks in a eugeosyncline is generally greater than in the adjacent miogeosyncline. Eugeosynclinal sedimentary rocks are almost entirely clastic and include shale, graywacke, and conglomerate. Well-sorted quartz sandstones and limestone are not abundant. Volcanic rocks within eugeosynclines include flows and tuffs as well as shallow dikes and sills. The volcanics are most commonly andesitic in composition, but basaltic and rhyolitic volcanics are abundant in some areas. Angular unconformities are common within eugeosynclinal sequences and indicate a tectonically unstable environment of deposition.

Modern depositional environments analo-gous to those in ancient eugeosynclines are found in the vicinity of island arcs, such as the Indonesian Island Arc (Fig. 6.15). These regions are typified by the presence of volcanoes, geanticlines (elongate upwarps), and structural troughs in which sediments are being deposited. In fact, the Indonesian Arc provides an excellent model in which most geosynclinal and plate margin features can be found. Sedimentation is especially rapid adjacent to volcanic and tectonic islands. Since modern island arcs are restricted to continental margins, it is likely that ancient eugeosynclinal deposits formed in similar environments. This relationship may be used in reconstructing the geography of ancient continents.

## MIOGEOCLINES AND EUGEOCLINES

Study of sediments along the eastern continental margin of North America by seismic refraction profiling has revealed the presence of thick sequences of sediments in two parallel, elongated zones. One of these is located under the continental shelf, and the other is located at the base of the continental shelf (Fig. 6.16). The rate of deposition of these sediments is more than 13 m (40 ft) per million years, but the basement does not seem to be downwarped into a syncline. Rather, the basement and the overlying sediments appear to dip uniformly seaward at a low angle. The structure in which they are located is more like that of a monocline than a syncline. Robert Dietz has proposed that the zone in which thick sediments underlie the continental shelf be called *miogeocline* and that the zone in which thick sediments underlie the base of the continental slope be called a *eugeocline* (**24**). Miogeosynclines and eugeosynclines are located near actively colliding plate margins where an oceanic or continental plate is being subducted. On the other hand, miogeoclines and eugeoclines are located on passive margins of spreading plates generally a considerable dis-

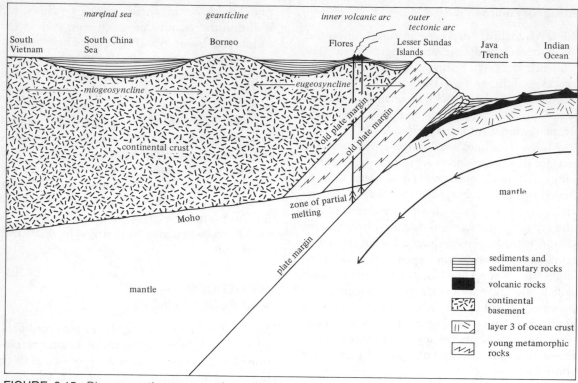

FIGURE 6.15 Diagrammatic cross section across the Indonesian Island Arc.

tance from the plate margin, as for example on the borders of the Atlantic and Indian oceans. The deposits of the miogeoclines are mainly unconsolidated sands, muds, and gravels where there is a nearby source of clastic sediments and limestone and chalk in areas away from a source of clastics. There are very few volcanics.

Eugeoclinal deposits are almost entirely clastic and include poorly sorted sands (often turbidites), muds, and minor gravels. Volcanic rocks are principally in the form of volcanic seamounts that extend from the basement into the surrounding sediments.

It is often very difficult to tell the difference between miogeosynclines and miogeoclines on the one hand and eugeosynclines and eugeoclines on the other. One of the reasons for this is that the original configuration of the basement rocks has often been considerably changed by folding. Miogeosynclines and eugeosynclines are thought to form adjacent to subduction zones where one plate descends beneath another plate, while miogeoclines and eugeoclines form on a passive continental margin adjacent to an expanding ocean. Andesites are erupted only in island arcs, and island arcs only form adjacent to subduction zones. Andesites are thought to form by partial melting of the rock near a plate boundary as the seafloor descends beneath a continent (Fig. 6.17). Therefore, the presence of andesite in a thick sequence of sedimentary and volcanic rock indicates that the sequence was probably deposited in a eugeosycline in the vicinity of an island arc. Miogeosynclines can form either adjacent to a eugeosyncline or adjacent to a mountain range in the center of a continent where there is a subduction zone beneath the mountains.

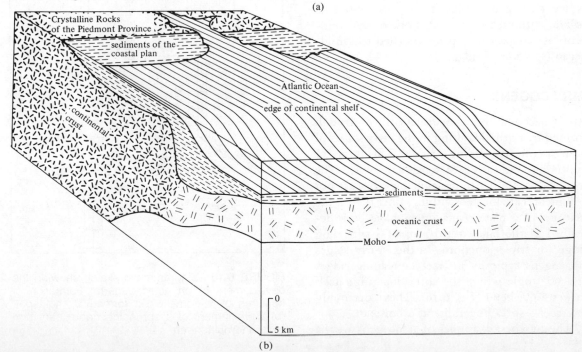

**FIGURE 6.16** Miogeocline and eugeocline on the East Coast of North America. **(a)** Isopach map of Mesozoic and Cenozoic deposits in the East Coast Geosyncline. The heavy lines are structure contours on the top of the crystalline basement. The dashed lines are submarine contours in fathoms. Thickness is given in feet. (After Eardley, Ref. 23.); **(b)** Block diagram showing the miogeocline under the continental shelf and the eugeocline at the base of the continental slope. Sediments in the miogeocline and the eugeocline are shown as short dashes.

(a)

Crystalline Rocks
of the Piedmont Province

sediments of the
coastal plan

Atlantic Ocean

edge of continental shelf

continental
crust

sediments

oceanic crust

Moho

0

5 km

(b)

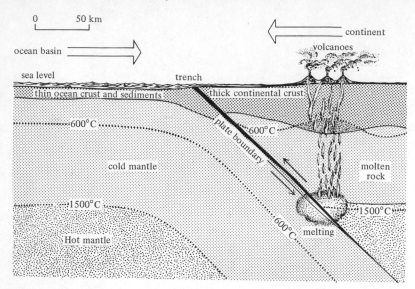

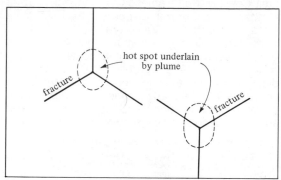

**FIGURE 6.17** Development of volcanoes as an oceanic plate descends beneath a continental plate. (Diagram courtesy of the National Science Foundation.)

Geoclines formed on the margins of an expanding ocean (formed by the separation of two continents) may be converted into a miogeosyncline-eugeosyncline pair bordering a subduction zone if the continents reverse direction and begin to approach each other. Furthermore, the eugeosyncline may be converted into a mountain range furnishing sediments to the adjacent miogeosyncline if the two continents collide with each other and if one of the continents is not part of a plate attached to a mid-ocean (spreading) ridge.

## AULACOGENS

An aulacogen is a sediment-filled trough formed by normal faulting in the early stages of a midocean (spreading) ridge. This type of faulting is indicative of tensional forces. The troughs are similar to geocynclines, but are not as long and are located within, rather than at the margins of a plate. Aulacogens have been described as the failed third arm of a spreading ridge (or triple junction). In their early stages aulacogens typically intersect spreading ridges at a 60° angle where the spreading ridge itself makes a 60° bend (Fig. 6.18). They commonly intersect a spreading ridge at a hot spot.

The two best examples of aulacogens are the

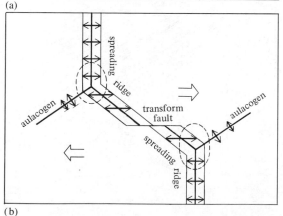

**FIGURE 6.18** Diagrammatic sketch showing the development of an aulacogen: **(a)** Formation of fractures over domal uplift formed by hot spot. **(b)** Development of a spreading ridge, transform fault, and aulacogen.

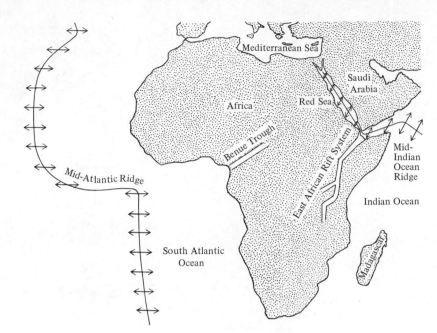

**FIGURE 6.19** Locations of two aulacogens in Africa.

East African Rift System and the Benue Trough (Fig. 6.19). The East African Rift System intersects the Mid-Indian Ocean Ridge where the ridge makes a 60° bend from the Gulf of Aden into the Red Sea. Active volcanoes and numerous earthquakes along the East African Rift System indicate that it is still active. The Benue Trough is now inactive and developed when Africa and South America began to split apart during the Mesozoic era. It is located in the bight of Africa where the Niger River enters the Atlantic Ocean. In this region, the Mid-Atlantic Ridge bends 60° to the west (Fig. 6.19).

Sedimentary deposits in aulacogens are generally very thick and are dominantly clastic. Shales, sandstones, and conglomerates are commonly interbedded with alkalic (rich in sodium and potassium) volcanic rocks. These rocks may be folded by compressional forces in the later stages of the evolution of an aulacogen.

## CRATONS

The interiors of continents are characterized by the presence of relatively thin sedimentary deposits or ancient crystalline rocks. In most areas, the sedimentary rocks are undeformed. The stable part of the continent, the *craton,* includes shields and platforms. *Shields* are extensive areas in which Precambrian rocks are exposed, while *platforms* contain sedimentary rocks. Areas of thicker-than-average sedimentary deposits on the platform are *basins;* areas of thinner-than-average deposits are *arches* or *domes.* The average rate of subsidence of platforms is generally less than 12 m (40 ft) per million years. However, higher rates have been calculated for the centers of some basins. Because these areas are not elongated, they would not be classified as geosynclines according to Hall's definition of the term. The deposits of the platform include abundant carbonates (limestone and dolostone) as well as clastics (shale and sandstone). Volcanic rocks are rare. The craton is generally bordered by geosynclines, geoclines, or mobile mountain belts (such as the Andes Mountains).

## CAUSE OF SUBSIDENCE

Geosynclines and geoclines commonly contain up to 10,000 m or more (30,000 ft) of sediments

**FIGURE 6.20** Development of a geosyncline on the margin of an ocean basin formed by continental drift and seafloor spreading: **(a)** Uplift and faulting over an upwelling convection current. **(b)** Erosion of continental margin to sea level before the underlying mantle cools to the temperature of the adjacent mantle. **(c)** Subsidence of continental margin and deposition of sediments as underlying mantle cools.

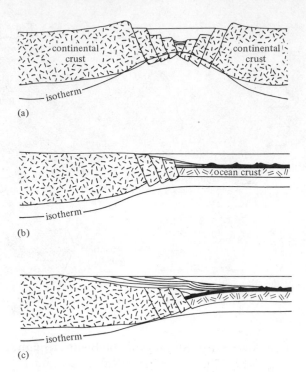

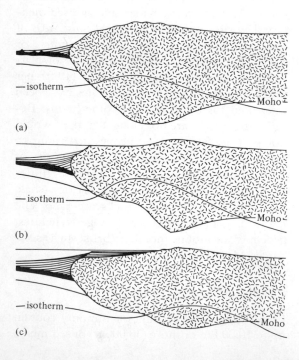

**FIGURE 6.21** Development of a geosyncline on the margin of a continent affected by compressional folding: **(a)** Beginning of erosion of mountain following its uplift. **(b)** Erosion of mountain belt to sea level before the underlying mantle cools to the temperature of the mantle beneath the stable interior of the continent. **(c)** Subsidence of margin and deposition of sediments as the underlying mantle cools.

or sedimentary and volcanic rocks, while platforms contain several thousands of feet of such deposits. Even the oldest of these deposits generally shows evidence of having been deposited in shallow water. Evidently the rate of subsidence within the geosyncline was approximately equal to the rate of accumulation of its sediments. Only in this way would the surface of deposition remain at approximately the same level throughout sedimentation. The cause of subsidence within geosynclines is the subject of considerable controversy. Some of the mechanisms that have been suggested are:

1. When sediments and volcanics are deposited in a geosyncline, a certain amount of subsidence results from the isostatic adjustment of the crust to the weight of the deposits. This adjustment takes place by a plastic flow within the mantle. The amount of subsidence is determined by the thickness of sediment deposited, the density of the sediments, and the density of the layer in which flow takes place. In most cases, isostatic adjustment can account for only about 75 percent of the observed subsidence. There must be another factor involved in the process.

2. It has been suggested that convection currents produce subsidence within geosynclines by dragging down the earth's crust as the currents descend and erode the crust from beneath geosynclines. Although convection currents may descend beneath miogeosynclines and eugeosynclines, the lack of seismic activity and folding associated with platforms and geoclines suggests that this mechanism is not applicable in such areas.

3. If the mantle beneath part of a continent is heated, it undergoes thermal expansion. As a consequence, the overlying continent is uplifted. The uplifted part of the continent is then subjected to increased subaerial erosion, resulting in a thinning of the earth's crust. If this erosion were followed by a cooling of the mantle, the mantle would contract and the overlying continent would subside. The sequence of heating - uplift - erosion - cooling - subsidence may occur over an upwelling convection current or in the vicinity of a mountain range. The miogeoclines bordering the Atlantic and Indian oceans may have been formed as a result of such a sequence of events (Fig. 6.20). Miogeosynclines and eugeosynclines commonly develop on the roots of ancient mountain systems. Mountains would initially have a high heat flow, and therefore the mantle under the mountains would be hotter than that under the stable interior of the continent (Fig. 6.21). If the mountainous region had been eroded before the underlying mantle cooled to the temperature obtaining under the stable interior, the crust could have been so reduced in thickness by erosion that it would be thinner than the crust of the stable interior. Further cooling of the mantle under the eroded mountain belt would cause subsidence of the mountain roots and would thus permit geosynclinal deposition.

## REFERENCES CITED

1. L. R. Sykes, 1969, *in* R. H. Hart, ed., *The Earth's Crust and Upper Mantle:* Geophysical Monograph No. 13, American Geophysical Union, Washington, D.C.
2. R. P. Von Herzen, 1965, *in* L. H. Ahren *et al.,* eds., *Physics and Chemistry of the Earth,* v. 6: Pergamon Press, London.
3. E. Orowan, 1964, *Science,* v. 146, p. 1003.
4. R. H. Dicke, 1969, *Jr. Geophys. Res.,* v. 74, p. 5895.
5. D. L. Anderson, 1965, *Science,* v. 151, p. 321. R. K. Mc Connel, Jr., 1968, *in* R. A. Phinney, ed., *The History of the Earth's Crust—A Symposium:* Princeton University Press, Princeton, N.J.
6. W. J. Morgan, 1968, *J. Geophys. Res.,* v. 73, p. 1959.
7. M. Barazangi and J. Dorman, 1969, *Bull. Seismol. Soc. Amer.,* v. 59, p. 369.
8. B. Isacks, J. Oliver, and L. R. Sykes, 1968, *Jr. Geophys. Res.,* v. 73, p. 5855.
9. J. T. Wilson, 1965, *Nature,* v. 207, p. 343.
10. B. Gutenberg and C. F. Richter, 1954, *Seismicity of the Earth,* 2nd ed.: Princeton University Press, Princeton, N.J.

11. B. Isacks and P. Molnar, 1969, *Nature,* v. 223, p. 1121.
12. D. P. McKenzie and W. J. Morgan, 1969, *Nature,* v. 224, p. 125.
13. F. A. Vening Meinesz, 1964, *Developments in Solid Earth Geophysics—1—The Earth's Crust and Mantle:* Elsevier, New York.
14. C. K. Seyfert, 1968, *Trans. Amer. Geophys. Union,* v. 49, p. 202; J. C. Maxwell, 1968, *Amer. Sci.,* v. 56, p. 35.
15. S. I. Van Andel and J. Hospers, 1968, *Tectonophysics,* v. 5, p. 273.
16. K. Horai and G. Simmons, 1969, *Earth Planet. Sci. Lett.,* v. 6, p. 386.
17. G. F. G. MacDonald, 1964, *Science,* v. 143, p. 921.
18. C. Wright, 1968, *Earth Planet. Sci. Lett.,* v. 5, p. 35.
19. W. J. Morgan, 1971, *Nature,* v. 230, p. 42.
20. W. J. Morgan, 1972, *Bull. Am. Assoc. Pet. Geol.,* v. 56, p. 203.
21. N. H. Sleep, *Geophys. Jr. Roy. Astronom. Soc.,* v. 24, p. 325.
22. L. L. Sloss, E. C. Dapples, and W. C. Krumbein, 1960, *Lithofacies Maps:* Wiley, New York.
23. A. J. Eardley, 1962, *Structural Geology of North America:* Harper and Row, New York.
24. R. S. Dietz, 1972, *Sci. Amer.,* v. 226, p. 30.

# Polar Wandering, Continental Drift, and Seafloor Spreading

**7**

Until the 1960s, most geologists believed in the permanence of continents and ocean basins. This concept did not provide answers to such problems as the origin of mountain ranges, the distribution of Paleozoic glacial deposits, the distribution of Paleozoic fauna and flora, and the rifting (splitting apart) of the earth's crust along midocean ridges (Fig. 7.1). Since about 1968, the idea that large scale movements of the earth's crust have taken place has gained widespread acceptance among geologists. Included in such movements are polar wandering, continental drift, and seafloor spreading.

During *polar wandering,* the earth's crust moves relative to the rotational poles of the earth. Because the spinning earth acts as a giant gyroscope that is stabilized by the earth's equatorial bulge, large changes in the angle of inclination of the earth's axis are considered to be unlikely. Rather, it is thought that the outer shell of the earth shifts slowly over the inner layers, while the axis remains at a relatively constant angle to the earth's plane of revolution about the sun (the ecliptic).

The movement of the continents relative to one another is termed *continental drift.* Unless continents move in a precisely east-west direction, continental drift will result in some degree of polar wandering. *Seafloor spreading,* which describes the movements of oceanic plates relative to one another, may operate independently or in association with continental drift and polar wandering. During polar wandering, continental drift, and seafloor spreading, the earth's crust and a significant portion of the upper mantle move together as a unit. The evidence which has led to the acceptance of continental drift, polar wandering, and seafloor spreading will be presented in this chapter.

## MATCHING CONTINENTAL OUTLINES

Although the shorelines of South America and Africa are similar in shape, they do not match exactly (Fig. 7.2a). Alfred Wegener was the first to propose that reconstructions of the continents should be made using submarine contours. He felt that the continents should be matched at "the margin of the continental slope in the deep sea" (1). This is at the edge of the continental shelf where it joins the continental slope. While the depth of this break in slope varies, it averages about 200 m (600 ft) below sea level. The matching of South America and Africa is better at this depth than at the shorelines, but the fit is still far from exact (Fig. 7.2b). S. Warren Carey has shown that the outlines of these two continents match almost exactly at the 2000-m (6500-ft) contour below sea level (Fig. 7.2c). Sir Edward Bullard, J. E. Everett, A. Gilbert Smith, and A. Hallam matched the continents bordering the Atlantic and Indian oceans with the help of a computer (3). This technique minimizes the overlaps and

(a)

(b)

FIGURE 7.1 Gaping fissures from the central Thingvellir Graben area of Iceland. The graben is bounded by normal faults. Rifting of this type occurs where seafloor spreading causes Iceland to be pulled apart: **(a)** Aerial view (courtesy of Sigurdir Thorarinsson). **(b)** Close up view (courtesy of Karlheinz Schafer).

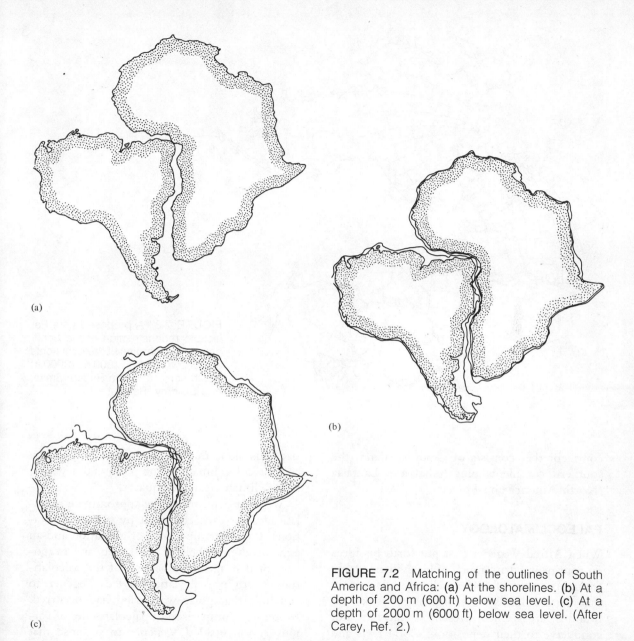

(a)

(b)

(c)

**FIGURE 7.2** Matching of the outlines of South America and Africa: **(a)** At the shorelines. **(b)** At a depth of 200 m (600 ft) below sea level. **(c)** At a depth of 2000 m (6000 ft) below sea level. (After Carey, Ref. 2.)

gaps between the continents. They found that the best fit between the continents could be obtained if the 1000-m (3300-ft) depth contour is considered to be the edge of the continent (Fig. 7.3). However, they found that it made

little difference whether the 1000-m or 2000-m depth contour were used. The reconstructions clearly show that the continents can be restored to form a supercontinent, "Pangaea," as it was named by Wegener. Pangaea is a composite

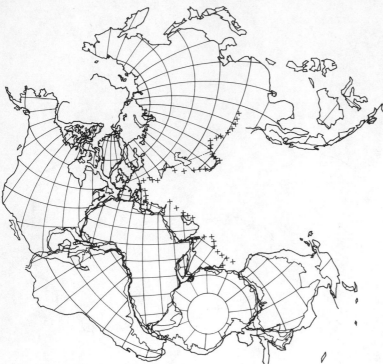

**FIGURE 7.3** A proposed Late Paleozoic reconstruction of the continents. The margins of the continents are shown at the 1000 m (3000 ft) depth contour. (From Briden, Smith, and Sallomy, Ref. 3.)

continent that consists of Gondwanaland (the southern continents plus India) and Laurasia (North America and Eurasia).

## PALEOCLIMATOLOGY

When Alfred Wegener first put forth his ideas on continental drift, much of his evidence came from his study of ancient climates. He noted that certain kinds of sedimentary rocks are found in areas where present climates are not conducive to their deposition. Coral reefs and coals derived from tropical plants in the Arctic and Antarctic are notable examples. Applying a uniformitarian approach, Wegener assumed that these rocks had been deposited under the same climatic conditions in the past as they are today and that the climatic zones of the earth have always been about as they are now (Fig. 7.4). Although climatic belts may shift and change in

size, it is likely that the earth has always had a cool or cold climate at the poles and a warm, moist climate near the equator.

Wegener determined the approximate paleolatitudes for each continent by placing a rotational pole near the center of glaciation and an equator along a belt of coral reefs and evaporites of that age. Comparison of the paleolatitudes with present latitudes led Wegener to conclude that polar wandering had occurred. Moreover, comparison of paleolatitudes of various continents led Wegener to propose that continental drift had taken place (Fig. 1.9).

### Further Paleoclimatic Observations

Late Ordovician glacial deposits have recently been discovered in the Sahara Desert in northern Africa and in several localities in southern

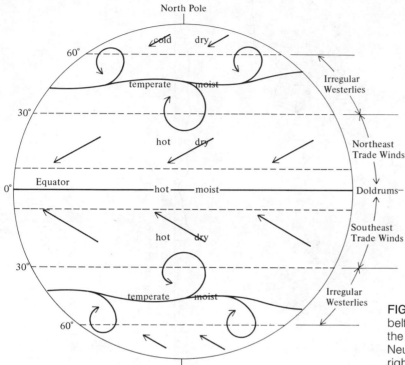

North Pole

South Pole

**FIGURE 7.4** Idealized climatic belts and prevailing directions for the present time. (Adapted from Neuberger and Cahir, Ref. 4. Copyright © 1969, Holt, Rinehart & Winston, Inc.)

Africa (**5**). These deposits are so extensive that they must have been laid down by continental rather than alpine glaciers. The orientation of *roches moutonnées,* glacial striations, chatter marks, and glacial valleys indicates that the ice moved *away* from the present equator (Fig. 10.2b).

Widespread glacial deposits of Permo-Carboniferous age occur in South America, Africa, India, Australia, and Antarctica (Fig. 11.29). Some of these deposits are on or near the present equator (Fig. 7.5). During this glacial episode, the ice moved away from what are now the Atlantic and Indian ocean basins. This relationship is puzzling, because the Pleistocene ice sheets almost always moved toward the ocean basins.

Many ancient evaporites and red beds are found in the middle and high latitudes where present climates are wet and/or cold. For exam-

ple, Paleozoic salt deposits are found within 20° of the present North Pole in northern Canada and Siberia, and Paleozoic red beds occur at 75°N latitude in Asia. Today, however, extensive red beds and evaporites are forming in hot, dry climates, at latitudes between 15° and 45°.

Ancient latitudes may also be approximated from the study of the direction of prevailing winds. At the present time, the prevailing winds blow from the west in the middle latitudes and from the east near the equator (Fig. 7.4). Changes in the direction of prevailing wind may indicate a change in latitude. Since cross beds in aeolian sandstones dip to the leeward, the average dip of such beds indicates the direction of prevailing winds at the time of deposition. Such measurements show that, in the southwestern United States and in the British Isles, the prevailing winds were easterly during the late Paleozoic (Fig. 7.6). Both these

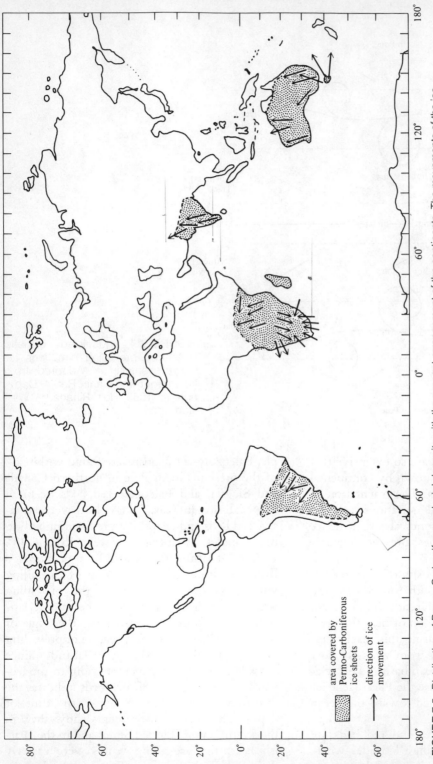

**FIGURE 7.5** Distribution of Permo-Carboniferous glacial deposits with the present arrangement of the continents. The movements of the ice sheets away from the present ocean basins is a strong argument for continental drift.

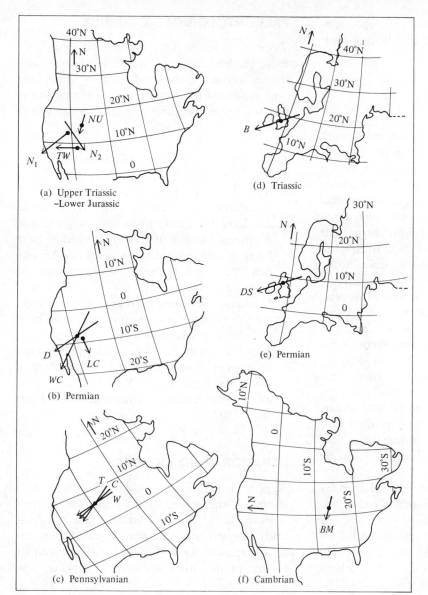

(a) Upper Triassic–Lower Jurassic

(b) Permian

(c) Pennsylvanian

(d) Triassic

(e) Permian

(f) Cambrian

FIGURE 7.6 Direction of prevailing winds during the Late Paleozoic and early Mesozoic for North America and Europe. Wind direction was determined from the average direction of dip of cross laminations in dune sandstones. (After Irving, Ref. 6.)

areas are now in the zone of prevailing westerlies.

Although modern coral reefs are restricted to warm water environments within about 30° of the equator, Paleozoic reefs are found at much higher latitudes in Greenland, northern Canada,

and Spitsbergen. Similarly, thick carbonates are generally deposited in warm waters, but extensive Paleozoic carbonates occur in areas bordering the Arctic Ocean. Carboniferous coal deposits contain fossil plants that lack growth rings. Such plants are typical of tropical or subtropical environments. However, these deposits are now located in the middle and high latitudes.

## Interpretation of Paleoclimatic Data

It is evident that, in many cases, rock units were deposited under climatic conditions quite different from those which exist presently in the regions where those rocks occur. The incidence of Paleozoic continental glaciers near the present equator, and evaporites, coral reefs, thick carbonates, coals, and red beds near the present poles, indicates that large-scale shifts in climatic belts have occurred since the Paleozoic. Some geologists have proposed worldwide climatic changes in order to account for this. However, such climatic changes do not provide a complete explanation for paleoclimatic data. For example, the Late Ordovician and Permo-Carboniferous glacial deposits are restricted to the southern hemisphere and India. Contemporaneously with the glaciation of the southern continents, evaporites, coral reefs, limestones, red beds, and coals were being deposited near the present north pole. These observations cannot be explained by worldwide climatic changes.

Paleoclimatic data can best be explained by polar wandering and continental drift. If the southern continents had been together in one large landmass during the Permian and Carboniferous, the ice sheets of that age would have radiated away from a center in southern Africa and eastern Antarctica (Fig. 11.26). The center of glacial accumulation was very close to the Permo-Carboniferous South Pole as determined from paleomagnetic data. Tropical coals, red beds, evaporites, and reefs of Permian and Carboniferous age would lie at low paleolatitudes. Britain and the southwestern part of the United States would fall within the belt of northeast trade winds, which would be consistent with the observed direction of winds during the Late Paleozoic in that area.

## PALEONTOLOGICAL EVIDENCE

Wegener believed that continental drift provided the best explanation for the striking similarities in fossil faunas and floras of continents now separated by deep oceans. Some examples that impressed Wegener and later investigators are:

1. The freshwater reptile *Mesosaurus* has been found only in South America and Africa (Fig. 7.7). The bone structure of this salamanderlike organism is such that it would not have been able to swim a large ocean.
2. Numerous bones of terrestrial reptiles have been discovered in Triassic rocks in Antarctica (**8**), which is now completely isolated from other continents.
3. In many cases early Paleozoic marine invertebrate assemblages are quite similar on both sides of the Atlantic Ocean.
4. Fossil leaves of seed ferns belonging to the *Glossopteris* flora (a distinctive assemblage of plants) of Permian age have been found on all of the southern continents and India (Fig. 7.8).

## STRATIGRAPHIC EVIDENCE

Wegener was also impressed by the strong similarity of the stratigraphic sequence of rock units in South America, Africa, India, and Australia. Recent studies have revealed that the stratigraphy of Antarctica is remarkably similar to that of the other southern continents. The correlation of Permo-Carboniferous tillites, Permian shales, Lower Triassic red beds, and Jurassic lavas is quite impressive (Fig. 7.7). In the northern hemisphere, the Paleozoic stratigraphic sequences of the Appalachian Geosyncline are similar to those of the Caledonian Geosyncline of Europe. Furthermore, the presence of oceanic rock, such as ophiolite sequences and turbidites, accreted onto many continental mountain ranges is evidence of seafloor spreading.

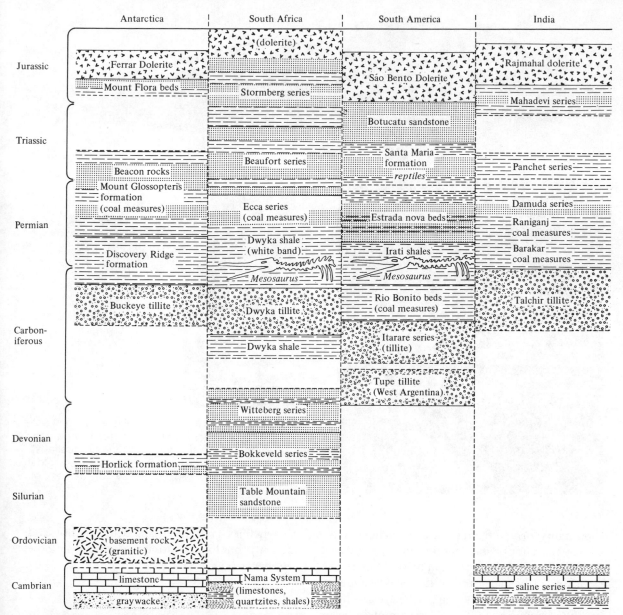

**FIGURE 7.7** Correlation of stratigraphic columns from the southern continents. (From Doumani and Long, Ref. 7. Copyright © 1962 by Scientific American, Inc., all rights reserved.)

## STRUCTURAL EVIDENCE

Modern geosynclines and geoclines are located on continental margins and encircle continents. Ancient geosynclines were probably distributed in a similar pattern. However, several geosynclinal belts appear to end abruptly. The Appalachian Geosyncline ends just northeast of Newfoundland, and the Caledonian Geosyncline is terminated just west of Ireland. The

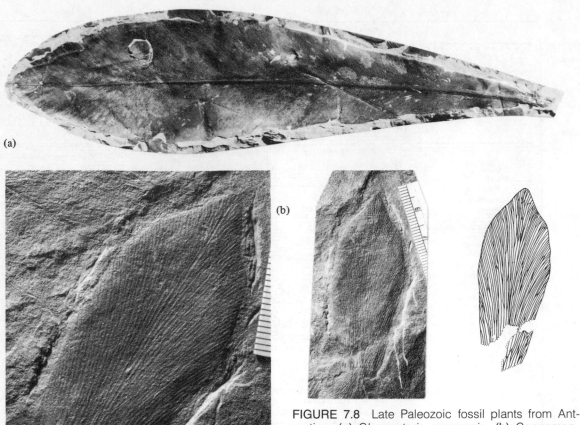

**FIGURE 7.8** Late Paleozoic fossil plants from Antarctica: **(a)** *Glossopteris communis,* **(b)** *Gangamopteris obovata* (scale at right is 0.1 in. graduations). (Print negatives 3099-1x and 2877-2x, U.S. Geological Survey, Columbus, Ohio, courtesy of James M. Schopf.)

mountains formed from these geosynclines also end at the continental margins and cannot be traced into the Atlantic Ocean basin. This relationship led Wegener and others (**9**) to suggest that the Appalachian Mountain belt was once continuous with the Caledonian Mountain belt of Europe (Fig. 7.9).

In the southern hemisphere, the Tasman Geosyncline ends south of Tasmania and the Transantarctic Geosyncline ends near Victoria Land. Geosynclines on the northwestern and southern borders of Africa and on the western border of South America are also abruptly terminated. If the continents are joined together,

geosynclines and mountain belts do not end suddenly, but form continuous belts circling the continents (Fig. 7.9).

## PALEOMAGNETISM

All rocks have magnetic fields owing to the presence of magnetic minerals, such as magnetite. By measuring the direction and inclination of this magnetic field, it is possible to calculate the approximate position of the magnetic poles of the earth at the time of crystallization or deposition of the rock. Thus rock magnetism furnishes a fossil "compass."

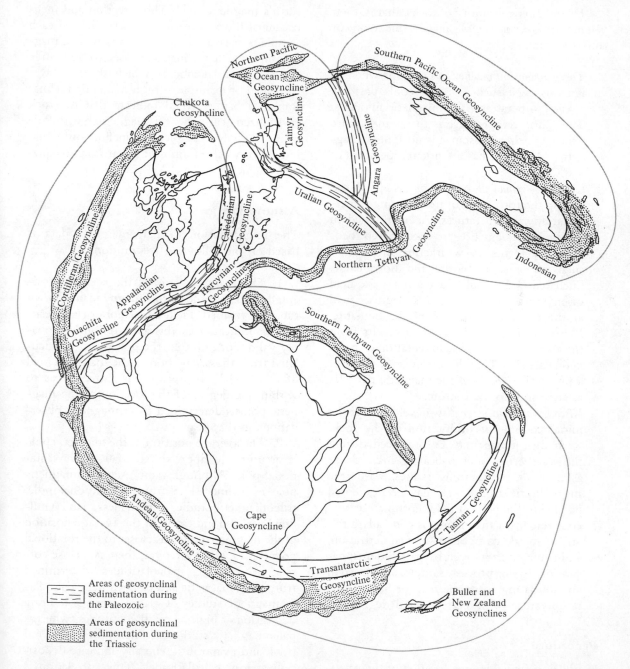

**FIGURE 7.9** Location of Paleozoic and Triassic geosynclines plotted on our reconstruction of the continents. Our reconstruction is essentially the same as that of Briden *et al.,* Ref. 3, except that Antarctica, Australia, and India have all been moved relative to Africa so that the Transantarctic Geosyncline is aligned with the Cape Geosyncline. The circled continents are five ancestral continents which existed during the Early Paleozoic prior to their joining into a single continent.

The magnetic memory of rocks, that is, their *natural remanent magnetism,* may be due to one or more factors:

1. *Thermoremanent magnetism* develops when an igneous rock such as basalt cools past the Curie temperature of its magnetic minerals. The Curie temperature is that temperature above which a substance is no longer magnetic. Magnetite has a Curie temperature of 578°C.
2. *Chemical remanent magnetism* is produced during the formation of iron-bearing minerals at low temperatures. As mentioned in Chapter 4, hematite in arkosic red beds probably forms by weathering in place of such minerals as biotite and hornblende. Such red beds would have a magnetic field oriented parallel to the earth's magnetic field at the time of formation of the hematite.
3. *Depositional remanent magnetism* results from the alignment during sedimentation of magnetic particles. This alignment is the result of a physical rotation of the magnetic particles as they sink to the bottom.
4. *Viscous remanent magnetism* develops after the initial cooling or lithification of the rock. Such magnetism may be produced during the weathering of iron-bearing minerals or as a result of lightning strikes. Also, the magnetic field of a rock may be reoriented parallel to the present field. Fortunately, viscous remanent magnetism is generally a relatively weak component of magnetization, and it may be removed by partial demagnetization. In this process, samples are placed in an alternating magnetic field or heated to a temperature of several hundred degrees C.

## Procedure

In order to determine the location of the paleomagnetic poles, oriented samples are collected from several horizons within a formation and at several different localities. This procedure serves to average out minor variations in the earth's magnetic field. The direction and inclination of the axis of the magnetic field of each sample is measured with a very sensitive magnetometer. The inclination is used to determine the magnetic latitude at which the rock was formed. For example, a high magnetic inclination in a basalt would indicate that the rock crystallized near either the magnetic North Pole or magnetic South Pole; low inclination would indicate crystallization near the magnetic equator (Fig. 7.10).

## Assumptions

The interpretation of paleomagnetic data is based on the following assumptions:

1. The magnetic axis of the rock was oriented parallel to the earth's magnetic field when the rock formed. Field and laboratory investigations indicate that this assumption is probably valid. For example, the pole position calculated from the orientation of the magnetic field of the 1669 lava flow on Mt. Etna agrees to within one degree of the location of the magnetic pole as determined from magnetic observations in that year.

2. The average position of the magnetic pole is assumed to approximate that of the rotational pole. Although the present geomagnetic pole is inclined 11.5° to the rotational pole, paleomagnetic studies of rocks less than 20 million years old indicate that the average position of the magnetic pole was closer to the rotational pole (Fig. 7.11). Moreover, there is a close correlation between paleolatitudes determined from paleomagnetic and paleoclimatic data. Paleomagnetic studies of glacial deposits indicate deposition at high or middle latitudes, whereas paleomagnetic studies of thick limestones, red beds, and evaporites generally indicate deposition at low paleolatitudes. This correlation is also shown by the agreement of the paleomagnetic pole with the center of the Late Ordovician and Permo-Carboniferous ice sheets (Fig. 10.36 and 11.1).

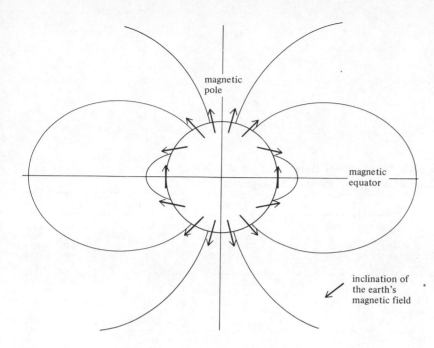

magnetic
pole

magnetic
equator

inclination of
the earth's
magnetic field

**FIGURE 7.10** Magnetic field of a uniformly magnetized sphere. The magnetic field of the earth is very similar, but is considerably distorted because of solar wind.

3. The earth's magnetic field is assumed to have always been dipolar—that is, with one north and one south magnetic pole. This assumption is consistent with the generally accepted *dynamo theory* of the origin of the earth's magnetic field. Briefly, the dynamo theory states that a conductor (in this case the liquid, nickel-iron core of the earth) moving in a pre-existing magnetic field produces an electric current, and an electric current flowing in a conductor generates a magnetic field. In this way, the earth's magnetic field is continuously being generated.

### Results

Thousands of paleomagnetic pole positions have been determined for formations of various ages and from various continents. However, only those poles that have been determined from samples that were tested for magnetic stability by partial demagnetization can be considered reliable, because partial demagnetization removes the unstable viscous remanent magnetism. The reliable pole positions have been

averaged for each continent and for each geological period and are shown in Table 7.1.

When paleomagnetic poles from North America are plotted on a map (Fig. 7.12), it is apparent that the positions of the rotational poles have changed relative to the continents. The early Paleozoic poles are located near the present equator, but late Paleozoic poles are generally in the present midlatitudes. Mesozoic and Cenozoic poles are located near the present geographic poles (Fig. 7.12). A line drawn through the paleomagnetic poles of any one continent for successive periods is the *apparent polar wandering (A.P.W.) curve* of that continent.

When A.P.W. curves are compared, it is evident that they do not coincide (Fig. 7.13). The North American A.P.W. curve lies to the west of the European curve, suggesting that North America has moved westward relative to Europe. The A.P.W. curves for Africa, India, and Australia lie to the east of the curve for Europe. Apparently these continents have moved eastward relative to Europe. An A.P.W. curve is due to either one or both of the following factors: (1) the movement of plates relative to each

TABLE 7.1 **Average Paleomagnetic North Poles Based on Samples Which Have Been Tested for Stability by Partial Demagnetization.**[a]

| Age | North America | | Europe | | Eurasia | | Siberia | | China | | Africa[b] | |
|---|---|---|---|---|---|---|---|---|---|---|---|---|
| Tertiary | 86N | 158W | 75N | 179E | 76N | 178W | 81N | 131W | | | 83N | 115E |
| Cretaceous | 69N | 168W | 85N | 176W | 77N | 175E | 70N | 168E | 70N | 155W | 71N | 109W |
| Jurassic | 68N | 147E | 77N | 139E | 74N | 144E | 70N | 148E | | | 67N | 109W |
| Triassic | 65N | 96E | 52N | 141E | 58N | 147E | 62N | 153E | | | 69N | 104W |
| Permian | 49N | 120E | 49N | 169E | 52N | 167E | 63N | 177W | | | 43N | 97W |
| Carbonif. | 37N | 125E | 37N | 166E | 41N | 168E | 45N | 170E | | | 40N | 145W |
| Devonian | 29N | 121E[c] | 32N | 162E | 31N | 162E | 28N | 162E | | | | |
| Silurian | | | 21N | 159E | 19N | 157E | 19N | 156E | | | | |
| Ordovician | 32N | 122E | 13N | 165E | | | 25S | 135E | | | 50S | 169E |
| Cambrian | 3N | 164E | 20 | 168E | | | 37S | 133E | 11S | 137E | 51S | 173E |
| 0.6 B.Y. | 0N | 159E | 8N | 163E | | | 30S | 145E | 3S | 127E | 85S | 25E |
| 0.7 B.Y. | 7S | 162E | 30S | 135E | | | 22S | 141E | 7N | 123E | 52S | 42W |
| 0.8 B.Y. | 8S | 168E | 2N | 122E | | | 13S | 130E | 16N | 116E | 26S | 74W |
| 0.9 B.Y. | 3S | 174W | 9N | 127E | | | 3S | 121E | 22N | 112E | 17S | 78W |
| 1.0 B.Y. | 3N | 179W | 17N | 136E | | | 6N | 114E | 28N | 106E | 9S | 82W |
| 1.1 B.Y. | 20N | 160W | 30N | 137E | | | 14N | 108E | 37N | 101E | 2N | 85W |
| 1.15 B.Y. | 34N | 165W | 35N | 137E | | | 20N | 117E | 40N | 96E | 7N | 87W |
| 1.2 B.Y. | 20N | 176W | 10S | 115E | | | 35N | 80E | | | 19S | 97W |
| 1.3 B.Y. | 1N | 169W | 4N | 161E | | | 33N | 51E | | | 66S | 135W |
| 1.4 B.Y. | 8S | 151W | 27N | 180E | | | 24N | 29E | | | 53S | 98W |
| 1.5 B.Y. | 11N | 144W | 12 | 134W | | | | | | | 32S | 89W |
| 1.6 B.Y. | 22N | 140W | 23N | 123W | | | | | | | 15S | 82W |
| 1.7 B.Y. | 40N | 124W | 30N | 120W | | | | | | | 8N | 72W |
| 1.8 B.Y. | 27N | 103W | | | | | | | | | 24N | 47W |
| 1.9 B.Y. | 14N | 98W[d] | | | | | | | | | 6N | 20W |
| 2.0 B.Y. | 4N | 99W | | | | | | | | | 15N | 42E |
| 2.1 B.Y. | 8S | 99W | | | | | | | | | 17S | 46E |
| 2.2 B.Y. | 30S | 103W | | | | | | | | | 47S | 40E |
| 2.3 B.Y. | 1N | 133W | | | | | | | | | 32S | 14W |
| 2.4 B.Y. | 47N | 137W | | | | | | | | | 23N | 62W |
| 2.5 B.Y. | 62N | 97E | | | | | | | | | 62N | 125W |
| 2.6 B.Y. | 53N | 69E | | | | | | | | | 73N | 163E |
| 2.7 B.Y. | 28N | 47E | | | | | | | | | 55N | 105E |
| 2.75 B.Y. | 5S | 20E | | | | | | | | | 11N | 67E |

[a] Poles from areas which have been rotated relative to the stable continent have been excluded.
[b] Location of some Precambrian paleomagnetic poles is based on the assumption that the continents which comprise Gondwanaland were together throughout the Precambrian
[c] Silurian and Devonian poles averaged together
[d] Poles older than 1.8 B.Y. are from the Superior Province

TABLE 7.1 (cont'd)

| Australia[b] | | South America | | India[b] | | Antarctica[b] | | Madagascar | | Greenland | |
|---|---|---|---|---|---|---|---|---|---|---|---|
| 70N | 56W |  |  | 35N | 82W | 86N | 178E |  |  | 61N | 170W |
| 53N | 31W |  |  | 15N | 59W | 81N | 157W | 71N | 132W |  |  |
| 45N | 37W | 87N | 110W |  |  | 55N | 36E | 74N | 93W | 54N | 169E |
| 53N | 28W | 84N | 102E | 11N | 54W |  |  | 65N | 66W | 68N | 160E |
| 41N | 41W | 75N | 152E | 15N | 60W |  |  | 48N | 96W |  |  |
| 62N | 29W | 58N | 157E | 32S | 46W |  |  |  |  |  |  |
| 47N | 136W[c] | 5N | 141E |  |  |  |  |  |  |  |  |
| 15N | 155W | 8S | 144E |  |  | 28N | 170W |  |  |  |  |
| 9N | 160E | 27S | 130E | 28N | 148W | 2N | 152W |  |  |  |  |
| 60S | 137W | 47S | 131W | 25S | 140E |  |  |  |  |  |  |
| 58S | 72W | 23S | 105W | 54S | 125E |  |  |  |  |  |  |
| 23S | 63W | 14N | 102W | 79S | 143W |  |  |  |  |  |  |
| 12S | 60W | 24N | 113W | 73S | 104W |  |  |  |  |  |  |
| 4S | 57W | 33N | 114W | 67S | 89W |  |  |  |  |  |  |
| 7N | 53W | 44N | 103W | 56S | 73W |  |  |  |  | 38N | 92W |
| 12N | 52W | 51N | 114W | 50S | 75W |  |  |  |  |  |  |
| 4S | 74W | 26N | 132W | 58S | 123W |  |  |  |  |  |  |
| 32S | 123W | 23S | 152W | 28S | 172E |  |  |  |  |  |  |
| 31S | 96W | 6S | 134W | 52S | 179E |  |  |  |  |  |  |
| 20S | 77W | 12N | 127W | 66S | 150W |  |  |  |  |  |  |
| 10S | 60W | 17N | 113W | 71S | 101W |  |  |  |  |  |  |
| 4S | 40W | 46N | 92W | 50S | 52W |  |  |  |  |  |  |
| 5N | 10W | 42N | 42W | 38S | 16E |  |  |  |  |  |  |
| 23S | 6E | 10N | 48W |  |  |  |  |  |  |  |  |
| 16S | 72E | 24S | 5E |  |  |  |  |  |  |  |  |
| 44S | 88E | 52S | 12W |  |  |  |  |  |  |  |  |
| 70S | 114E |  |  |  |  |  |  |  |  |  |  |
| 60S | 11W |  |  |  |  |  |  |  |  |  |  |
| 11N | 22W |  |  |  |  |  |  |  |  |  |  |

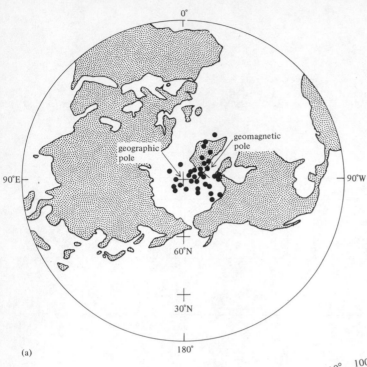

(a)

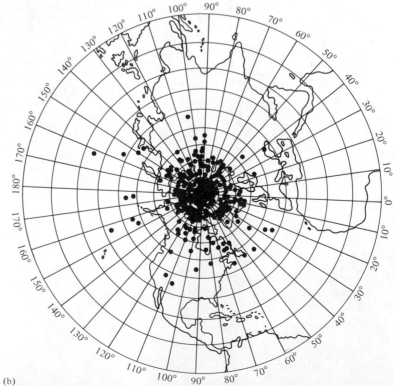

FIGURE 7.11 Virtual geomagnetic poles: **(a)** Determined from various stations on the earth measuring the direction and inclination of the present field; note the clustering of the poles about the geomagnetic pole. (After Cox and Doell, Ref. 10.) **(b)** Determined from measurements of the direction and inclination of the magnetic field in rocks less than 30 million years old; note the clustering of the poles about the earth's geographic pole. (After D. H. Tarling, Ref. 6.)

(b)

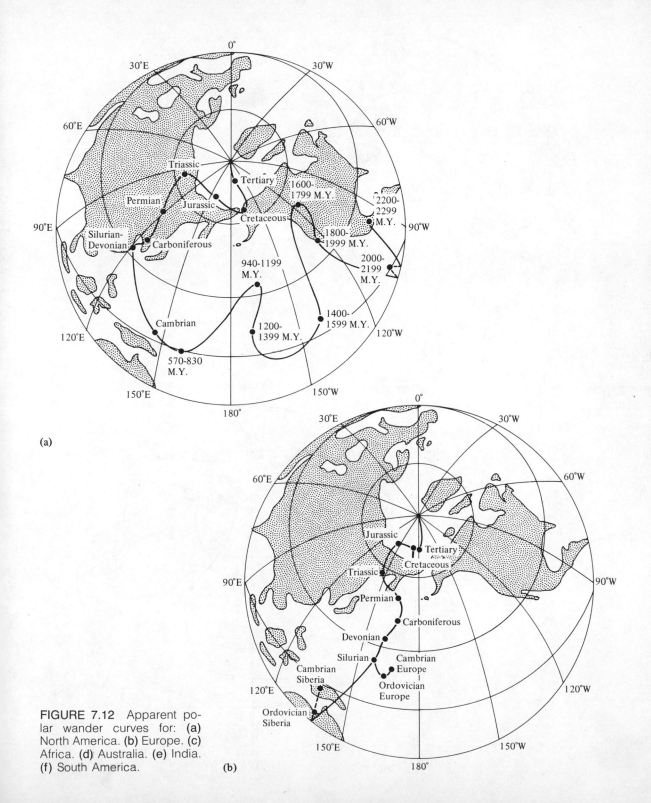

FIGURE 7.12 Apparent polar wander curves for: **(a)** North America. **(b)** Europe. **(c)** Africa. **(d)** Australia. **(e)** India. **(f)** South America.

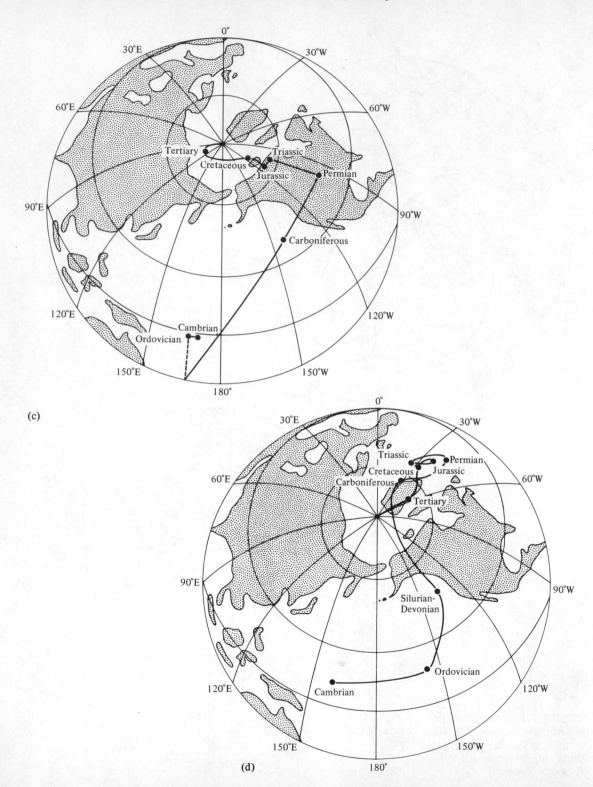

(c)

(d)

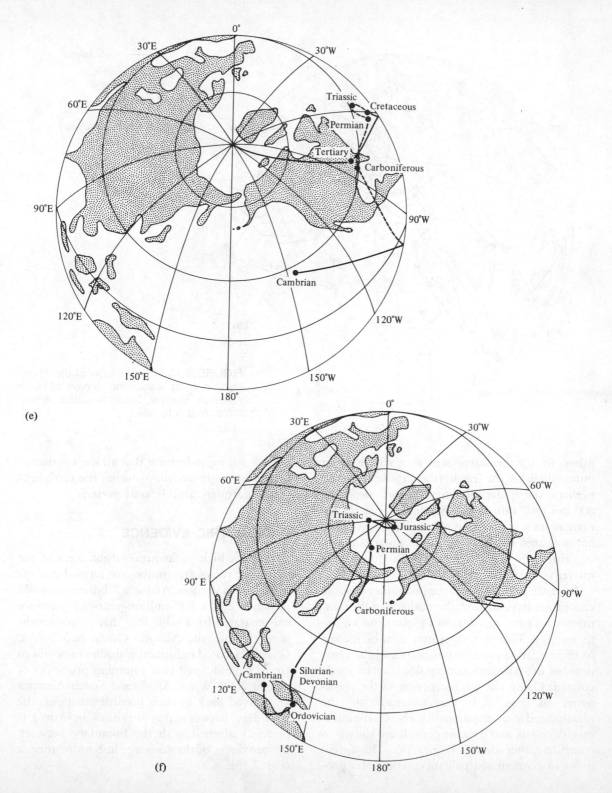

(e)

(f)

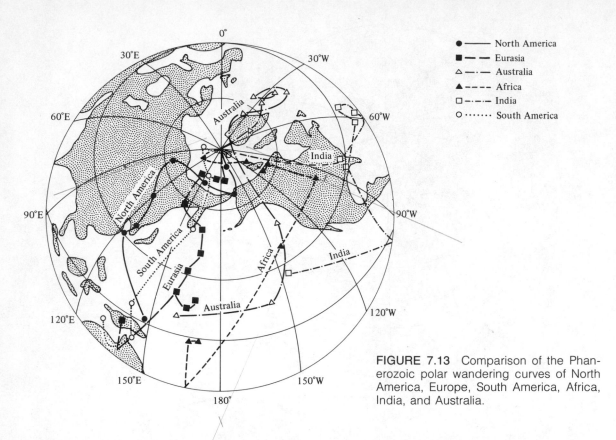

**FIGURE 7.13** Comparison of the Phanerozoic polar wandering curves of North America, Europe, South America, Africa, India, and Australia.

other, or (2) the movement as a whole of an outer shell of the earth (the lithosphere and perhaps the underlying mantle to a depth of 800 km (500 mi) or so). Studies of paleomagnetic poles and direction of movement of plates has indicated that most apparent polar wandering since the Early Cretaceous is due to the movement of plates (**11**).

In order to further test the possibility that the continents have drifted, the outlines of the continents may be traced from a globe onto a plastic overlay. The continents may then be moved to their relative predrift positions and the positions of the paleomagnetic poles may be recalculated. Using the reconstruction of the continents of Fig. 7.9, the locations of the recalculated pole positions for the Carboniferous, Permian, and Triassic periods are shown to coincide rather closely (Fig. 7.14). The agreement of recalculated paleomagnetic poles pro-

vides a strong indication that all the continents formed a single landmass during the Carboniferous, Permian, and Triassic periods.

## RADIOMETRIC EVIDENCE

Areas in which potassium—argon ages of the basement rocks are similar are termed *geologic provinces*. In western Africa, a 2-billion-year-old province and a 600-million-year-old province are separated by a boundary that trends southwesterly into the Atlantic Ocean near Accra, Ghana. Detailed radiometric studies in northern Brazil have defined two adjoining provinces of similar ages. When Africa and South America are moved back to their predrift positions, the boundary between the provinces in Africa is perfectly aligned with the boundary between the provinces of the same age in South America (Fig. 7.15).

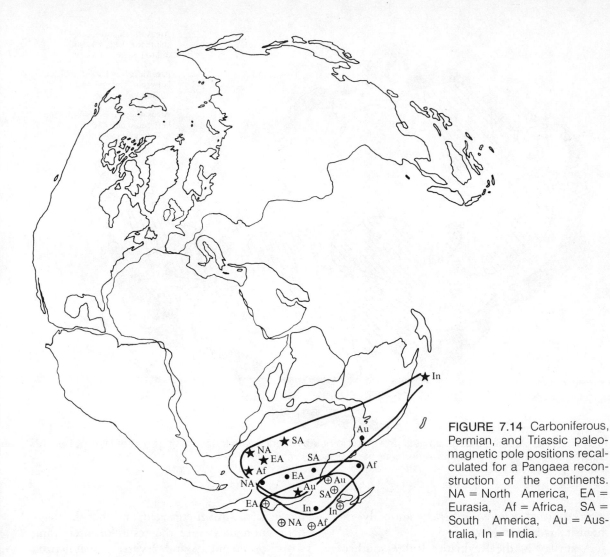

FIGURE 7.14 Carboniferous, Permian, and Triassic paleomagnetic pole positions recalculated for a Pangaea reconstruction of the continents. NA = North America, EA = Eurasia, Af = Africa, SA = South America, Au = Australia, In = India.

## SEAFLOOR SPREADING

Harry Hess was the first to propose that blocks of seafloor are moving relative to one another in response to the motion of convection currents within the mantle. The evidence in support of the concept of seafloor spreading has come from many sources, including the magnetism of the ocean floor, the age of oceanic volcanic rocks, and the age of deep-sea sediments.

## Marine Magnetics

Since 1950, there have been extensive surveys of the ocean floor. Measurements of the intensity of the earth's magnetic field in these areas have revealed the presence of a series of north-south trending magnetic anomalies. A *magnetic anomaly* is a deviation from the average intensity of the earth's magnetic field. In an area with a positive anomaly, the earth's magnetic field has a greater than average intensity, whereas in

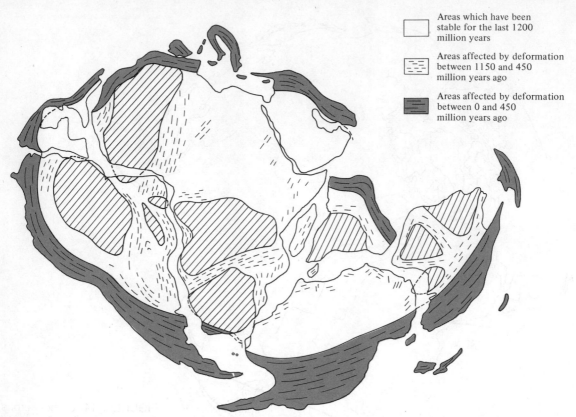

Areas which have been stable for the last 1200 million years

Areas affected by deformation between 1150 and 450 million years ago

Areas affected by deformation between 0 and 450 million years ago

FIGURE 7.15  Matching of geological provinces in South America with those in Africa. (After Hurley, Ref. 12. Copyright © 1968 by Scientific American, Inc., all rights reserved.)

an area with a negative magnetic anomaly, the intensity is below average.

A survey of the Reykjanes Ridge south of Iceland revealed a symmetrical pattern of magnetic anomalies (Fig. 7.16). Numbers have been assigned to distinctive anomalies. The broad positive anomaly over the crest of the Mid-Atlantic Ridge is Anomaly Number 1, and the two broad positive anomalies equidistant on either side of the ridge are each numbered 5 (Fig. 7.17).

The origin of the linear magnetic anomalies in the oceans remained a mystery for several years after their discovery. Then in 1963, Fred Vine and D. H. Matthews proposed that the anomalies had resulted from periodic reversals

in the earth's magnetic field, and that the magnetic anomalies were the result of alternating strips of normally and reversely magnetized lavas that had been extruded at the crests of midoceanic ridges (**15**). The normally magnetized lavas were extruded during a time of normal polarity of the earth's magnetic field, and they cause a positive anomaly, since their magnetism is added to the magnetism produced in the core of the earth. The reversely magnetized lavas were extruded during a time of reversed polarity of the earth's magnetic field and cause a negative anomaly, since their magnetism would cancel some of the earth's magnetism.

The Vine-Matthews hypothesis provides an explanation for the striking magnetic symmetry

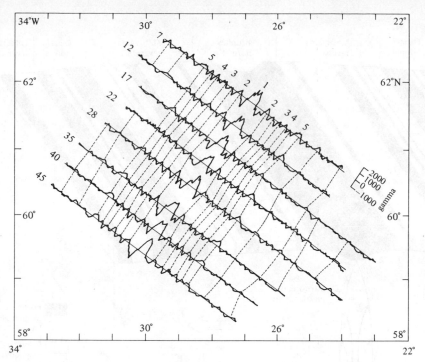

**FIGURE 7.16** Eight magnetic profiles taken across the Reykjanes Ridge south of Iceland. The profiles show the total intensity of the earth's magnetic field as measured by a magnetometer towed behind a low-flying aircraft. Flight paths were parallel to each other and in a WNW-ESE direction and are numbered 7 through 45. Magnetic anomalies are shown as deviations from the average magnetic field. Positive anomalies (such as anomaly #1 on the ridge crest) have a greater than average intensity of the magnetic field, and negative anomalies a less than average intensity of the magnetic field. Notice the symmetry of the anomalies 2 through 5 about anomaly #1. The strength of the anomalies is measured in gammas. (After Heirtzler *et al.,* Ref. 13. Used with permission of Maxwell International Microfilms Corp.)

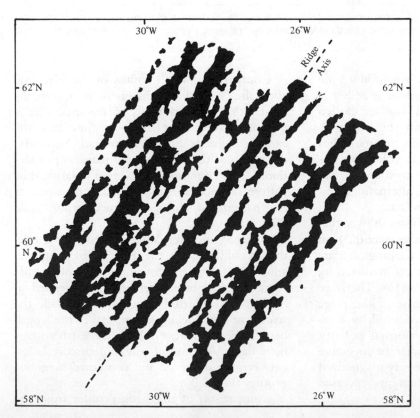

**FIGURE 7.17** Pattern of magnetic anomalies on the Reykjanes Ridge south of Iceland. The anomalies were determined by using the data from the magnetic profiles shown in Fig. 7.16 (and other profiles as well). Positive magnetic anomalies are colored in black and negative anomalies are shown in white. Notice how the anomalies parallel the ridge crest. (From F. J. Vine, "Magnetic Anomalies Associated with Mid-Ocean Ridges," in *The History of the Earth's Crust* by Robert A. Phinney (ed.). Copyright © 1968 by Princeton University Press, Fig. 6, facing p. 82. Reprinted by permission of Princeton University Press, Ref. 14.)

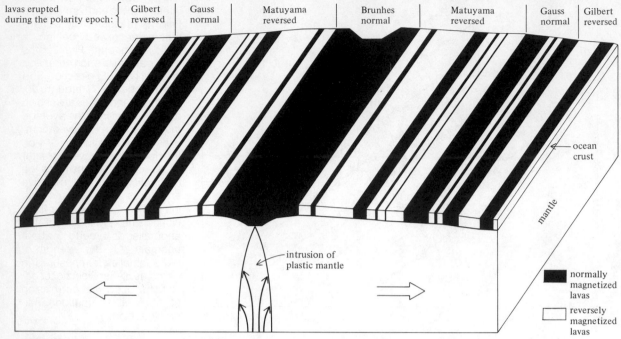

lavas erupted during the polarity epoch: { Gilbert reversed | Gauss normal | Matuyama reversed | Brunhes normal | Matuyama reversed | Gauss normal | Gilbert reversed

ocean crust

mantle

intrusion of plastic mantle

■ normally magnetized lavas

□ reversely magnetized lavas

**FIGURE 7.18** Block diagram of a midoceanic ridge showing the inferred distribution of normally and reversely magnetized lavas according to the Vine-Matthews hypothesis.

that appears on profiles taken across midoceanic ridges (Fig. 7.16). As new seafloor is added at ridge crests, older magnetized lavas are carried in both directions away from the crests. A symmetrical pattern of magnetic anomalies would be produced only if (1) volcanism occurred principally at the ridge crest and (2) the seafloor was spreading during alternating periods of normal and reversed polarity.

In 1965, Vine and J. Tuzo Wilson showed that the pattern of magnetic anomalies over the Juan de Fuca Ridge off the coast of Washington State conformed exactly to the pattern predicted by the Vine-Matthews hypothesis (**16**). The large positive anomaly over the crest of the ridge (Anomaly Number 1) was produced by lavas erupted during the Brunhes normal polarity epoch, whereas the two adjacent negative anomalies were caused by reversely magnetized lavas erupted during the Matuyma reversed polarity epoch (Fig. 7.18).

Comparison of the widths of the magnetic anomalies with the geomagnetic time scale indicates that the width of positive anomalies is proportional to the length of time that the earth's magnetic field was normal. Similarly, the width of the negative anomalies is proportional to the duration of reversed polarity. The pattern of anomalies is essentially the same as the pattern of the geomagnetic time scale (Fig. 7.19).

Using a computer, Vine constructed a model that would predict the pattern of magnetic anomalies for a series of thin blocks of oceanic crust that had been alternately magnetized in normal and reversed directions. He made the widths of the blocks proportional to the length of time between reversals of the earth's magnetic field. Vine found an excellent correlation between the observed and computed magnetic profiles (Fig. 7.20).

Comparison of magnetic profiles from the

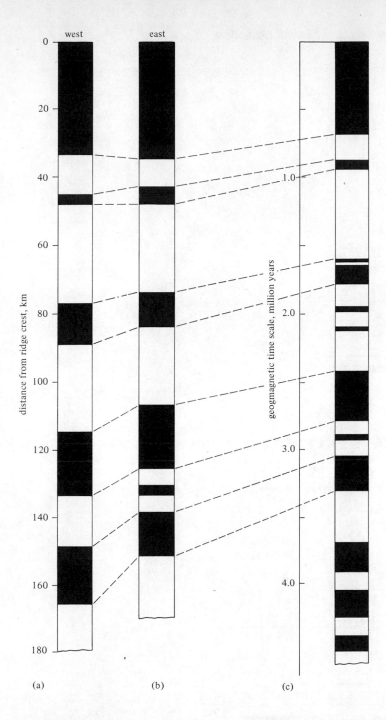

FIGURE 7.19 Correlation of magnetic anomalies with the geomagnetic time scale: **(a)** Magnetic anomalies on the west side of the East Pacific Rise measured on the magnetic profile shown in Fig. 7.20c. Positive anomalies are shown in black, negative in white. **(b)** Magnetic anomalies measured on the east side of the East Pacific Rise. **(c)** The geomagnetic time scale with the times of normal polarity shown in black and times of reversed polarity shown in white. Comparison of the magnetic anomaly patterns in **(a)** with those in **(b)** shows that the anomalies are symmetrical. Comparison of the anomaly patterns in **(a)** and **(b)** with the geomagnetic time scale **(c)** shows that positive magnetic anomalies correlate with times of normal polarity of the earth's magnetic field, and negative magnetic anomalies correlate with times of reversed polarity of the magnetic field. This indicates that seafloor spreading and volcanism at the crest of the East Pacific Rise occurred at times when the earth's magnetic field alternated between normal and reversed polarity.

Atlantic, Pacific, and Indian oceans shows that the anomalies may be correlated from one ocean to another (Fig. 7.21). This would be expected because changes in the earth's polarity affect the entire earth.

As discussed earlier, it is possible to date magnetic anomalies by comparing the pattern of anomalies with the geomagnetic time scale. If the age of an anomaly is known, the average rate of seafloor spreading may be determined by measuring the distance of that anomaly from the crest of the adjacent midoceanic ridge, since

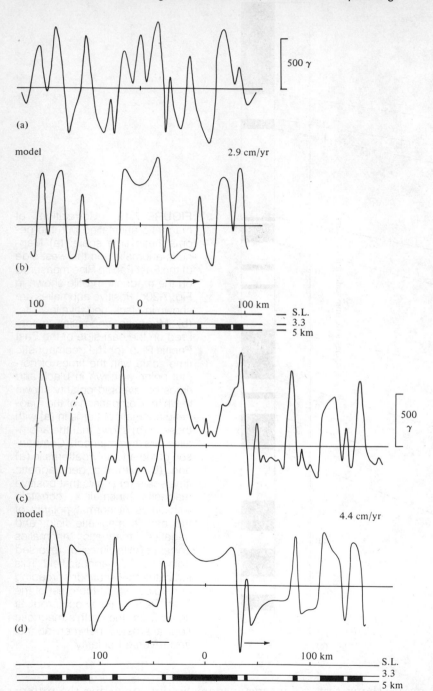

**FIGURE 7.20** Comparison of observed and computed magnetic profiles from the Pacific Ocean: **(a)** Observed magnetic profile taken across the Juan de Fuca Ridge at 46°N latitude. **(b)** Computed magnetic profile for the above area determined by using alternate strips of normally and reversely magnetized lavas with a width proportional to the length of time of normal and reversed polarity of the earth. **(c)** Observed magnetized profile taken across the East Pacific Rise at 51°S latitude. **(d)** Computed magnetic profile for the above area. (From Vine, Ref. 16. Copyright © 1966 by the American Association for the Advancement of Science.)

distance equals rate × time. The rate of seafloor spreading is usually expressed as a *spreading half rate* which is the rate of movement of one block of seafloor relative to the adjacent midocean ridge. Blocks of seafloor along with the attached continents move apart from other blocks of seafloor and continents at twice the spreading half rate. The half rate of seafloor spreading in the Atlantic Ocean along the Mid-Atlantic Ridge varies from approximately 1 to 2 cm (½ to 1 in.) per year, while that in the Pacific Ocean along the East Pacific Rise varies from about 2 to 6 cm (1 to 3 in.) per year. The continents are moving apart at approximately the same rate as the rate a person's fingernails grow. They will have drifted apart a distance approximately equal to a person's height during the period of a lifetime.

Correlation of magnetic anomalies within a single ocean basin reveals that the distance between anomalies varies with latitude (Fig. 7.21b). This is an indication of a change in the rate of seafloor spreading along a midocean ridge. The rate of seafloor spreading decreases toward the poles of spreading (Fig. 6.7). Finally, magnetic anomaly lineations may be used to map the age of ocean basins (Fig. 7.22).

## Oceanic Volcanoes

One of the earliest indications of seafloor spreading was the general increase in age of volcanic islands and seamounts away from the Mid-Atlantic Ridge (Fig. 7.23). Several islands with active volcanoes, such as Tristan da Cunha and Iceland, are located on or very close to the

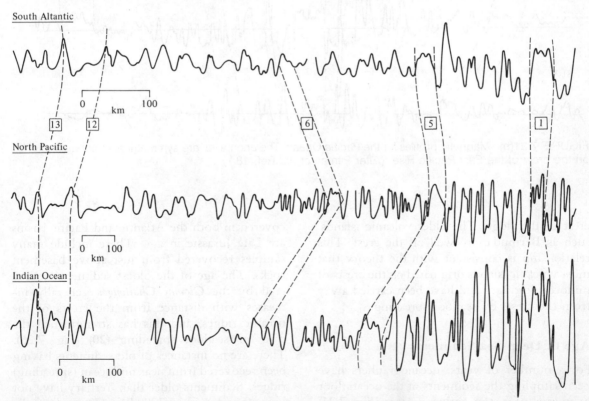

**FIGURE 7.21(a)** Comparison of magnetic profiles from various oceans. The correlation of anomalies indicates that they are due to a worldwide rather than a local cause, as would be a reversal in the earth's magnetic field. (After Heirtzler *et al.,* Ref. 17.)

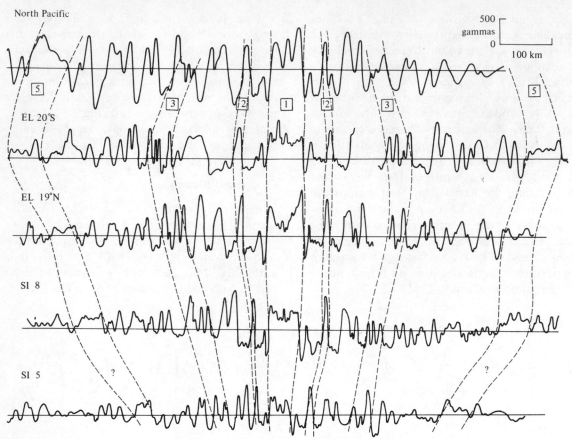

**FIGURE 7.21(b)** Magnetic profiles for the Pacific Ocean. The anomalies are symmetrical about anomaly #1 on the crest of the East Pacific Rise. (After Pitman *et al.*, Ref. 18.)

crest of the ridge. The older volcanic islands, such as Bermuda, are well off the crest. This relationship is consistent with the theory that most volcanic islands originated on the crests of midoceanic ridges and have been carried away from the crests by seafloor spreading.

### Age of Deep-Sea Sediments

For a number of years, oceanographers have been sampling the sediments of the ocean floor by means of piston coring devices (Fig. 7.24) and a drilling rig mounted on the ship, *Glomar Challenger* (Fig. 7.25). The oldest sediments re-

covered in both the Atlantic and Pacific basins are Late Jurassic in age. These include many samples recovered from just above basement rocks. The age of the oldest sediments recovered by the *Glomar Challenger* generally increases with distance from the crests of the oceanic ridges, and this has strengthened the case for seafloor spreading (**20**) (Fig. 7.26). There are no instances of old sediments having been recovered from near midocean (spreading) ridges. Sediments older than Tertiary have not been found within 1200 km (700 mi) of the crest of the Mid-Atlantic Ridge and 4000 km (2400 mi) of the crest of the East Pacific Rise.

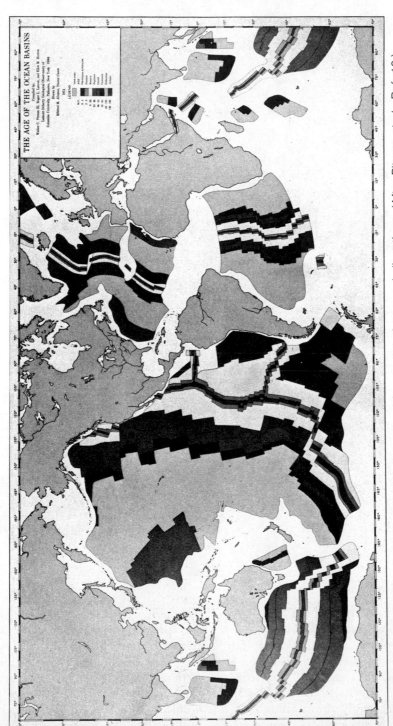

FIGURE 7.22  Map of the age of ocean basins determined from magnetic anomaly lineations. (After Pitman *et al.*, Ref. 18.)

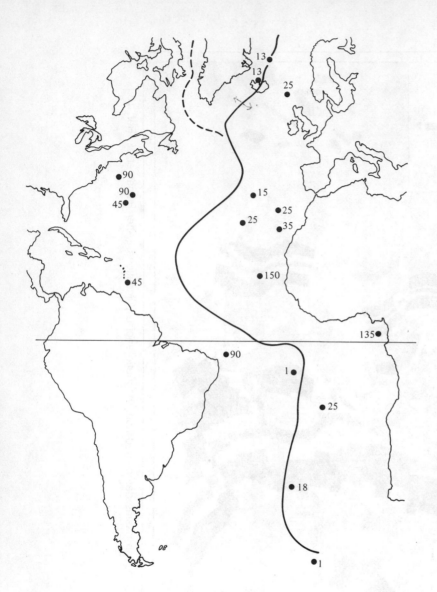

**FIGURE 7.23** Variation in the age of volcanoes in the Atlantic Ocean with distance from the crest of the Mid-Atlantic Ridge. (From Wilson, Ref. 19.)

## Undeformed Sediments

There is almost no deformation of the sediments of the deep ocean floor (**21**). However, seismic reflection studies have indicated the presence of deformed sedimentary units landward of some trenches (Fig. 7.27). The sediments in this area are intensely folded (**22, 23**) and cut by faults, probably thrust faults (**24**). Seismic refraction measurements in these areas show that the sediment layer is relatively thin. The absence of large accumulations of deformed sediments associated with oceanic trenches may be due to the conversion of sediments into sedimentary and metamorphic rocks as the lithosphere descends beneath a continent or island arc. During this process, the sediments would be folded, faulted, and thickened, which in turn would result in their being dewatered,

**FIGURE 7.24** Piston corer used for the sampling of sediments from the deep ocean. (Lamont-Doherty Geological Observatory, Columbia University photo.)

**FIGURE 7.25** The drilling ship *Glomar Challenger* which has been used to take core drillings from the deep oceans. (Scripps Institution of Oceanography photo.)

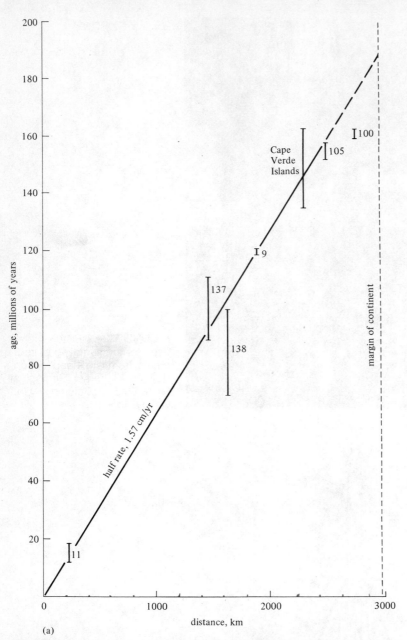

(a)

**FIGURE 7.26** Age of oldest sediment recovered by the *Glomar Challenger* plotted against distance from the nearest midocean ridge: **(a)** For the North Atlantic Ocean on either side of the Mid-Atlantic Ridge. **(b)** For the South Atlantic Ocean on either side of the Mid-Atlantic Ridge. **(c)** For the Pacific Ocean west of the crest of the East Pacific Rise.

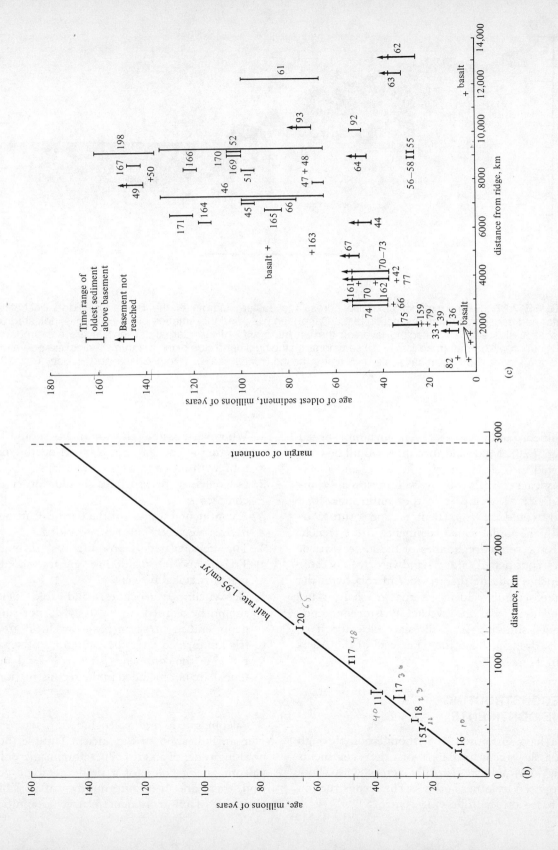

(c)

**Legend (in figure c):**
- Time range of oldest sediment above basement
- Basement not reached

age of oldest sediment, millions of years

distance from ridge, km

basalt +

+ basalt

198
167
166
171
164
165
170
169 52
46
45
51
66
49
+50
47 + 48
+163
93
92
64
44
55
56–58
61
63
62
67
70
74
161
162
70–73
+42
77
75 66
+159
+79
33 +39
36
82
+
+
+ basalt

(b)

age, millions of years

distance, km

margin of continent

half rate, 1.95 cm/yr

20
17
17
11
18
15
16

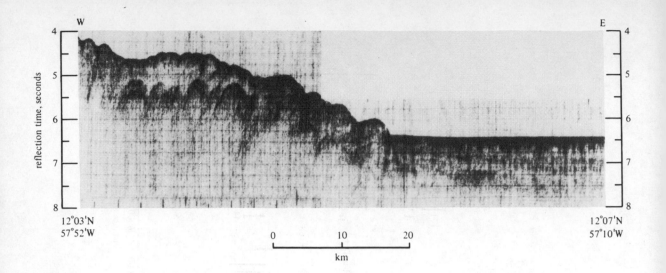

**FIGURE 7.27** Seismic reflection profile across the eastern margin of the Barbados Ridge illustrating deformed sediments near the junction of the Caribbean plate and the America plate. Folding is indicated by a scallop-shaped dark reflecting layer on the left hand side of the diagram. These layers are probably continuous with undeformed (flat-lying) sediments on the right hand side of the diagram. Vertical exaggeration is approximately 5:1 (vertical scale is 5 times the horizontal scale). (From Chase and Bunce, Ref. 22.)

lithified, and perhaps even metamorphosed (Fig. 7.28). Muds and turbidites would be converted into shale, graywacke, slate, and meta-graywacke. Such rocks would presumably underlie the lower part of the continental slope and would be equivalent to and continuous with the undeformed sediments in the trench. Seismic refraction studies of the lower part of the continental slope landward of oceanic trenches indicate the presence of rock with the same seismic velocity as that of shale, slate, mudstone, and graywacke. Pleistocene mudstones showing incipient slaty cleavage have been dredged from the inner wall of the Aleutian Trench (**26**).

## RECONSTRUCTING THE CONTINENTS

We have proposed a Carboniferous through Triassic reconstruction of the continents (Fig. 7.9) which generally agrees with those proposed by other authors. The reconstruction is based on the following controls:

1. Wherever possible, the continents should fit together at the 2000-m (6000-ft) depth contour with no large gaps or overlaps.
2. Geosynclines or geoclines should border all continents.
3. A continental source for the Permo-Carboniferous ice sheets must be provided.
4. The recalculated Carboniferous through Triassic polar wandering curves should match reasonably well.
5. Geosynclines, geoclines, and fold belts should be aligned so that they do not suddenly end in an ocean basin or shield area.
6. Islands formed after the beginning of separation of the continents (such as Iceland and Cuba) are not included in the reconstruction.

Paleomagnetic and paleoclimatic data provide an indication of the ancient latitude, but not longitude, of an area. Therefore, using only paleomagnetic poles for a single geologic period, there are an infinite number of possible positions of one continent relative to another

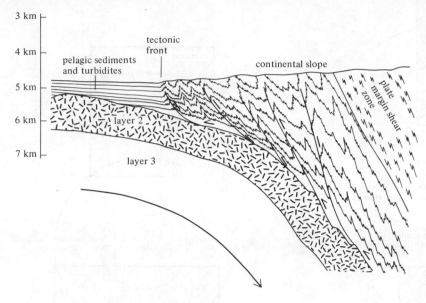

FIGURE 7.28 Diagrammatic cross section through the Peru-Chile Trench at approximately 36°S latitude. Notice how undeformed sediments in the trench are continuous with folded and faulted sediments of the continental slope. Vertical exaggeration is approximately 2:1. (From Seyfert, Ref. 25.)

(Fig. 7.29). However, if two or more points on a polar wandering curve are known for two continents, and if no relative movements took place between them during the time of polar wandering, the relative positions of the two continents may be established (**27**) (Fig. 7.30).

## Dating the Initial Contact of Two Continents

A collision between two continents will produce a mountain belt along the zone of contact between the two continents. Because geosynclines and geoclines border continents, mountains formed in this way will generally contain deposits laid down in geosynclines or geoclines. Mountains may also form on a continental margin bordering a deep ocean where the seafloor descends beneath a continent (in a subduction zone). This is an area where geosynclines occur, and therefore the presence of geosynclinal deposits in a mountain range does not necessarily indicate an area of collision between two continents. The time of collision of two continents may be dated as the beginning of the following events associated with mountain building:

1. Folding
2. Reverse faulting
3. Intrusion of granitic rocks and emplacement of ophiolite sequences
4. Metamorphism
5. Uplift
6. Erosion of the mountains and deposition of the sediments derived from those mountains in regions adjacent to the mountains

The time of collision of two continents may also be deduced from determination of the time of joining of the apparent polar wander (A.P.W.) paths of the two continents. Two parts of a continent which are now joined but were once separated will have A.P.W. paths which will coincide with each other during that period when they were joined together. However, the A.P.W. curves will be separate from each other during that time when the continents were not together. The time at which the two A.P.W. paths join is the time at which the two continents initially contacted (collided with) each other (Fig. 7.31). In the case of a collision between two continents which were subsequently separated by rifting, the A.P.W. curves must be first recalculated for the prerifting configuration

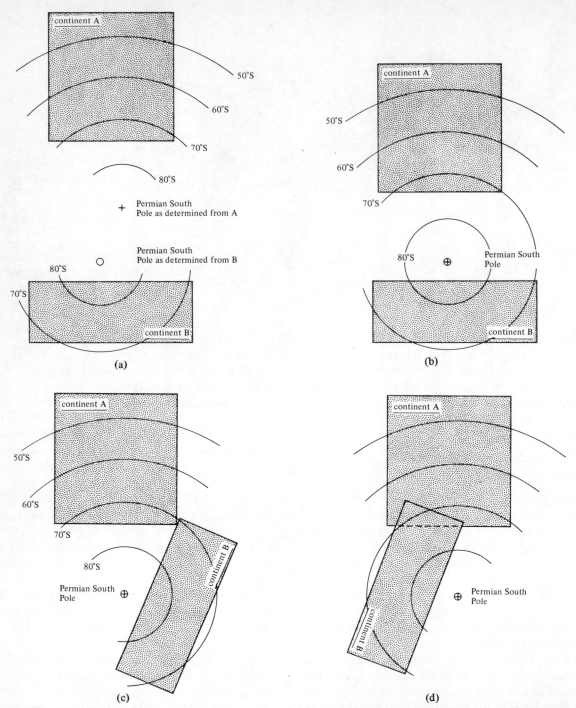

**FIGURE 7.29** Using paleomagnetic data to reconstruct the continents: **(a)** Before reconstruction, two continents with separated paleopoles. **(b)** One of many possible reconstructions in which palepoles coincide. **(c)** Another reconstruction in which paleopoles coincide. **(d)** Reconstruction not permitted because of overlap of the continents, even though paleopoles agree.

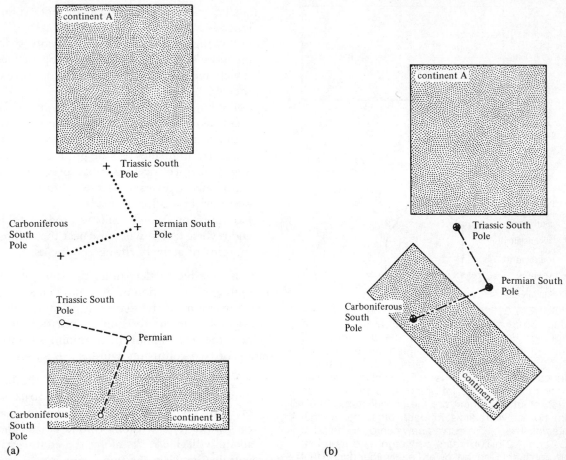

**FIGURE 7.30** Using two apparent polar wander (A.P.W.) curves for making a reconstruction of the continents: **(a)** Before reconstruction of the continents, two continents have two separate A.P.W. curves. **(b)** After reconstruction of the continents, the two A.P.W. curves agree. This is the only reconstruction in which the two curves will agree.

of the continents. If a continent is moved to some new position, its A.P.W. path must move with it. A *recalculated A.P.W. curve* is an A.P.W. curve which has been moved along with a continent to some new position, usually the position of that continent (relative to another continent) at some previous time. The recalculated A.P.W. curves will coincide with each other during that period when the continents were joined together, and they will be separate from each other when the continents were not together.

gether. The time of joining of the two recalculated A.P.W. paths will be the time of initial contact of those two continents (Fig. 7.32).

## Dating the Beginning of Separation of Two Continents

The *rifting* (separating by pulling and tearing apart) of two continents will leave its mark on the continental margins between the two continents:

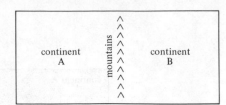

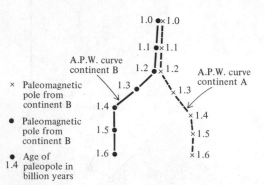

**FIGURE 7.31** A.P.W. curves for two continents which were once separated, but are now joined together. Paleomagnetic poles dated between 1.6 and 1.3 billion years do not agree with each other, and therefore continents A and B were separated during that interval. The A.P.W. curves between poles dated at 1.6 to 1.4 billion years are parallel to each other, and therefore the continents probably did not move *relative to each other* during that interval. The A.P.W. curves between poles dated at 1.4 and 1.2 billion years converge, and therefore the continents moved toward each other during that interval. Collision of the two continents occurred 1.2 billion years ago. The A.P.W. curves between poles dated at 1.2 to 1.0 billion years coincide with each other, and therefore the two continents were joined together in their present arrangement during that interval.

1. As two continents begin to separate, they will be stretched as a result of the tensional forces operating on them. Consequently, normal faults will be expected near the continental margins. Tension also tends to result in mafic (basaltic) igneous activity, which generally includes intrusion of diabase sills and dikes and extrusion of flood basalts. Furthermore, tension (especially that associated with the formation of aulacogens) tends to result in the intrusion and extrusion of alkalic (rich in sodium and/or potassium) igneous rocks.

2. Following separation of the continents, sediments will be deposited in miogeoclines which underly the newly formed continental shelves.

3. Following separation of the continents, sediments will be deposited on newly formed oceanic crust between the two continents.

4. Magnetic anomalies will be produced in the newly formed oceanic crust between the continents if reversals of the magnetic field occur as the seafloor spreads.

5. Compression (which tends to produce folds and reverse faults) will cease near the trailing margins of actively rifting continents.

It is possible to determine the time of the beginning of separation of two continents from a comparison of recalculated paleomagnetic poles. Based on the foregoing discussion, it is evident that the time of the beginning of separation of two continents will be:

1. Approximately equal to the age of the beginning of (a) normal faulting, (b) intrusion of mafic dikes and sills and alkalic igneous rocks, and (c) extrusion of flood basalts and alkalic volcanic rocks along the continental margins between the two continents.

2. Equal to or greater than the age of the oldest sediments known in miogeoclines bordering the continents. This may be established by deep drilling from sites on or near the continental shelf, by drilling into the continental slope from drilling ships such as the *Glomar Challenger,* or by dredging samples from the continental slope. One of the problems with this method of determining the age of continental separation is that it is not always evident whether a sample from the continental margin was deposited in a miogeocline or whether it was deposited prior to the formation of the miogeocline while the continents were still together. Evidence of a deep water environment deposition would support a miogeoclinal origin.

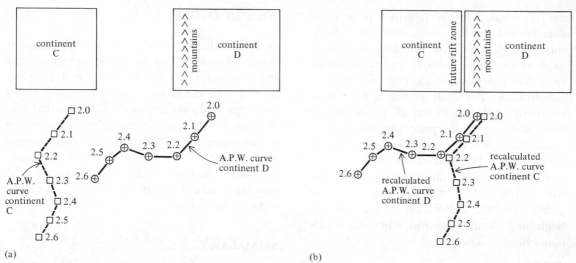

**FIGURE 7.32** A.P.W. curves for two continents which were separate, then joined together, and finally rifted apart (separated) from each other. **(a)** The A.P.W. curves do not agree with each other in the interval from 2.0 to 2.6 billion years ago. This indicates that the continents have moved relative to each other sometime since 2.0 billion years ago. **(b)** The continents and their A.P.W. curves have been moved back together to their relative positions before they rifted apart from each other. In this position, the recalculated A.P.W. curves agree with each other between poles dated at 2.0 to 2.2 billion years ago. This indicates that the continents were joined together during that interval. Paleomagnetic poles dated between 2.3 and 2.6 billion years do not agree with each other, and therefore continents C and D were separated during that interval. The A.P.W. curves join together at poles dated at 2.2 billion years old, and therefore this was the time of joining of the two continents.

3. Equal to or greater than the age of the oldest sediments known on oceanic crust between the two continents. The *Glomar Challenger* has drilled more than a hundred holes in the bottom of the Atlantic and Indian oceans, many of these down to *basement* (mafic volcanic rocks underlying sediments or sedimentary rocks). The oldest sediment in a rifted ocean basin will be closest to the continents, but unfortunately this is the area where the sediment is the thickest. Consequently, drilling sites are generally more than several hundred kilometers from the edge of the continent. The age of the oldest sediments in a given area will provide a minimum age for the underlying ocean crust. Generally the age of these sediments will be very close to the age of the crust, but their age may be significantly younger if later volcanic activity covered the oldest

sediment or if a period of erosion and/or nondeposition occurred between the time of formation of the ocean crust and deposition of the oldest sediments. Generally, drilling is terminated soon after basement is reached, and therefore there is the possibility of older sediment being present below the bottom of the hole.

4. Equal to or greater than the age of the oldest magnetic anomaly in the deep ocean between two continents. Magnetic anomalies may be dated either by comparing the pattern of anomalies with the geomagnetic time scale or by paleontological dating of the oldest sediment overlying a given magnetic anomaly.

5. Less than the age of the youngest compressive folding near the continental margins between the continents.

6. At the time of the beginning of separation of

the recalculated A.P.W. (apparent polar wander) curves for the two continents. Using this technique, the two continents are moved to their predrift positions and their paleomagnetic poles are recalculated for that reconstruction. For each continent, the paleomagnetic poles are joined in order of decreasing age. This procedure produces a recalculated A.P.W. curve for each continent. Ideally, the A.P.W. curves for the two continents will coincide approximately for the time during which the continents were together, and the A.P.W. curves will diverge beginning during the time when the continents began to separate.

Methods 3 and 4 provide only a minimum age for the time of separation of the continents. The actual age of separation may be approximated by first measuring the distance from the edge of the continent (usually the 2000-m depth contour) to the magnetic anomaly or drilling site where the age of the underlying ocean crust is known, and then calculating the time of separation using some assumed rate of seafloor spreading. For example, one might use the average rate of seafloor spreading from the anomaly or drilling site to the crest of the midocean ridge, or one might use the average rate of seafloor spreading just before the formation of the site of the dated ocean crust. The time between the separation of the continents and the formation of the dated oceanic crust may be determined using the formula

$$D = (r)(t)$$

where

$D$ = distance
$r$ = rate
$t$ = time

However, the time of separation may be in error by a factor of several tens of millions of years if the actual rate of seafloor spreading significantly differs from the assumed rate.

## WILSON CYCLE

Numerous lines of evidence of the type thus far discussed indicate that the continents have rifted apart and collided a number of times in the past. As the continents and their associated plates converge and diverge, oceans are created and destroyed. The opening and closing of an ocean in this manner is called the *Wilson cycle,* named after J. Tuzo Wilson, one of the pioneers in plate tectonics. The evidence for and specific times of openings and closings of oceans are discussed in subsequent chapters.

## SUMMARY

We have presented evidence from a large variety of sources that supports the concepts of polar wandering, continental drift, seafloor spreading, and plate tectonics.

The evidence for movement of the poles relative to the continent includes:

1. Thick limestones, red beds, evaporites, and coral reefs are often found at high latitudes, whereas ancient glacial deposits commonly occur near the present equator.
2. The direction of movement of Late Ordovician glaciers was away from the present equator in Africa.
3. Both the southwestern part of the United States and the British Isles were in the belt of prevailing easterlies during the late Paleozoic; however, these areas are now in the zone of prevailing westerly winds.
4. Paleomagnetic data indicate a large shift in the position of the magnetic poles. Presumably this is associated with a shift in the geographic poles.

The evidence for movement of the continents relative to each other includes:

1. The outlines of the continents bordering the Atlantic and Indian oceans can be matched with only small areas of overlap or gap.

2. The Paleozoic and early Mesozoic fauna and flora of Europe are similar to those of North America. Moreover, the fauna and flora of the same age from each of the southern continents are also quite similar to those of the other southern continents.

3. The Permo-Carboniferous ice sheets appear to have moved from the site of the present Atlantic and Indian oceans. A continental source for the ice sheets is provided if the continents are joined.

4. Geosynclines, geoclines, and mountain belts often end abruptly at the continental margins. They form continuous belts encircling the continents if the continents are joined into a single landmass.

5. Apparent polar wandering (A.P.W.) curves do not agree with one another. However, recalculated Carboniferous, Permian, and Triassic paleomagnetic poles agree reasonably well for a Pangaea reconstruction of the continents.

6. Provinces based on radiometric dating may be traced from one continent to another if the continents are joined.

The evidence indicating that the seafloor is moving away from midoceanic ridges includes:

1. The age of volcanic islands generally increases away from midoceanic ridges.

2. The age of the oldest sediment recovered at a given site generally increases away from midoceanic ridges.

3. Magnetic anomalies are symmetrical about the crest of midoceanic ridges.

4. The pattern of magnetic anomalies is the same as the pattern of the geomagnetic time scale.

## REFERENCES CITED

1. A. Hallam, 1975, *Sci. Amer.,* v. 232, no. 2, p. 88; E. T. Drake, 1976, *Geology,* v. 4, p. 41.

2. S. W. Carey, 1958, *Continental Drift: A Symposium:* Geology Department University of Tasmania, Hobart.

3. E. C. Bullard, J. E. Everett, and A. G. Smith, 1965, *in* P. M. S. Blacket, E. C. Bullard, and S. K. Runcorn, eds., *Phil. Trans. Roy. Soc.* (London), v. 1088, p. 41; A. G. Smith and A. Hallam, 1970, *Nature,* v. 225, p. 139; J. C. Briden, A. G. Smith and J. T. Sallomy, 1970, *Geophys. J. Roy. Astron. Soc.,* v. 23, p. 101.

4. H. Neuberger and J. Cahir, 1969, *Principles of Climatology:* Holt, Rinehart and Winston, New York.

5. R. W. Fairbridge, 1970, *Geotimes,* v. 15, no. 6, p. 18.

6. D. H. Tarling, 1971, *Principles and Applications of Paleomagnetism:* Chapman and Hall, London.

7. G. A. Doumani and W. E. Long, 1962, *Sci. Amer.,* vol. 207, no. 3, p. 168.

8. J. Lear, 1970, *Sat. Rev.,* vol. 53, no. 6, p. 46; A. Hallam, 1972, *Sci. Amer.* v. 227, no. 5, p. 56.

9. J. F. Dewey and G. M. Kay, 1968, *in* R. A. Phinney, ed., *The History of the Earth's Crust—A Symposium:* Princeton University Press, Princeton, N.J., p. 161.

10. A. Cox and R. R. Doell, 1960, *Geol. Soc. Amer. Bull.,* v. 71, p. 645.

11. D. M. Jurdy and R. Van Der Voo, 1975, *Science,* v. 187, p. 1193.

12. P. M. Hurley, 1968, *Sci. Amer.,* v. 218, no. 4, p. 53.

13. J. R. Heirtzler, X. Le Pichon, and J. G. Baron, 1965, *Deep-Sea Res.,* v. 13, p. 427.

14. F. J. Vine, 1968, *in* R. A. Phinney, ed., *The History of the Earth's Crust—A Symposium:* Princeton University Press, Princeton, N.J., p. 73, Fig. 6 facing p. 82.

15. F. J. Vine and P. M. Matthews, 1963, *Nature,* v. 199, p. 947.

16. F. J. Vine, 1966, *Science,* v. 154, p. 1405, Dec. 16.

17. J. R. Heirtzler, G. O. Dickson, E. M. Herron, W. C. Pitman and X. Le Pichon, 1968, *Jr. Geophys. Res.,* v. 73, p. 2119.

18. W. C. Pitman, E. M. Herron, and J. R. Heirtzler, 1968, *Jr. Geophys. Res.,* v. 73, p. 2069; W. C. Pitman, R. L. Larson, and E. M. Herron, 1974, *Geol. Soc. America* map and chart, MC-6.

19. J. T. Wilson, 1965, *in* P. M. S. Blacket, E. C. Bullard, and S. K. Runcorn, eds., *Phil. Trans. Roy. Soc.* (London), v. 1088, p. 145.

**20.** A. E. Maxwell *et al.,* 1970, *Science,* v. 168, p. 1047.

**21.** D. W. Scholl, R. von Huene, and J. B. Ridlon, 1968, *Science,* v. 159, p. 869.

**22.** R. L. Chase and E. T. Bruce, 1969, *Jr. Geophys. Res.,* v. 74, p. 1413.

**23.** M. L. Holmes, R. von Heune, and D. S. McManus, 1972, *Jr. Geophys. Res.,* v. 77, p. 959.

**24.** B. C. Heezen and M. Rawson, 1977, *Science,* v. 196, p. 423.

**25.** C. K. Seyfert, 1969, *Nature,* v. 222, p. 70.

**26.** J. C. Moore and J. E. Geigle, 1974, *Science,* v. 183, p. 509.

**27.** K. W. T. Graham, C. E. Helsley, and A. L. Hales, 1964, *Jr. Geophys. Res.,* v. 69, p. 3895.

# The Origin of the Earth and Its Early History

# 8

THE EARTH IS APPROXIMATELY 4.6 billion years old, while the oldest known terrestrial rocks are about 3.8 billion years old. This leaves 800 million years of earth history unaccounted for in the rock record. Thus the origin of the earth and the early evolution of its crust, atmosphere, and oceans must be reconstructed from indirect evidence. By observing stars in the early stage of their evolution and applying the principal of uniformitarianism, we can suggest processes by which our own solar system was formed. The origin and development of the earth's crust, mantle, core, atmosphere, and oceans may be deduced from studies of the moon, meteorites, asteroids, other planets, and our sun.

## THE ORIGIN OF THE ELEMENTS

As was discussed in Chapter 6, the earth is composed principally of oxygen, iron, silicon, magnesium, sulfur, nickel, aluminum, calcium, hydrogen, sodium, titanium, and potassium. These elements came from several sources. Essentially all of the hydrogen and most of the helium was produced in a primeval fireball but most of the elements which make up the earth were produced within the interior of stars. A very small percentage of the earth's elements were produced within the solar nebula and within the earth by the process of radioactive decay.

### Primeval Fireball

George Gamow was one of the first to postulate that the universe as we know it was produced in a gigantic explosion which resulted in a *primeval fireball* (1). This theory, often referred to as the "big bang theory," states that at one time all of the matter of the universe was contained in a relatively small, incredibly dense mass. When this mass exploded, matter was ejected outward in all directions. An important line of evidence supporting this theory is that the universe appears to be expanding. Indications of this expansion come from the study of distant galaxies (Fig. 8.1). The lines in the electromagnetic spectra of distant galaxies show a systematic shift toward the red end of the spectrum (Fig. 8.2). Such a red-shift would be produced in the spectrum of light from an object receding at a great speed from the point of observation. Astronomers have thus reached the conclusion that the galaxies they observed were receding from our own galaxy. Edwin Hubble estimated the distance between the Milky Way and a number of galaxies and found that the apparent rate of their recession is proportional to their distance (Fig. 8.3). In other words, the more distant galaxies appear to be receding at a faster rate than nearby galaxies. This relationship has been interpreted as evidence that the universe is expanding.

The rate of expansion has decreased somewhat with time due to the mutual gravitational

FIGURE 8.1 Spiral galaxy M 74 in the constellation *Pisces*. Galaxies are assemblages of billions and billions of stars. They may be spiral, elliptical, or irregular in shape. Our sun is located in one of the arms of a spiral galaxy like this one. (Mount Wilson and Palomar Observatories.)

attraction of all of the matter in the universe. The very rapid expansion of this matter is thought to have produced a fireball with temperatures in excess of 1,000,000,000,000°C (approximately 1,800,000,000,000°F). Using a very sensitive radio antenna, Arno Penzias and Robert Wilson discovered a form of microwave radiation that was unusual in that it was isotropic. Isotropic radiation strikes the earth from all directions with equal intensity. It has been suggested that this radiation is the remnant of the primeval fireball that has now cooled down nearly to absolute zero (3). It is interesting to note that just prior to the observation of this isotropic microwave radiation, physicist Robert Dicke had predicted the existence of such radiation on the assumption that the universe began as a primeval fireball.

As the fireball expanded, protons (the nuclei of hydrogen atoms), electrons, and neutrons were formed. Virtually all of the hydrogen, as well as 90 percent of the helium in the universe,

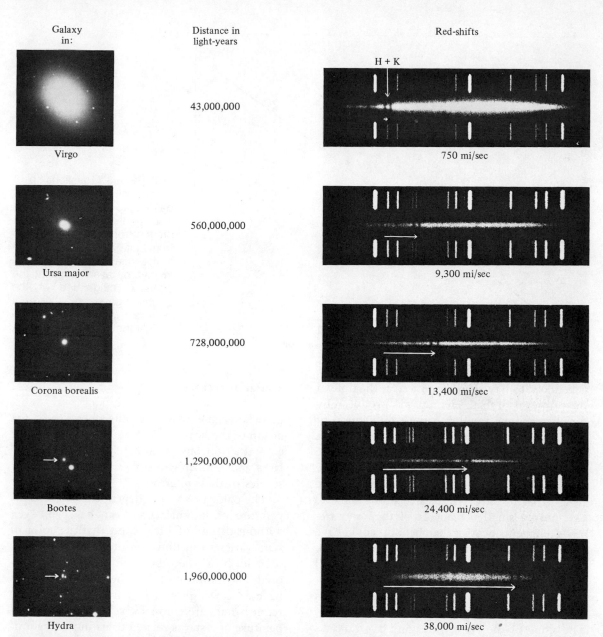

| Galaxy in: | Distance in light-years | Red-shifts |
|---|---|---|
| Virgo | 43,000,000 | 750 mi/sec |
| Ursa major | 560,000,000 | 9,300 mi/sec |
| Corona borealis | 728,000,000 | 13,400 mi/sec |
| Bootes | 1,290,000,000 | 24,400 mi/sec |
| Hydra | 1,960,000,000 | 38,000 mi/sec |

**FIGURE 8.2**  Relation between red-shift and distance for galaxies at various distances. The spectra of the galaxies is shown on the right side of the diagram as a fuzzy, pencil-shaped bright patch which is crossed by a series of dark lines. The bright lines on either side of the spectrum of the galaxy are spectral lines used for comparison purposes. Spectra are obtained either by passing the light from the galaxy through a prism or reflecting that light from a grating (a surface on which numerous, closely spaced, parallel lines have been scratched). The spectrum of a galaxy shows a continuous gradation in color from red through orange, yellow, green, and blue to violet, like the colors of a rainbow. Dark lines in the spectrum are caused by selective absorption of the light by different elements. Arrows on the right side of the diagram indicate the shift of two dark lines produced by the element calcium. (Mount Wilson and Palomar Observatories.)

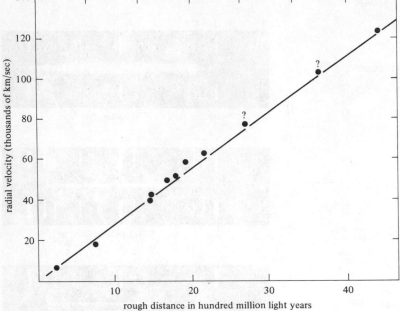

FIGURE 8.3 Relationship of the velocity of recession of galaxies to their distance. The velocity of recession is based on the red-shift of the light from that galaxy. The distance to the galaxy was determined in part by assuming that the absolute magnitude of the brightest galaxy in a cluster of galaxies is always approximately the same. (After Abell, Ref. 2.)

is believed to have been produced at the time of the primeval fireball. Helium was produced by fusion reactions involving the combination of hydrogen nuclei. By weight, hydrogen makes up 72 percent and helium 27 percent of the matter in the universe (**4**), so that most of the atoms in the universe originated during the "Big Bang." Theoretical calculations indicate that only a small percentage of the *heavy elements* (elements heavier than helium) could have been formed in the fireball. Most of the heavy elements and some of the helium were formed at a later time.

The concept of an expanding universe allows us to get some idea as to the age of the universe. The slope of the curve in Figure 8.3 indicates that the expansion of the universe began less than 18 billion years ago. The exact time of the beginning of expansion is not known because the rate of expansion is decreasing. However, the universe could be much, much older than 18 billion years since it is possible that the expansion may have been preceded by an earlier phase of contraction.

## Stellar Interiors

In seeking the answer to the question of the origin of the heavy elements, astronomers have embarked on observational studies of the surfaces and atmospheres of stars and theoretical studies of their interiors.

The color of a star is dependent on the temperature of its surface. It can be seen, even without the aid of a telescope, that the color of stars varies from bluish-white on one extreme to reddish on the other, with our yellow-colored sun in the middle. The blue-white stars have a very high surface temperature, but the reddish stars are comparatively cool. The temperature of a star is related to its intrinsic luminosity (actual brightness). When the intrinsic luminosity of stars is plotted against their surface temperature on an H-R (Hertzsprung-Russell) diagram, most of the stars fall along a straight line known as the main sequence (Fig. 8.4). Intrinsically bright stars have high surface temperatures while stars that are intrinsically dim generally have low surface tempera-

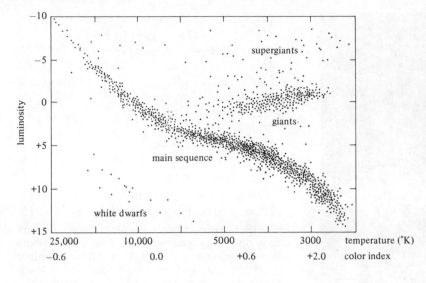

FIGURE 8.4 The Hertzsprung-Russell diagram on which a star's intrinsic (absolute) luminosity is plotted against its surface temperature. Intrinsic luminosity was determined by measuring the apparent luminosity of a star and its distance and then using an equation to calculate intrinsic luminosity. Temperature was calculated from the color of the star. To convert temperature in degrees K (Kelvin) to degrees C (centigrade), subtract 273 degrees. (After Abell, Ref. 2.)

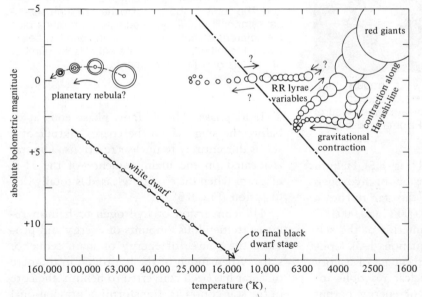

FIGURE 8.5 Evolutionary track of a star of 1.2 solar masses on a Hertzsprung-Russell diagram. The size of the circles is proportional to the size of the star at various stages in its evolution. (After Abell, Ref. 2.)

tures. The H-R diagram may be used to illustrate the evolution of a star (Fig. 8.5).

It is believed that stars originate as slowly contracting masses of gas and dust. Some astronomers have suggested that globules—dark, spherical bodies associated with certain irregular clouds of gas and dust—represent an early stage in the formation of stars (Fig. 8.6). Many globules have approximately the same mass as

our sun, but their radii are thousands of times that of the entire solar system.

As the cloud of gas and dust in a globule contracts, it heats up due to the release of gravitational energy. The phase of the evolution of a star in which heating is due only to the release of gravitational energy is known as the *Hayashi phase*. In this phase, named for the Japanese astrophysicist Hayashi (**5**), contraction and

FIGURE 8.6 A nebula showing irregular and spherical dust clouds. The small spherical dust clouds are globules of approximately the mass of our sun. (Mount Wilson and Palomar Observatories.)

$$\begin{align}
{}_{1}^{1}H + {}_{1}^{1}H &\to {}_{1}^{2}H + e + \text{neutrino} \tag{1}\\
{}_{1}^{2}H + {}_{1}^{1}H &\to {}_{2}^{3}He + \text{gamma ray} \tag{2}\\
{}_{2}^{3}He + {}_{2}^{3}He &\to {}_{2}^{4}He + 2{}_{1}^{1}H \tag{3}\\
{}_{2}^{4}He + {}_{2}^{4}He &\to {}_{4}^{8}Be \tag{4}\\
{}_{4}^{8}Be + {}_{2}^{4}He &\to {}_{6}^{12}C \tag{5}\\
{}_{6}^{12}C + {}_{2}^{4}He &\to {}_{8}^{16}O \tag{6}\\
{}_{6}^{12}C + {}_{2}^{4}He &\to {}_{10}^{20}Ne \tag{7}\\
{}_{10}^{20}Ne + {}_{2}^{4}He &\to {}_{12}^{24}Mg \tag{8}\\
{}_{12}^{24}Mg + {}_{2}^{4}He &\to {}_{14}^{28}Si \tag{9}\\
{}_{14}^{28}Si + {}_{2}^{4}He &\to {}_{16}^{32}S \tag{10}\\
{}_{14}^{28}Si + {}_{14}^{28}Si &\to {}_{28}^{56}Ni \tag{11}\\
{}_{28}^{56}Ni &\to {}_{27}^{56}Co + \text{positron} \tag{12}\\
{}_{27}^{56}Co &\to {}_{26}^{56}Fe + \text{positron} \tag{13}
\end{align}$$

FIGURE 8.7 Equations showing nuclear reactions occurring in the interiors of stars. Chemical symbols are H—hydrogen; He—helium; Be—beryllium; C—carbon; O—oxygen; Ne—neon; Mg—magnesium; Si—silicon; Ni—nickel; Co—cobalt; Fe—iron. The number above and to the left of the chemical symbol is equal to the sum of the protons plus the neutrons of that element, and the number below and to the left of the chemical symbol is the number of protons only in that element.

heating of a star is very rapid (Fig. 8.5). Helium begins to form through fusion of hydrogen nuclei when the interior of the star reaches a temperature of about 10,000,000°C (20,000,000 °F) and the density of the interior of the star reaches 100 g/cc. (Fig. 8.7, equations 1–3). Contraction continues as the percentage of energy output of the star from nuclear reactions increases and the percentage of energy output from the release of gravitational energy decreases. In the *T Tauri phase* of a star's evolution, the energy of a star comes from both gravitational and nuclear energy. The T Tauri phase of stellar evolution has been observed in a number of stars, including its namesake T Tauri. T Tauri stars are all characterized by rapidly expanding clouds of gas and dust surrounding them. While dust obscures a star during the Hayashi phase, a star may be seen in the T Tauri phase. The T Tauri phase ends at or before the stage when the energy output of a star is due entirely to nuclear reactions. The star is located on the main sequence of the H-R diagram when the energy released is totally due to fusion (Fig. 8.5).

The conversion of hydrogen to helium releases tremendous amounts of energy and provides the source of energy of main sequence stars. When much of the hydrogen in the core of a star has been converted to helium, the core of the star contracts, transforming gravitational energy into heat, which again causes the core to heat up. This in turn makes the outer layers of the star expand and causes the surface temperature of the star to decrease. At this stage, the star moves off the main sequence and becomes a *red giant,* which is a relatively large star with a comparatively cool, red surface (Fig. 8.5). Betelgeuse, the brightest star in the constellation of Orion, is such a star. At this stage the star burns hydrogen in the outer layers of its interior, rather than in its core. When the hydrogen around the core of the star is exhausted, the star contracts suddenly and the temperature of the stellar interior rises still further. When the temperature reaches approximately 100,000,000°C (200,000,000°F), helium in the core begins to fuse to produce carbon (Fig. 8.7, equations 4, 5). At this point, the luminosity of the star increases briefly (for about 1 hour) in what is termed a *helium flash*. Its luminosity decreases somewhat, and it again becomes a red giant continuing to "burn" helium to produce carbon. When the helium in the core is exhausted, the star will contract again and its internal temperature will rise while its outer surface will expand and cool. Large quantities of hydrogen, helium, and carbon will be thrown off into space in this phase of the star's evolution. Eventually the helium in the core will become largely depleted, and in a star the size of our sun, the core will continue to contract. However, since the star is no longer fueled by nuclear reactions, it will cool and contract to the next stage in its development, a *white dwarf*.

If the red giant is between 4 and 8 times the mass of our sun, it may explode when most of the helium in the core is used up, forming a supernova. The explosion releases hydrogen, helium, and carbon into space. If the red giant has a mass of more than 8 times that of the sun, it will contract and the temperature of the core will rise again when most of the helium in the core is used up. The temperature will increase to the point where oxygen is produced by fusion (Fig. 8.7, equation 6). The star may go through a series of contractions and increases in temperature within the core during which time neon, magnesium, silicon, sulfur, nickel, cobalt, and iron are produced (Fig. 8.7, equations 7–13).

The star is then enriched in the elements out of which the earth is made. After most of the material in the core of the star is converted to iron, the star is no longer able to produce energy by fusion in its core. The star collapses catastrophically, crushing the matter in its core

**FIGURE 8.8** The spiral galaxy NGC 7331 **(a)** before and **(b)** during the eruption of a supernova. Arrow in **(b)** indicates the supernova. (Lick Observatory photograph.)

into an extremely dense sphere. Due to subsequent rebound of this matter collapsing, the star may explode in a supernova much more violent than the type discussed earlier. The star may flare up to hundreds of millions of times its former brightness (Fig. 8.8). In fact, the energy output of a single supernova may equal the energy output of an entire galaxy for a short period of time. It is believed that the elements heavier than iron in our solar system were produced during or immediately preceding supernova explosions (6).

A supernova will result in the dispersal of most of a star's mass into space (7), where other stars will form from the matter ejected (Fig. 8.9). Some of the material in our sun and its planets was made and released by such supernova explosions.

FIGURE 8.9 The Crab Nebula is the remains of a supernova that occurred in 1054. The bright filaments are excited gas which is moving outward from the center of the explosion. (Mount Wilson and Palomar Observatories.)

## Quasars and the Origin of the Heavy Elements

In 1963, Maarten Schmidt discovered a new class of starlike objects which emit very large amounts of energy in the form of radio waves. These have been called quasi-stellar radio sources, or quasars (**8**). The spectra of quasars indicate that they are receding from our galaxy at very large velocities. Using Hubble's velocity-distance relation (Fig. 8.3) it has been calculated that many quasars are probably more distant than any known galaxy. If these calculations are correct, then the energy output of some quasars must be greater than that of the most luminous known galaxy.

Recent studies show that the red-shifts of quasars do indicate their distance and therefore that they are indeed at great distances. One such study has shown that some quasars which appear to be part of a cluster of galaxies have the same red-shift as that of the galaxies in the cluster (**9**).

A large percentage of known quasars is located between 7 and 8 billion light years from our galaxy (**9**). It is quite possible that quasars were very abundant 7 or 8 billion years ago and that their numbers have decreased with time. The apparent lack of quasars beyond a distance of 9 billion light years may indicate that there were no quasars prior to 9 billion years ago.

It has been suggested that quasars are the extremely luminous nuclei of galaxies the outer portions of which we are not able to observe because they are so far away. Many of the distinctive characteristics of quasars were observed by Carl Seyfert (Senior) in an interesting class of spiral galaxies named *Seyfert galaxies* after their discoverer (**10**). These galaxies are distinctive in that they possess highly luminous nuclei (Fig. 8.10). Seyfert galaxies may be an evolutionary stage between quasars and normal galaxies, since their luminosities overlap those of quasars and normal galaxies (Fig. 8.11).

It has been suggested that the enormous amounts of energy released by both quasars and Seyfert galaxies originate in the simultaneous occurrence of large numbers of novae and supernovae in the nuclei of galaxies. In our galaxy, supernovae occur once every 30 years or so. In an early stage in the history of this galaxy, the rate of occurrence of supernovae may have been much greater than it is today.

Very massive stars burn hydrogen at a very rapid rate and therefore evolve rapidly to supernovae. Large numbers of very massive stars would result in large numbers of supernovae. The rapid variations in energy output of quasars and Seyfert galaxies may be due to differences in the number of supernovae occurring at any one time.

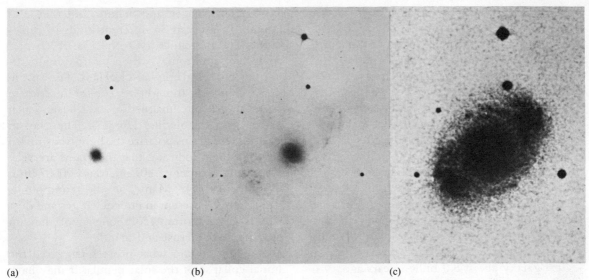

(a)                              (b)                              (c)

**FIGURE 8.10**  Photographs of negative plates of a Seyfert galaxy with increasing time: **(a)** Only the nucleus is visible (at this point, the nucleus resembles a quasar). **(b)** and **(c)** With a longer time of exposure, the stars in the spiral arms of the galaxy become visible. (Copyrighted by the National Geographic Society-Palomar Observatory Sky Survey: original photographs from the Hale Observatories; montage prepared by W. W. Morgan, Yerkes Observatory.)

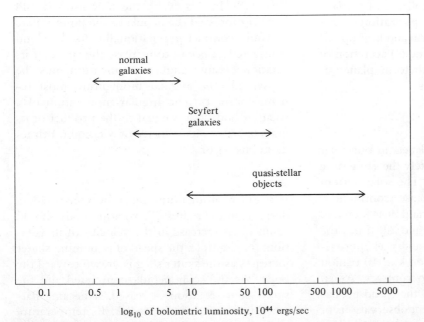

$\log_{10}$ of bolometric luminosity, $10^{44}$ ergs/sec

**FIGURE 8.11**  Comparison of the luminosities of normal galaxies, Seyfert galaxies, and quasars. (Modified after Colgate, Ref. 11.)

In our own galaxy (the Milky Way Galaxy), there is a wide variation in the age of the stars. However, most of the main sequence stars have essentially the same percentage of heavy elements, regardless of the age of the star (**12**). Evidently, most of the heavy elements were created at an early stage in the history of our galaxy, and the rate of heavy element formation has since been drastically reduced. C. M. Hohenburg estimates that most of the heavy elements in our galaxy were produced between 8.0 and 8.8 billion years ago (**13**). A high rate of heavy element formation would require large numbers of massive, rapidly evolving stars and therefore large numbers of supernovae. Perhaps our galaxy was a quasar that evolved into a Seyfert galaxy about 8.0 billion years ago.

## FORMATION OF THE SOLAR SYSTEM

The formation of our solar system probably occurred in four stages: (1) collapse of a *nebula* (cloud of gas and dust), (2) vaporization of that nebula as the sun began to form and heat up, (3) condensation of the nebula, and (4) accretion of the particles in the nebula to form planetesimals, and, eventually, planets.

### Collapse of the Nebula

Scientists theorize that the planets in our solar system formed at approximately the same time as the formation of the sun. The solar system probably formed by the collapse (contraction) of an interstellar cloud of gas and dust (nebula). This cloud was between 1.2 and 2.0 times the present mass of the sun (**14**) and was approximately 2 light years (17 trillion km, 10 trillion mi) across when it began to contract. At an early stage in its contraction, the cloud probably resembled a globule. Recent observations of interstellar nebulae have helped to determine the probable composition of this gas and dust cloud. Interstellar nebulae are principally composed of hydrogen and helium, but they also contain more than 36 different kinds of molecules. These include hydroxyl ($^2/_3$ of a water molecule), ammonia, carbon monoxide, methane, cyanide, and ethyl alcohol (**15**). The dust in these clouds is thought to consist in part of grains of graphite, magnetite, and silicates (such as the mineral olivine) (**16**, **17**). The way in which these grains polarize the light transmitted through them indicates that the grains are very small, on the order of 0.07 microns (.00007 mm, .000003 in) (**16**). Many of the compounds thought to be present in interstellar gas and dust clouds are important components of the sun and the earth's crust and mantle.

It is not certain what process triggered the initial collapse of the solar nebula. It may have been due to more or less random motions of gas and dust which at some point resulted in a concentration of material sufficient for the nebula to collapse due to the mutual gravitational attraction of its components. Alternatively, a nearby supernova may have triggered the collapse (**18**). In this event, the supernova would have compressed the nebula to the point where it would contract gravitationally. As the cloud contracted, it began to increase the speed of its rotation because angular momentum must be conserved (i.e., angular momentum must remain constant). The angular momentum of a rotating body ($w$) is equal to the product of its mass ($m$) times its velocity of rotation ($v$) times its radius ($r$) or

$$w = mvr$$

If angular momentum must be conserved, a decrease in the radius of a rotating body should result in an increase in the velocity of its rotation; for example, the speed of a spinning skater increases as the skater's leg is drawn toward the body. As the solar nebula contracted and the speed of its rotation increased, it became flattened into the shape of a disk. Its temperature increased as a result of the increase in the number and the violence of the collisions of the particles within the cloud.

## Vaporization and Condensation of the Solar Nebula

The solar nebula became luminous at its center as the sun began to take shape during the Hayashi phase of its evolution. Heating of the nebula surrounding the sun caused dust near the sun to become vaporized. A sphere of vaporized dust grew around the sun and probably reached its maximum extent during the T Tauri phase of the sun's evolution. After the sun entered the T Tauri phase of its evolution, the nebula began to cool and condense.

The evidence for vaporization and condensation within the solar nebula comes from the study of *meteorites,* which are fragments of rock or metal from space which have fallen on the earth, moon, or other planet. They are grouped by composition into stony, stony-iron, and iron meteorites. *Stony meteorites* are composed of rock, while *stony-iron meteorites* are a mixture of rock and nickel-iron, and *iron meteorites* are composed of nickel-iron with small amounts of iron sulfide (troilite). The most abundant of these are stony meteorites constituting 92.8 percent of the falls while stony-irons make up 1.5 percent and irons 5.7 percent of the observed falls (**19**). Most (94 percent) stony meteorites belong to a class called *chondrites,* and most of these contain small, more or less spherical bodies called *chondrules.* Typical chondrules are 0.5 to 8 mm ($^1/_{32}$″ to $^1/_3$″) in diameter.

There are two principal types of chondrites, ordinary and carbonaceous. As their names imply, *carbonaceous chondrites* contain a significant percentage of carbon, while *ordinary chondrites* do not. The chondrules in ordinary chondrites are composed of either glass or a mixture of glass and crystals. A number of different theories have been proposed for the origin of these chondrules, but all of them involve rapid cooling of liquid droplets. One recently proposed theory suggests that chondrules are formed by remelting of earlier condensed materials (**20**), as might occur in the sun's T Tauri phase of evolution. This phase is characterized by sudden increases and decreases in a star's luminosity. A sudden increase in the sun's luminosity might cause some earlier condensates to melt.

Chondrules in carbonaceous chondrites include glassy chondrules and completely crystalline chondrules (**21**). The latter provide the strongest evidence for vaporization within the solar nebula. Completely crystalline chondrules contain a mixture of finely crystalline minerals such as melilite, spinel, and calcium-rich plagioclase (anorthite). Experimental studies on these minerals indicate that they condensed at high temperatures (around 1500°K, 1227°C, 2240°F) (**21**). Carbonaceous chondrites also contain (between the chondrules) significant quantities of water and hydrocarbons (mixtures of carbon, hydrogen, and nitrogen). These substances condense from the solar nebula at about 350°K (77°C, 170°F) (**21**). Thus, carbonaceous chondrites contain a mixture of high and low temperature condensates and accretion of the parent bodies from which these meteorites were derived occurred after condensation was largely completed.

Accretion of the earth also probably occurred for the most part after, rather than during, the condensation of its components. This process is referred to as *homogeneous accretion* as opposed to *inhomogeneous accretion,* in which accretion begins before completion of condensation of a planet's materials.

In order for meteorites to be used in the determination of conditions within the solar nebula, it must be first established where they originated. Most meteorites come from the *asteroid belt,* a zone located between Mars and Jupiter in which there are a large number of small- to intermediate-size bodies called asteroids which are in orbit about the sun (Fig. 8.12). Observations of the surface reflectivity of asteroids indicates that many are similar in composition to meteorites (**22**). Furthermore, the orbits of meteorites as determined from their trajectories as they pass through the earth's

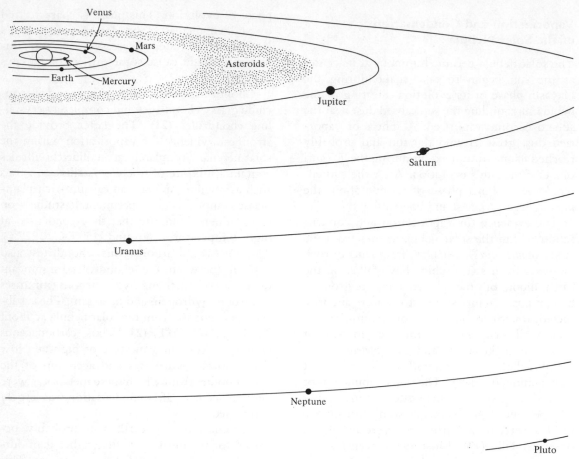

**FIGURE 8.12** Diagram of the solar system.

atmosphere indicate that meteorites originate in the asteroid belt (**23**). It is therefore likely that asteroids are the parent bodies for most if not all meteorites. The size distribution of fragments in the asteroid belt indicates that asteroids are the remnants of approximately 100 bodies that have been subjected to collisional fragmentation (**22**).

Asteroids in the main part of the asteroid belt appear to have a composition similar to the carbonaceous chondrites while those inside the inner edge of the main part of the asteroid belt at a distance of less than 2.3 astronomical units (356,000,000 km or 214,000,000 mi) appear to have a composition similar to ordinary chon-

drites (**22**). Carbonaceous chondrites show evidence of having been formed from materials condensed from a hot solar nebula. It is likely that the earth, which is closer to the sun, also formed from materials condensed from a hot solar nebula.

## Accretion of the Planets

In order to understand the events that occurred during the accretion of the planets, it is necessary to look at their composition. Comparison of the uncompressed densities (densities at standard temperatures and pressures) of various planets gives an indication of their compositional differences. The uncompressed density of the terrestrial planets (Mercury, Venus, Earth, and Mars) is dependent on the percentage of nickel-iron (density 7.5 g/cc) within them. Mercury's uncompressed density is high in comparison to the other planets, and there is a general decrease in density away from the sun (Table 8.1). Mercury has a higher nickel-iron content than the other planets. Thus, Mercury may have begun to accrete after iron condensed, but before olivine had completely condensed from the solar nebula (**24**). This would place the temperature of the beginning of accretion of Mercury at about 1300°K (1027°C, 1880°F) (**21**).

The temperature of accretion of the earth would have been less than Mercury. The earth's core probably contains significant amounts of troilite (**25**) which condensed at about 650°K (377°C, 710°F) (**26**), and therefore the earth probably began to accrete at a temperature below 650°K. The earth is strongly depleted in lead, and because lead condenses at temperatures of about 520°K (247°C, 476°F) (**26**), the

earth probably condensed at a temperature above 520°K. Estimates place the temperature of accretion of the earth at about 600°K (327°C, 620°F) (**27**). As discussed earlier, the temperature of accretion of carbonaceous chondrites (and therefore at least some of the asteroids) was about 350°K. Comparison of the temperatures of accretion of Mercury, Earth, and the asteroids indicates that a rather high temperature gradient existed within the solar system at the time of accretion of the planets.

A comparison of the compositions of the terrestrial planets and the asteroids with that of the sun indicates that they are all deficient in *volatile elements* (elements which vaporize at relatively low temperatures). This deficiency includes not only hydrogen and helium, but carbon, nitrogen, and certain other elements. Mercury is also deficient in magnesium and silicon (the components of olivine). These observations indicate that considerable amounts of material must have been lost from the solar nebula following accretion of the planets, perhaps as a result of the material having been

TABLE 8.1 **Solar system data**

| PLANET | AVERAGE DISTANCE FROM SUN, MILLIONS OF KM | EQUATORIAL DIAMETER, KM | DENSITY, G/CC | UNCOMPRESSED DENSITY | MASS (EARTH = 1) | ECCENTRICITY OF ORBIT | INCLINATION OF AXIS (OBLIQUITY) | INCLINATION OF ORBIT TO ECLIPTIC |
|---|---|---|---|---|---|---|---|---|
| Mercury | 58 | 4,840 | 5.42 | 5.2 | 0.055 | 0.0206 | 1° | 7° |
| Venus | 108 | 12,200 | 5.25 | 4.0 | 0.815 | 0.007 | 3° | 3.4° |
| Earth | 150 | 12,756 | 5.51 | 4.0 | 1.0 | 0.017 | 23°27′ | 0° |
| Moon | 150 | 3,476 | 3.34 | 3.31 | 0.012 | 0.055 | 6°42′ | 5.15° |
| Mars | 228 | 6,760 | 3.96 | 3.7 | 0.108 | 0.093 | 23°59′ | 1.9° |
| Jupiter | 778 | 142,700 | 1.33 | ? | 317.9 | 0.048 | 3°05′ | 1.3° |
| Saturn | 1426 | 121,000 | 0.68 | ? | 95.2 | 0.056 | 26°44′ | 2.5° |
| Uranus | 2868 | 47,000 | 1.60 | ? | 14.6 | 0.047 | 82°5′ | 0.8° |
| Neptune | 4494 | 45,000 | 1.65 | ? | 17.2 | 0.009 | 28°48′ | 1.8° |
| Pluto | 5896 | 5,000 | 3 ?? | ? | 0.1? | 0.25 | ? | 17.2° |

**FIGURE 8.13** A painting showing planetesimals orbiting the sun prior to their coalescing into planets. (From *Moons and Planets: An Introduction to Planetary Science* by William K. Hartmann, © 1972 by Wadsworth Publishing Company, Inc. Reprinted by permission of the publisher, Ref. 30.)

blown away by an intense solar wind. The sun may have been at this time in the T Tauri phase of its evolution, a time of considerable mass loss.

It is not known with certainty how the planets actually accreted. Several possibilities have been suggested. One is that the planets grew as a result of gravitational instabilities within the solar nebula. In order for particles and/or gas to collapse gravitationally to form a body, the material must be concentrated into a critical density. In one proposed model, particles settled rapidly into a thin disk which then broke up into many rings. These rings then collapsed into small bodies called *planetesimals* a few kilometers or a few tens of kilometers in diameter (**29**) (Fig. 8.13). Bodies of this size would have been large enough to attract and hold small particles that might collide with them. One of the necessary conditions for growth of a small body is that the particles colliding with it must be in similar orbits so that encounter velocities

would be low. Accretion would occur as larger bodies swept up smaller bodies in their path as they orbited the sun.

Alternatively, initial accretion of the planets may have occurred as a result of electrical and/or magnetic forces attracting small particles to each other. In the asteroid belt and beyond, particles may have been stuck together by ices (frozen water, ammonia, or methane) or by carbonaceous compounds. As with the terrestrial planets, the outer planets would have continued to grow after reaching some critical size due to gravitational forces.

## Time of Formation

The time of formation of the earth and the rest of the solar system has been placed at about 4.6 billion years ago. Uranium-lead, potassium-argon, and rubidium-strontium dates fall between 4.5 and 4.7 billion years for most meteorites (**31**). This is in good agreement with the age of the oldest lunar samples (4.66 billion years) and the age of the earth based on lead-isotopic ratios in oceanic basalts (4.53 billion years) (**32**).

## ORIGIN OF THE EARTH-MOON SYSTEM

Recent investigations of the moon during manned landings in the Apollo Space Program have rekindled interest in the origin of the earth-moon system. The theories of origin fall into three categories:

1. *Fission Theory:* The moon was once part of the earth and separated from it as a result either of rapid rotation of the earth or of the impact of a Mars-size body (Fig. 8.13)
2. *Binary Theory:* The moon formed while orbiting the earth, either from materials derived from the earth or from material orbiting the sun (Fig. 8.14)
3. *Capture Theory:* The moon formed outside of the earth's orbit and was subsequently captured by the earth (Fig. 8.15)

In order to evaluate these theories, let us look at some relevant data concerning the moon:

1. For the past 360 million years, the moon has been receding from the earth at a rate such that the moon must have become a satellite of the earth between 0.5 and 2.0 billion years ago, assuming that the rate of recession was not significantly lower prior to 360 million years ago (**33**).
2. The uncompressed density of the moon is 3.31 g/cc, while that of the earth is 4.0 g/cc.
3. Seismic studies of the moon indicate that the moon has no core or at best a very small core. The maximum possible size for the core is 360 km (210 mi) (**34**) which is less than 1 percent of its volume. By contrast, the earth's core is 16 percent of its volume. Thus the earth has a much greater percentage of nickel-iron than the moon.

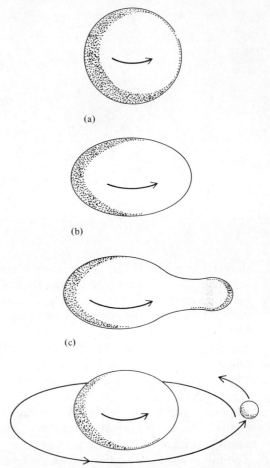

(a)

(b)

(c)

**FIGURE 8.14** Diagrammatic sketch illustrating the formation of the moon by fission from the earth.

4. Samples brought back from the moon and geophysical studies made on the moon indicate that the entire satellite is probably very low in volatile elements (**35**). For example, the moon has only the slightest traces of water and is very low in sodium and potassium. The earth is much richer in these substances than the moon. The moon is also rich in refractory elements (those elements which vaporize at high temperatures) such as titanium, zirconium, and chromium (**35**). The low volatile and high refractory element content of the moon in comparison to that of

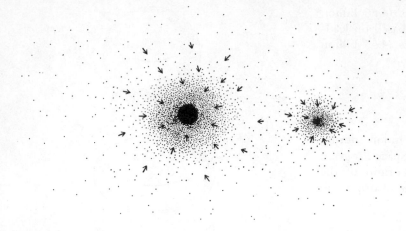

FIGURE 8.15 Diagrammatic sketch illustrating the formation of the moon near the earth at the same time as the formation of the earth.

the earth is probably due to the moon's having accreted at a higher temperature than the earth. Depletion of the moon in gold, copper, sodium, and potassium (**26**, **36**) probably indicates that the moon accreted at a temperature above 1230°K (957°C, 1754°F) (**26**, **37**).

5. The moon's crust is approximately 75 km (45 mi) thick and therefore constitutes about 10 percent of its volume (**38**). As mentioned earlier, the earth's crust constitutes only about 1 percent of its volume.

6. The moon shows abundant evidence of volcanic activity (Fig. 8.16 and 8.17). Radiometric dating of lunar samples indicates that this activity largely ceased by about 3.2 billion years ago (**35**).

7. The plane of the moon's orbit is inclined at 5° to the ecliptic (the plane about which the earth orbits the sun), while the earth's axis is inclined at 23.5° to the ecliptic.

8. The mass of the moon is approximately 1.2 percent of the mass of the earth.

There are four very important objections to both the Fission and the Binary theories. The first is that the rate of recession of the moon seems to indicate that it became a satellite of the earth less than 2 billion years ago. The Binary Theory requires that the moon has been a satellite of the earth since the earth was formed 4.6 billion years ago. Furthermore, if the moon formed from the earth by fission, this event must have occurred more than 3.8 billion years ago because there is evidence for the continuous existence of oceans since that time and a fission would probably result in the loss of the earth's oceans. However, it is possible that the rate of recession was significantly slower during the Precambrian than it was later, and therefore this objection may not be a serious one.

Secondly, the chemistry of the earth and moon are very different. The Fission Theory easily explains why the moon has so little nickel-iron because the fission would have taken place after separation of the earth's core and mantle. Infall of liquid nickel-iron toward the center of the earth could have promoted fission since the infall of dense material would have increased the speed of rotation of the earth (due to the conservation of angular momentum) (**39**). However, the Fission Theory does not adequately explain the high content of refractory elements and the low content of volatile elements and water on the moon, although attempts have been made to do so (**38**, **40**).

The third objection to these theories is that the moon does not revolve around the earth in the earth's equatorial plane. In general, moons of other planets, particularly those which rotate

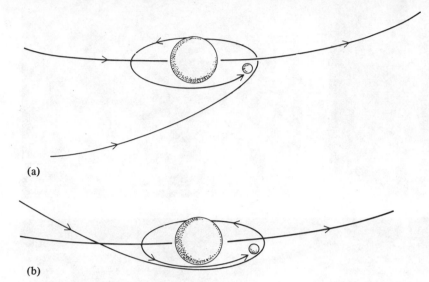

(a)

(b)

**FIGURE 8.16** Diagrammatic sketch illustrating the capture of the moon in a prograde orbit from **(a)** outside the earth's orbit and **(b)** inside the earth's orbit.

around the planets in a prograde sense (counterclockwise when viewed from above the plane of the ecliptic) do so in the equatorial plane of that planet. However, our moon's orbit is inclined at about 18.5° to the earth's equatorial plane.

Lastly, the mass of the moon is 1.2 percent the mass of the earth. Fission by rotation should have produced a body between 10 and 40 percent of the mass of the parent body (**41**). Furthermore, the moon is much larger than other moons in the solar system in comparison to the size of the planet about which it is rotating. The moons of Jupiter and Saturn, for example, are less than 0.01 percent of the mass of their planets.

Many scientists believe that the earth's moon was captured. A variety of capture models have been proposed: (1) capture from an orbit near the earth, (2) capture from an orbit outside of the earth's orbit, and (3) capture from an orbit inside the earth's orbit. Capture from an orbit near the earth's orbit would not be difficult from the standpoint of dynamics, but the chemical differences between the earth and moon are difficult to explain, although attempts have been made here also (**42**).

There are major difficulties in explaining the moon's origin by capture from an orbit outside the earth's orbit. First, the iron content of the moon is probably too low to have been formed outside the earth's orbit. Not only does the moon have only a very small nickel-iron core at best, but its total iron (combined and uncombined) is between 6 and 12 percent of its mass (**43**). Differences between the moon and Mars indicate that it is unlikely that the moon originated in that vicinity. Mars has an uncompressed density of 3.7 g/cc (the moon's is 3.3) and therefore Mars probably has considerably more nickel-iron than the moon. It is also unlikely that the moon originated in or near the asteroid belt. Most meteorites have significant amounts of iron. Ordinary chondrites contain about 13 percent nickel-iron on the average and have a total iron content averaging 25 percent (**44**). Carbonaceous chondrites have a total iron content averaging 23 percent (**44, 45**). Secondly, the moon is too low in volatiles to have been formed outside the earth's orbit. Mars is thought to have fairly abundant water, both chemically combined in its soil and as ice in the polar regions (**46**). Furthermore, meteorites generally contain water. Ordinary chondrites

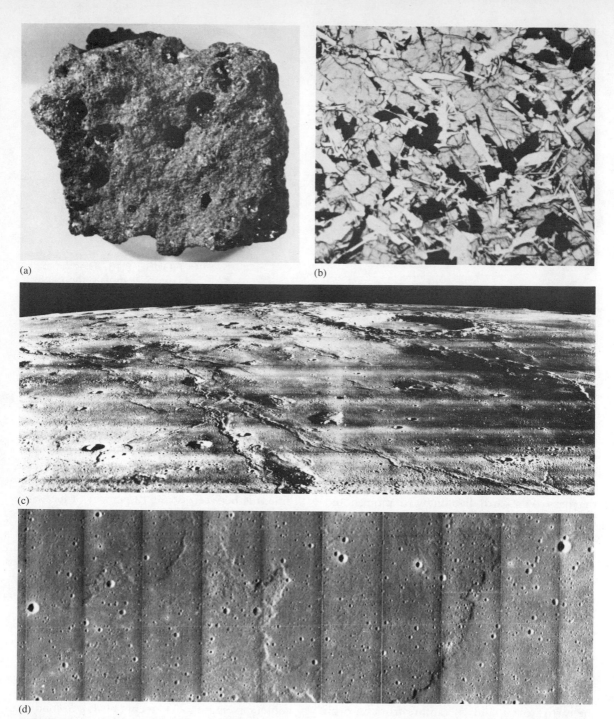

FIGURE 8.17 Evidence of volcanic activity on the moon: (a) Vesicular lava collected during the *Apollo eleven* flight to the moon. (Courtesy of NASA.) (b) Photomicrograph of thin sections of a lunar rock taken in polarized light with uncrossed nicol prisms. The colorless mineral is plagioclase, the light gray is pyroxene, and the black is ilmenite. Magnification approximately 250X. (c) Oblique photograph made by *Lunar orbiter 11* showing domes on the lunar surface. The domes are from 2 to 10 mi in diameter and from 1000 to 1500 ft high. They are shaped much like terrestrial volcanic domes. For this reason, it has been suggested that lunar domes have volcanic origins. (d) Lobate margin of a volcanic flow. (Courtesy of NASA.)

contain about 0.3 percent water (**43**) and carbonaceous chondrites average between 1 and 20 percent water (depending on the type) (**44**). Planets beyond the asteroid belt contain even more volatiles, so that is an unlikely place of origin of the moon. Similarity of oxygen isotope composition of the moon and the earth indicate that the moon originated within the solar system (**47**). It is unlikely that the moon formed near Venus because Venus has abundant volatiles in its atmosphere. It would be expected that the moon would be rich in refractory elements and low in volatiles if the moon formed inside the orbit of Venus, but the low content of nickel-iron of the moon would be difficult to explain. Mercury has an uncompressed density of 5.2 g/cc and therefore contains even more nickel-iron than the earth (**48**).

The moon may have formed inside the orbit of Mercury (**49**). This would explain the moon's high content of refractory elements and low content of volatile elements, and would explain the moon's low content of iron if the moon accreted before nickel-iron began to condense from the solar nebula. In this case the moon would also be low in olivine and that olivine which was present would be very poor in iron. However, the moon probably is too rich in iron oxide (FeO) to have formed in this manner. The surface rocks on the moon have an FeO content similar to that of the surface rocks of the earth (**35**). Basalts from the lunar mare have an FeO content of about 20 percent, even higher than that of most terrestrial basalts. Rocks from the lunar highlands have an FeO content of about 6.5 percent. The average composition of terrestrial crust is about 6 percent FeO (all iron calculated as FeO). Early condensates would probably be too low in FeO to produce a lunar crust with such a high FeO content. Furthermore, there are severe dynamical difficulties in capturing a body such a great distance from the earth.

From the foregoing discussion, it is evident that there are severe problems with the capture theory if the moon was formed as a separate body anywhere within the solar system. It is possible that the moon was captured, but did not form as a separate body. It may have formed by fission from the planet Mercury (**50**). There are several features about Mercury which support this hypothesis:

1. Since Mercury formed near the sun, it is almost certainly rich in refractory elements and low in volatile elements, like the moon. It is estimated that Mercury began its accretion at about 1400°K (1127°C, 2060°F) (**27**). If Mercury began to accrete before olivine had completely condensed, as Anderson has suggested (**43**), its temperature of accretion would have been about 1300°K (1027°C, 1880°F). As discussed earlier, the moon probably accreted at a temperature of more than 1230°K.
2. Earth-based observations of Mercury indicate that its surface composition is very similar to that of the moon (**51**). Reflectivity measurements made by *Mariner 10* confirm this similarity (**52**).
3. Television pictures sent back to earth from *Mariner 10* show that the surface of Mercury consists of light-colored, heavily cratered terrains similar to the moon's highlands, and less cratered, darker plains which are similar to the lunar mare.
4. The volume of the moon is approximately 26 percent of that of Mercury. This is well within the permissible limits of a body produced by rotational fission (**41**).
5. Mercury almost certainly has a large nickel-iron core.

The lack of significant amounts of nickel-iron in the moon could be explained by fission of the moon from Mercury after the formation of Mercury's core. It is thought that Mercury's core formed at an early stage in its history (**53**) and therefore fission of the moon probably took place about 4.5 billion years ago.

On the initial separation of the moon from Mercury, the velocity of the moon may have been sufficient to break free of the gravitational attraction by Mercury. Alternatively, gravitational interaction between the moon and sun following separation of the moon from Mercury may have resulted in the moon's breaking free of Mercury's gravitational field at a later time. The moon would then have an elliptical orbit which may have crossed the orbit of Venus. Once Mercury's orbit crossed that of Venus, these two bodies would interact in such a way that the moon would accelerate each time it passed close to Venus. After repeated passes near Venus, sufficient energy may have been transferred from Venus to the moon so that the moon achieved an earth-crossing orbit. At this point, the moon would begin to interact gravitationally with the earth and repeated close passes of the moon and earth could place the moon into an orbit which would permit the earth to capture the moon. Two features about Venus support the idea that interactions between the moon and Venus played a role in capture of the moon by the earth:

1. Venus rotates more slowly than any other planet and it rotates in a retrograde sense. (**54**). Uranus is the only other planet that rotates in a retrograde sense, and it probably does so because its axis has been tilted more than 90° to the ecliptic. The retrograde rotation of Venus could have been caused by repeated close passes between it and the moon. Other attempts to explain the retrograde orbit of Venus have not been successful (**55**).

2. Every time Venus and the earth come into superior conjunction (i.e., every time they approach each other at minimum distances), Venus presents the same face to the earth (**56**). This observation might be explained if the earth had captured the moon from an orbit which was interacting with Venus.

Capture of the moon by the earth may have been a very violent event for both the moon and the earth if the moon approached the earth closely following capture. Widespread igneous activity would have occurred on both the earth and moon and tides on the earth would have been huge. However, if capture occurred at some distance from the earth, there would have been less effect on both the earth and moon. If capture was a violent event, it had to have occurred more than about 3.2 billion years ago, the time of the last widespread volcanic activity on the moon. If the earth and moon achieved a resonant orbit before the time of closest approach, they would be approximately 7.5 earth radii (100,000 km, 60,000 mi) apart at the time of their closest approach (**58**). In this case, the moon would not have produced any great upheavals on the earth at the time of its closest approach. Capture could have been either in a prograde or retrograde sense. Several moons in the solar system rotate in a retrograde sense, but most, like our moon, rotate prograde. Those with a retrograde rotation are thought to have been captured (**57**). These include Triton, a moon of Neptune, which is 6000 km (3600 mi) in diameter. For comparison, our moon is 3476 km (2090 mi) in diameter.

The moon may have been captured in a retrograde orbit, spiraled in toward the earth, and then changed to a prograde orbit after its orbital plane was inverted (**58**). In a prograde orbit, the moon would spiral away from the earth, as it is doing today. If capture was in a retrograde orbit, a considerable amount of time may have elapsed between the time of capture and the time of closest approach of the moon to the earth. Capture may have been more than 4 billion years ago if it occurred in a retrograde orbit.

If the moon's orbit did change from retrograde to prograde, the time of closest approach would almost certainly have left its mark in the rock record. The period between 1.4 and 1.5 billion years ago was unique in several respects, and this may have been the time of closest approach of the moon to the earth:

1. Many large anorthosite bodies were intruded at that time (**59**). These plagioclase-rich intrusions may have been formed by partial melting of the lower crust during a time of an unusually high geothermal gradient (**59**).
2. Many large charnockites were intruded at this same time (**60**). These pyroxene-bearing granitic rocks give evidence of having been intruded into areas with unusually high geothermal gradients (**60**).
3. A very large number of granites were intruded during this time interval, and yet there is no evidence of the folding or regional metamorphism usually associated with them (**61**). Widespread granitic intrusion is almost always accompanied by folding, faulting, and mountain building.
4. There is no firm evidence on earth of sedimentary rocks deposited during this time interval.

The high geothermal gradient and widespread melting of the earth's crust which produced the anorthosites, charnockites, and granites may have been due to heating of the earth (perhaps due to earth tides) when the moon was near the earth. The apparent lack of sedimentary rocks between 1.4 and 1.5 billion years old may be due to huge tides at the time of closest approach. Such tides might have prevented deposition of sediment on the continents. All deposits of this age may have been carried into the deep oceans.

## SEPARATION OF THE EARTH'S CORE, MANTLE, AND CRUST

Separation of the earth's core, mantle, and crust required that the earth be heated to the melting point of a mixture of nickel, iron and sulfur (about 5000°C) in the core and to the melting point of the silicates (about 1000°C) in the earth's crust. As mentioned earlier, the temperature of accretion of the earth was probably about 600°K (327°C, 620°F). Since the earth probably formed by the accretion of planetesimals ranging in size from tiny particles to fragments up to several hundreds of kilometers in diameter, the earth may have been heated significantly by the impact of these bodies. The rate of heating depends primarily on the rate of growth of the earth. Estimates of this rate vary considerably, but if most of the material of the earth was added rapidly, the earth would have reached the melting point of nickel-iron before accretion was completed.

Another heating process which would have operated within the earth is the decay of radioactive isotopes. The decay of short-lived isotopes (such as aluminum 26) may have played a role in the initial heating of the earth (**19**). Such isotopes may have been produced by intense bombardment of material within the solar nebula by protons or neutrons (**63**) or they may have been produced in a supernova explosion just prior to the formation of the solar system (**18**). If initial heating of the earth took place over a relatively long period of time, the decay of long-lived radioactive isotopes (such as potassium 40, uranium 235, and uranium 238) would have been important.

## Formation of the Core

Radiometric dating of meteorites suggests that at least some of the asteroids were heated up to a temperature above that of the melting point of nickel-iron between 4.5 and 4.7 billion years ago. Investigators believe that when the asteroids formed, they were similar in size to the largest asteroids in the asteroid belt (600 to 900 km, 360 to 540 mi). If the earth and the parent bodies of meteorites were subjected to the same heating processes, the earth, a much larger body, would probably have melted at approximately the same time or even before the asteroids. Thus the earth's core probably formed very soon after the formation of the solar system. It must have formed before about

2.75 billion years ago, because rocks of this age have a strong thermoremnant magnetism (**64**). A liquid nickel-iron core is necessary for the formation of such magnetism.

## Formation of the Crust

The earth's crust may have been produced largely at a very early stage in its history, or it may have been added at a relatively uniform rate throughout geologic time. Study of the moon and other planets suggests that much of the material in the earth's crust was probably formed soon after the earth formed. The moon's crust formed before about 4.4 billion years ago (**35**) and that of Mercury probably also formed at an early stage in its development (**65**). It is likely that most of the earth's crust formed at the same time as the earth's core. The infall of nickel-iron into the earth's core would have resulted in a temperature increase of about 2000° to 3000°C (4000° to 6000°F) due to the release of gravitational energy (**66**). This would have generated sufficient heat to begin melting the mantle. The melt formed would have risen to the surface because its density would have been lower than that of the mantle, and it would have formed either a succession of lava flows or a great pool of molten rock. Some scientists have suggested that the lunar crust formed from such a great molten pool and that the earth may have had a similar history. Alternatively, the earth's crust may have formed as a result of volcanic activity triggered by the impact of large meteorites (**67**). However, all remnants of the primitive crust on the earth and moon have been destroyed by meteorite impacts, metamorphism, remelting, and erosion.

## EVOLUTION OF THE EARTH'S ATMOSPHERE AND HYDROSPHERE

The earth's atmosphere is greatly depleted in hydrogen, helium, and the noble gases (e.g., argon, neon, and xenon) relative to the sun. In the earth today, the ratio of neon to silicon is only 1/10,000,000,000 of that existing in the sun (**68**). Therefore, it is evident that the earth's atmosphere was not solely derived from the gaseous part of the solar nebula. The earth's early atmosphere has often been assumed to have been composed of methane with small amounts of hydrogen, ammonia, and water (**73**). This assumption is based largely on the presumed composition of the atmosphere of Jupiter and the other giant planets. However, these planets probably accreted at a much lower temperature than the earth and therefore they cannot be used as a model for the earth's early atmosphere. If the earth accreted from planetesimals at temperatures around 600°K, it would be expected that its atmosphere postdates accretion. It is widely believed that the atmosphere and hydrosphere (oceans) formed from materials released through volcanic activity (**69**). However, the earth may have contained little water when it accreted because water condenses from the solar nebula at temperatures around 350°K (**21**). The source of the earth's water and its atmosphere may have been the asteroid belt (**70**). Carbonaceous chondrites are quite rich in water and other volatiles present in the earth's atmosphere. The asteroid belt originally contained much more material than it does today and much of this material was lost through collisions with other planets (Fig. 8.18) (**71**). Countless craters on Mercury, Moon, Mars, and Venus indicate intense bombardment by meteorites presumably derived largely from the asteroid belt (Fig. 8.18). The earth must have also been intensely bombarded at an early stage in its history. With a water content of about 10 percent in carbonaceous chondrites, as much as 0.5 percent of the earth may be composed of carbonaceous chondrites from the asteroid belt. Some of the atmosphere and hydrosphere may have formed from materials released during the impact of these meteorites (**72**).

Large quantities of water vapor would be released during the impact of large carbonaceous chondrites. Some would remain in the

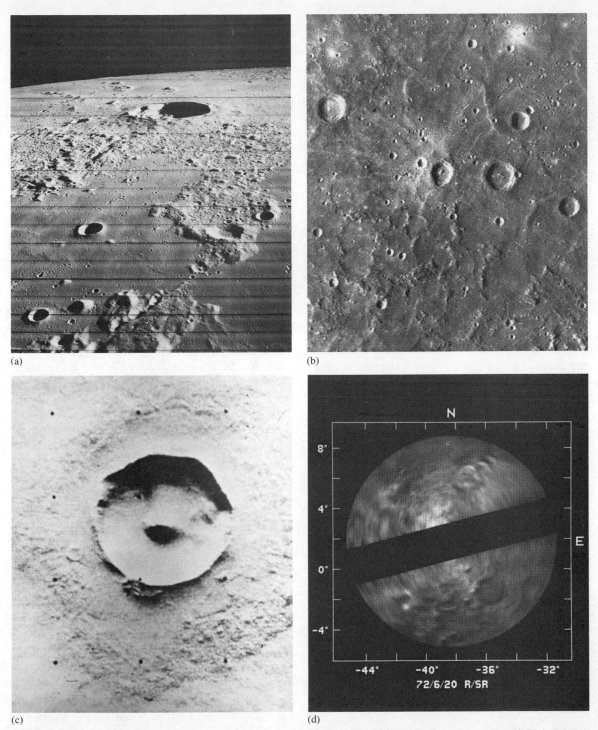

**FIGURE 8.18** Craters on: **(a)** the moon, **(b)** Mercury, **(c)** Mars, and **(d)** Venus (radar observations). [**(a)**, **(b)**, **(c)** From NASA; **(d)** courtesy of R. M. Goldstein].

atmosphere, but most would condense out to become water which would form streams, rivers, lakes, and eventually the primitive oceans. Nitrogen might also be released as a result of heating during the impact. Complex hydrocarbons within the meteorites might break down to produce simple hydrocarbons such as methane and ethane. They might also react with water to produce such gases as carbon monoxide, hydrogen, ethyl alcohol, and ammonia. Additional hydrogen and oxygen would also be produced by the hydrolysis of water due to the energy of ultraviolet radiation and electrical discharges in the atmosphere. Some ammonia and methane may have come to the earth in a frozen state from meteorites or cometary debris (comets may also have been more abundant at an early stage in earth history) or from planetesimals formed near Jupiter or Saturn.

Part of the earth's atmosphere was derived from volcanic activity. Partial melting during the formation of the crust and mantle would have released large quantities of water and other gases from the earth's interior. These gases were probably rich in carbon monoxide, hydrogen, methane, and ammonia like those released during meteorite impacts. With time, the volcanic gases became more like modern volcanic gases. In Hawaii these gases are composed principally of water (79 percent), carbon dioxide (12 percent), sulfur dioxide (6 percent), and nitrogen (1 percent).

The earth's present atmosphere is largely composed of nitrogen and oxygen. The water and sulfur dioxide of the early volcanic gases went into the developing ocean basins. The sulfur dioxide eventually formed sulfuric acid, which reacted with materials in the earth's crust and ocean bottom sediments to form sulfates. After life developed, free oxygen was released from carbon dioxide by photosynthesis.

## THE ORIGIN OF LIFE

Interest in the origin of life on earth gained momentum in the first half of the twentieth century. In 1936, the Russian biochemist A. I. Oparin proposed the theory that the early atmosphere of the earth was chemically reducing and was rich in methane and ammonia (**74**). Oparin suggested that the first step in development of living organisms was the formation of organic molecules. According to his theory, under reducing conditions the organic molecules were not destroyed by oxidation, as they would be in the modern atmosphere. These organic molecules were concentrated in the primitive ocean in a dilute "organic soup." Some of the molecules then combined chemically to form more and more complex molecules. Those molecules which had a favorable composition or favorable internal arrangement acquired new molecules more readily than others. Ultimately, organic molecules capable of reproduction were formed.

In 1953 Stanley L. Miller produced organic molecules experimentally from the components that had been suggested by Oparin as common to the earth's primitive atmosphere (**75**). Miller passed an electric spark through a closed chamber filled with a mixture of ammonia, methane, water vapor, and hydrogen. The spark produced the necessary energy to bring about the reactions. In the earth's primitive atmosphere, this energy may have been supplied by lightning or ultraviolet rays. The resulting organic molecules were amino acids, the basic components of proteins.

Philip Abelson, in 1957, added carbon dioxide, carbon monoxide, and nitrogen to Miller's original mixture and was able to produce all the amino acids commonly found in living matter, plus some proteins. Not only have organic molecules (those molecules containing carbon) been found in interstellar space (as discussed earlier in this chapter), but they have also been found in carbonaceous chondrites (**76**). This indicates that the inorganic synthesis of organic molecules is a common process. Furthermore, the organic molecules in meteorite fragments may have reacted directly with water in the formation of more complex organic molecules

(**77**) and ultimately self-replicating molecules. It should be remembered, however, that even the simplest forms of life are far more complex than the proteins that have been synthesized thus far, and scientists are still a long way from creating living molecules in the laboratory.

Although scientists disagree about where and how molecules capable of reproduction were formed, most theorists consider an aqueous environment to be most probable. Since fluids of plants and animals are chemically close to the composition of seawater, a chemically similar body of water would appear to be a likely environment (**78**).

In a warm lagoonal environment, clay minerals and quartz particles could have concentrated organic molecules through the process of adsorption. However, P. K. Weyl has theorized that organic molecules might have concentrated at a depth of several hundred feet in the open ocean at the top of a thermocline (**79**). This would have allowed life to develop and evolve within a zone protected from harmful ultraviolet radiation. When this all might have happened has not yet been revealed in the geologic record. However, the oldest rocks containing probable fossils are approximately 3.4 billion years old (see Chapter 9). Therefore, the oceans, atmosphere, and life all had to develop in the first billion years of the earth's existence.

## SUMMARY

According to the prevailing theory, the universe originated between 9 and 18 billion years ago in a gigantic explosion. As the matter in the universe moved away from the center of expansion, some of it coalesced into galaxies. Within the galaxies there were secondary centers of contraction, and these become stars. The first stars were probably composed almost entirely of hydrogen and helium. During this early stage in the history of our galaxy (8–9 billion years ago), there were probably large numbers of very massive, rapidly evolving stars in which the heavier elements were generated.

At this time, numerous supernovae would have dispersed heavy elements into the interstellar medium as gas and dust.

About 5 billion years ago, a cloud of gas and dust which was to become our solar system began to contract. Most of the material of the cloud went into its center to form the sun. The sun was initially heated as a result of the release of gravitational energy, but when the core heated sufficiently, hydrogen began to fuse to form helium. Energy from the sun vaporized the solar nebula, which then condensed and accreted into planets.

Heating of the earth resulted in the melting of nickel-iron which sank to form the earth's core. The earth's crust was probably produced by partial melting of the mantle. As the magma rose to the earth's surface, widespread volcanic activity occurred. Gases released during the eruption of the lavas and during the impact of large meteorites formed the earth's primitive atmosphere and hydrosphere. Ultimately, the action of lightning, ultraviolet radiation, or both in the presence of the carbon, oxygen, hydrogen, and nitrogen compounds produced organic molecules of increasing complexity, and these were eventually joined into molecules capable of reproducing themselves. Life as we know it evolved in the intervening eras from such self-reproducing molecules.

## REFERENCES CITED

1. G. Gamow, 1949, *Rev. Mod. Phys.,* v. 21, p. 367.
2. G. Abell, 1969, *Exploration of the Universe:* Holt, Rinehart and Winston, New York.
3. R. B. Partridge, 1969, *Amer. Sci.,* v. 57, p. 37.
4. R. Wagoner, W. A. Fowler, and F. Hoyle, 1967, *Astrophys. J.,* v. 148, p. 3.
5. C. Hayashi, 1961, *Publ. Astron. Soc. Japan,* v. 13, p. 450.
6. W. A. Fowler, 1964, *Chem. Eng. News,* v. 42, no. 1, p. 90.
7. G. Fritz *et al.,* 1969, *Science,* v. 164, p. 709.
8. G. Burbidge, 1967, *Amer. Sci.,* v. 55, p. 282.
9. W. D. Metz, 1973, *Science,* v. 181, p. 1154.
10. J. H. Oort, 1970, *Science,* v. 170, p. 1363; R. J. Weymann, 1969, *Sci. Amer.,* v. 220, no. 18, p. 28.

11. S. A. Colgate, 1969, *Phys. Today,* v. 22, p. 27.
12. A. O. J. Unsold, 1969, *Science,* v. 163, p. 1015.
13. C. M. Hohenberg, 1969, *Science,* v. 166, p. 212.
14. W. Ward, 1975, *Rev. Geophys. Space Phys.,* v. 13, p. 422.
15. E. Stephen-Sherwood and J. Oro, 1973, *Space Life Sciences,* v. 4, p. 5.
16. F. Hoyle, and N. C. Wickramasinghe, 1970, *Nature,* v. 226, p. 62.
17. D. R. Huffmann, 1973, *in* J. M. Greenberg and H. C. van de Hulst, eds., *Interstellar Dust and Related Topics:* D. Reidel, Dordrecht.
18. A. G. W. Cameron, 1976, *Icarus,* v. 3, p. 447.
19. J. A. Wood, 1968, *Meteorites and the Origin of Planets:* McGraw-Hill, New York.
20. A. G. W. Cameron, 1975, *in* G. B. Rield and A. G. W. Cameron, eds., *The Dusty Universe:* Neale Watson Academic Publications, New York, p. 1.
21. L. Grossman and J. W. Larimer, 1974, *Rev. Geophys. Space Phys.,* v. 12, p. 71; L. Grossman, 1975, *Sci. Amer.,* v. 232, no. 2, p. 30.
22. C. R. Chapman, 1976, *Geochim. Cosmochim. Acta,* v. 40, p. 701.
23. E. L. Krinov, 1960, *Principles of Meteoritics:* Pergamon Press, New York.
24. D. L. Anderson, 1974, *Phys. Today,* March, p. 44.
25. V. R. Murthy and H. T. Hall, 1970, *Phys. Earth Planet. Interiors,* v. 2, p. 276; T. M. Usselman, 1975, *Amer. J. Sci.,* v. 275, p. 278; T. M. Usselman, 1975, *Amer. J. Sci.,* v. 275, p. 291.
26. A. E. Ringwood and S. E. Kesson, 1976, *Basaltic Magmatism and the Composition of the Moon, Part II:* Research School of Earth Sciences, Australia National Univ., Canberra; E. Anders and T. Owen, 1977, *Science,* v. 190, p. 453.
27. W. M. Kaula, 1976, *Icarus,* v. 28, p. 429.
28. E. Anders, R. Hayatsu, and M. H. Studier, 1973, *Science,* v. 182, p. 781.
29. P. Goldreich and W. R. Ward, 1973, *Astrophys. J.,* v. 183, p. 1051.
30. W. K. Hartmann, 1972, *Moons and Planets:* Wadsworth Publishing, Belmont, Ca.
31. W. Compston *et al.,* 1970, *Science,* v. 167, p. 474; S. K. Kanushal and G. W. Wetherill, 1970, *J. Geophys. Res.,* v. 75, p. 463; L. T. Silver, 1970, *Science,* v. 167, p. 468.
32. V. M. Oversby and P. W. Gast, 1968, *Science,* v. 162, p. 925.
33. D. L. Lamar and P. M. Merifield, 1967, *Bull. Geol. Soc. Amer.,* v. 78, p. 1359.
34. Y. Nakamura *et al.,* 1974, *Geophys. Res. Lett.,* v. 1, p. 137.
35. S. R. Taylor, 1975, *Lunar Science: A Post-Apollo View:* Pergamon Press, New York.
36. H. T. Hall and V. R. Murthy, 1971, *Earth Planet. Sci. Lett.,* v. 11, p. 239.
37. R. Ganapathy and E. Anders, 1974, *Proc. Lunar Sci. Conf.,* 5th, p. 1181; W. M. Kaula and A. W. Harris, 1975, *Rev. Geophys. Space Phys.,* v. 13, p. 363.
38. M. N. Toksoz, 1975, *Rev. Geophys. Space Phys.,* v. 13, p. 306.
39. D. U. Wise, 1969, *J. Geophys. Res.,* v. 74, p. 6034.
40. S. P. Clark, Jr., *et al.,* 1975, *Nature,* v. 258, p. 219; W. K. Hartman and D. R. Davis, 1975, *Icarus,* v. 24, p. 504; H. E. Miller, 1975, *Icarus,* v. 24, p. 256; J. A. O'Keefe, 1970, *J. Geophys. Res.,* v. 75, p. 6565; A. E. Ringwood, 1970, *Earth Planet. Sci. Lett.,* v. 8, p. 131; J. A. Wood, 1975, *Proc. IAU Colloq.,* v. 28.
41. J. A. O'Keefe, 1969, *J. Geophys. Res.,* v. 74, p. 2758.
42. D. L. Anderson, 1972, *Nature,* v. 239, p. 263.
43. D. L. Anderson, 1975, *Geotimes,* v. 20, no. 5, p. 26.
44. J. A. Wood, 1963, *in* B. M. Middlehurst and G. Kuiper eds., *The Moon, Meteorites, and Comets:* The University of Chicago Press, Chicago, p. 337.
45. J. M. Herndon *et al.,* 1976, *Earth Planet. Sci. Lett.,* v. 29, p. 283.
46. J. A. Cutts, 1973, *J. Geophys. Res.,* v. 78, p. 4231; R. L. Huguenin, 1976, *Icarus,* v. 28, p. 203.
47. R. N. Clayton, 1973, *Science,* v. 182, p. 485.
48. B. C. Murray, 1975, *J. Geophys. Res.,* v. 80, p. 2342.
49. A. G. W. Cameron, 1973, *The Moon,* v. 7, p. 377; A. G. W. Cameron, 1972, *Nature,* v. 240, p. 299.
50. C. K. Seyfert, 1977, *Trans. Am. Geophys. Union,* v. 58, p. 989.
51. T. McCord and J. Adams, 1972, *Science,* v. 178, 745.
52. M. J. S. Belton *et al.,* 1974, *Science,* v. 185, p. 169; B. Hapke *et al.,* 1975, *J. Geophys. Res.,* v. 80, p. 2431.
53. R. W. Siegfried II and S. C. Solomon, 1974, *Icarus,* v. 23, p. 192.

**54.** A. Young and L. Young, 1975, *Sci. Amer.*, v. 233, no. 3, p. 70.

**55.** S. F. Singer, 1970, *Science,* v. 170, p. 1196.

**56.** A. Young and L. Young, 1975, *Sci. Amer.*, v. 233, p. 70.

**57.** H. Alfven and G. Arrhenius, 1976, *Evolution of the Solar System:* National Aeronautics and Space Administration, Washington, D.C.

**58.** H. Gerstenkorn, 1955, *Z. Astrophys.*, v. 36, p. 245.

**59.** N. Herz, 1969, *Science,* v. 164, p. 944.

**60.** S. K. Saxena, 1977, *Science,* v. 198, p. 614.

**61.** L. T. Silver *et al.,* 1977, *Geol. Soc. Amer. Abstr. Programs* v. 9, p. 1176.

**62.** P. M. Merifield and D. L. Lamar, 1970, *in* S. K. Runcorn, ed., *Proceedings of NATO Conference on Paleogeophysics:* Academic Press, New York.

**63.** W. A. Fowler, *et al.,* 1962, *Geophys. J. Roy. Astron. Soc.,* v. 6, p. 148; D. Heymenn and M. Dziczkaniec, 1976, *Science,* v. 191, p. 79; G. J. Wasserburg, *et al.,* 1977, *Geophys. Res. Lett.,* v. 4, p. 299.

**64.** H. W. Bergh, 1970, *in* S. K. Runcorn, ed., *Paleogeophysics,* Academic Press, New York, p. 143.

**65.** B. C. Murray *et al.,* 1975, *J. Geophys. Res.,* v. 80, p. 2508.

**66.** G. C. Kennedy, 1969, *in The Primitive Earth: A Symposium:* Miami University Department of Geology, Oxford, Ohio, p. K-1.

**67.** A. M. Goodwin, 1974, *Amer. J. Sci.,* v. 274, p. 987; P. D. Lowman, Jr., 1976, *J. Geol.,* v. 84, p. 1; J. W. Salisbury and L. B. Ronce, 1966, *Nature,* v. 210, p. 669.

**68.** H. Brown, 1952, *in* G. P. Kuiper, ed., *The Atmospheres of the Earth and Planets:* University of Chicago Press, p. 258.

**69.** W. W. Rubey, 1955, *in* A. Podervarrt, ed., *Crust of the Earth:* Geological Society of America, New York, p. 631.

**70.** E. Anders and T. Owen, 1977, *Science,* v. 198, p. 453.

**71.** C. R. Chapman and D. R. Davis, 1975, *Science,* v. 190, p. 553.

**72.** F. P. Fanale, 1976, *IEEE Trans. on Geos. Electronics,* v. GE-14, p. 183.

**73.** H. D. Holland, 1962, *in* A. E. Engel *et al.,* eds., *Petrologic Studies: A Volume to Honor A. F. Buddington:* Geological Society of America.

**74.** A. I. Oparin, 1962, *Life: Its Nature, Origin, and Development:* Academic Press, New York.

**75.** S. L. Miller, 1953, *Science,* v. 117, p. 528.

**76.** G. Jungclaus *et al.,* 1976, *Nature,* v. 261, p. 126.

**77.** H. E. Seuss, 1975, *Origin of Life,* v. 6, p. 9.

**78.** A. Banin and J. Navrot, 1975, *Science,* v. 189, p. 550.

**79.** P. K. Weyl, 1968, *in* R. Siever, ed., *Science,* v. 161, p. 712.

# The Precambrian 9

T HE PRECAMBRIAN includes all of geologic time prior to the beginning of the Cambrian period and thus covers approximately 87 percent of geologic time. Rocks of this age are often metamorphosed to gneisses and schists, but in some areas they have not been metamorphosed. Precambrian rocks underlie stable areas of low relief called *shields*. Shields are the nuclei of the continents around which the younger rock systems have developed (Fig. 9.1). In the older portions of these areas, the roots of very ancient mountains are exposed. Shields generally show evidence of several episodes of sedimentation, mountain building, and erosion. Rocks of Precambrian age are also found in the cores of young mountain ranges.

## SUBDIVIDING THE PRECAMBRIAN

Except for algal stromatolites, very few fossils have been found in deposits of Precambrian age. For this reason, the degree of metamorphism served as a basis for subdividing the Precambrian before the widespread use of radiometric dating. Intensely metamorphosed rocks were thought to be the oldest units and were assigned to the Archean (Archeozoic) era, whereas rocks of lower metamorphic grade were assigned to the Proterozoic era. In some instances, however, radiometric dating has demonstrated that slightly metamorphosed

rocks are actually older than intensely metamorphosed rocks in nearby areas. Thus the intensity of metamorphism has proven to be an unsatisfactory method of subdividing the Precambrian. At the present time, reconstruction of Precambrian geologic history relies heavily on radiometric dating.

Potassium-argon, rubidium-strontium, and uranium-lead dating of Precambrian basement rocks has shown that several major *orogenic episodes* (periods of mountain building, folding, faulting, metamorphism, uplift, and intrusion) occurred simultaneously on all continents (1). The frequency distribution of radiometric dates for Precambrian rocks indicates that there were peaks of metamorphism and intrusion at 1.0, 1.7, and 2.5 billion years ago (Fig. 9.2). Using these dates as boundaries, the Precambrian may be subdivided into four eras from oldest to youngest: the Archean, the Early Proterozoic, the Middle Proterozoic, and the Late Proterozoic. Other classification systems for the Precambrian are also in use (Fig. 9.3).

## PRECAMBRIAN GEOGRAPHY

Some geologists believe that the continents formed soon after the formation of the earth, and that only a relatively small fraction of the earth's crust was added at a later date (3). Others believe that a large part of the crust was added slowly throughout geological time (4, 5).

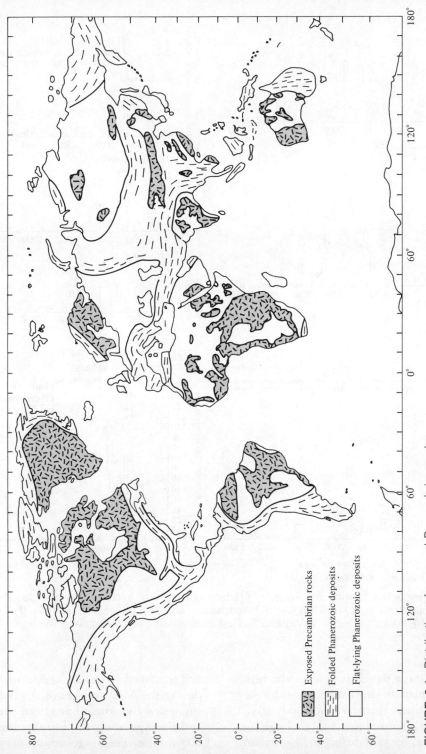

**FIGURE 9.1** Distribution of exposed Precambrian rocks.

Exposed Precambrian rocks

Folded Phanerozoic deposits

Flat-lying Phanerozoic deposits

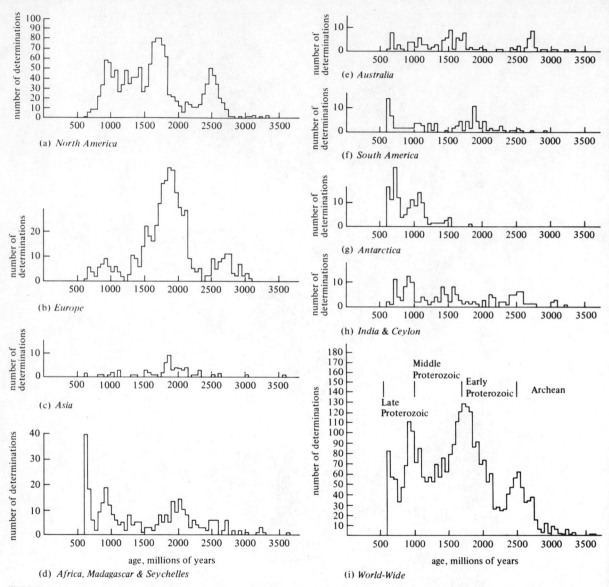

**FIGURE 9.2** Frequency distribution of ages of Precambrian plutonic and metamorphic basement rocks: **(a)** North America. **(b)** Europe. **(c)** Asia. **(d)** Africa, Madagascar, and Seychelles. **(e)** Australia. **(f)** South America. **(g)** Antarctica. **(h)** India and Ceylon. **(i)** Worldwide. Notice the smaller vertical scale for the North American and worldwide data.

Study of the crust development on the moon and on other planets lends support to the view that most continental material formed early.

It has been suggested that a single large continent (Pangaea) formed during the Proterozoic and remained as a single unit until it broke apart during the Mesozoic era (**6, 7**). It has also been suggested that there have been some openings and closings of oceans, but that this began only 600 million years ago at the very end of the

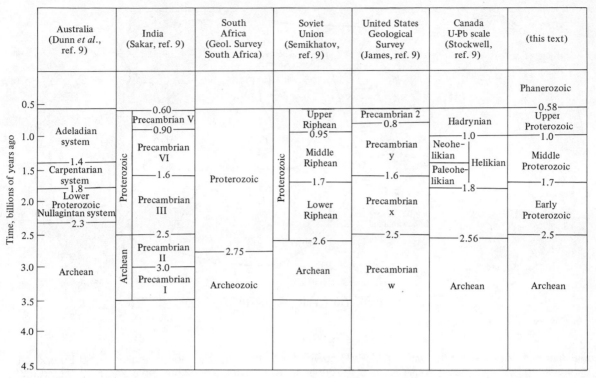

**FIGURE 9.3** Comparison of some of the proposed Precambrian classification systems.

Precambrian (**8**). However, paleomagnetic, structural, lithological, and stratigraphic data indicate that continents moved relative to each other throughout the Proterozoic, and probably during the Archean as well (**9–13**).

In order to determine the relative positions of the continents during the Precambrian, apparent polar wander (A.P.W.) curves for North America, Gondwanaland, Europe, and Asia were recalculated for a Pangaea reconstruction of the continents. This was done by moving these continents along with their A.P.W. paths to the positions which they occupied prior to the beginning of separation of the continents during the Mesozoic (Fig. 9.4). The recalculated A.P.W. based on samples 1.0 to 0.6 billion years old from North America, Gondwanaland, Europe, and Asia do not agree with each other (Fig. 9.4a and 9.4d). Therefore, Pangaea probably did not exist during that interval (the Late

Proterozoic). Paleomagnetic data indicate that there were five separate continents during that time. These continents are Ancestral North America, Ancestral Europe, Gondwanaland, Ancestral Siberia, and Ancestral China (Fig. 9.5). Ancestral North America includes most of the present continent of North America except for parts of the southeastern United States and Mexico, and it also includes northwestern Britain, western Norway, Spitsbergen, and northeastern Siberia. Most of central and northern Europe are included in Ancestral Europe. Ancestral Siberia is comprised of most of Siberia, northernmost China, and Outer Mongolia. Ancestral China includes most of China, all of Japan, North Korea, South Korea, Southeast Asia, and portions of Siberia, India, Pakistan, and the Middle East. Gondwanaland consists of Africa, South America, Antarctica, Australia, most of the Middle East and India,

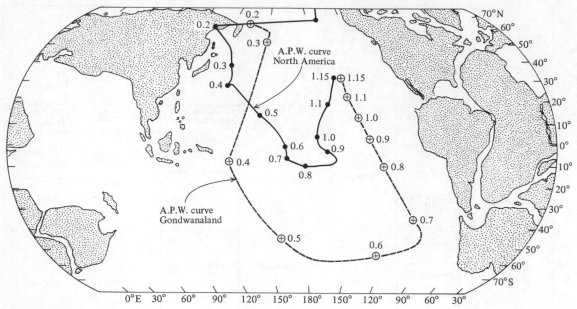

⊕ Gondwanaland   • North America

(a)

FIGURE 9.4   Apparent polar wander (A.P.W.) curves for North America, Gondwanaland, and Europe are recalculated for a Pangaea reconstruction of the continents by moving these continents along with their A.P.W. curves to the positions which they occupied prior to the beginning of separation of the continents during the Mesozoic: **(a)** Comparison of the recalculated A.P.W. curves based on samples 1.15 to 0.2 billion years old from North America and Gondwanaland indicates that these continents were together (or at least were not widely separated) 1.15 billion years ago, separated soon thereafter, and were rejoined about 400 million years ago. **(b)** Comparison of the recalculated A.P.W. curves based on samples 1.7 to 1.15 billion years ago from North America and Gondwanaland indicates that these continents were joined together (or at least were not widely separated) 1.7 billion years ago, separated soon thereafter, and were rejoined about 1.2 billion years ago. **(c)** Comparison of recalculated A.P.W. curves based on samples 2.75 to 1.7 billion years old from North America and Gondwanaland indicates that these continents were joined together (or at least were not widely separated) between 2.75 and 2.4 billion years ago, then separated from each other, were rejoined about 1.85 billion years ago, and remained together (or at least were not widely separated) between 1.85 and 1.7 billion years ago. **(d)** Comparison of recalculated A.P.W. curves based on samples 1.15 to 0.1 billion years old from North America and Europe indicates that these continents were joined together (or at least were not widely separated) between 1.15 and 0.9 billion years ago, then separated from each other, and were rejoined about 600 million years ago. **(e)** Comparison of recalculated A.P.W. curves based on samples 1.7 to 1.15 billion years old from North America and Europe indicates that these two continents were joined together (or at least were not widely separated) 1.7 billion years ago, separated soon thereafter, and then were rejoined between 1.2 and 1.15 billion years ago.

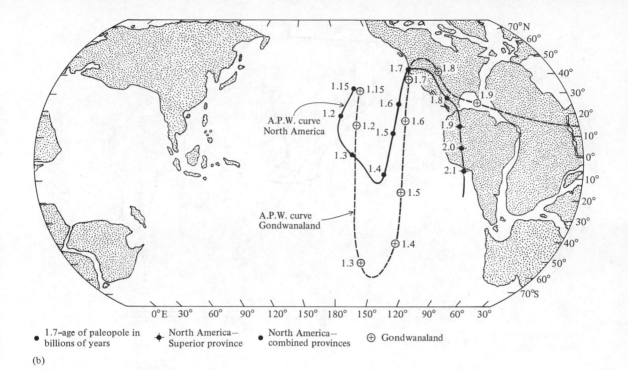

**(b)**

- 1.7–age of paleopole in billions of years
- North America—Superior province
- North America—combined provinces
- ⊕ Gondwanaland

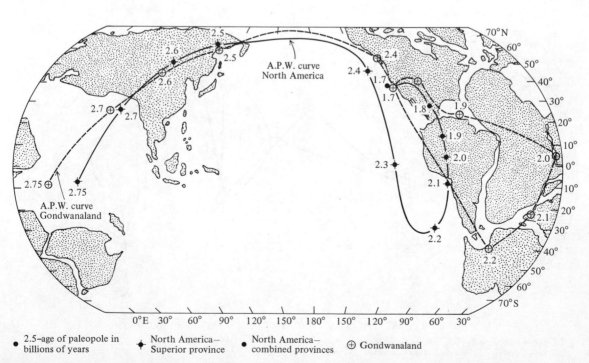

**(c)**

- 2.5–age of paleopole in billions of years
- North America—Superior province
- North America—combined provinces
- ⊕ Gondwanaland

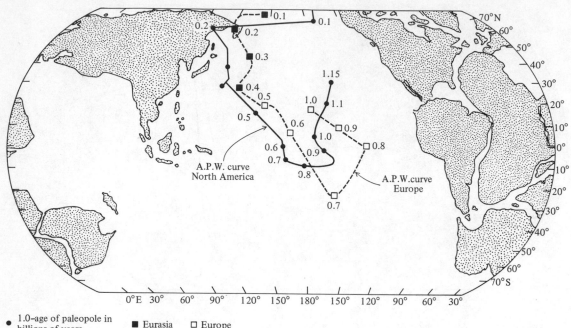

(d)

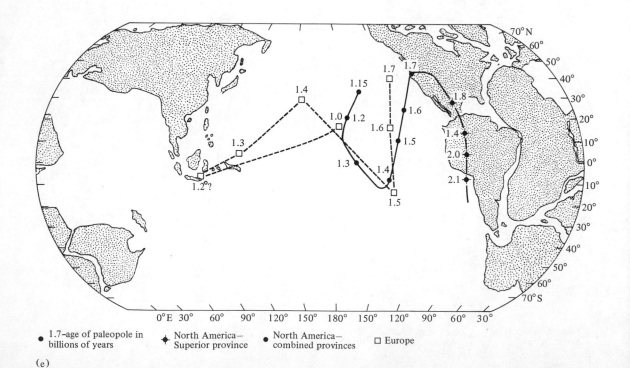

(e)

(a)

**FIGURE 9.5** Ancestral continents: **(a)** Ancestral North America. **(b)** Ancestral Europe. **(c)** Ancestral Siberia. **(d)** Ancestral China. **(e)** Gondwanaland.

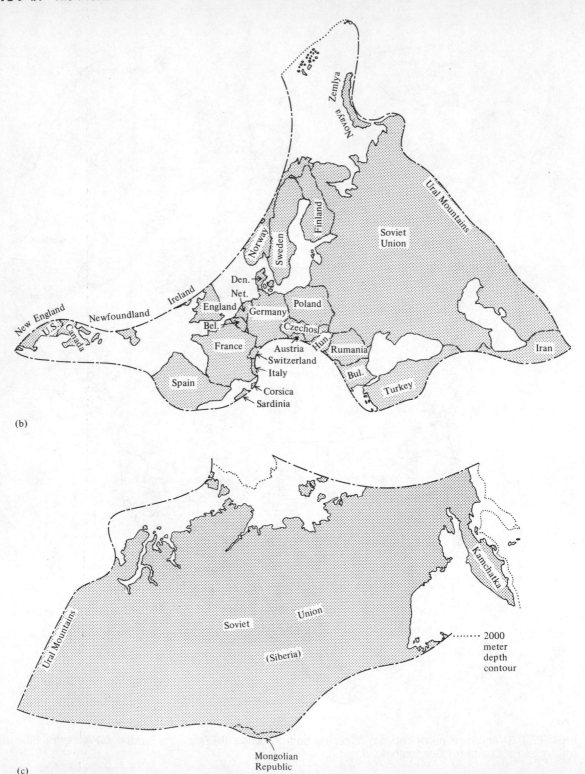

(b)

(c)

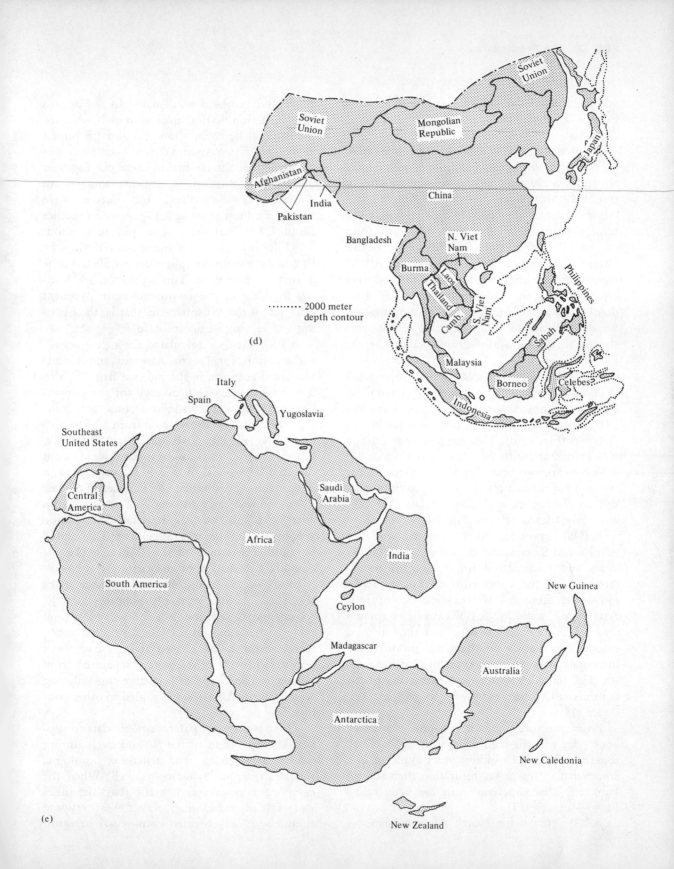

Soviet
Union

Mongolian
Republic

Soviet
Union

Japan

Afghanistan

India

China

Pakistan

Bangladesh

N. Viet
Nam

Burma

Laos

Thailand

S. Viet
Nam

Philippines

Camb.

Sabah

Malaysia

Borneo

Celebes?

Indonesia

········· 2000 meter
depth contour

(d)

Italy

Spain

Yugoslavia

Southeast
United States

Saudi
Arabia

Central
America

Africa

India

South America

New Guinea

Ceylon

Madagascar

Australia

Antarctica

New Caledonia

New Zealand

(e)

and segments of southern Europe, the south-eastern United States, Mexico, and Central America.

Recalculated A.P.W. curves based on samples 1.2 to 1.15 billion years old from North America and Gondwanaland coincide rather closely (Fig. 9.4a). It is likely that North America and Gondwanaland were then in approximately the same positions relative to each other as during the late Paleozoic and early Mesozoic before Pangaea separated (Fig. 9.6b). The A.P.W. curves diverge after 1.15 billion years and then rejoin during the Paleozoic (Fig. 9.4a). Evidently, North America and Gondwanaland were together 1.15 billion years ago, split apart shortly thereafter, and then rejoined during the Paleozoic.

The paleomagnetic pole from Europe dated at about 1.0 billion years and recalculated for a Pangaea reconstruction agrees well with the 1.05 billion years paleomagnetic pole for North America (Fig. 9.4d). This suggests that Europe was in contact with North America about 1.0 billion years ago and that the two continents were in approximately the same positions relative to each other as during the late Paleozoic and early Mesozoic eras (Fig. 9.6b).

A.P.W. curves for North America and Asia (China and Siberia) dated at about 1.15 billion years and recalculated for a Pangaea reconstruction of the continents do not coincide. However, these A.P.W. curves do coincide if Asia along with its A.P.W. curve is rotated about 90° counterclockwise and then moved eastward so that the southern margin of Asia is in contact with the west coast of North America (Fig. 9.6b). A similar reconstruction of the continents has been proposed on geologic evidence (**14**).

From the foregoing discussions, it may be concluded that the continents were probably all together about 1.15 billion years ago, but in a somewhat different configuration than that of Pangaea. This supercontinent has been called Proto-Pangaea (**11**).

A.P.W. curves for North America, Europe,

and Gondwanaland recalculated for a Pangaea reconstruction do not agree with each other for the interval between about 1.3 and 1.5 billion years ago (Fig. 9.4b and 9.4e). Therefore, it is probable that the continents were not together during that interval. It appears that North America, Gondwanaland, and Europe were apart 1.5 billion years ago, then joined together about 1.2 billion years ago, split apart about 1.15 billion years ago, rejoined again during the Paleozoic, and finally split apart for the last time during the Mesozoic. This separation and joining together of the continents is an excellent example of the Wilson cycle, that is, the opening and closing of ocean basins.

A.P.W. curves recalculated for a Pangaea reconstruction for North America and Gondwanaland coincide in the interval from 2.75 to 2.40 billion years ago, diverge for the interval between 1.90 and 2.35 billion years, and coincide rather well in the interval from 1.85 to 1.70 million years (Fig. 9.4b and 9.4c). Evidently, North America and Gondwanaland were joined together between 2.75 to 2.40 billion years ago in an arrangement much like Pangaea in the late Paleozoic and early Mesozoic. They then split apart and subsequently rejoined 1.85 billion years ago in a similar arrangement. This would be another example of the Wilson cycle, the opening and closing of an ocean.

Study of the composition of volcanic rocks about 2.75 billion years old indicates that the continents went through a short-lived Wilson cycle at that time. In the last 2.8 billion years, there appear to have been at least five times during which North America separated from other continents and at least five times during which North America was joined to other continents.

There is not only paleomagnetic data to support the operation of the Wilson cycle during the Precambrian era, but structural, lithologic, and stratigraphic evidence as well. When the A.P.W. curves diverged, mafic (basaltic) dikes were intruded and mafic lavas were erupted along the newly formed continental margins.

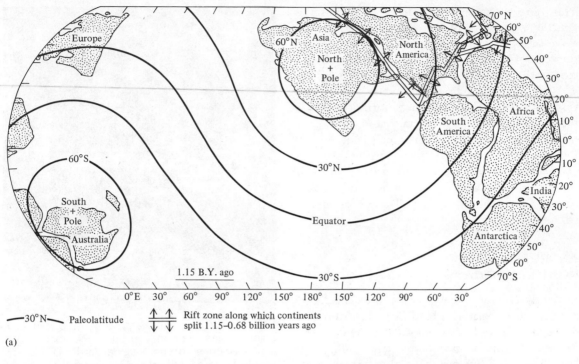

(a)

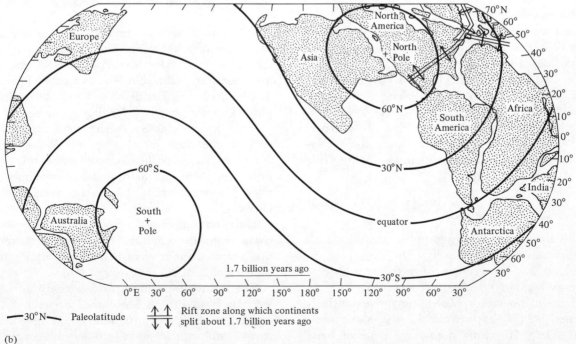

(b)

**FIGURE 9.6** Proposed reconstruction of the continents for: **(a)** 1.15 billion years ago, **(b)** 1.7 billion years ago.

This igneous activity is indicative of crustal tension (stretching of the earth's crust). The times of mountain building that occurred along the zone of junction of the two continents coincide with the dates where A.P.W. curves join. This mountain building is generally accompanied by folding, faulting, metamorphism, and intrusion affecting sedimentary and volcanic deposits on continental margins. Moreover, geoclines are formed on the margins of the continents following their separation. During the formation of a geocline, sedimentation may change from shallow water to deep water environments.

## ANCESTRAL NORTH AMERICA

Precambrian rocks are exposed in the Canadian Shield, the Appalachian Fold Belt, the Adirondacks, the Rocky Mountains of both the United States and Canada, and in Greenland and Scotland. Rocks of this age have also been penetrated in oil wells in the central United States and western Canada where they are covered by a relatively thin sequence of Phanerozoic sedimentary rocks.

The Canadian Shield, which covers most of eastern and central Canada, is one of the largest and most extensively studied areas of Precambrian rocks in the world (Fig. 10.4). Where Pleistocene glaciation has removed the overlying soil, the rocks of the shield are exceptionally well exposed (Fig. 9.7).

### Geological Provinces

Potassium-argon dating and structural studies indicate that the Canadian Shield may be divided into a number of distinct geological provinces in which the rocks were last deformed, metamorphosed, and/or extensively intruded at approximately the same time (Fig. 9.8) (**15**). Potassium-argon dates on basement rocks in the United States and western Canada show that the rest of Ancestral North

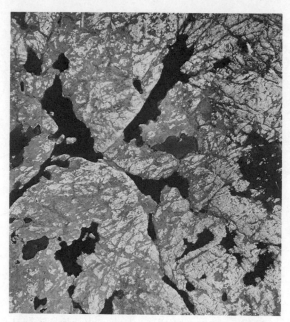

**FIGURE 9.7** Precambrian rocks of the Canadian Shield exposed by the action of Pleistocene glaciers. [Courtesy of the National Air Photo Library, Dept. of Energy, Mines and Resources (Canada).]

America may also be divided into geologic provinces (Fig. 9.9) (**5**). The oldest rocks in Ancestral North America occur in the Superior, Slave, eastern Nain, and Wyoming provinces. These rocks were last deformed and metamorphosed approximately 2.7 billion years ago.

The Churchill and Southern provinces nearly encircle the Superior, Slave, and Wyoming provinces. Rocks of the Churchill and Southern provinces were last deformed and metamorphosed approximately 1.7 billion years ago. However, whole-rock rubidium-strontium isochron and uranium-lead concordia dating of rocks from the Churchill Province indicates that some of them were first metamorphosed more than 2.5 billion years ago (**16**).

The last widespread intrusive activity in the western Nain Province occurred between 1.4 and 1.5 billion years ago. Widespread igneous rocks with similar potassium-argon ages are also found in the Central Province (**16**) (Figs. 9.8 and 9.9).

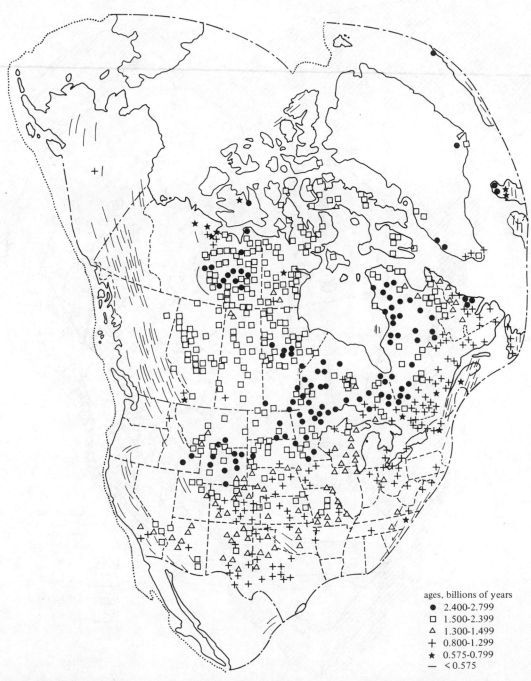

**FIGURE 9.8** Radiometric dates on basement rocks from Ancestral North America.

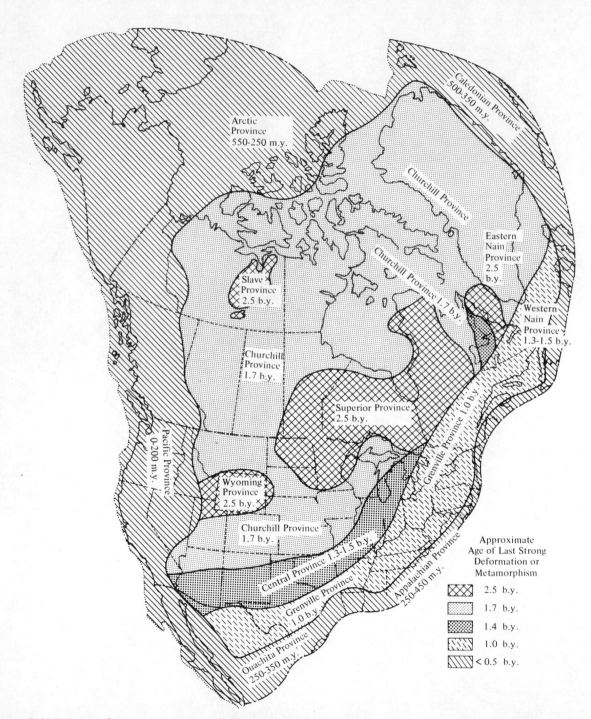

**FIGURE 9.9**   Geologic provinces of Ancestral North America.

The Grenville Province is the youngest Precambrian province in Ancestral North America. It cuts across the eastern borders of the Central, Churchill, Superior, and Nain provinces. Rocks of the Grenville Province were last deformed and metamorphosed approximately 1.2 billion years ago. Whole-rock rubidium-strontium and uranium-lead dates of samples from the Grenville Province in southeastern Ontario indicate that the first metamorphic episode in this province occurred at least 2.4 billion years ago, and a second orogenic event occurred 1.8 billion years ago (**17**). Iron formation in Labrador which was first folded about 1.8 billion years ago has been traced into the Grenville Province, where it has been subjected to at least one additional deformation (**18**). Thus parts of older provinces may have been incorporated into younger provinces during subsequent deformation.

## The Archean Era (More Than 2.5 Billion Years Ago)

Archean rocks occur in the Rocky Mountains, in the Superior, Slave, and eastern Nain provinces, in southwestern Greenland and in northern Scotland. These rocks include an older, highly metamorphosed sequence and a younger, less metamorphosed sequence. The chemical compositions of the rocks and their primary structures indicate that they were sandstones, shales, iron formation, and volcanic rocks before metamorphism. Radiometric dating of Archean rocks indicates that major orogenic episodes (times of mountain building) occurred at about 3.6, 3.0, and 2.7 billion years ago. Some, or perhaps all, of these episodes may be times of plate collisions which alternate with periods of continental rifting.

3.83 TO 3.6 BILLION YEARS.   The oldest known rocks belong to the Isua sequence in the Godthaabsfjord region of southwestern Greenland (Fig. 9.10). These rocks consist of 3000 m (10,000 ft) of metamorphosed volcanic

**FIGURE 9.10**   Isua supracrustal rocks from the Gothaabsfjord region of southwestern Greenland. (Photo: Geologic Survey of Greenland. Copyright).

and sedimentary rocks including iron formation and rhyolite-bearing conglomerates (**6**). The sedimentary rocks in this sequence contain zircons that were eroded from an older terrain. These zircons have been dated by the uranium-lead concordia method at about 3.83 billion years (**19**). This is a maximum age for the deposition of the Isua sequence because it dates the time of formation of the rocks from which the Isua sediments were derived.

The Isua sequence is intruded by granitic igneous rocks that have been deformed and metamorphosed to a gneiss (the Amîtsoq Gneiss). Uranium-lead concordia and whole-rock rubidium strontium isochron dating of these gneisses gives ages of about 3.59 to 3.70 billion years (**19**). This is approximately the time of intrusion of the granitic rocks and de-

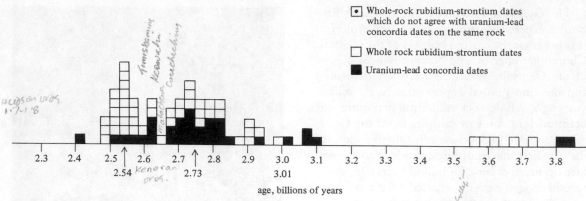

FIGURE 9.11  Frequency distribution of uranium-lead concordia and whole-rock rubidium-strontium age dates for North America and Greenland for the interval between 3.9 to 2.3 billion years ago.

formation and metamorphism of the Isua sequence rocks. This deformation, metamorphism, and intrusion is the oldest known orogenic activity, and may have been caused by collisions between plates.

Another area which contains very ancient rocks is the Minnesota River Valley in southwestern Minnesota. Here metamorphosed sedimentary and volcanic rocks have a whole-rock rubidium-strontium date of 3.8 billion years (**20**). Uranium-lead and whole-rock rubidium-strontium isochron dating of granitic rocks which cut the metamorphic rocks indicates that the granites were intruded between 3.55 and 3.65 billion years ago (**21**). The Hebron Gneiss and the Uivak Gneiss of Labrador (eastern Canada) have been dated at 3.6 billion years by these same methods (**22**). The similarity of radiometric dates from Greenland, Minnesota, and Labrador suggests that a major intrusive and metamorphic event occurred approximately 3.6 billion years ago.

3.60 TO 2.85 BILLION YEARS.  The period between 3.60 and 3.04 billion years ago was relatively quiet from the standpoint of orogenic activity. There are no known intrusive or metamorphic events in Ancestral North America during that interval (Fig. 9.11). This was a time of deposition of sedimentary and volcanic

rocks that were subsequently metamorphosed to gneisses and amphibolites.

In several areas granitic and metamorphic rocks have been dated at between 3.04 and 2.85 billion years by the uranium-lead concordia and whole-rock rubidium-strontium isochron methods. These include southwestern Ontario, southeastern Manitoba, southwestern Wyoming, northeastern Wyoming, northwestern Wyoming in the Grand Tetons, central Wyoming, District of Mackenzie in northwestern Canada, and Godthaabsfjord, Greenland (Fig. 9.12) (**19**, **23–30**). Radiometric dates in this range are common elsewhere in the world, and it may be that this was a time of worldwide collisions of plates. Because 3.5 to 3.6 billion years ago also appears to have been a time of plate collisions, it is possible that continents rifted apart between about 3.5 and 3.1 billion years ago prior to their collision. This may be the first Wilson cycle.

2.85 TO 2.73 BILLION YEARS.  In many parts of the world, including Ancestral North America, there are thick sequences of slightly metamorphosed Archean rocks consisting of greenstones and metasedimentary rocks. These occur in elongated, discontinuous zones termed *greenstone belts* (Fig. 9.13). A part of one of these belts in the Rainy Lake region of northern

(a)

(b)

FIGURE 9.12 Precambrian rocks which were metamorphosed about 3.0 billion years ago: **(a)** Precambrian rocks exposed in the Grand Tetons National Park, Wyoming. **(b)** Paragneiss in the Superior Province near Nipigon, Ontario. Sedimentary parent is indicated by the strongly layered character of the gneiss.

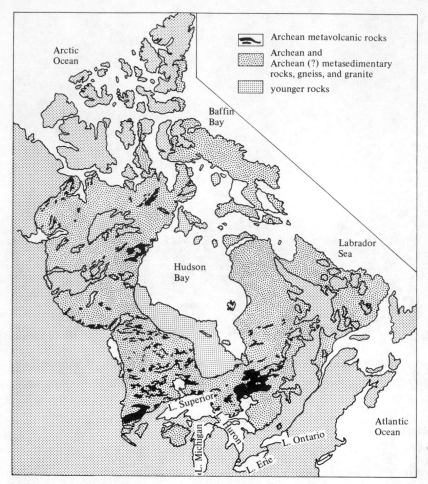

Archean metavolcanic rocks

Archean and
Archean (?) metasedimentary
rocks, gneiss, and granite

younger rocks

Arctic
Ocean

Baffin
Bay

Labrador
Sea

Hudson
Bay

L. Superior

L. Michigan

Huron

L. Ontario

L. Erie

Atlantic
Ocean

**FIGURE 9.13** Geologic map showing the distribution of metavolcanic rocks in Archean sequences of the Canadian Shield. (From *The Geological Map of Canada*, Geological Survey of Canada, Ottawa.)

Minnesota and southwestern Ontario was studied by A. C. Lawson around the turn of the century (**31**). Lawson mapped a sequence of metamorphosed sedimentary and volcanic rocks that were intruded by granites. The oldest sequence of metasedimentary rocks, which he named the "Coutchiching series," was primarily sandstones, siltstones, and shales prior to metamorphism. Uranium-lead dating of detrital zircons from these rocks indicates that at least part of the Coutchiching was deposited less than 2.75 billion years ago (**32**). In the Rainy Lake region, the Coutchiching is overlain by a sequence of metavolcanic rocks that Lawson called the "Keewatin series." The Keewatin

has been dated at between 2.73 and 2.77 billion years by the uranium-lead concordia and whole-rock rubidium-strontium isochron methods (**24, 31, 22**). The youngest metasedimentary sequence Lawson called the "Timiskaming series." In nearby areas, metasedimentary and metavolcanic equivalents of the Coutchiching, Keewatin, and Timiskaming series are not superimposed on one another, but rather interfinger as a result of facies changes.

The metavolcanic rocks in greenstone belts include a wide variety of rock types such as flows, breccias, pillow lavas, and tuffs (Fig. 9.14). These range in composition from mafic (basalt) to felsic (rhyolite). Relative pro-

**FIGURE 9.14** Pillow greenstone from the Superior Province near Terrace Bay, Ontario, Canada. The outcrop has been striated and polished by the action of Pleistocene glaciers. The shape of the pillows indicates that the younger beds are on the left.

portions are approximately 57 percent basalt, 30 percent andesite, and 13 percent dacite and rhyolite (**33**). This may be compared to modern island arcs which contain 50 percent basalt, 35 percent andesite, and 15 percent dacite and rhyolite (**28**). Olivine–rich basalts (komiitites) are present at the base of some volcanic sequences in the greenstone belts and these grade up into flood basalts (tholeiites) (**34**). Flood basalts commonly form during times of continental rifting, and it may well be that volcanism in the greenstone belts was the result of tension caused by separation of an older continental terrain into small plates (microplates) which separated from each other.

It has been suggested that the mafic volcanics in greenstone belts were extruded on the crests of midocean ridges (**35**). Andesites, dacites, and rhyolites overlie the flood basalts and the similarity in composition between these rocks and rocks found in modern island arcs suggest that this part of the sequence in greenstone belts was extruded in an island arc environment (**33**, **36**, **37**). In modern island arcs, there is a systematic change in the composition

of volcanic rocks in the direction of dip of the Benioff zone beneath the island arc (**38**). Changes in the composition of volcanic rocks across the Abitibi Greenstone Belt in southeastern Ontario and southwestern Quebec indicate that the dip of the Benioff zone was toward the north at the time of extrusion of the andesitic to rhyolitic volcanic rocks (**33**).

Island arcs are areas in which plates are converging, and yet the flood basalts lower in greenstone belts indicate separation of plates. These seemingly contradictory observations may be reconciled by postulating that rifting resulted in the opening of a series of small ocean basins between microplates and that shortly thereafter these small ocean basins began to close as the pattern of convection changed.

The metasedimentary rocks in greenstone belts were sandstone, shale, iron formation, and conglomerate before metamorphism (Fig. 9.15). Most of the sandstone is poorly sorted graywacke and much of it was derived from the erosion of nearby volcanic islands. Study of the distribution of iron formation within greenstone belts has suggested that there were at least

FIGURE 9.15 Archaen conglomerate near Wawa, Ontario. Pebbles and cobbles in the conglomerate were deformed during the Kenoran orogeny.

ten separate basins of deposition in Canada (33). These may have been between or part of eight to ten separate island arcs.

### 2.73 TO 2.50 BILLION YEARS—THE KENORAN OROGENY.

Rocks of the Superior, Slave, Wyoming, and Eastern Nain provinces were folded, metamorphosed, intruded, uplifted, and eroded between 2.73 and 2.50 billion years ago during the Kenoran orogeny. Northern Scotland was part of Ancestral North America during the Archean, and the Scourian sequence in this area was metamorphosed at about the same time (46). The Kenoran orogeny had two peaks of activity, one at 2.73 billion years (Phase 1) and a second peak at 2.54 billion years (Phase 2) (Fig. 9.11).

Granitic and metamorphic rocks dated by the uranium-lead concordia and whole-rock rubidium-strontium isochron methods at between 2.65 and 2.78 billion years have been found in southwestern Quebec, southern Ontario, northern Minnesota, northern Michigan, southeastern Manitoba, southern Minnesota, and southeastern Wyoming (20, 24, 26, 28, 31, 32, 39–45). Granitic rocks from Rainy Lake in northern Minnesota are dated at between 2.75 and 2.72 billion years. These dates are surprisingly close to those on the Keewatin and Coutchiching sequences into which the granites are intruded. Evidently, little time elapsed between extrusion of the Keewatin, deposition of the Coutchiching, and metamorphism and intrusion of these sequences.

The cause of this phase of the Kenoran orogeny may have been arc-arc or arc-continent plate collisions. As the seafloor descends beneath a series of island arcs, they move closer together. If the greenstone belts were formed in some eight to ten island arcs, their deformation and metamorphism may have been caused by the joining together of the island arcs about 2.72 billion years ago (Fig. 9.16). A modern analog would be the island arcs in the western part of the Pacific Ocean. An eastern arc extends southward from Japan through the Mariana Islands, while a western arc extends southwestward from Japan to the Philippine Islands. The eastern island arc is moving toward the western arc as the seafloor on which it is located descends beneath the western arc. Presumably these two arcs will someday collide.

The Precambrian island arcs of the greenstone belts may have been originally built on oceanic crust in the same manner as some modern island arcs (such as the western Aleutian Islands). Alternatively, the island arcs may have formed during a fragmentation of one or more large continents. The principal evidence of such a fragmentation is the composition of the older volcanics in the greenstone belts. Other evidence may have been obliterated by folding and metamorphism during Phase 1 of the Kenoran orogeny.

A somewhat younger series of granitic intrusives has been dated at about 2.54 billion years by uranium-lead concordia and whole-rock rubidium-strontium isochron methods (Fig. 9.11). These intrusives have been found in southwestern Quebec, southwestern Ontario, southeastern Manitoba, northern Michigan,

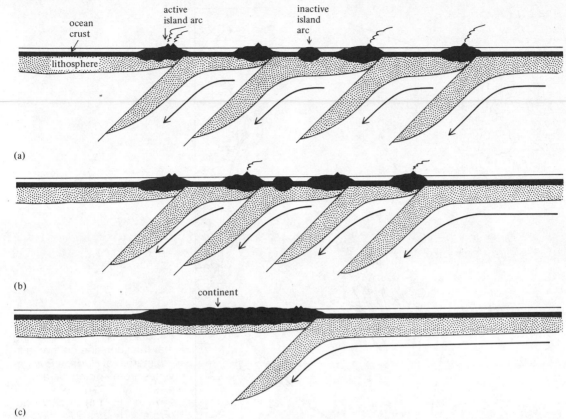

FIGURE 9.16 Diagrammatic sketch illustrating a possible method of formation of a continent. As the seafloor moves under several adjacent island arcs, they are brought together, forming a single continent.

northern Minnesota, much of Wyoming, Yellowknife in the Northwest Territories, and Godthaabsfjord, Greenland (**19, 23, 25, 27, 28, 31, 32, 39, 40, 47, 48**).

MATACHEWAN DIKE SWARM. A large swarm of north-south trending diabase dikes is present in southeastern Ontario south of James Bay. This swarm, known as the Matachewan dikes, has been dated at 2.69 billion years by the whole-rock rubidium-strontium isochron method (**49**). The Matachewan dikes are unfolded and unmetamorphosed. If the age of these dikes is accurate, Phase 2 of the Kenoran orogeny (which occurred about 2.54 billion years ago) was probably not accompanied by

folding and metamorphism, at least in southeastern Ontario. Potassium-argon dates of 2.4 to 2.6 billion years are common in the rocks of the Superior Province, but apparently these dates reflect a time of uplift and cooling rather than metamorphism.

There is a large circular structure on the eastern margin of Hudson Bay which may be a fossil impact crater created by a giant meteorite (asteroid) (**50**). This structure is defined by an almost perfectly circular arc of about 153° having a radius of 237 km (192 mi) (Fig. 9.17). The circularity of the arc indicates that the proposed impact occurred after the folding in Phase 1 of the Kenoran orogeny (2.72 billion years ago). The Belcher Islands, near the center of the arc

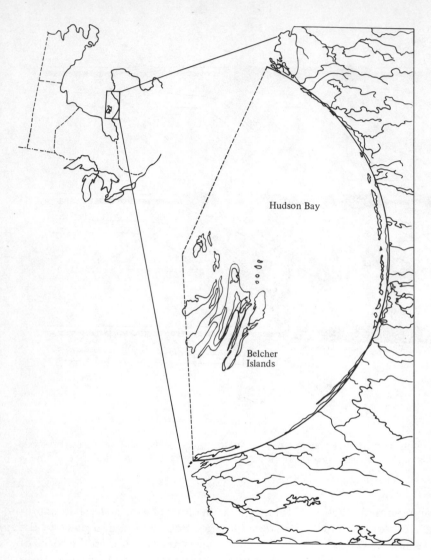

Hudson Bay

Belcher
Islands

FIGURE 9.17 Map showing
that the shoreline on the eastern margin of Hudson Bay de-
viates from a circular arc by
less than 7 km (4 mi) for
237 km (162 mi) of shoreline.
(From Thomas A. Mutch,
*Geology of the Moon: A Strati-
graphic View,* rev. ed. copyright © 1970, 1972 by Princeton University Press, Fig.
IV-32, p. 93. Reprinted by Permission of Princeton University
Press, after C. S. Beales, Ref.
50.)

(Fig. 9.17), may be a central uplift which has
been covered by sedimentary and volcanic
rocks. Subsequent compressive forces resulted
in the folding of these rocks and may have
displaced the central uplift somewhat. Sedimentary and volcanic rocks filling the crater are
of Early Proterozoic age, more than 2.0 billion
years old. Therefore the proposed impact probably took place between 2.7 and 2.0 billion
years ago.

## Early Proterozoic Era
## (2.5 to 1.7 Billion Years Ago)

Deposits of Early Proterozoic age are found in
many parts of the Canadian Shield, in the
Rocky Mountains, in southwestern United
States, in Greenland, and in northern Scotland.
As discussed earlier, it is likely that the continents of Ancestral North America and Gondwanaland were part of a single landmass during

the later part of the Archean era and that this landmass broke apart near the beginning of the Early Proterozoic era (Fig. 9.4c). The continent of Ancestral North America itself may have been fragmented into three or more plates at about the same time. During the middle of the Early Proterozoic, the continents began to move back towards each other and collided about 1.85 billion years ago near the end of the Early Proterozoic. The rifting and collision of plates had a profound effect on the depositional history of Ancestral North America during the Early Proterozoic.

2.50 TO 2.16 BILLION YEARS. Archean rocks are unconformably overlain by Early Proterozoic rocks of the Huronian sequence in southeastern Ontario. The lower Huronian sequence consists of quartzite, shale, limestone, dolostone, tillite, and volcanics. It is about 5000 m (16,000 ft) thick north of Lake Huron (**51**). A similar sequence of Early Proterozoic rocks has been found in Wyoming (**51**). It is possible that both sequences were deposited in a geocline on the southeastern border of the Superior Province following separation of Gondwanaland from North America about 2.4 billion years ago. As discussed earlier, paleomagnetic data support this suggestion (Fig. 9.4c). The Thessalon volcanics at the base of the sequence are mafic (basaltic) flood basalts similar to those formed as a result of continental rifting, and they give a date of about 2.33 billion years (**52**).

There is evidence of glaciation during the Early Proterozoic in several parts of Ancestral North America. Three tillites occur within the Lower Huronian sequence north of Lake Huron (**51**). The youngest of these is found in the Gowganda Formation. This tillite contains striated pebbles, and locally it rests on a striated pavement (Fig. 9.18). The direction of ice movement was apparently toward the southwest (**53**). A whole-rock rubidium-strontium isochron date of 2.29 billion years has been

(a)

(b)

(c)

FIGURE 9.18 Glacial deposits from the Gowganda Formation: (a) Striated boulder in till from near Cobalt, Ontario (collected by M. E. Wilson, courtesy the Geological Survey of Canada, Ottawa). (b) Tillite showing the very poorly sorted nature of the deposit from north of Elliot Lake, Ontario. (c) Probable glacial varves showing a pebble which was probably dropped into the deposit by floating ice. [(b) and (c) Courtesy of Grant M. Young.]

obtained on Gowganda sedimentary rocks. However, rubidium-strontium dating of sedimentary rocks gives minimum ages, and the actual time of deposition may have been earlier. Early Proterozoic tillite has also been found in southeastern Wyoming, northern Utah, and the central part of the Churchill Province (Hurwitz Group) (**51**).

Paleomagnetic investigation of the Gowganda Formation indicates that the North Pole was located close to the present North Pole relative to Ancestral North America (**54**). This would place the areas glaciated within about 42° of the North Pole. For comparison, glaciation occurred within 55° of the North Pole during the last ice ages (Pleistocene).

A flat-lying sequence of Early Proterozoic mafic volcanics (tholeiitic flood basalts) is also present on the coast of Labrador near the town of Okhakh. This sequence, known as the Mugford Group Volcanics, is about 1000 m (3000 ft) thick and has been dated at 2.39 billion years by the whole-rock rubidium-strontium isochron method (**55**). The paleomagnetic poles of the volcanics in the Mugford Group and the Thessalon Volcanics are reasonably close, and therefore they were probably extruded at about the same time. The cause of volcanism in these two areas may have been a rifting apart of the continents. Tholeiitic flood basalts are commonly formed during continental rifting. It is possible that the Eastern Nain Province may have separated from the Superior Province beginning with the extrusion of the Mugford Group volcanics about 2.39 billion years ago. Rifting in this area is indicated by a 3000 m (10,000 ft) thick sequence of sediments in a graben that heads eastward in the general direction of the Mugford Group volcanics. Kevin Burke and John Dewey have suggested that this graben is an aulacogen which was part of a triple junction that formed in response to an underlying plume (**10**). The triple junction may have formed as an ocean opened between the Eastern Nain and Superior provinces, in which case the aulacogen would be a failed third arm of the rift.

The Hurwitz Group in the Churchill Province of western Canada is a sequence of sandstone, shale, volcanics, and tillite of Early Proterozoic age. This sequence is remarkably similar to the Huronian sequence of southeastern Ontario. It is about 5000 m (16,000 ft) thick, and it may have been deposited in a geocline formed as a result of a separation of a plate including the Slave Province from a plate including the Superior Province. If the tillite in the Hurwitz Group correlates with the Gowganda tillite (**51**) (which has been dated at 2.29 billion years), and if the Hurwitz Group was deposited in a geocline after the beginning of separation of the Superior and Slave Provinces, the ocean probably began to open about 2.4 billion years ago. However, as will be discussed below, there is some evidence that the separation of these two provinces did not begin until about 2.16 billion years ago.

**2.16 BILLION YEARS.** A number of mafic (basaltic) sills and dikes were emplaced in southeastern Ontario, southwestern Quebec, northern Wyoming, and central Wyoming about 2.16 billion years ago. These intrusions include the Nipissing Diabase, the northeast-trending Abitibi dikes, Beartooth dikes 7 and 8, and dikes in the Owl Creek Mountains of Wyoming (**49, 57, 58**). Dikes and sills of this age have also been found in the Slave Province (**49**). The emplacement of these dikes and sills may have been associated with the beginning of convergence between Ancestral North America and Gondwanaland and between the Superior and Slave Provinces 2.16 billion years ago. There is some evidence for deformation and intrusion on the southeastern margin of Ancestral North America between 2.0 and 2.1 billion years ago, and it is possible that this margin was a subduction zone at that time. Alternatively, the Nipissing Diabase and dikes in the Beartooth and Owl Creek mountains may have been caused by the beginning of separation of North America and Gondwanaland. Paleomagnetic data (Fig. 9.4c) is consistent with this possibility, but additional data from these two areas is needed to determine whether these two continents began to separate 2.4 or 2.16 billion years ago. Still another possibility is that these two continents separated 2.4 billion years ago, rejoined about 2.2 billion years ago, and separated again 2.16 billion years ago.

The Ungava dikes in northeastern Canada have been dated at about 2.1 billion years old, and Kevin Burke and John Dewey consider them to be an ocean-opening dike swarm formed when the Superior and Slave provinces began to separate. Paleomagnetic data are consistent with this suggestion (Fig. 9.22), but additional paleomagnetic studies of samples of Middle Proterozoic age from the Superior and Slave provinces are needed to test this possibility of separation of these two provinces between 2.10 and 2.16 billion years ago.

**2.16 TO 1.85 BILLION YEARS.** The Animikie Group, which is the youngest group in the Huronian sequence, is found in Northern Michigan, northern Wisconsin, and northwestern Minnesota. It was deposited after intrusion of the Nipissing Diabase. The Animikie Group contains quartzite, siltstone, limestone, an iron formation, and minor amounts of coal (presumably from algae). The iron formation is mined in the Mesabi and Cuyuna districts of Minnesota and the Gogebic, Marquette, and Menominee districts of Michigan. These deposits are some of the most important sources of iron in North America. The Hemlock Volcanics of northern Michigan overlie the iron formation. Uranium-lead concordia measurements on rhyolite from these volcanics give a date of about 2.00 billion years (**59**), and therefore the iron formation is probably more than 2.0 billion years old.

North of Lake Superior, the Huronian sequence is only a few thousand feet thick, nearly horizontal, and almost unmetamorphosed. Toward the southeast, it thickens to several tens of thousands of feet, and is folded and metamorphosed. The marked thickening suggests a transition from platform to geosynclinal conditions of deposition. Most of the clastic deposits in the Huronian sequence were probably derived from the Superior Province, but in the outer parts of the basin, clastics may have been supplied by island arcs bordering the continent.

Thick sequences of Early Proterozoic sedimentary and volcanic rocks are found in a wide belt extending from the Great Lakes northeastward to Labrador and thence westward to the Ungava Peninsula and southward to the Belcher Islands. Thick Early Proterozoic sequences are also found in the Black Hills of South Dakota, Wyoming, and Arizona. In northern Arizona in the vicinity of the Grand Canyon, schists (Vishnu Schist) and amphibolites are intruded by granites (Zoroaster Granite) dated at up to 1.7 billion years (**60**). The metamorphic rocks were originally shale, siltstone, sandstone, and mafic (basaltic) volcanics (Fig. 9.19). Metamorphosed sandstone, shale, iron formation, and limestone of the Loch Maree series in Scotland was intruded by pegmatites between 1.7 and 1.6 billion years ago, and therefore these deposits are probably of Early Proterozoic age.

In many areas, Early Proterozoic deposits are more than 5000 m (16,000 ft) thick. Their distribution suggests that they may have been deposited in a geosyncline or geocline, perhaps the one in which the Lower Huronian deposits were laid down. This geosyncline may have bordered the Superior and Wyoming Provinces during much of the Early Proterozoic age (Fig. 9.20).

**1.85 TO 1.70 BILLION YEARS—THE HUDSONIAN OROGENY.** A major episode of folding, metamorphism, and intrusion known as the Hudsonian orogeny occurred near the end of the Early Proterozoic (Fig. 9.20). This event created a mountain range approximately coincident with the Churchill, Southern, and Bear provinces (Fig. 9.9). Uranium-lead concordia and whole-rock rubid-

(a)

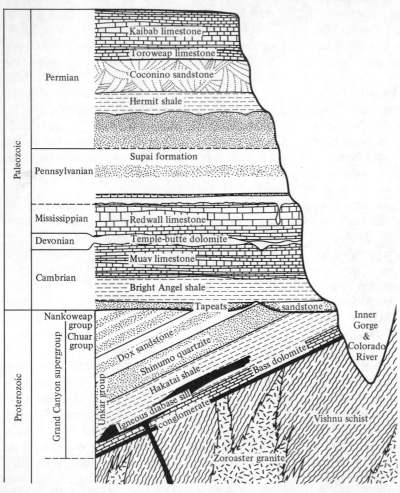

(b)

FIGURE 9.19 (a) Vishnu Schist in the Inner Gorge of the Grand Canyon. Note the granitic dikes cutting the schist and the sandstone unconformably overlying the schist. (Courtesy of Phil von Stade.) (b) Generalized stratigraphy of the Grand Canyon. (Adapted from Breed and Roat, 1974.)

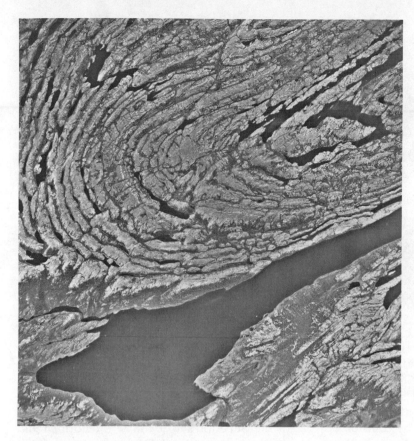

**FIGURE 9.20** Aerial view of a plunging fold in eastern Quebec, Canada. These Early Proterozoic rocks were deformed during the Hudsonian orogeny between 1.7 and 1.8 billion years ago. [Courtesy of the National Air Photo Library, Dept. of Energy, Mines and Resources (Canada).]

ium-strontium isochron dating of granitic and metamorphic rocks from these provinces indicates that the intrusion and metamorphism began about 1.85 billion years ago (Fig. 9.11). Dates between 1.83 and 1.75 billion years have been obtained on rocks from southeastern Ontario, eastern and central Wisconsin, northern Minnesota, southwestern Saskatchewan, northern Colorado (Fig. 9.21), and central Colorado (**61–69**). This orogenic episode is referred to as Phase 1 of the Hudsonian orogeny. Whole-rock rubidium-strontium isochron dates of 1.72 to 1.65 billion years have been obtained on granitic and metamorphic rocks from southeastern Ontario, eastern and central Wisconsin, southwestern Saskatchewan, northern Colorado, central Colorado, and northwestern Arizona (**61, 66, 67, 70–73**). This event appears

to be a separate event from Phase 1 (Fig. 9.11), and it is referred to as Phase 2 of the Hudsonian orogeny.

Paleomagnetic data indicate that the Hudsonian orogeny was probably caused by a collision of plates. The Churchill Province separates the Superior Province from the Slave Province (Fig. 9.9). A.P.W. paths have been constructed for these provinces by connecting (in order of decreasing age) paleomagnetic poles from each province (Fig. 9.22). The A.P.W. path determined on rocks older than 1.75 billion years from the Suprior and southeastern Churchill provinces is located consistently to the west of the A.P.W. path of the same age from the Slave and northwestern Churchill provinces (Fig. 9.22). The two A.P.W. curves join together in the vicinity of paleomagnetic poles

FIGURE 9.21 Precambrian metamorphic rocks in Rocky Mountain National Park. These rocks were metamorphosed approximately 1.8 billion years ago. (Courtesy of Mark Miller.)

dated at about 1.75 billion years. This date agrees well with the age of the Hudsonian orogeny. Therefore it is very likely that the folding and metamorphism which accompanied the Hudsonian orogeny in a part of the Churchill Province was caused by collision between: (1) a plate containing the Superior Province and the southeastern Churchill Province and (2) a second plate containing the Slave Province and the northwestern Churchill Province (**74**). The suture which marks the former boundary between the two plates is probably located in the Churchill Province approximately midway between the Superior and Slave provinces (Fig. 9.23). If an ocean did open and close between these two provinces, it would be another example of the Wilson cycle.

A part of the Churchill Province separates the Eastern and Western Nain provinces from the Superior Province (Fig. 9.8). The folding in the Churchill Province between these provinces occurred about 1.75 billion years ago and it may have been caused by the collision between a plate containing the Superior Province and a second plate containing the Eastern and Western Nain provinces and possibly Europe as well. Again, there is a possibility that an ocean opened before this closing, and this may provide another example of the Wilson cycle.

The Southern Province is located southeast of the Superior and it too was deformed and metamorphosed during the Hudsonian orogeny (called the Penokean orogeny in this area). The deformation and metamorphism may have been caused by the collision of North America with Gondwanaland. Recalculated A.P.W. curves from North America and Gondwanaland are separated between 2.4 and 1.9 billion years and join together at paleomagnetic poles dated at 1.85 billion years (Fig. 9.4c). This is the time of the beginning of the Penokean orogeny which occurred between North America and Gondwanaland.

The Wyoming Province is separated from the Superior Province by a narrow zone containing rocks that were deformed and metamorphosed during the Hudsonian orogeny. Paleomagnetic data suggests that the Wyoming Province was on the same plate as the Superior Province during most of the Precambrian.

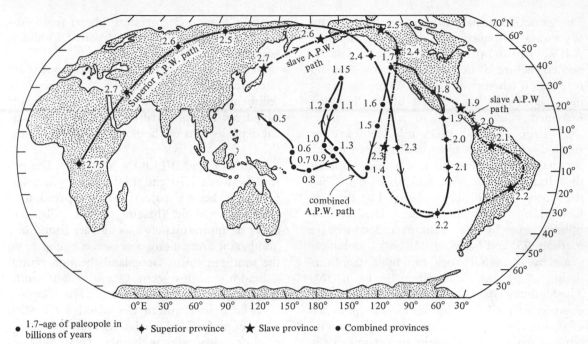

**FIGURE 9.22** Comparison of the A.P.W. paths of the Superior, Slave, and Eastern Nain provinces. (After Cavanaugh and Seyfert, Ref. 74.)

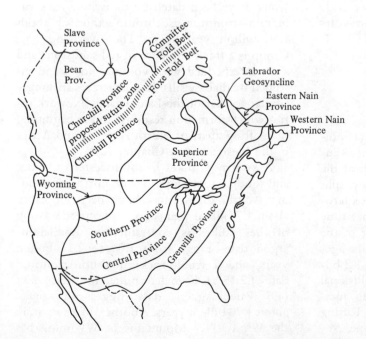

**FIGURE 9.23** Map showing geological provinces in the Canadian Shield along with the proposed suture between Early Paleozoic plates. (After Cavanaugh and Seyfert, Ref. 74.)

Three out of four paleomagnetic poles from the Wyoming Province fall along the Superior A.P.W. path (**74**). However, more data is needed before a definite conclusion can be reached on whether or not the Wyoming Province has always been a part of the Superior Province.

A large elliptical basin near Sudbury in southeast Ontario is filled with igneous rocks, breccia, and lake deposits. It was folded during the Hudsonian orogeny, and prior to folding the basin was probably circular. The time of formation of the basin has been dated at 1.84 billion years by the uranium-lead concordia method (**75**) and between 1.68 and 1.96 billion years by the whole-rock rubidium-strontium dating of igenous rocks within the basin (**76**). The Sudbury Basin contains some of the largest deposits of nickel in the world. The presence of shatter cones and breccias in the basin and its circular shape have led to the suggestion that it was formed by the impact of a giant meteorite (**77**). The present diameter of the basin is about 70 km (42 mi), but the original diameter of the crater may have been several times as wide because erosion in the last 1.85 billion years may have removed much of the basin.

## Middle Proterozoic Era (1.7 to 1.0 Billion Years Ago)

As discussed earlier, paleomagnetic data indicate that the continents of North America, Europe, and Gondwanaland were together at the end of the Early Proterozoic, but they split apart at the beginning of the Middle Proterozoic and then rejoined at the end of that time (Fig. 9.4b, 9.4e, and 9.5). Rifting apart of the continents began about 1.7 billion years ago, and the continents were rejoined about 1.2 billion years ago. This joining was a collisional event accompanied by mountain building, metamorphism, folding, and intrusion. Rifting apart of Gondwanaland, Ancestral Europe, An-

cestral China, and Ancestral Siberia from Ancestral North America began about 1.15 billion years ago.

In many areas, Early Proterozoic deposits are unconformably overlain by Middle Proterozoic sedimentary and volcanic rocks. These deposits vary from thin platform deposits to geosynclinal deposits tens of thousands of meters thick.

1.70 TO 1.50 BILLION YEARS. The period between 1.70 and 1.60 billion years ago may have been a major time of separation of continents. Mafic (basaltic) dikes, sills, and flows of approximately this age are found in a number of areas along a zone extending from the southern tip of Greenland through central Labrador, southwestern Quebec, and southeastern Ontario to Wyoming. The Gardner Lake lavas in southern Greenland are a thick series of mafic flood basalts. Radiometric dating of rocks associated with these lavas indicates that the lavas were extruded between 1.7 and 1.3 billion years ago (**78**). The Croteau Group of central Labrador contains mafic volcanics which have been dated by the whole-rock rubidium-strontium isochron method at about 1.56 billion years (**79**). The overlying Seal Group is a thick sequence of red sandstone and shale interbedded with volcanic flows and intruded by diabase sills. Its character is strikingly similar to that of the Late Triassic Newark series which formed in response to the beginning of a rifting apart of North America and Africa in the Mesozoic (see Chapter 12).

A swarm of dikes in southwestern Quebec and southeastern Ontario (east–northeast trending Abitibi dikes) may have been emplaced about 1.7 billion years ago. Radiometric dating of these dikes is contradictory. Potassium-argon dates range from 1.09 to 2.14 billion years, and a whole-rock rubidium-strontium date of 2.15 has been determined for these rocks (**49**). Paleomagnetic data suggest an age of about 1.70 billion years. Another dike swarm in the Wind River Mountains of Wyoming has been dated at 1.68 to 1.88 billion years by the potassium-argon method, and the Marathon dikes north of Lake Superior have been dated at 1.22 billion years by the whole-rock potassium-argon method (**57**). Paleomagnetic data suggest that all of these dikes, sills, and flows may have been intruded at the same time, perhaps 1.70 billion years ago. They may have been emplaced during a period of crustal tension (stretching) as Gondwanaland began to separate from Ancestral North America about 1.7 billion years ago.

1.50 TO 1.40 BILLION YEARS—THE ELSONIAN OROGENY. The period between 1.5 and 1.4 billion years ago was a time of widespread igneous activity (the Elsonian orogeny) and little if any sedimentation on the continents. The area affected by the igneous activity was primarily the Central Province and Western Nain Province (Fig. 9.9). This activity included intrusion of anorthosites, granites, charnockites (defined in Appendix C), and related igneous rocks, and the extrusion of rhyolites. This igneous activity is unusual in that it was not accompanied by folding, faulting, or metamorphism, at least not in the areas which have been studied (**80**). Widespread granitic igneous activity almost always accompanies folding, faulting, and metamorphism.

Anorthosites were intruded along a wide belt extending from southern Greenland through central Labrador, southern Quebec, the Adirondack Mountains, and southeastern Wyoming to southern California (**81**). Radiometric dating indicates that most anorthosites were emplaced between 1.50 and 1.40 billion years ago (**81**). Whole-rock rubidium-strontium and uranium-lead concordia dating indicate that granitic rocks (including charnockites) were intruded 1.50 and 1.40 billion years ago along a wide belt which more or less parallels the belt of anorthosites. Granitic rocks of this age have been found in eastern and central Wisconsin, eastern Minnesota, central Missouri, northern

New Mexico, and southern Nevada (**61, 64, 66, 82–84**). Very large volumes of rhyolite were extruded at about the same time in the St. François Mountains of central Missouri (**83**).

The widespread intrusion of large bodies of anorthosites and charnockites is unique in the geological record. S. K. Saxena has suggested that these intrusive igneous rocks formed during a time of unusually high geothermal gradient (**81**). Norman Herz has suggested that a cataclysmic event, such as the birth of the earth-moon system, could be responsible for such a high geothermal gradient (**81**). It is possible that this was the time of capture of the moon in a prograde orbit, or that this was the time of closest approach of the moon following capture in a retrograde orbit. The tides resulting from either of these events would have been immense, and this would be consistent with the lack of observed sedimentary rocks of this age on the continents.

1.40 TO 1.30 BILLION YEARS. The Belt Supergroup is a 11,000 m (35,000 ft) thick sequence of limestone, sandstone, shale, and volcanic rocks in northern Idaho, western Montana, and southern British Columbia (Fig. 9.24). The Purcell Group of the Canadian Rockies in southeastern British Columbia is very similar in thickness and lithology to the Belt Supergroup and is thought to correlate with it. The deposits of the Belt and Purcell groups are of geosynclinal thickness, but they do not have a significant lateral extent. It is therefore probable that they were deposited in a basin rather than in a geosyncline.

The basement on which the Belt Supergroup rests was last metamorphosed about 1.7 billion years ago, and therefore the Belt must be younger than 1.7 billion years (**89**). The Neihart Quartzite at the base of the Belt has been dated at 1.36 to 1.27 billion years by the whole-rock rubidium-strontium method (**89**). Furthermore, the paleomagnetic pole for the Spokane formation (**90**), which is in the lower part of the Belt,

FIGURE 9.24 Gently dipping sedimentary rocks of the Belt sequence of Middle Proterozoic age in Glacier National Park, Montana. (Courtesy of the Montana Highway Commission.)

lies between paleomagnetic poles dated at 1.45 and poles dated at 1.30 billion years. The Purcell Basalt is found near the middle of the Belt sequence in Glacier National Park. These lavas and associated sills have been dated at between 1.4 and 1.1 billion years (**89**). It is probable that they were formed during the Mackenzie igneous event (discussed in the next section) about 1.30 billion years ago. Thus, that part of the Belt below the Purcell Lavas was probably deposited between 1.40 and 1.30 billion years ago.

A Middle Proterozoic geosyncline or geocline may have bordered Ancestral North America on the southeast during the Middle Proterozoic (Fig. 9.25). Some of the deposits in this geosyncline include the Grenville and

Hastings sequences of southeastern Ontario, the Sibley Group on the north shore of Lake Superior (**91**), metasedimentary rocks in the Adirondacks of New York State and in the Llano area of Texas, and sedimentary rocks of the Grand Canyon sequence.

Tens of thousands of feet of intensely metamorphosed carbonate, clastic, and volcanic rocks comprise the Grenville and Hastings series in the northern part of the Grenville Province. Volcanics interbedded with these deposits have been dated at between 1.2 and 1.3 billion years by the uranium-lead method (**85**), but they may be older than this because of the possibility of resetting of ages during intense metamorphism.

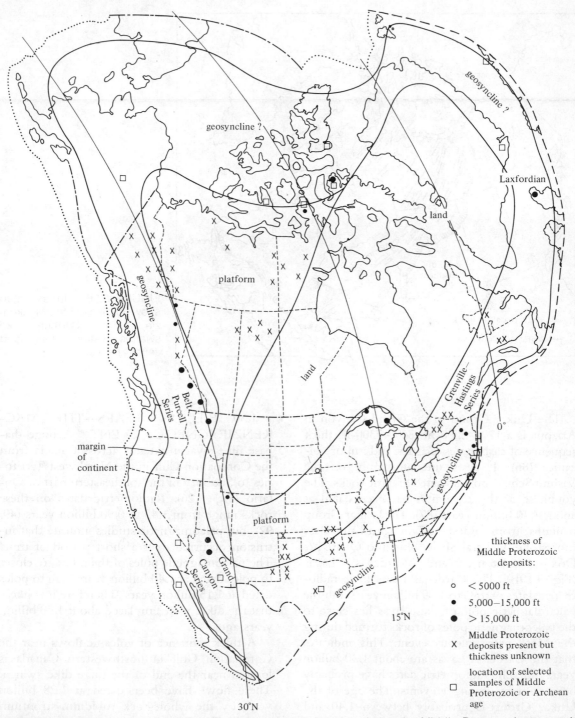

**FIGURE 9.25** Paleogeographic map of Ancestral North America during the Middle Proterozoic, approximately 1.3 billion years ago.

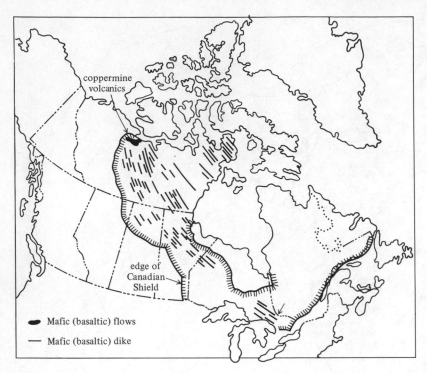

coppermine
volcanics

edge of
Canadian
Shield

⬮ Mafic (basaltic) flows

— Mafic (basaltic) dike

**FIGURE 9.26** Index map showing dikes of the Mackenzie III and Sudbury dike swarms. (After Fahrig and Jones, Ref. 92.)

The Unkar Group in the Grand Canyon in Arizona is a 1700–2200-m (5600–7300-ft) thick sequence of carbonates, clastics, and minor volcanics (**86**). It rests unconformably on the Vishnu Schist and associated granitic rocks. The youngest of these granites has been dated at about 1.40 billion years (**87**). The Unkar Group contains from oldest to youngest: the Bass Limestone, Hakatai Shale, Shinumo Quartzite, Dox Sandstone, and Cardenas Lavas (Fig. 9.19B). The Cardenas Lavas have radiometric dates of .895 to 1.09 billion years, but the paleomagnetic pole of these lavas lies close to the paleomagnetic poles of rocks formed during the Mackenzie igneous event. This indicates that the Cardenas Lavas are about 1.30 billion years old. The radiometric dates have probably been reset by younger events. The age of the Unkar Group is probably between 1.40 and 1.30 billion years.

**1.30 BILLION YEARS—THE MACKENZIE IGNEOUS EVENT.** A huge diabase dike swarm extends across Canada from the Coronation Gulf in the Northwest Territories to Sudbury in the southeastern part of Ontario (Fig. 9.26). Radiometric dates on these rocks range from 1.43 to 0.90 billion years (**49, 92**), but paleomagnetic studies indicate that intrusion occurred over a short period of time. The paleomagnetic poles of the dikes are closer to poles dated at 1.40 billion years than to poles dated at 1.15 billion years. Therefore it is likely that the dikes were emplaced about 1.30 billion years ago.

A thick sequence of volcanic flows near the Coronation Gulf in northwestern Canada is located near the end of the huge dike swarm. These flows have been dated at 1.28 billion years by the whole-rock rubidium-strontium method (**93**). The paleomagnetic pole of these flows is close to that of the dike swarm, and therefore it is probable that they were extruded at about the same time as the intrusion of the dike swarm.

As discussed previously, volcanics were extruded and sills and dikes intruded during the Mackenzie igneous event in central Arizona and in the northwestern United States. Therefore, this activity was very widespread in Ancestral North America. The Mackenzie igenous event may have been associated with the change from opening to closing oceans as Ancestral North America, Gondwanaland, and Ancestral Europe reversed direction and began to move toward each other.

1.30 TO 1.20 BILLION YEARS. The Nankoweap and Chuar groups, which unconformably overlie the Cardenas lavas in the Grand Canyon region, are 2400 m (8000 ft) thick and are composed dominantly of shale and sandstone (**86**). The Troy Quartzite in central Arizona unconformably overlies basalt flows and is therefore probably correlative with the Nankoweap and/or Chuar group. The Troy quartzite has been intruded by diabase sills dated at 1.14 to 1.15 billion years by the uranium–lead concordia method on zircons and the potassium–argon method on biotite (**94**). Thus, the Troy Quartzite and probably the Nankoweap and Chuar groups were deposited prior to 1.15 billion years ago. The Belt series contains a thick sequence of quartzite, shale, and minor limestone above the Purcell Lavas. It is likely that these deposits also correlate with the Troy Quartzite and were therefore also deposited before 1.15 billion years ago. It is likely that parts of the Grenville and Hastings sequences and related units were deposited at the same time.

1.20 TO 1.15 BILLION YEARS—THE GRENVILLE OROGENY. The Grenville orogeny was a period of intense metamorphism, folding, and intrusion within the Gren-

FIGURE 9.27 Granitic dikes of different ages in the Grenville Province near Parry Sound, Ontario, Canada.

ville Province (Fig. 9.27). As a result of the deformation, a lofty mountain range was produced on the southeastern margin of Ancestral North America. The mineralogy of the rocks produced during the Grenville orogeny suggests that the rocks were metamorphosed at depth in excess of 20 km (12 mi). At least three generations of folds have been observed in the Grenville marbles near Bancroft, Ontario (**95**) and similar relations have been observed elsewhere in the Grenville Province. Large folds with subhorizontal axial planes (*nappes*) are common in the Grenville Province. Thus it is likely that the Grenville orogeny consisted of more than one phase of deformation and it was one of the most intense orogenies to have affected Ancestral North America. The deformed belt is 5000 km (3500 mi) long and 1000 km (600 mi) wide.

Unfortunately, the age of the beginning of the Grenville orogeny is not well established.

Numerous potassium-argon dates place the time of last uplift and cooling at about 0.95 to 1.00 billion years ago (**96**). However, the time of the initial folding and metamorphism of the Middle Proterozoic rocks in the Grenville Province was probably between 1.15 and 1.25 billion years ago (**97**). Uranium-lead and whole-rock rubidium-strontium dating of intrusives into rocks of probable Middle Proterozoic age range from 1.25 to 1.05 billion years (**98**).

It has been suggested on the basis of paleomagnetic data that the Grenville orogeny was caused by a collision between a plate consisting of the southwestern part of the Grenville Province and another consisting of the rest of Ancestral North America (**99**). However, because the recalculated A.P.W. curves from Gondwanaland and Ancestral North America join in the vicinity of poles dated at about 1.2 billion years (Fig. 9.4b), it is more likely that the Grenville orogeny was caused by a collision between Ancestral North America and Gondwanaland. During this collision, Ancestral North America may have slid beneath Gondwanaland forming a double thickness of continent. It has been suggested that the leading edge of India has been sliding beneath Asia throughout much of the Cenozoic and that the 70 km (42 mi) thickness of crust beneath the Himalaya Mountains and the Tibet Plateau is due to a double thickness of continent. The collision between Ancestral North America and Gondwanaland may have occurred in a similar fashion. This collision may have produced a 70 km (35 mi) thick continent which would have been reduced to 35 km (21 mi) thick by subsequent erosion. In this case, the Grenville Province would represent that part of Ancestral North America that was beneath Gondwanaland.

The character of the rocks of the Grenville Province support the model proposed here. The sliding of Ancestral North America under Gondwanaland could easily account for the intense metamorphism during the Grenville Orogeny in the Grenville Province, and it would also account for the formation of nappes. Nappes cannot be produced by simple compression between two blocks because this will tend to produce folds with steeply inclined axial planes. Nappes are readily produced by a shearing motion of one block riding over another block. The nappes in the Alps are thought to have formed as a result of the sliding of Europe beneath a small plate of which Italy is now a part (see Chapter 13).

**1.15 BILLION YEARS.** Mafic (basaltic) igneous activity occurred in many parts of Ancestral North America about 1.15 billion years ago. These areas include the Grand Canyon region, the Lake Superior region, southern Greenland, and western Texas (Fig. 9.28) (**100**).

Approximately 15,000 m (50,000 ft) of volcanics, shale, sandstone, and conglomerate were deposited in the vicinity of Lake Superior beginning about 1.15 billion years ago. These deposits belong to the Keweenawan Group. Keweenawan volcanics are for the most part flood basalts that are up to 6000 m (20,000 ft) thick. Some of the volcanics and the associated sedimentary rocks contain mineable quantities of native copper. This copper was once used by Indians to make tools. Whole-rock rubidium-strontium isochron and uranium-lead concordia dating of rhyolites interlayered with Middle Keweenawan mafic volcanic rocks give dates of 1.10 to 1.15 billion years (**101**). Radiometric dating of the Duluth Gabbro near the western end of Lake Superior indicates that it was intruded about 1.12 billion years ago (**102**), at about the same time as the extrusion of the Keweenawan lavas. Normal faulting accompanied or followed shortly after the Keweenawan volcanism. In central Arizona, displacement occurred along north-south trending normal faults at about the same time as the intrusion of mafic sills and dikes in this area.

It has been suggested that the faulting and volcanism associated with the Keweenawan sequence occurred in response to a major period of rifting in Ancestral North America (**12**).

**FIGURE 9.28** Thick mafic (diabase) sill (at river level) intruding sedimentary rocks of the Grand Canyon Supergroup. Note the columnar jointing in the diabase.

Gravity and magnetic measurements and also sampling by deep drilling indicate that the rift zone (midcontinent gravity high) extends southwestward from Lake Superior to central Kansas, southeastward into central Michigan, and from there southward into Tennessee (**103**) (Fig. 9.25). The rifting in the Lake Superior region and the normal faulting and mafic igneous activity in northern Arizona may have occurred along aulacogens associated with the beginning of separation of Gondwanaland from North America about 1.15 billion years ago. The rift zones in the midcontinent region head in the general direction of a possible triple junction in central Louisiana (Fig. 9.25). This triple junction is located at an embayment in the coastline of Ancestral North America (Fig. 9.25). Such embayments commonly occur at the site of triple junctions that form when continents separate. The Gulf of Guinea bordering Africa at the mouth of the Niger River is a more recent example of such a case.

The Torridonian series, a thick sequence of nonmarine sandstone and shale in northern Scotland, was probably deposited during the later part of the Middle Proterozoic and the early part of the Late Proterozoic. The paleomagnetic pole from the Lower Torridonian (**104**) agrees well with paleomagnetic poles 1.15 billion years old from North America if Scotland (along with this pole) is rotated back to its predrift position adjacent to Ancestral North America. Rubidium-strontium whole-rock dating on shale from the Lower Torridonian give a date of 885 million years (**46**). However, such dates are generally minimum ages, and the paleomagnetic pole position may give a more accurate age than this method. The Upper Torridonian has a rubidium-strontium date of 815 million years (**46**) which agrees with the position of its paleomagnetic pole. The Moinian series of central Scotland, a thick sequence of metamorphosed sandstone, shale, and volcanics, may correlate with the Torridonian (**46**).

Thick layers of sedimentary and volcanic rocks were deposited in the Cordilleran Geosyncline during the Late Proterozoic in a belt extending from eastern Alaska to southern California (Fig. 9.29). John Stewart has suggested that this geosyncline formed as a result of the separation of an unspecified continent from western North America about 850 million years ago (**13**). This continent was probably Asia, based on paleomagnetic data described earlier, and rifting may have occurred 1.15 billion years ago.

1.15 TO 1.0 BILLION YEARS. There are no Late Proterozoic sedimentary or volcanic rocks which are known to have been deposited in Ancestral North America after the Keweenawan sequence. The end of the Late Proterozoic was probably a time of erosion of the continent. Sedimentation was probably limited to the deep oceans and outermost margins of Ancestral North America.

## Late Proterozoic Era (1.0 to 0.57 Billion Years Ago)

Thick sequences of sedimentary and volcanic rocks of Late Proterozoic age are found in the Appalachian region, the Cordilleran region, along the east coast of Greenland, in northern Scotland, and in Spitsbergen, all of which were part of Ancestral North America at that time. These deposits underlie thick sequences of Paleozoic sedimentary rocks in the Appalachian, Cordilleran, East Greenland, and Caledonian geosynclines (Fig. 10.3).

The oldest deposits in the Northern Appalachian geosyncline in the Taconic region of New York State are arkosic red beds of the Rensselaer Graywacke. Jack Bird and John Dewey have suggested that these deposits may have accumulated in a graben formed during the initial rifting of Ancestral Europe from Ancestral North America, and that the rifting initiated deposition in the Northern Appalachian geosyncline (**11**). Flood basalts of Late Protero-

zoic age in Newfoundland may have been extruded as a consequence of crustal tension associated with rifting (**11**). A number of mafic (basaltic) dikes were emplaced in the Grenville Province about 675 million years ago and in southern Britain about 650 million years ago (**46**, **105**). It is possible that these were emplaced during separation of the continents, which would mean that separation began about 675 million years ago. A diamictite unit of probable glacial origin was deposited following the rifting. This area was at about 5°S latitude at the time.

The Caledonian and East Greenland geosynclines were continuous during the Late Proterozoic. As much as 1700 m (5600 ft) of shale, sandstone, limestone, dolostone, and diamictite were deposited in these geosynclines during this interval (Fig. 9.30 and 9.31a) (**106**). There are at least two separate diamictite horizons in these sequences, and striated pebbles and boulders occur within them (**107**). Although none of the diamictites rests on a striated and polished basement, they are probably of glacial origin, perhaps deposited by floating ice. The diamictites are estimated to be about 600 million years old on the basis of sedimentation rates in the overlying Paleozoic units (**108**). The paleolatitude of East Greenland was 15°S and of southern Norway was 35°S 600 million years ago. Deposition in the East Greenland and Caledonian geosynclines may have been initiated by separation of Ancestral Europe from Ancestral North America.

The oldest deposits in the Southern Appalachian geosyncline belong to the Mount Rogers Group. The lower part of this unit consists of conglomerate, diamictite, and varved shale. The diamictites and varved shales are probably of glacial origin (**107**). A 1500-m (5000-ft) thick volcanic unit overlies the sedimentary rocks. This unit contains rhyolite and basalt in approximately equal proportions (**109**). Rhyolites from this unit have been dated at between 850 and 950 million years by the uranium-lead method on zircons (**110**). The underlying

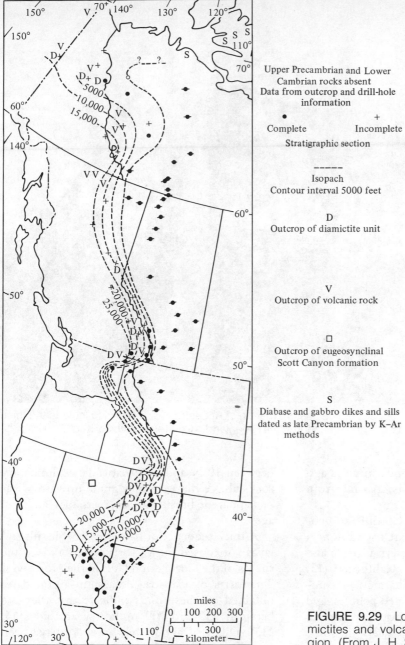

Upper Precambrian and Lower
Cambrian rocks absent
Data from outcrop and drill-hole
information

●           +
Complete     Incomplete
Stratigraphic section

- - - - -
Isopach
Contour interval 5000 feet

D
Outcrop of diamictite unit

V
Outcrop of volcanic rock

□
Outcrop of eugeosynclinal
Scott Canyon formation

S
Diabase and gabbro dikes and sills
dated as late Precambrian by K–Ar
methods

**FIGURE 9.29** Location of Late Proterozoic diamictites and volcanic rocks in the Cordilleran region. (From J. H. Stewart, Ref. 13.)

diamictites may have been deposited about 875
million years ago at the same time that extensive glaciation occurred in Africa (**111**). Diamictite of probable glacial origin also occurs
above the volcanics. It may have been deposited
740 or 610 million years ago during a time of
widespread glaciation in other parts of the
world (**111**). The Southern Appalachian geosyncline was approximately on the equator 875
million years ago, and at a latitude of about 20°S
between 740 and 610 million years ago.

Deposition in the Southern Appalachian geosyncline appears to have begun before deposition in the Northern Appalachian geosyncline.

**FIGURE 9.30** Late Proterozoic deposits on the eastern margin of Ancestral North America include the thick sedimentary rocks of the Agardhsbjerg Formation in East Greenland. (Courtesy of H. R. Katz).

It may have begun about 1.15 billion years ago with the separation of Gondwanaland from Ancestral North America.

The oldest deposits in the Cordilleran geosyncline are diamictites which are found along a zone extending from northwestern Canada and eastern Alaska to southern California (**13**) (Fig. 9.29 and 9.31b). The diamictites commonly contain striated and faceted pebbles, and in northern Utah they rest on grooved, striated, and polished basement (**107, 112**). In some areas, the diamictites are overlain by volcanics which have been dated at 830 to 920 million years by the whole-rock potassium-argon method (**113**) (Fig. 9.29). The underlying diamictites may have been deposited about 875 million years ago at a time of widespread glaciation in Africa and elsewhere. Paleomagnetic data indicate that the diamictites were deposited between 10°N and 20°N latitude. Sedimentary rocks above the volcanics grade upward without apparent break into deposits of Cambrian age.

A thick sequence of Late Proterozoic sills and lavas is exposed on Victoria Island and on the shores of the Coronation Gulf in the Northwest Territories of Canada. Potassium-argon dates obtained on these units indicate that they are between 650 and 700 million years old (**114, 115**). Dikes of about the same age have been found in northern Wyoming (**58**).

## ANCESTRAL EUROPE

Precambrian rocks are exposed in the Baltic Shield, the Ukrainian Shield, and in southern Europe. Precambrian rocks are also exposed in a belt extending from eastern Newfoundland to

(a)                                              (b)

**FIGURE 9.31** Late Proterozoic diamictites: **(a)** The Sveanor Tillite from Spitsbergen. (Photo by A. Hjelle, courtesy of S. Winsnes, Norsk Polarinstitutt.) **(b)** The Rapitan Group from the Mackenzie Mountains, Northwest Territories. (Photo by Grant Young.)

eastern Massachusetts. These areas were part of Ancestral Europe during the Precambrian. Throughout most of eastern and central Europe, Precambrian rocks are covered by flat-lying sedimentary rocks of Paleozoic age.

## Archean Era (More Than 2.5 Billion Years Ago)

The oldest known rocks in Ancestral Europe are gneisses which crop out in the northeastern part of the Baltic Shield and in the Ukrainian Shield (Fig. 9.32). Radiometric dates show that these rocks were metamorphosed between 3.0 and 3.1 billion years ago (**116**). These ancient rocks are overlain unconformably by a sequence of sedimentary and volcanic rocks that were deformed, metamorphosed, and intruded about 2.6 billion years ago (**117**). It is interesting to note that metamorphism and intrusion occurred in Ancestral North America about 3.1 and 2.6 billion years ago.

## Early Proterozoic Era (2.5 to 1.7 Billion Years Ago)

Early Proterozoic sedimentary and volcanic rocks occur in the Baltic and Ukrainian shields and have been encountered in deep wells in eastern Europe. A sequence of quartzite, dolomitic marble, iron formation, and schist more than 5000 m (15,000 ft) thick is exposed in Scandinavia, in the northwestern corner of the Soviet Union, and in the eastern Ukraine. To the west is a eugeosynclinal sequence consisting of tens of thousands of feet of Early Proterozoic metamorphosed shale and volcanics with few limestones or quartzites.

The rocks in the western part of the Baltic and Ukrainian shields were deformed, metamorphosed, and intruded at the end of the Early Proterozoic age, 1.7 billion years ago. This deformation may have been caused by a collision between Ancestral Europe and Ancestral North America.

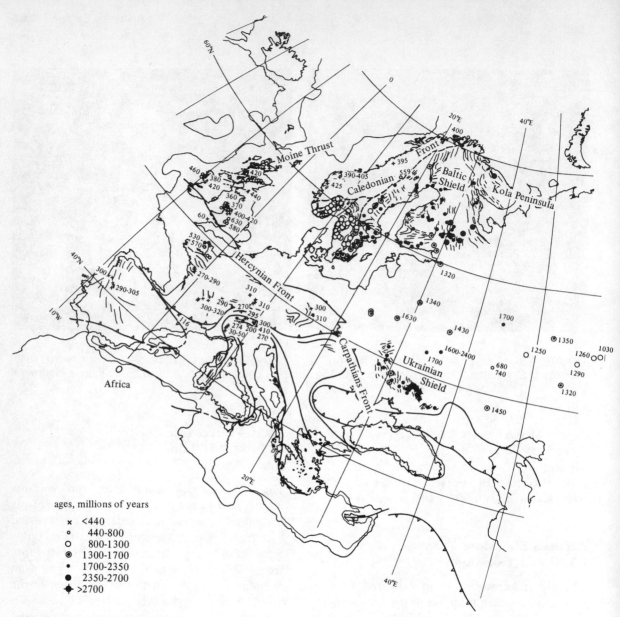

**FIGURE 9.32** Radiometric dates for Europe. (From Hurley and Rand, Ref. 7. Copyright © 1969 by American Association for the Advancement of Science.)

### Middle Proterozoic Era (1.7 to 1.0 Billion Years Ago)

Mafic (basaltic) dikes which were intruded in Sweden and Finland during the early part of the Middle Proterozoic (**118**) may have been associated with a beginning of separation of Ancestral North America and Ancestral Europe. In Scandinavia, deposition of quartzite, conglomerate, shale, and felsic volcanics seem to have followed the proposed rifting. The thickness of these deposits generally increases toward the margins of the Baltic Shield, which would have been the site of the rifting.

Widespread mafic (basaltic) igneous activity occurred in many parts of the Baltic Shield about 1.3 billion years ago at about the same time as the Mackenzie igneous event in Ancestral North America. The Jotnian Dolerites, Vaasa Dolerites, and Jotnian Basalts were emplaced at that time (**119**).

A period of deformation, metamorphism, and intrusion occurred in the western part of the Baltic Shield between 1.2 and 1.0 billion years ago (**120**). This activity resulted in the formation of a mountain range, and it may have been caused by a collision of Ancestral Europe with Gondwanaland and/or Ancestral North America.

### Late Proterozoic Era
### (1.0 to 0.58 Billion Years Ago)

Great thicknesses of sedimentary and volcanic rocks were deposited during the Late Proterozoic in geosynclines bordering Ancestral Europe. These include the Southeastern Caledonian, Northern Appalachian, the Hercynian, and Uralian geosynclines (Fig. 10.27).

The Southeastern Caledonian geosyncline extends from southern England through Norway to Spitsbergen. Late Proterozoic deposits in this geosyncline consist of up to 5000 m (15,000 ft) of sandstone, shale, limestone, dolostone, and diamictites (two rather closely spaced horizons that are probably of glacial origin). These deposits are remarkably similar to deposits of the same age from the East Greenland geosyncline (**109**). Both the Southeast Caledonian and East Greenland geosynclines may have formed as a result of the separation of Ancestral North America from Ancestral Europe beginning about 675 million years ago.

The eastern part of the Northern Appalachian geosyncline extends from eastern Massachusetts through Maine and Nova Scotia to eastern Newfoundland. The Hercynian geosyncline extends from southern England through Normandy (France) to southern Germany. Before the breakup of Pangaea during the Mesozoic era, these two geosynclines were part of a continuous belt (Fig. 10.27). Sedimentary rocks in these geosynclines are dominantly shale, but sandstone, conglomerate, and diamictite are also present. Volcanic rocks include basalt, andesite, and rhyolite. Chemical composition of these rocks indicates that they were deposited in an island arc environment (**121**). The sedimentary and volcanic rocks in these geosynclines were folded and intruded by granitic rocks between 575 and 600 million years ago during the Avalonian orogeny. Granitic rocks of this age are found in eastern Massachusetts, eastern Newfoundland, southeastern Ireland, southern England, the Channel Islands between England and France, Normandy, and southern Germany (**122**). The beginning of volcanism and deformation may mark the beginning of closure of the ocean between Gondwanaland and Ancestral Europe as these two continents began to move toward each other.

Late Proterozoic diamictites in Normandy, Norway, and eastern Newfoundland are probably glacial in origin. The diamictites in Normandy were deposited between 560 and 630 million years ago on the basis of radiometric dates on associated units (**110**). Paleomagnetic data indicate that the areas where the diamictites are found were at a latitude of between 15°S and 45°S at the time of deposition of the diamictites.

## GONDWANALAND

Precambrian rocks are exposed on all of the continents that comprised Gondwanaland. Paleomagnetic, lithological, and stratigraphic data from Gondwanaland support the idea that the Wilson cycle operated between Gondwanaland and other continents, and perhaps within Gondwanaland throughout the Precambrian.

### Archean Era (More Than
### 2.5 Billion Years Ago)

One of the remarkable features of the Archean history of Gondwanaland is that it is so similar

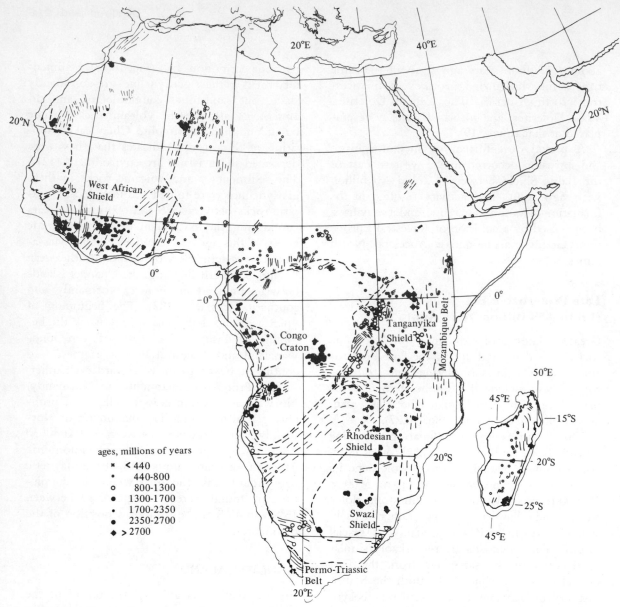

**FIGURE 9.33** Radiometric dates for basement rocks in Africa. Data were obtained mainly by the potassium-argon method. (From Hurley and Rand, Ref. 7. Copyright © 1969 by American Association for the Advancement of Science.)

to that of Ancestral North America. Both areas had orogenic episodes at about 3.6, 3.0, and 2.7 billion years ago, and both areas have a number of greenstone belts in which volcanic activity occurred about 2.75 billion years ago.

The oldest known rocks in Gondwanaland are gneisses in southern Africa. Whole-rock rubidium–strontium dates on granitic rocks intrusive into these gneisses range from 3.42 to 3.58 billion years (**123**, **124**). This same method of dating also indicates that a second orogenic episode occurred at about 3.0 billion years in southern Africa, central Africa, India, Australia, and Antarctica (**124–126**).

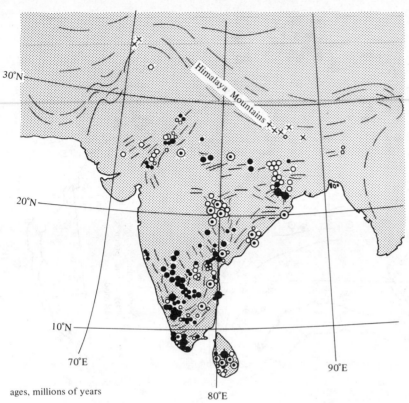

ages, millions of years

× &lt;440
∘ 440–800
○ 800–1300
◉ 1300–1700
• 1700–2350
● 2350–2700
⬢ &gt;2700

**FIGURE 9.34** Radiometric dates for basement rocks in India. (Data from Hurley and Rand, Ref. 7 and Crawford, Ref. 126. Copyright ©️ 1969 by American Association for the Advancement of Science.)

Greenstones approximately 2.7 billion years old are found in greenstone belts in India, Australia, and southern Africa (**123**). Following volcanism in these belts, there was a very extensive period of metamorphism and intrusion which lasted from about 2.7 to 2.5 billion years ago (**125**). Radiometric dates in this range are so widely distributed that it is very likely that much of Gondwanaland was transformed into mountains at that time (Figs. 9.33 to 9.37). The formation and deformation of the greenstone belts may have been related to the opening and closing of small oceans between microplates.

There were, however, some parts of Africa that were not affected by the folding and meta-morphism 2.7 to 2.5 billion years ago. In southern Africa, relatively undeformed and unmetamorphosed lavas and quartzites of the Dominion Reef sequence are about 2.85 billion years old (**128**). The Witwatersrand sequence overlies the Dominion Reef sequence and it is composed of quartzite, shale, conglomerate, and minor volcanics. The Ventersdorp sequence, which overlies the Witwatersrand sequence has been dated at between 2.75 and 2.6 billion years (**128**). The reason that these ancient rocks are not deformed may be that they were near the center of a moderately large, stable block which has not been fragmented or compressed in the last 2.85 billion years. Rifting and

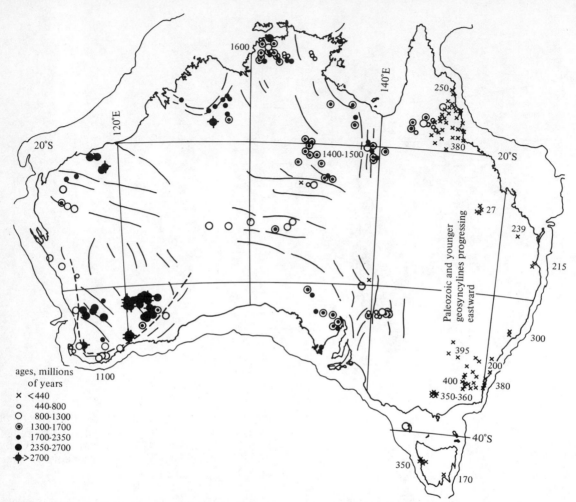

**FIGURE 9.35** Radiometric dates for basement rocks in Australia. (Data from Hurley and Rand, Ref. 7. Copyright © 1969 by American Association for the Advancement of Science.)

compression would have been restricted to the margins of this block during that interval.

### Early Proterozoic Era
### (2.5 to 1.7 Billion Years Ago)

Early Proterozoic sedimentary and volcanic rocks are widely exposed in Gondwanaland. Many of these may have been initiated by the breakup of blocks within Gondwanaland and by the separation of Gondwanaland from An-

cestral North America and Ancestral Europe. Arkosic sandstones and basaltic lavas from the base of the Transvaal sequence (Wolkberg Group) in southern Africa may have formed during a period of rifting about 2.40 billion years ago. Moreover, a large, east-west trending, mafic dike swarm (Widgiemooltha dikes) was intruded in southwestern Australia and a swarm of mafic dikes (Obusi dolerites) was intruded in Ghana at about the same time (2.42 billion years ago) (**129, 130**). The emplacement

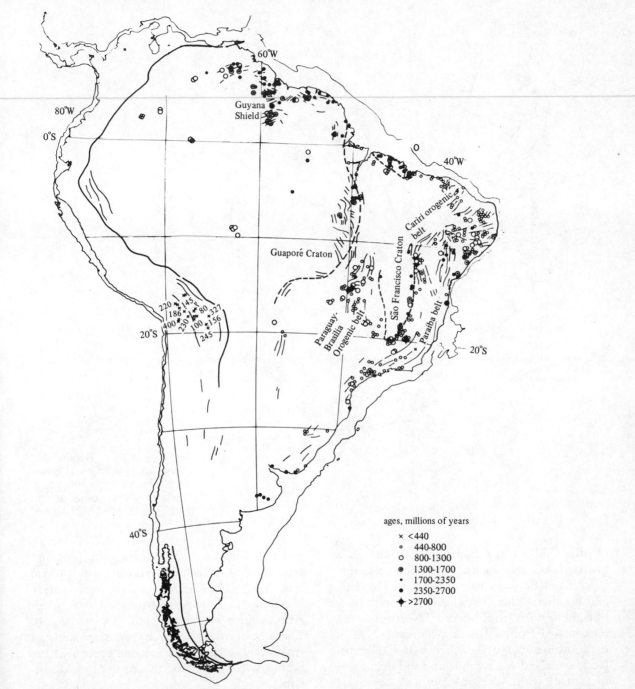

**FIGURE 9.36** Radiometric dates for basement rocks in South America. (Data from Hurley and Rand, Ref. 7. Copyright © 1969 by American Association for the Advancement of Science.)

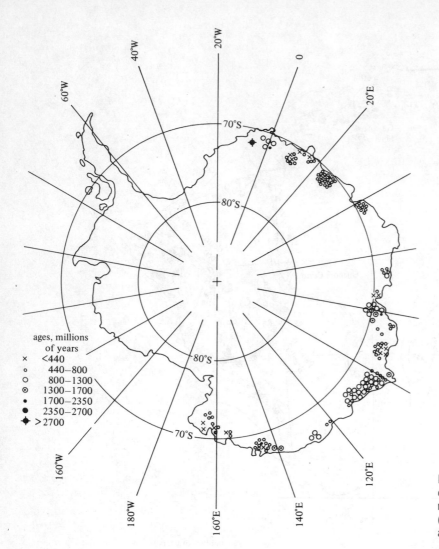

ages, millions
of years
× <440
∘ 440–800
○ 800–1300
⊙ 1300–1700
• 1700–2350
● 2350–2700
✦ >2700

**FIGURE 9.37** Radiometric dates for Precambrian basement rocks in Antarctica. (Data primarily from Piaciotto and Coppez, Ref. 127.)

of these dikes may have been related to a breakup of Gondwanaland 2.4 billion years ago. Paleomagnetic poles from the Obusi Dolerites and Widgiemooltha dikes agree well with paleomagnetic poles from Ancestral North America recalculated from a Pangaea reconstruction of the continents, and therefore it is likely that Gondwanaland was a single unit and that it was in contact with Ancestral North America 2.4 billion years ago. The recalculated A.P.W. curves from North America and Gondwanaland diverge between 2.4 and 2.0 billion years,

and therefore these continents were probably separating from each other during that interval (Fig. 9.4c). The character of the sedimentary rocks in the Anti-Atlas Mountains of North Africa suggests rifting 2.4 billion years ago (**9**). The zone of rifting between Gondwanaland and Ancestral North America would probably have been just northwest of the Anti-Atlas Mountains.

The Transvaal sequence contains dolostone, iron formation, shale, volcanics, and a tillite near its base. This tillite (Griquatown Glacials) is exposed over an area of 32,000 sq km (11,500 sq mi), and contains striated pebbles. Radiometric dating indicates that the tillite formed between 2.30 and 2.22 billion years ago at approximately the same time as the tillites in the Gowganda Formation (**131**). Paleomagnetic data (**132**) indicate that these glacial deposits accumulated within 30° of the South Pole. The iron formation is located near the middle of the Transvaal sequence, and it is underlain by dolostone and overlain by shale and volcanics. It has been intruded by rocks dated at about 1.95 billion years and is therefore between 2.3 and 1.95 billion years old. In central and western Australia, felsic (granitic in composition) volcanics in a thick sequence of dolostone, sandstone, iron formation, and shale have been radiometrically dated at between 2.2 and 2.0 billion years (**133**). Similar sequences have also been found in South America and in India where they have been dated at between 2.1 and 1.8 billion years (**134**).

Whole-rock rubidium-strontium dating has indicated that widespread deformation and metamorphism took place 2.0 billion years ago in South America and Africa (**135**). This orogenic episode is called the Trans-Amazon orogeny, and it may have been caused by collisions between microplates within Gondwanaland. Alternatively, this may have been the time when subduction began as the continents moved back toward each other.

The Bushveld Complex, an extensive sequence of mafic to felsic igneous rocks in southern Africa, has been dated at about 1.95 billion years (**136**). Its origin has been attributed to the impact of three giant meteorites (**136**), but the evidence for this is rather weak.

An episode of folding, metamorphism, intrusion, and mountain building affected many parts of Gondwanaland near the end of the Early Proterozoic, beginning about 1.85 billion years ago (**137, 138**). As was discussed earlier, paleomagnetic data (Fig. 9.4c) indicate that Ancestral North America collided with Gondwanaland at this time. The Penokean fold belt, which may have been formed as a result of this collision, is located at the place where Ancestral North America would have been joined to Gondwanaland.

## Middle Proterozoic Era (1.7 to 1.0 Billion Years Ago)

In the Anti-Atlas Mountains of North Africa, shallow water deposits are overlain by a sequence of volcanic rocks and shale which appear to have been deposited in a deep water environment about 1.7 billion years ago (**9**). Paleomagnetic data (Fig. 9.4) indicate that Ancestral North America began to separate from Gondwanaland at that time. Therefore the change from a shallow water to a deep water environment may have occurred in response to rifting on the continental margin of Africa. Diabase intrusions from the Ivory Coast and from Surinam, South America (in the Roraima Formation) are about 1.7 billion years old (**139**) and they may have been deposited during a time of fragmentation of Gondwanaland.

Diabase dike swarms were intruded and mafic lavas extruded between 1.25 and 1.30 billion years ago in several parts of Africa. These include the Waterberg Dolerites, Guperas Formation lavas, and the Barby Formation lavas. This mafic igneous activity may have been part of a rift system (**140**). It is interesting to note that this igneous activity occurred at the same time as the Mackenzie igneous event in North America and a similar event in Europe. Apparently mafic igneous activity was very widespread at this time. This activity could have been associated with the beginning of the closing of the oceans between North America, Europe, and Gondwanaland.

There are a number of orogenic belts in Gondwanaland in which the rocks were deformed, metamorphosed, and intruded between 1.2 and 1.0 billion years ago. Some of these orogenic belts may have been caused by a collision between Ancestral North America and Gondwanaland, while others may have been caused by collisions between plates within Gondwanaland.

Mafic dikes were also emplaced in several parts of Gondwanaland about 1.15 billion years ago. These include the Kabaledo dikes in South America, the Kleindrans dikes in Africa, and the Lakeview Dolerites in Australia. These dikes may have been intruded during a time of fragmentation of Gondwanaland, or they may have been emplaced during the separation of Gondwanaland from Ancestral North America.

## Late Proterozoic Era (1.0 to 0.58 Billion Years Ago)

Red beds, volcanic rocks, and deep water sedimentary rocks of Late Proterozoic age are exposed in the Atlas Mountains of Africa (**9**). These deposits may have formed in response to the separation of North America from Gondwanaland 1.15 billion years ago (Fig. 9.4a).

Late Proterozoic sequences containing diamictites are present in many parts of Gondwanaland including central, northwest, and southern Africa, Antarctica, northeast South America, and in Australia in New South Wales, Western Australia, Northern Territories, and Tasmania (**140**) (Fig. 9.38). Three different diamictite horizons have been reported in Africa and Australia (**111, 141**). Radiometric dating indicates that the youngest diamictite is about 610 million years old, the middle diamictite is about 740 million years old, and the oldest diamictite is about 875 million years old (**110, 111**). The diamictites contain striated pebbles, and in one locality in Africa and three localities in Australia, diamictites rest on a striated basement (**141**). The diamictites are associated with thick limestones and dolostones and with stromatolite reefs (**142**), all indicators of a warm climate. Paleomagnetic data indicate that:

1. Central Africa and northeastern South America were on the equator 740 million years ago when diamictites in these areas were deposited.

2. Almost all of Africa was within 20° of the equator about 875 million years ago when diamictites in these areas were deposited.
3. Several areas in Africa which were glaciated 610 million years ago were very close to the equator when the diamictites were deposited.

Thus, these glacial deposits were laid down in very low latitudes.

Deformation, metamorphism, and intrusion occurred in many parts of Gondwanaland between about 560 and 620 million years ago (**144**) (Fig. 9.33 to 9.38). This deformation apparently affected primarily the center rather than the margins of the continents.

## Cause of Deformations

The Precambrian fold belts within the continents comprising Gondwanaland may have been due to plate collisions, or they may have been due to stresses operating within plates. Paleomagnetic data have been cited to support the proposal that Proterozoic deformations within and between the Gondwanaland continents is not due to the convergence of widely separated continents (**143**). We believe that this has been established for some Late Proterozoic deformations, but not for the Early and Middle Proterozoic deformations. In order to establish that two areas (provinces or continents) have *not* moved relative to each other, it must be shown that a paleomagnetic pole determined from rocks of some given age from one area agrees (coincides) with a paleomagnetic pole *of the same age* from another area. Even when this has been done, it establishes only that the two areas were then (at the time of the paleomagnetic pole) in the same positions relative to each other as they are today. Relative movements could have occurred before the time marked by the coincidence of paleomagnetic poles or, if the continent returned to its previous position, after the time of the coincidence of paleomagnetic poles.

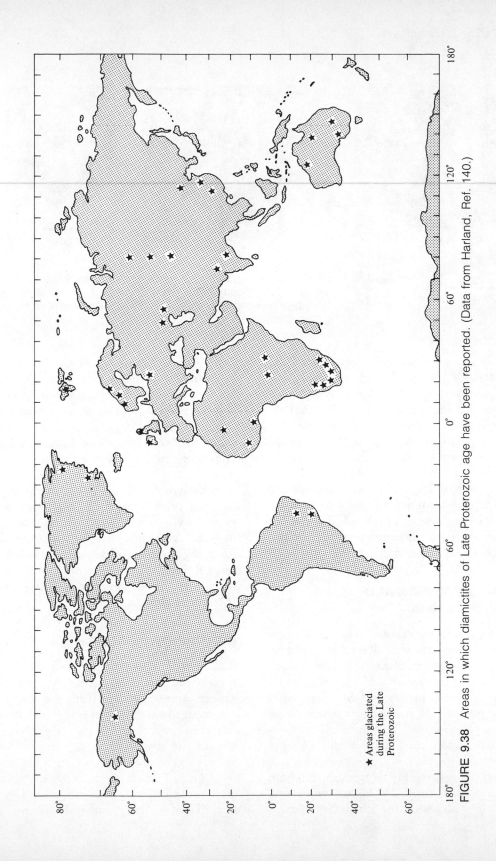

FIGURE 9.38  Areas in which diamictites of Late Proterozoic age have been reported. (Data from Harland, Ref. 140.)

★ Areas glaciated
during the Late
Proterozoic

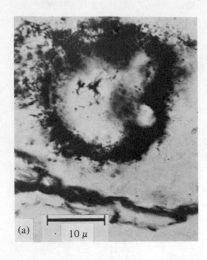

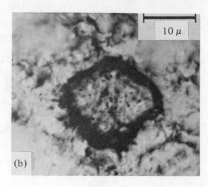

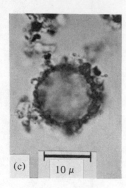

**FIGURE 9.39** Algalike microfossils of Archean age from the Onverwacht Series of South Africa. (From Engel *et al.*, Ref. 148. Copyright ⓒ 1968 by American Association for the Advancement of Science.)

Late Proterozoic paleomagnetic poles from Africa agree rather well with paleomagnetic poles of the same age from Australia and India if the poles are recalculated for a Pangaea reconstruction of the continents. Therefore, it is likely that there were no large-scale movements between or within these continents. However, comparison of Early and Middle Proterozoic paleomagnetic poles within or between the continents which comprise Gondwanaland can not be done because there are no two poles of the same age from different provinces which can be compared to each other.

## ANCESTRAL SIBERIA AND ANCESTRAL CHINA

Crystalline rocks of Precambrian age are exposed in three relatively small shields in Siberia. Gneisses of the Anabar Shield in northern Siberia are dated at more than 3.55 billion years by the uranium-lead method (**145**). Somewhat younger gneisses of the Angara Shield were folded, metamorphosed, and intruded approximately 2.6 billion years ago. A still younger sequence of metasedimentary and metavolcanic rocks that unconformably overlies the Archean sequence was intruded 1.8 billion years ago. The youngest sequence, the Sinian Complex, is

composed of shale, sandstone, limestone, and dolomites. This sequence is conformably overlain by Lower Cambrian sedimentary rocks, and therefore it is in part Late Proterozoic.

Precambrian rocks occur at the surface in eight relatively small shield areas in China as well as one in South Korea and another in Vietnam.

Although few radiometric dates are available, the relative ages of these sequences are known. The oldest rocks are gneisses, which may be Archean. These gneisses are unconformably overlain by schists. Relatively undeformed limestones, quartzites, and shales of Middle or Late Proterozoic age lie unconformably above the schists.

## CONTINENTAL GROWTH

Radiometric dates determined on basement rocks from Ancestral North America show a general decrease in age from the center of the continent toward the margins (Fig. 9.8). In Ancestral Europe, there is a general decrease in age toward the west (Fig. 9.32), and in Australia there is a general decrease in age toward the east (Fig. 9.35). This relationship has led to the suggestion that the continent has grown by accretion of sedimentary and volcanic rocks onto older continental nuclei (**4**). However, in Ancestral North America, rocks 2.9 billion years old or older are found in four different areas separated by zones of younger rocks (Fig. 9.8). Furthermore, there are some relatively ancient rocks near the margins of some of the continents. In southern California, rocks 1.0 billion years old crop out only 330 km (200 mi) from the continental margin, and in California and Washington State, rocks 1.7 billion years old crop out only 500 km (300 mi) from the margin of the continent (Fig. 9.8). In Scotland, rocks 2.6 billion years old occur less than 330 km (200 mi) from the margin of Ancestral North America (Fig. 9.32), and, in Peru, 2.0 billion-year-old rocks are found less than 50 km (30 mi) from the edge of the continent (**146**).

The occurrence of ancient rocks near the margins of the continents suggests that the rate of continental growth was probably more rapid during the early part of the Archean than later or that Precambrian rifting has affected these areas.

## PRECAMBRIAN LIFE

There is no direct evidence in the fossil record of just when and how life first evolved on the earth. The oldest known fossils are in rocks which are about 3.8 billion years old, but life may have evolved long before then.

### Archean

The oldest rocks which contain possible fossils are in the Isua sequence in southwestern Greenland. These deposits contain spherical structures 5 to 30 microns (.0002 to .0012 in.) in diameter which may be from primitive algae (**147**). As discussed earlier, the Isua sequence is probably between 3.83 and 3.60 billion years old. Similar microstructures have been found in carbonaceous cherts of the Onverwacht Group in South Africa (**148**) (Fig. 9.39). These rocks have been dated at approximately 3.35 billion years old (**149**). Although the organic nature of these microstructures has not been conclusively es-

tablished, their association with carbonaceous matter supports an organic origin. These structures may have been produced by algae or bacteria. The overlying rocks of the Fig Tree series have been dated at about 3.1 to 3.2 billion years, and they contain microscopic structures that are almost certainly fossils (**150**). Some of the cells show evidence of binary fission (Fig. 9.40). These probably represent the remains of primitive, *procaryotic* (without distinct nuclei) algae and bacteria.

The earliest known stromatolites occur in the 3.0 to 3.1 billion-year-old Bulawayan series in Rhodesia (**151**). The organisms that built the stromatolites were probably similar to the modern blue-green algae, and if so they are the oldest organism known to be photosynthetic.

### Early Proterozoic

Fossil stromatolite reefs are common in Early Proterozoic units throughout the world (Fig. 9.41). Thus, blue-green algae were probably abundant in Early Proterozoic seas. *Microfossils* (fossils too small to be seen with the naked eye) from the Transvaal System are the oldest undoubted remains of blue-green algae, and the oldest known fossils with diversified cells (**152**). The Transvaal System is between about 2.3 to 2.0 billion years old (**133, 136**). Microfossils from the Gunflint Formation, which is part of the 2.0 billion-year-old Anamikie Group, include algal filaments resembling those of modern blue-green algae and bacteria (**153**) (Fig. 9.42). Blue-green algae and bacteria are procaryotic. Some of the microstructures from the Gunflint Formation may be from *protoeukaryotes,* organisms which are transitional between procaryotic and *eukaryotic* (possessing a well-defined nucleus) organisms (**154**). Possible protoeukaryotes have also been reported from the Belcher Islands in rocks of about the same age (**154**). However, some investigators have questioned all reports of eukaryotic fossils older than 900 million years (**155**). Fossil bacteria have been found in the

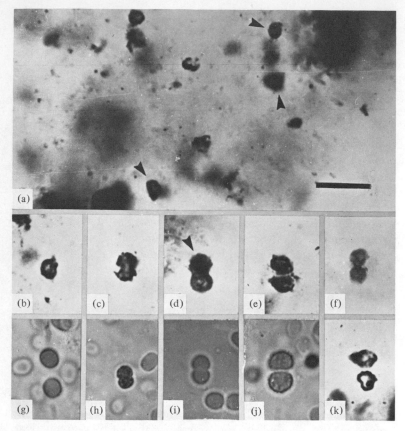

FIGURE 9.40 Microfossils from the Fig Tree Series of South Africa. Arrows indicate discernible individuals. **(a)** Several single microfossils. **(b)** to **(f)** Stages of cell division preserved. **(g)** to **(j)** Stages in cell division of a modern prokaryotic organism. (From Knoll and Barghorn, Ref. 150. Copyright © 1977 by American Association for the Advancement of Science.)

FIGURE 9.41 Early Proterozoic stromatolites viewed from the top on a surface parallel to bedding at Great Slave Lake, northern Canada. (Photo by P. F. Hoffman, courtesy of the Geological Survey of Canada, Ottawa.)

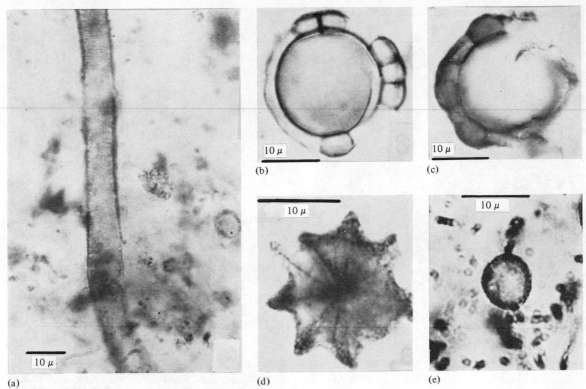

**FIGURE 9.42** Early Proterozoic microfossils from the Gunflint chert of the Anamikie Group: **(a)** *Anamikies septata,* an algal filament. **(b)** and **(c)** *Eosphaera tyleri,* an organism of uncertain affinity. **(d)** *Kakabekia umbellata,* an organism of unknown affinity. **(e)** *Huroniospora microreticulata.* (Courtesy of E. S. Barghoorn, Harvard University, Ref. 153. From Barghoorn and Tyler, copyright © 1965 by American Association for the Advancement of Science.)

Huronian sequence in the Gowganda Formation (**156**), and bacteria are thought to have played an important role in the precipitation of the iron in the Anamikie iron formation.

## Middle Proterozoic

Stromatolite reefs are common in the Belt Supergroup and in the Grand Canyon Supergroup. Possible worm trails and burrows have also been reported from these units (**157, 158**) (Fig. 9.43). Furthermore, structures resembling impressions of jellyfish have been found in the Grand Canyon Supergroup (**159**) (Fig. 9.44a). The rocks in which the structures have been found are between about 1.3 and 1.15 billion years old. These structures may be the oldest known fossils from multicellular animals. However, inorganic origins for these structures have been proposed and it is possible that they are not fossils. Unequivocal worm trails have been found in southern Africa in rocks about 1.0 billion years old (**160**). Microfossils from the Beck Springs area of southern California are almost certainly from eukaryotic organisms (**161**). This is probably the oldest known occurrence of true eukaryotic organisms. The rocks in which they have been found are about 1.4 billion years old (**161**). Fossil green algae from the 1.0 to .9 billion-year-old Bitter Springs Limestone of Australia are thought to be eukaryotic on the basis of the presence of structures within the cells which resemble a nucleus

FIGURE 9.43 Probable worm trails from the Middle Proterozoic rocks of the Belt Series, Dawson Pass, Glacier National Park, Montana. (Courtesy of Fenton and Fenton, Ref. 157.)

(**162**). Some of the cells are preserved in the process of division.

The oldest unequivocal fossil animals are one-celled chitinozoans from the Kwagunt Formation in the Chuar Group of the Grand Canyon Supergroup (**163**) (Fig. 9.45). These are probably about 1.2 billion years old. The oldest megascopic (visible without the aid of a microscope) plant fossil is the 2 to 3 mm (.08 to .12 in.) diameter *Chuaria* (Fig. 9.46) from rocks of about the same age.

## Late Proterozoic

The oldest fossils which show evidence of sexual reproduction are found in rocks which are about 900 million years old (**164**) (Fig. 9.47). These units have been radiometrically dated at more than 790 million years old (**165**). Worm trails have also been found in Late Proterozoic rocks in Nevada 3000 ft below the base of the Cambrian (**166**) and in North Carolina in a

(a)

(b)

FIGURE 9.44 Possible impressions of jellyfish from the Grand Canyon Series: (a) *Brooksella canyonensis* from the Nankoweap Group. (From Bassler, Ref. 159, courtesy of the Smithsonian Institution.) (b) From the Unkar Group south of Bright Angel Creek. (Courtesy of Raymond M. Alf Museum, Webb School of California, Claremont.)

sequence of rocks that have been dated by the uranium-lead method on zircon at 620 million years old (**167**) (Fig. 9.47).

The most famous of the Late Proterozoic metazoan (multicellular animal) fossils are those found in the Edicaran Hills of Australia (**168**). The fossils found here include impressions of jellyfish, worms, worm trails, sea pens,

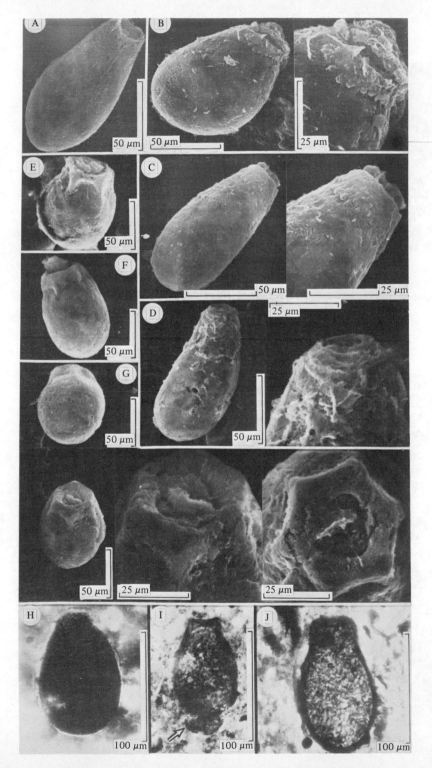

**FIGURE 9.45** Chitinozoans from the Kwagunt Formation of the Grand Canyon Supergroup. (From Bloesser *et al.* Ref. 163. Copyright © 1977 by American Association for the Advancement of Science.)

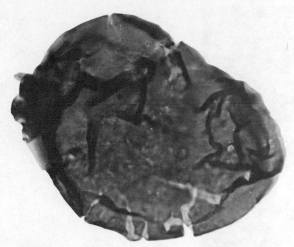

FIGURE 9.46 *Chuaria* from the Chuaria Shale, Grand Canyon supergroup. *Chuaria* is a fossil of uncertain affinity, but it is probably from a plant. (Courtesy of C. Downie.)

sponges, coral-like forms, and arthropods (Fig. 9.48). This fauna is probably about 680 million years old (**167**). Fossils similar or identical to the Edicaran fauna have also been found in a number of other localities around the world (**168**) including southern England and eastern Newfoundland (**169**) (Figs. 9.49 and 9.50).

Although fossils are generally not used in correlating Precambrian strata, the distinctive morphology of Late Proterozoic stromatolites makes them useful in correlating beds of this age (**170**). Spores have been used in the determination of relative ages within Late Proterozoic deposits in the Soviet Union and China (**171**).

## OXYGEN IN THE ATMOSPHERE

As discussed in Chapter 8, the origin of life probably required a reducing atmosphere. The

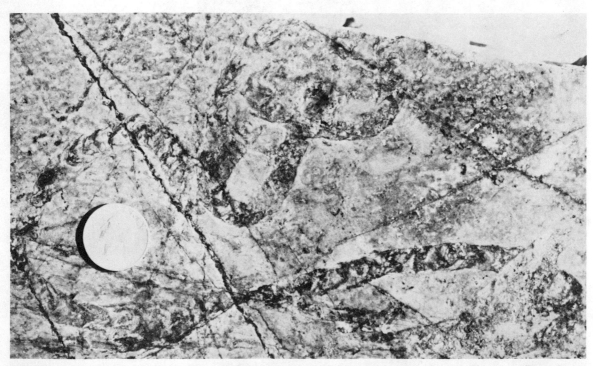

FIGURE 9.47 Worm trails (*Vermiforma antiqua*) from Late Proterozoic rocks of North Carolina. (From Cloud *et al.,* Ref. 142.)

**FIGURE 9.48** Late Proterozoic fossils from the Ediacaran Hills of Australia: **(a)** *Dickinsonia costata.* **(b)** *Spriggina floundersi,* a segmented worm. **(c)** *Ranges grandis,* a sea pen. **(d)** *Medusinites asteroides,* a possible jellyfish. (Courtesy of Prof. M. F. Glaessner, Ref. 168. Copyright © 1961 by Scientific American, Inc., all rights reserved.)

time of the transformation from a reducing atmosphere to one with abundant free oxygen is the subject of some controversy. Preston Cloud proposed that the time of this transition was about 2 billion years ago at about the time of the formation of the last thick sequences of iron formation and the first thick sequences of red beds (**172**). Cloud links the appearance of the first eukaryotic organisms with the appear-

ance of abundant free oxygen. However, others have placed the time of the appearance of abundant free oxygen before the end of the Archean era, more than 2.5 billion years ago (**178**). It is generally agreed that the oxygen in the atmosphere was originally released as a result of photosynthesis by blue-green algae, and therefore this process has probably been going on for more than 3.0 billion years (since the time of the

FIGURE 9.49 Late proterozoic fossils from south-eastern Newfoundland: Round lobate organisms probably jellyfish (**a, f**); spindle-shaped organisms (**c, e**); leaf-shaped organism (**d**); stalk (**g**); and base (**h**) of the leaf-shaped organisms. (Courtesy of S. B. Misra, Ref. 169.)

FIGURE 9.50 *Charnia masoni*, a spindle-shaped organism from Late Proterozoic deposits of England, three-quarters natural size. This impression is possibly of a primitive pennatulid type of coelenterate or of a fucoid seaweed. It was found in the Charnian rocks of Charnwood Forest near Leicester. (Courtesy of Trevor D. Ford.)

oldest known stromatolites). At first, the oxygen released would go into various oxygen-consuming reactions such as oxidation of iron in iron formation. A billion years or more does seem like a long time between the beginning of release of free oxygen and the beginning of abundant free oxygen in the atmosphere.

## PRECAMBRIAN CLIMATE

The reconstruction of ancient climates should be made within the context of polar wandering and continental drift, because climatic indicators must be compared with paleolatitudes rather than with modern latitudes. Paleoclimates may be investigated by plotting such data as the locations of glacial deposits, red beds, evaporites, reefs, and thick limestones on reconstructions of the continents.

Precambrian rock sequences provide evidence of wide fluctuations in climate, but generally the ranges of climates are similar to that indicated for the Phanerozoic. Thick limestones are generally found near the paleoequator, and glacial deposits are found within 40° of the poles except in the late Proterozoic.

## Archean

Unfortunately, little can be determined about the climatic conditions of the Archean, an era covering almost half of geologic time. One reason for this lack of data is that most Archean deposits have been metamorphosed to such an extent that the original character of the rock is not evident. However, even if paleoclimatic indicators were preserved in these rocks, it would be difficult to make any inferences about Archean climates, owing to the scarcity of Archean paleomagnetic data.

## Early Proterozoic

The presence of tillites indicates cold climatic conditions near the poles approximately 2.3 billion years ago in several parts of Ancestral North America and in Africa. Paleomagnetic measurements show that glaciers moved to about 50°N latitude in North America and 60°S latitude in South Africa. Much of the Superior Province was presumably at a relatively high elevation following the Kenoran orogeny and this may have helped cause the cooling which preceded the glaciation.

The thick limestones, dolomites, and iron formation of late Early Proterozoic age in North America, Africa, Australia, and India are indicative of deposition under warm climates. The iron concentrations in these deposits may have resulted from deep weathering in a tropical environment. During such lateritic weathering, large amounts of iron and silica would have been carried in solution from the continents to the oceans, where they may have been precipitated as iron oxide and chert. Paleomagnetic

data indicate that most deposits of iron formation were laid down within 45° of the equator. That latitude is consistent with the suggestion that these deposits were the result of lateritic weathering.

The iron formation and carbonates of the Anamikie Group are significantly younger than the tillites. The change in climate between the times of deposition of these two sequences may have been due in part to movement of the continents away from polar regions between the beginning and the end of the Early Proterozoic. Red beds have been found in the upper part of the Transvaal sequence in Africa (**128**). These are the oldest known red beds, and they suggest that southern Africa was in an arid belt about 2.0 billion years ago (**128**).

## Middle Proterozoic

The thick limestones of the Belt Supergroup and the thick marbles of the Grenville series suggest deposition in a warm climate. The presence of salt and gypsum in Grenville deposits suggests the climate may have been somewhat dry in these areas (**174**). Paleomagnetic data indicate that both units were deposited within 30° of the Middle Proterozoic equator.

## Late Proterozoic

Diamictites of Late Proterozoic age have been found on all continents except Antarctica (Fig. 9.34). The presence of striated and polished basement beneath the diamictites, and the presence of striated and faceted pebbles within the diamictites establishes that they are of glacial origin. Stratigraphic studies have shown that there are three distinctly different horizons which contain diamictites. Radiometric dating on units associated with the diamictites indicates that glaciation occurred approximately 875, 740, and 610 million years ago. Paleomagnetic studies indicate that glaciation often occurred on or very near the equator. Thus the worldwide climate during these intervals must have been much colder than that of today and even colder than that of the Pleistocene (last ice ages). One explanation for the equatorial glaciations is that the angle of inclination of the earth's axis to the plane of the ecliptic (the plane about which the earth orbits the sun) may have been significantly greater during the Proterozoic than it is today (**175**).

## Overview of Precambrian Climates

The Proterozoic lasted 2 billion years, and there are four known glaciations within it. This contrasts with the Phanerozoic (the last 575 million years) during which time there were three major glaciations. Thus, Proterozoic climates may have been slightly warmer on the average than they were during the Phanerozoic. There are no known glaciations between about 2.25 and 0.88 billion years ago. It is possible that this period was warmer than average due either to the capture of the moon in a prograde orbit at that time, or that this was the time of closest approach of the earth and moon as the moon's orbit changed from retrograde to prograde.

## SUMMARY

The oldest known rocks, which are in Greenland and Minnesota, have been dated at about 3.8 billion years old. The oldest known period of extensive deformation, metamorphism, and intrusion occurred about 3.6 billion years ago and affected North America, Greenland, Africa, and Siberia. These areas were affected by a second orogenic episode about 3.0 billion years ago. All continents went through a period of extensive mafic (basaltic) volcanic activity about 2.8 to 2.75 billion years ago and the volcanic rocks were folded, metamorphosed to greenstones, and intruded by granitic rocks near the end of the Archean era, about 2.73 billion years ago. A second period of granite intrusion occurred about 2.53 billion years ago.

The first geosynclines or geoclines formed during the Early Proterozoic. Iron formation was an important deposit in them. Two periods of deformation, metamorphism, and intrusion affected all continents near the end of the Early Proterozoic, the first about 1.85 billion years ago and the second about 1.70 billion years ago.

Middle Proterozoic geosynclines or geoclines developed on continental margins following deformation, metamorphism, and intrusion about 1.50 to 1.40 billion years ago. Still another orogenic episode occurred about 1.2 billion years ago followed by the development of new geoclines during the Late Proterozoic.

It is likely that the Wilson cycle was in operation during the Archaen. The orogenic activity near the end of the Archean may have been caused by the joining together of island arcs. It is probable that several continents were joined together near the end of the Early Proterozoic about 1.75 billion years ago, and it is possible that another joining of continents occurred 1.2 billion years ago near the end of the Middle Proterozoic. Rifting apart of the continents in Ancestral North America may have occurred 2.8, 2.3, 1.4, and 1.15 billion years ago. Rifting during these times may have occurred on other continents as well.

The earliest evidence we have of life is microscopic spherical bodies that are found in rocks about 3.8 billion years old. These structures may be the remains of primitive algae and bacteria. The oldest stromatolites indicate that blue-green algae probably appeared at least 3.0 billion years ago. The oldest probable eukaryotic organisms are in rocks about 1.4 billion years old, but possible eukaryotic organisms have been found in rocks as old as 2.0 billion years. The oldest possible metazoa are in rocks about 1.2 billion years old and the oldest unequivocal fossil from an animal, a microscopic chitinozoan, was found in rocks of about the same age. The oldest known unequivocal fossil remains of a metazoan (multicellular animal) are worm burrows in rocks about 1.0 billion years old. There are abundant fossil remains of soft-bodied metazoans in rocks about 680 million years old.

While little is known about the climate during the Archean era, the Proterozoic may have been slightly warmer than the average for the rest of geological time. Four episodes of glaciation have been recorded for the Proterozoic.

## REFERENCES CITED

1. G. Gastil, 1960, *Amer. J. Sci.,* v. 258, p. 1.
2. P. R. Dunn, D. A. Plumb, and H. G. Roberts, 1966, *Geol. Soc. Aust. J.,* v. 13, p. 593; H. L. James, 1972, *Amer. Assoc. Petrol. Geol. Bull.,* v. 56, p. 1026; C. H. Stockwell, 1973, *Geol. Survey Can. Pap. 72-52;* S. N. Sarkar, 1972, *Proc. 24th Int. Geol. Cong.,* Sec. 1, p. 260; M. A. Semikhator, 1974, *Acad. Sci. USSR, Geol. Inst. Trans.,* v. 256, p. 302.
3. R. Dearnley, 1965, *Phys. and Chem. of the Earth,* v. 7, p. 3.
4. R. S. Dietz and J. C. Holden, 1966, *Amer. Assoc. Petrol. Geol. Bull.,* v. 50, p. 351; P. M. Hurley and J. R. Rand, 1969, *Science,* v. 164, p. 1229; J. T. Wilson, 1948, *Trans. Amer. Geophys. Union,* v. 29, p. 691.
5. A. E. J. Engel, 1963, *Science,* v. 140, p. 143.
6. S. Moorbath, 1977, *Sci. Amer.,* v. 236, p. 92.
7. P. M. Hurley and J. R. Rand, 1938, *Science,* v. 164, p. 1238.
8. R. B. Hargraves, 1976, *Science,* v. 193, p. 363.
9. P. E. Schenk, 1971, *Can. J. Earth Sci.,* v. 8, p. 1218.
10. K. C. A. Burke and J. R. Dewey, 1973, *J. Geol.,* v. 1, p. 406.
11. J. M. Bird and J. F. Dewey, 1970, *Geol. Soc. Amer. Bull.,* v. 81, p. 1031.
12. F. J. Sawkins, 1976, *Geology,* v. 4, p. 427.
13. J. H. Stewart, 1972, *Geol. Soc. Amer. Bull.,* v. 83, p. 1345.
14. J. W. Sears and R. A. Price, 1978, *Geology,* v. 6, p. 267.
15. C. H. Stockwell, 1965, *Amer. Assoc. Petrol. Geol. Bull.,* v. 49, p. 887.
16. R. J. W. Douglas, 1969, *Geological Map of Canada:* Geol. Survey Can.
17. J. A. Grant, 1964, *Science,* v. 146, p. 1049; T. E. Krogh and G. L. Davis, 1971, *Carnegie Inst. Geophys. Lab., Ann. Rept. of Director, 1969–1970,* p. 339.
18. H. R. Wynne-Edwards, 1969, *Geol. Assoc. Can. Spec. Pap. 5,* p. 163.
19. H. Baadsgaard, 1976, *Earth Planet. Sci. Lett.,* v. 33, p. 261; S. Moorbath and R. J. Pankhurst, 1976, *Nature,* v. 262, p. 124; H. Baadsgaard, R. St. J. Lambert, and J. Krupicka, 1976, *Geochim. Cosmochim. Acta,* v. 40, p. 513.
20. J. L. Wooden, S. S. Goldich, and G. N. Ankenbauer, 1975, *Geol. Soc. Amer. Abstr. Programs,* v. 7, p. 1322.
21. S. S. Goldich, C. E. Hedge, and T. W. Stern, 1970, *Geol. Soc. Amer. Bull.,* v. 81, p. 3671.
22. I. M. Barton, 1975, *Earth Planet. Sci. Lett.,* v. 27, p. 427; R. W. Hurst and G. R. Tilton, 1976, *Geol. Soc. Amer. Abstr. Programs,* v. 8, p. 933.
23. Z. E. Peterman and R. A. Hildreth, 1977, *U.S. Geol. Survey Rept. 77-140,* p. 1.
24. T. E. Krogh and G. L. Davis, 1971, *Carnegie Inst. Geophys. Lab., Ann. Rept. of Director, 1969–1970,* p. 339.
25. T. E. Krogh, N. B. W. Harris, and G. L. Davis, 1976, *Can. J. Earth Sci.,* v. 13, p. 1212.
26. T. E. Krogh, I. F. Ermanovics, and G. L. Davis, 1974, *Carnegie Inst. Geophys. Lab., Ann. Rept. of Director, 1973–1974,* p. 573.
27. J. C. Reed, Jr. and R. E. Zartman, 1973, *Geol. Soc. Amer. Bull.,* v. 84, p. 561.
28. R. C. Johnson and F. A. Hills, 1976, *Geol. Soc. Amer. Bull.,* v. 87, p. 809.
29. R. A. Heimlich and P. O. Banks, 1968, *Amer. J. Sci.,* v. 266, p. 180.
30. Z. Nikie, H. Baadsgaard, R. E. Folinsbee, and A. P. Leech, 1975, *Geol. Soc. Amer. Abstr. Programs,* v. 7, p. 1213.
31. Z. E. Peterman, S. S. Goldich, C. E. Hedge, and D. H. Yardley, 1972, *Geochronology of the Rainy Lake Region, Minnesota-Ohio:* Geol. Soc. Amer. Mem. 135, p. 193.
32. S. R. Hart and G. L. Davis, 1969, *Geol. Soc. Amer. Bull.,* v. 80, p. 595.
33. A. M. Goodwin, 1974, *in* C. A. Burk and C. L. Drake, eds., *The Geology of Continental Margins:* Springer-Verlag, New York, p. 767.
34. W. T. Jolly, 1975, *Geol. Soc. Amer. Abstr. Programs,* v. 7, p. 793.
35. B. F. Windley, 1973, *Phil. Trans. Roy. Soc. (London) Ser. A* v. 273, p. 321.

36. R. E. Folinsbee, 1968, *in* L. Knopoff, C. L. Drake, and P. J. Hart, eds., *The Crust and Upper Mantle of the Pacific Area:* Amer. Geophys. Union Mon. 12, p. 441.

37. B. Jahn, C. Shih, and V. R. Murthy, 1974, *Geochim. Cosmochim. Acta,* v. 38, p. 611.

38. W. R. Dickinson and T. Hatherton, 1967, *Science,* v. 157, p. 801.

39. A. Turek and Z. E. Peterman, 1971, *Can. J. Earth Sci.,* v. 8, p. 572.

40. R. S. Naylor and R. H. Steiger, and G. J. Wasserburg, 1970, *Geochim. Cosmochim. Acta,* v. 34, p. 1133.

41. G. N. Hanson, S. S. Goldich, J. G. Arth, and D. H. Yardley, 1971, *Can. J. Earth Sci.,* v. 8, p. 1110.

42. G. R. Tilton and R. H. Steiger, 1969, *J. Geophys. Res.,* v. 74, p. 2118.

43. R. H. Steiger and G. J. Wasserburg, 1969, *Geochim. Cosmochim. Acta,* v. 33, p. 1213.

44. P. K. Sims, Z. E. Peterman, and W. C. Prinz, 1977, *J. Res. U.S. Geol. Survey,* v. 5, p. 185.

45. L. T. Aldrich, G. L. Davis, and H. L. James, 1965, *J. Petrol.,* v. 6, p. 445.

46. G. M. Bennison and A. E. Wright, 1969, *The Geological History of the British Isles:* St. Martin's Press, New York.

47. P. K. Sims and Z. E. Peterman, 1976, *J. Res. U.S. Geol. Survey,* v. 4, p. 405.

48. F. A. Hills, P. W. Gast, R. S. Houston, and I. G. Swainbank, 1968, *Geol. Soc. Amer. Bull.,* v. 79, p. 1757; D. C. Green and H. Baadsgaard, 1971, *J. Petrol.,* v. 12, p. 177.

49. R. VanSchmus, 1965, *J. Geol.,* v. 73, p. 755; T. M. Gates and P. M. Hurley, 1973, *Can. J. Earth Sci.,* v. 10, p. 900.

50. C. S. Beals, 1968, *in* C. S. Beals and D. A. Shenstone, eds., *Science, History and Hudson Bay,* v. 2, p. 985; T. A. Mutch, 1970, *Geology of the Moon:* Princeton University Press, Princeton, N.J.

51. K. C. Condie, 1967, *Geol. Soc. Amer. Bull.,* v. 78, p. 1317; G. M. Young, 1970, *Paleogeography, Paleoclimatology, Paleoecology,* v. 7, p. 85.

52. D. T. A. Symons and R. J. O'Leary, 1975, *Geol. Soc. Amer. Abstr. Programs,* v. 7, p. 867.

53. D. A. Lindsey, 1966, *Science,* v. 154, p. 1442.

54. W. A. Morris, 1977, *Geology,* v. 5, p. 137; J. L. Roy and P. L. Lapointe, 1976, *Can. J. Earth Sci.,* v. 13, p. 749; D. T. A. Symnos, 1975, *Geology,* v. 3, p. 303.

55. P. Hoffman, J. F. Dewey, and K. Burke, 1974, *in* R. H. Dott, Jr. and R. H. Shaver, eds., *Modern and Ancient Geosynclinal Sedimentation:* Society of Economic Paleontologists and Mineralogists, Special Publication No. 19. p. 38; J. M. Barton, Jr., 1975, *Can. J. Earth Sci.,* v. 12, p. 1196.

56. G. D. Jackson and F. C. Taylor, 1972, *Can. J. Earth Sci.,* v. 9, p. 1650; W. R. Van Schmus, 1976, "The Penokean Orogeny of the Great Lakes Region," a paper presented at Stockwell Symposium on the Hudsonian Orogeny and Plate Tectonics, Ottawa, March 4–5, 1976, organized by Canadian Geodynamics Subcommittee, Geol. Survey of Can.

57. H. Spall, 1973, *Earth Planet. Sci. Lett.,* v. 18, p. 1.

58. E. E. Larson, R. Reynolds, and R. Hoblitt, 1973, *Geol. Soc. Amer. Bull.,* v. 84, p. 3231.

59. W. R. VanSchmus, personal communication.

60. P. Pasteels and L. T. Silver, 1966, *Geol. Soc. Amer. Spec. Pap. 87,* p. 124.

61. Z. E. Peterman, C. E. Hedge, and W. A. Braddock, 1968, *J. Geophys. Res.,* v. 73, p. 2277.

62. C. E. Hedge, Z. E. Peterman, and W. A. Braddock, 1967, *Geol. Soc. Amer. Bull.,* v. 78, p. 551.

63. J. N. Rosholt, Z. E. Peterman, and A. J. Bartel, 1970, *Can. J. Earth Sci.,* v. 7, p. 184.

64. Z. E. Peterman, 1966, *Geol. Soc. Amer. Bull.,* v. 77, p. 1031; J. M. Barton, Jr. and R. Doig, 1973, *Amer. J. Sci.,* v. 273, p. 376.

65. F. Barker, Z. E. Peterman, and R. A. Hildereth, 1969, *Contr., Mineral, and Petrol.,* v. 23, p. 271.

66. W. R. Van Schmus, L. G. Medaris, Jr., and P. O. Banks, 1975, *Geol. Soc. Amer. Bull.,* v. 86, p. 907.

67. T. E. Krogh and G. L. Davis, 1970, *Carnegie Inst. Geophys. Lab., Ann. Rept. of Director, 1968–1969,* p. 308.

68. T. E. Krogh and G. L. Davis, 1971, *Carnegie Inst. Geophys. Lab., Ann. Rept. of Director, 1969–1970,* p. 339.

69. T. E. Krogh and G. L. Davis, 1973, *Carnegie Inst. Geophys. Lab., Ann. Rept. of Director, 1972–1973,* p. 601.

70. F. Barker, Z. E. Peterman, W. T. Henderson, and R. E. Hildreth, 1974, *J. Res. U.S. Geol. Survey,* v. 2, p. 705.

71. Z. E. Peterman and C. E. Hedge, 1964, *U.S. Geol. Survey Prof. Pap. 475-D,* p. D100.

72. T. E. Krogh, G. L. Davis, L. T. Aldrich, and S. R. Hart with A. Stueber, 1968, *Carnegie Inst.*

*Geophys. Lab., Ann. Rept. of Director, 1966–1967,* p. 528.

73. W. R. Van Schmus and L. L. Woolsey, 1975, *Can. J. Earth Sci.,* v. 12, p. 1723.

74. M. D. Cavanaugh and C. K. Seyfert, 1977, *Geology,* v. 5, p. 207.

75. T. E. Krogh and G. L. Davis, 1974, *Carnegie Inst. Geophys. Lab., Ann. Rept. of Director, 1973–1974,* p. 567.

76. W. A. Gibbons and R. H. McNutt, 1975, *Can. J. Earth Sci.,* v. 12, p. 1970.

77. R. S. Dietz, 1964, *J. Geol.,* v. 72, p. 412.

78. J. D. A. Piper, 1977, *Earth Planet. Sci. Lett.,* v. 34, p. 247.

79. J. L. Roy and W. F. Fahrig, 1973, *Can. J. Earth Sci.,* v. 10, p. 1279.

80. L. T. Silver *et al.,* 1977, *Geol. Soc. Amer. Abstr. Programs,* v. 9, p. 1176.

81. N. Herz, 1969, *Science,* v. 164, p. 944; S. K. Saxena, 1977, *Science,* v. 198, p. 614.

82. L. E. Long, 1972, *Geol. Soc. Amer. Bull.,* v. 83, p. 3425.

83. M. E. Bickford and D. G. Mose, 1975, *Geol. Soc. Amer. Spec. Pap. 165,* p. 1.

84. G. J. Wasserburg and M. A. Lanphere, 1965, *Geol. Soc. Amer. Bull.,* v. 76, p. 735.

85. L. T. Silver and S. B. Lumbers, 1966, *Geol. Soc. Amer. Spec. Pap. 87,* p. 156.

86. D. P. Elston and C. S. Grommé, 1974, *in* T. N. V. Karlstrom, G. A. Swann, and R. L. Eastwood, *Geology of Northern Arizona, Part I—Regional Studies,* for Geol. Soc. Amer. Rocky Mt. Section Mtg.

87. B. J. Giletti and P. E. Damon, 1961, *Geol. Soc. Amer. Bull.,* v. 72, p. 639.

88. J. H. Stewart, 1976, *Geology,* v. 4, p. 11.

89. J. D. Obradovich and Z. E. Peterman, 1968, *Can. J. Earth Sci.,* v. 5, p. 737.

90. I. Vitorello and R. Van der Voo, 1977, *Can. J. Earth Sci.,* v. 14, p. 67.

91. W. A. Robertson, 1973, *Can. J. Earth Sci.,* v. 10, p. 180.

92. W. F. Fahrig and D. L. Jones, 1969, *Can. J. Earth Sci.,* v. 6, p. 679; W. R. Van Schmus, 1975, *Can. J. Earth Sci.,* v. 12, p. 1690.

93. R. K. Wanless and W. D. Loveridge, 1972, *Geol. Survey Can., Pap. 72–23,* p. 21.

94. C. E. Helsley and H. Spall, 1972, *J. Geophys. Res.,* v. 77, p. 2115.

95. R. R. Divi and W. K. Fyson, 1973, *Geol. Soc. Amer. Bull.,* v. 84, p. 1607.

96. R. K. Wanless, R. D. Stephens, G. R. Lachance, and C. M. Edmonds, 1968, *Geol. Survey Can., Pap. 67-2, Part A.*

97. R. Doig, 1977, *Geol. Soc. Amer. Bull.,* v. 88, p. 1843.

98. G. R. Tilton, G. W. Wetherill, G. L. Davis, and M. N. Bass, 1960, *J. Geophys. Res.,* v. 65, p. 4173.

99. E. Irving, R. F. Emslie, and H. Ueno, 1974, *J. Geophys. Res.,* v. 79, p. 5491.

100. A. B. Blaxland, O. van Breemen, C. H. Emeleus, and J. G. Anderson, 1978, *Geol. Soc. Amer. Bull.,* v. 89, p. 231.

101. W. R. Van Schmus, 1971, *Geol. Soc. Amer. Bull.,* v. 82, p. 3221.

102. G. Faure, S. Chandhuri, and M. D. Fenton, 1969, *J. Geophys. Res.,* v. 74, p. 720.

103. L. C. Ocola and R. P. Meyer, 1973, *J. Geophys. Res.,* v. 78, p. 5173.

104. E. Irving, 1964, *Paleomagnetism and Its Application to Geological and Geophysical Problems:* Wiley, New York.

105. G. S. Murthy, 1971, *Can. J. Earth Sci.,* v. 8, p. 802.

106. T. S. Winsnes, 1965, *in* K. Rankama, ed., *The Precambrian:* Wiley, New York, v. 2, p. 1.

107. L. J. G. Schermerhorn, 1974, *Amer. J. Sci.,* v. 274, p. 673.

108. J. Steiner and E. Grillmair, 1973, *Geol. Soc. Amer. Bull.,* v. 84, p. 1003.

109. D. W. Rankin, G. H. Espenshade, and K. W. Shaw, 1973, *Amer. J. Sci.,* v. 273-A, p. 1.

110. D. W. Rankin, T. W. Stern, J. C. Reed, Jr., and M. F. Newell, 1969, *Science,* v. 166, p. 741.

111. A. Kroner, 1977, *J. Geol.,* v. 85, p. 289.

112. R. W. Ojakangas and C. L. Matsch, 1976, *Geol. Soc. Amer. Abstr. Programs,* v. 8, p. 1035.

113. F. K. Miller, E. H. McKee, and R. G. Yates, 1973, *Geol. Soc. Amer. Bull.,* v. 84, p. 3723.

114. W. A. Robertson and W. R. A. Baragar, 1972, *Can. J. Earth Sci.,* v. 9, p. 123.

115. H. C. Palmer and A. Hayatsu, 1975, *Can. J. Earth Sci.,* v. 12, p. 1439.

116. S. E. Zykov, A. I. Tugarinov, I. V. Belkov, and E. V. Bilikova, 1964, *Geochem. Internat.,* No. 2, p. 262.

117. O. Kouvo and G. R. Tilton, 1966, *J. Geol.,* v. 74, p. 421.

118. K. J. Neuvonen, 1973, *Bull. Geol. Soc. Finland,* v. 45, p. 23; K. J. Neuvonen, 1974, *Bull. Geol. Soc. Finland,* v. 46, p. 75.

119. H. N. A. Priem, F. G. Mulder, N. A. I. M. Boelrijk, E. H. Hebeda, R. H. Verschure, and E. A. Th. Verdurmen, 1968, *Phys. Earth Planet. Int.,* v. 1, p. 373.

120. J. L. Kulp and H. Neumann, 1961, *in* J. L. Kulp, ed., *Geochronology of Rock Systems:* New York Academy of Sciences, v. 91, p. 469; K. Rankama, ed., 1963, *The Precambrian:* Wiley, New York, v. 1.

121. C. J. Hughes and W. D. Bruckner, 1971, *Can. J. Earth Sci.,* v. 8, p. 899.

122. W. Krebs and H. Wachendorf, 1973, *Geol. Soc. Amer. Bull.,* v. 84, p. 2611; R. St. J. Lambert and D. C. Rex, 1966, *Nature,* v. 209, p. 605; A. E. Wright, 1969, *Amer. Assoc. Petrol. Geol. Mem. 12,* p. 93; S. J. Cribb, 1975, *J. Geol. Soc.,* v. 131, p. 203.

123. S. Moorbath, J. F. Wilson, and P. Cotterill, 1976, *Nature,* v. 264, p. 536; H. Allsopp, H. Roberts, G. Schreiner, and D. Hunter, 1962, *J. Geophys. Res.,* v. 67, p. 5307.

124. R. D. Davies and H. L. Allsopp, 1976, *Geology,* v. 4, p. 553.

125. R. A. Old and D. C. Rex, 1971, *Geol. Mag.,* v. 108, p. 353; M. Halpern, 1970, *Science,* v. 169, p. 977; A. L. Hales, 1961, *in* J. L. Kulp, ed., *Geochronology of Rock Systems:* New York Academy of Sciences, v. 91, p. 524.

126. A. R. Crawford, 1969, *Nature,* v. 223, p. 380.

127. E. Picciotto and A. Coppez, 1961, *Extrait des Annales de la Société Geologique de Belgique,* v. 85, p. 263.

128. P. Cloud, 1976, *Major Features of Crustal Evolution:* Geol. Soc. South Africa, Annex to v. 79.

129. M. E. Evans, 1968, *J. Geophys. Res.,* v. 73, p. 3261.

130. J. D. A. Piper and K. Lomax, *Geophys. J. Roy. Ast. Soc.,* v. 34, p. 434.

131. J. N. J. Visser, 1974, *Geology,* v. 2, p. 279.

132. J. D. A. Piper, J. C. Briden, and K. Lomax, 1973, *Nature,* v. 245, p. 244.

133. A. F. Trendall, 1968, *Geol. Soc. Amer. Bull.,* v. 79, p. 1527.

134. U. Aswathanarayana, 1964, *in* B. C. Roy, ed., *Proc. 22nd Int. Geol. Cong.:* New Delhi, p. 1; L. O. Nicolaysen, *et al.,* 1958, *Geol. Soc. S. Afr. Trans.,* v. 48, p. 161; C. B. van Niekerk and A. J. Burger, 1964, *Union of S. Afr. Geol. Survey Annals,* v. 3, p. 75.

135. P. M. Hurley, *et al.,* 1967, *Science,* v. 157, p. 495.

136. R. C. Rhodes, 1975, *Geology,* v. 3, p. 549.

137. R. W. Page, 1976, *Carnegie Inst. Geophys. Lab., Ann. Rept. of Director, 1975–1976,* p. 813; P. M. Hurley, H. W. Fairbairn, A. Boudda, and W. H. Kanes, 1974, *Geol. Soc. Amer. Abstr. Programs,* v. 6, p. 803.

138. P. W. G. Tanner, 1973, *Geol. Soc. Amer. Bull.,* v. 84, p. 2839.

139. J. D. A. Piper, 1976, *Phil. Trans. Roy. Soc.,* v. 280, p. 469.

140. M. R. Cooper, 1978, *Nature,* v. 272, p. 810.

141. D. A. Brown, K. S. W. Campbell, and K. A. W. Crook, 1968, *The Geological Evolution of Australia and New Zealand:* Pergamon Press, Elmsford, N.Y.; A. Kroner, 1977, *J. Geol.,* v. 85, p. 289.

142. L. Cahen and J. Lepersonne, 1967, *in* K. Rankama, ed., *The Precambrian:* Wiley, New York, v. 3, p. 143.

143. M. W. McElhinny and M. O. McWilliams, 1977, *Tectonophysics,* v. 40, p. 137.

144. R. J. Fleck *et al.,* 1976, *Geol. Soc. Amer. Bull.,* v. 87, p. 9; L. Glover III and A. K. Sinha, 1973, *Amer. J. Sci.,* v. 273-A, p. 234; M. Halpern, 1972, *Proc. Internat. Symp. Carboniferous and Permian Systems S. Amer.:* Academia Brasileiro de Ciencias, Sao Paulo, Brazil, p. 77.

145. A. Ya. Krylov *et al.,* 1963, *Geochemistry,* no. 12, p. 1193.

146. E. J. Cobbing, J. M. Ozard, and N. J. Snelling, *Geol. Soc. Amer. Bull.,* v. 88, p. 241; B. Dalmayrac, J. R. Lancelot, and A. Leyreloup, 1977, *Science,* v. 198, p. 49.

147. D. H. Tarling, 1975, *Nature,* v. 255, p. 12.

148. A. E. J. Engel, B. Nagy, and L. A. Nagy, 1968, *Science,* v. 161, p. 1005.

149. P. M. Hurley *et al.,* 1972, *Earth Planet. Sci. Lett.,* v. 14, p. 360.

150. A. H. Knoll and E. S. Barghorn, 1977, *Science,* v. 198, p. 397.

151. J. W. Schopf, D. Z. Oehler, R. I. Horodyski, and K. A. Kvenvolden, 1971, *J. Paleont.,* v. 45, p. 477.

152. L. A. Nagy, 1973, *Science,* v. 183, p. 514.

153. E. S. Barghorn and S. A. Tyler, 1965, *Science,* v. 147, p. 563.

154. Anonymous, 1974, *Nature,* v. 248, p. 730.

155. V. Ramanathan, 1975, *Science,* v. 190, p. 52.

156. T. A. Jackson, 1967, *Science,* v. 155, p. 1003.

157. C. L. Fenton and M. A. Fenton, 1937, *Geol. Soc. Amer. Bull.,* v. 48, p. 1873.

**158.** M. F. Glaessner, 1969, *Lethaia,* v. 2, p. 369.

**159.** R. S. Bassler, 1941, *Proc. U.S. Nat. Mus.,* v. 48, p. 519.

**160.** H. Clemmey, 1976, *Nature,* v. 261, p. 576.

**161.** P. E. Cloud, Jr., G. R. Licari, L. A. Wright, and B. W. Troxel, 1968, *Proc. Nat. Acad. Sci. U.S.A.,* v. 61, p. 779; R. J. Horodyski and B. Bloeser, 1978, *Science,* v. 199, p. 682.

**162.** J. W. Schopf, 1968, *J. Paleont.,* v. 42, p. 651.

**163.** B. Bloeser, J. W. Schopf, R. J. Horodyski, and W. J. Breed, 1977, *Science,* v. 195, p. 676.

**164.** J. W. Schopf *et al.,* 1973, *J. Paleont.,* v. 47, p. 1.

**165.** L. C. Ranford, P. J. Cook, and A. T. Wells, 1965, *Bur. Min. Resour. Aust.,* v. 86, p. 1.

**166.** S. A. Kirsch, 1971, *Geol. Soc. Amer. Bull.,* v. 82, p. 3169.

**167.** P. Cloud, J. Wright, and L. Glover III, 1976, *Amer. Scientist,* v. 64, p. 396.

**168.** M. F. Glaessner, 1961, *Sci. Amer.,* v. 204, no. 3, p. 72; M. F. Glaessner, 1971, *Geol. Soc. Amer. Bull.,* v. 82, p. 509.

**169.** S. B. Misra, 1969, *Geol. Soc. Amer. Bull.,* v. 80, p. 2133.

**170.** M. E. Raaben, 1969, *Amer. J. Sci.,* v. 267, p. 1.

**171.** D. V. Nalivkin, 1960, *The Geology of the U.S.S.R.,* trans. S. I. Tomkeieff: Pergamon Press, New York.

**172.** P. E. Cloud, Jr., 1968, *Science,* v. 160, p. 729.

**173.** L. Margulis, J. C. Walker, and M. Rambler, 1976, *Nature,* v. 264, p. 620.

**174.** M. J. de Wit, 1970, *Nature,* v. 277, p. 829.

**175.** G. E. Williams, 1975, *Geol. Mag.,* v. 112, p. 441.

# The Early Paleozoic Era 10

T HE EARLY PALEOZOIC includes the Cambrian, Ordovician, and Silurian periods. The abundant and complex life represented in its fossil record contrasts markedly with the extreme scarcity of evidence for advanced organisms in Precambrian deposits. Early Paleozoic deposits may be easily correlated not only within a continent, but between continents where sufficient paleontological data are available. Such long-range correlations are invaluable in the reconstruction of ancient continents.

In many parts of the world, the boundary between Cambrian sedimentary rocks containing abundant fossils with hard parts and Precambrian deposits is marked by an unconformity (Fig. 10.1). In other areas, thick, apparently conformable sequences underlie the lowest fossiliferous beds. The base of the Cambrian is generally placed just below the lowest fossil bearing strata. However, this means that the base of the Cambrian may have to be shifted downward locally if new finds of fossils are made lower down. Other problems with this method are discussed in the section on early Paleozoic life. Another method is to locate the base of the Cambrian at the first unconformity below the lowest beds containing fossils with hard parts. This criterion is somewhat unsatisfactory because beds that are conformable in one area may be unconformable in another. Thus strata designated as Cambrian in one area could be the time-stratigraphic equivalent of Precambrian deposits elsewhere.

## EARLY PALEOZOIC GEOGRAPHY

Paleomagnetic and structural data have been used to reconstruct movements of the continents during the early Paleozoic. Such data indicate that the there were five separate continents at the beginning of the Cambrian. When early Paleozoic paleomagnetic poles are recalculated for a Pangaea reconstruction of the continents, they do not coincide. For example:

1. The recalculated early Paleozoic apparent polar wander (A.P.W.) path for Gondwanaland diverges significantly from those of both North America and Europe (Fig. 9.4). This indicates that Gondwanaland was separated from North America and Europe by a wide ocean during the early Paleozoic.
2. The recalculated Cambrian paleomagnetic pole for North America does not agree very well with that of Europe, but the Ordovician through Triassic paleomagnetic poles are in reasonable agreement (Fig. 9.4). Thus it may be inferred that these two continents were separated during the Cambrian and were joined or were reasonably close from the Ordovician through the Triassic (1).
3. The paleomagnetic poles from Ancestral Siberia do not agree with the Cambrian and Ordovician paleomagnetic poles for Europe, but the Silurian and later paleomagnetic poles for these two continents are in reason-

**FIGURE 10.1** The unconformity between the Cambrian Flathead Sandstone and the Archean gneiss is plainly visible on the highway between Cody, Wyoming, and Yellowstone National Park. This unconformity represents 2 billion years of erosion. The tilting of the Cambrian sandstone occurred during the Laramide orogeny in the early Tertiary.

able agreement (Fig. 7.13). Therefore, it is probable that Ancestral Siberia and Ancestral Europe were separated until the Silurian and that they have been joined together since that time.

4. The Cambrian paleomagnetic pole of Ancestral China is located relatively close to that of Ancestral Siberia (Fig. 7.13). This suggests that these two continents were probably in close proximity to each other during the Cambrian. Alternatively, they could have been widely separated from each other if the pole of spreading about which the continents moved coincided approximately with the paleomagnetic pole.

Based on this paleomagnetic data, it is likely that five separate continents were present at the beginning of the Cambrian and that joining of these continents began in the Middle Ordovician (Fig. 10.2). There seems to be a general agreement among geologists who work on Paleozoic reconstructions of the continents that there were five separate continents during the Cambrian and that they had approximately the size and shape of the ancestral continents shown in Figure 9.5. Opinions differ as to the relative position of the continents during the Early Paleozoic (**3**) and as to the times when the continents initially collided (**3, 4**).

A major difference between continental reconstructions is the relative positions of Ancestral North America and Gondwanaland during the early Paleozoic. Our reconstruction is consistent with the available paleomagnetic data. Cambrian and Ordovician paleomagnetic poles

(a)

**FIGURE 10.2** Reconstruction of the continents during the Early Paleozoic with paleoclimatic indicators. Paleolatitudes are based on paleomagnetic data from samples which were tested for magnetic stability. The average paleomagnetic pole for each continent (Table 7.1) was used to compute an average pole position for the continents as a whole. (a) Cambrian. (b) Ordovician (directions of ice movement are from R. W. Fairbridge, Ref. 2). (c) Silurian.

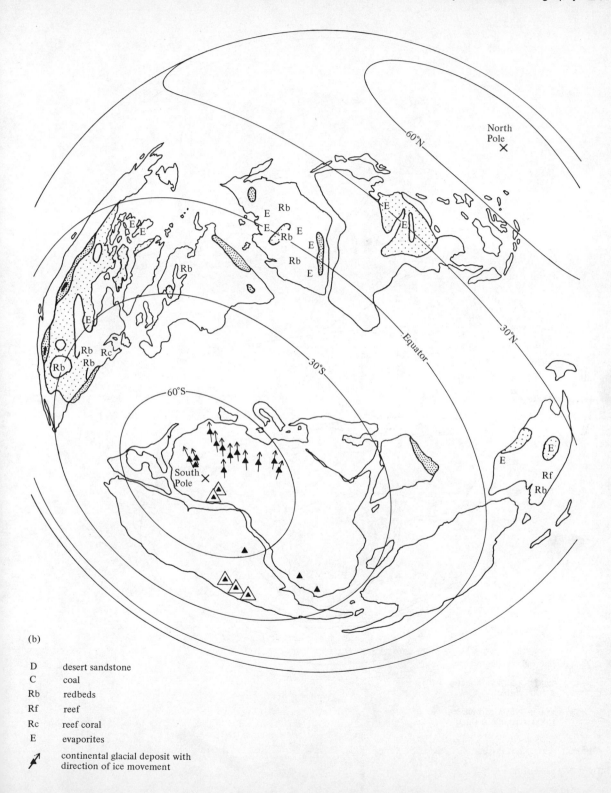

(b)

| | |
|---|---|
| D | desert sandstone |
| C | coal |
| Rb | redbeds |
| Rf | reef |
| Rc | reef coral |
| E | evaporites |
| ⚡ | continental glacial deposit with direction of ice movement |

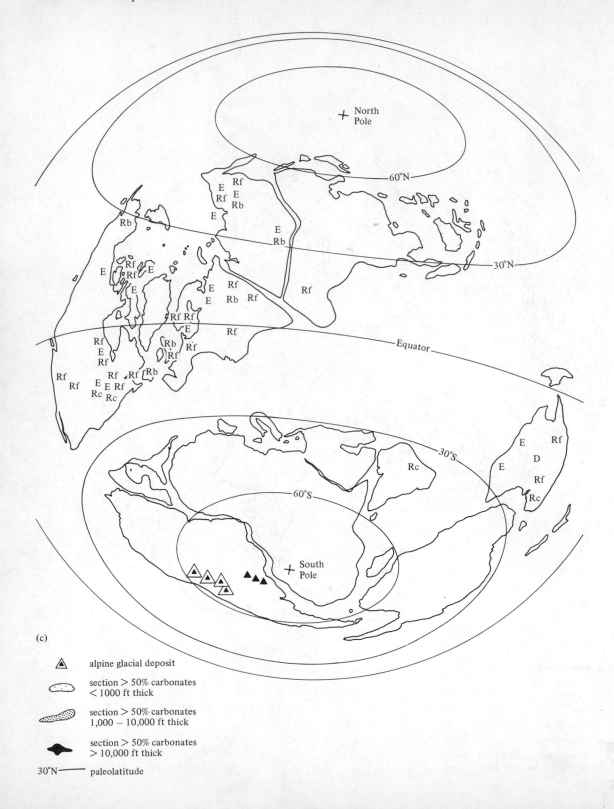

(c)

| | |
|---|---|
| ⬚ (triangle) | alpine glacial deposit |
| ⬭ | section > 50% carbonates < 1000 ft thick |
| ⬭ (stippled) | section > 50% carbonates 1,000 – 10,000 ft thick |
| ◆ | section > 50% carbonates > 10,000 ft thick |
| 30°N—— | paleolatitude |

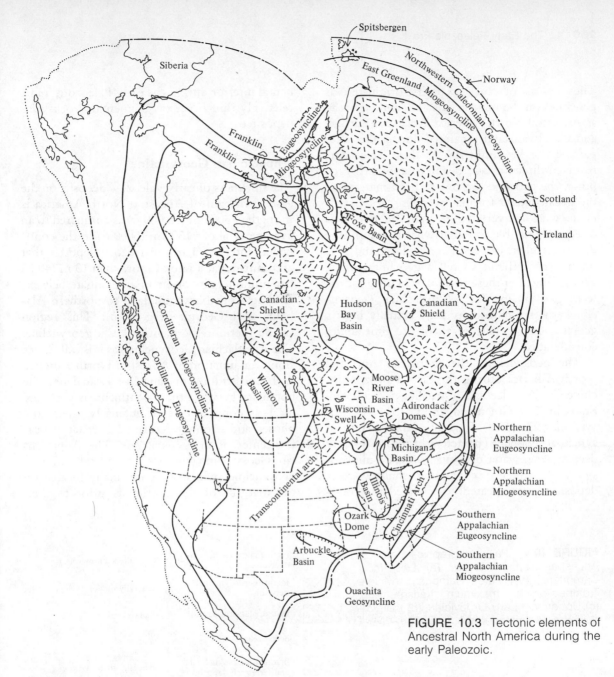

**FIGURE 10.3** Tectonic elements of Ancestral North America during the early Paleozoic.

for North America coincide with the paleo-magnetic poles of the same age for Gondwana-land if North America (along with these poles) is rotated away from Gondwanaland about a spreading pole just west of South America (Fig. 10.2a). It is likely that this was the spread-ing pole during the early Paleozoic and that rotation about this pole resulted in the closure of the ocean between the two continents.

## ANCESTRAL NORTH AMERICA

The continent of Ancestral North America is also known as proto–North America and Laurentia, and the areas which comprise it are described in Chapter 9. During the early Paleo-zoic, thick sequences of sedimentary and vol-canic rocks were deposited on the margins of Ancestral North America (Figs. 10.3 and 10.4).

These areas of thick deposits originated as geoclines on the margin of an expanding ocean as Ancestral North America, Ancestral Europe, and Gondwanaland separated during the latter part of the Precambrian. As these continents reversed direction and began to approach each other, the geoclines on one or both margins of the ocean which formed between these continents were converted into geosynclines. It is often not possible to tell at what point in time this reversal in direction took place, and it is often very difficult to tell a geocline from a geosyncline. For these reasons, when we use the term geosyncline (i.e., Appalachian Geosyncline), the structure may have actually been a geocline during part of all of the time under consideration.

The geosynclines that bordered Ancestral North America during the early Paleozoic include the Appalachian, Ouachita, Cordilleran, Franklin, East Greenland, and Caledonian geosynclines (Fig. 10.4). In most of these, a eugeosynclinal sequence characterized by moderately deep to deep water deposits borders a miogeosynclinal sequence containing shallow-water deposits. Seas repeatedly inundated the continental interior during the early Paleozoic, and a relatively thin cover of sedimentary rocks was deposited.

## Appalachian Geosyncline

The thickness of early Paleozoic deposits on the eastern margin of Ancestral North America is more than 5000 m (15,000 ft) as compared to an average of about 1500 m (5000 ft) in the continental interior. Early Paleozoic deposits that accumulated at a rate of more than 13 m (40 ft) per million years occur in an elongate belt extending from Newfoundland to northern Alabama and Georgia (Fig. 10.5). This region would therefore be identified as a geosyncline. The Appalachian Geosyncline, as it is called, lies along the junction of Ancestral North America with Ancestral Europe and Gondwanaland. The boundary between these continents is probably located along a belt of intensely metamorphosed and strongly deformed rocks at or near the center of the geosyncline. The Appalachian Geosyncline consists of several different geosynclines which were originally located on the margins of the continents prior to their

FIGURE 10.4 Paleogeographic maps of Ancestral North America: **(a)** Lowermost Cambrian. **(b)** Late Cambrian (Middle Trempealeauan) maximum transgression. **(c)** Uppermost Early Ordovician. **(d)** Late Ordovician, Richmond stage.

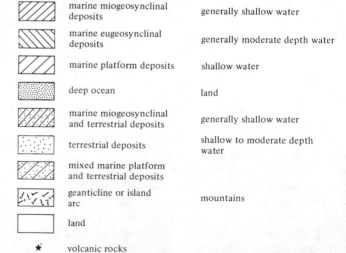

| | |
|---|---|
| marine miogeosynclinal deposits | generally shallow water |
| marine eugeosynclinal deposits | generally moderate depth water |
| marine platform deposits | shallow water |
| deep ocean | land |
| marine miogeosynclinal and terrestrial deposits | generally shallow water |
| terrestrial deposits | shallow to moderate depth water |
| mixed marine platform and terrestrial deposits | |
| geanticline or island arc | mountains |
| land | |
| ★ volcanic rocks | |

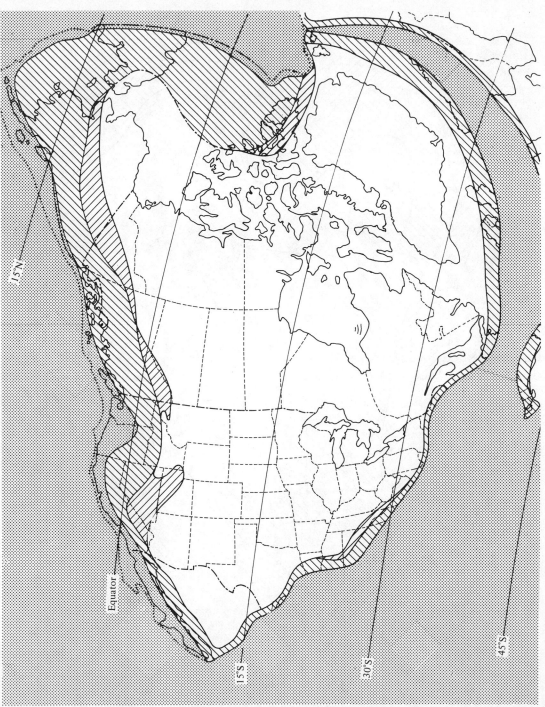

(a)

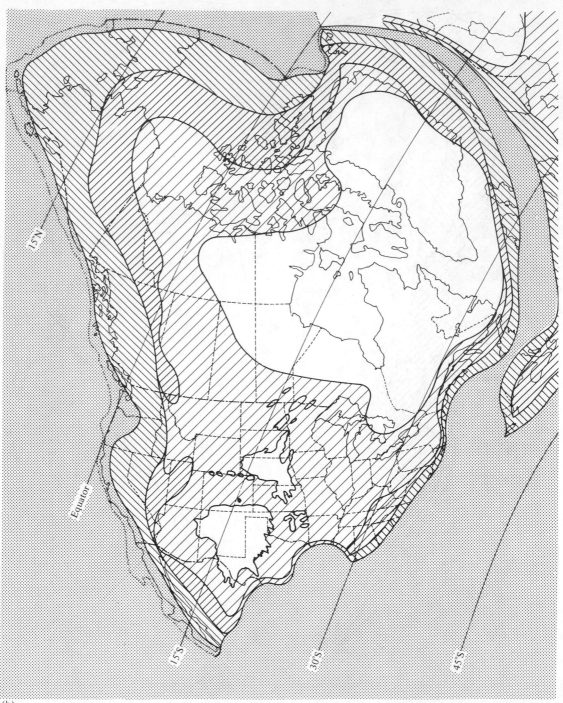

(b)

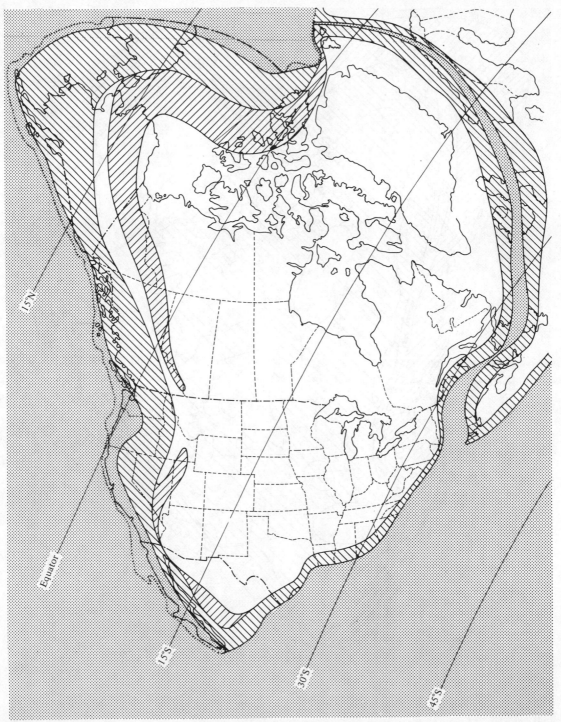

15°N

Equator

15°S

30°S

45°S

(c)

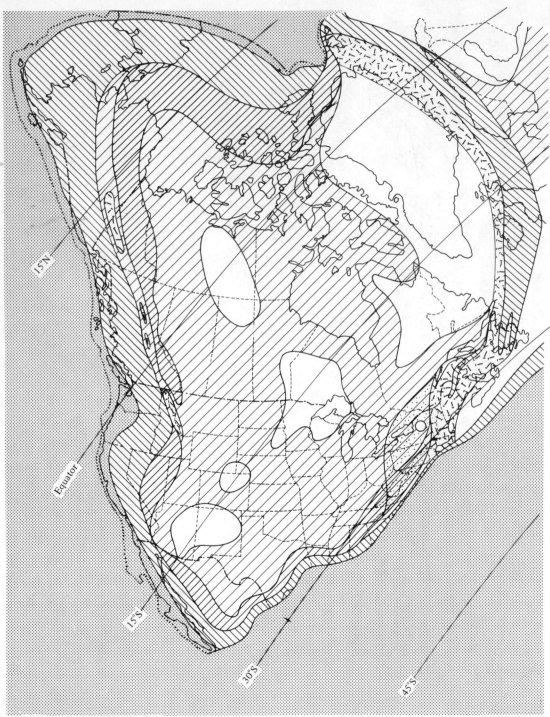

(d)

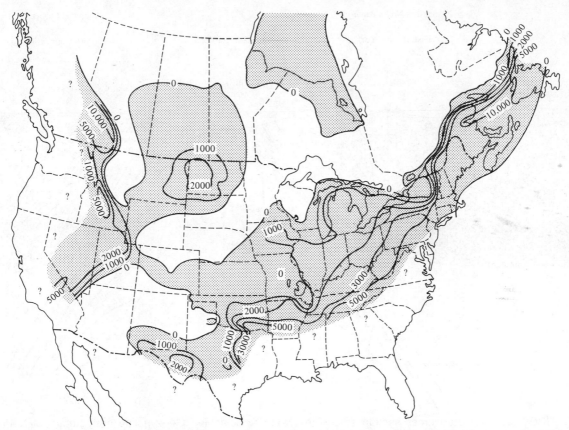

**FIGURE 10.5** Isopach map showing the thickness of Ordovician deposits in the United States and southern Canada. (Compiled by R. Macomber, *in* Sloss, Dapples, and Krumbein, Ref. 5.)

joining together into a single landmass (Fig. 10.6). The Northwest Appalachian Geosyncline bordered Ancestral North America opposite Ancestral Europe, and the Southwest Appalachian Geosyncline bordered Ancestral North America opposite Gondwanaland. The Northeast Appalachian Geosyncline bordered Ancestral Europe opposite Ancestral North America (see Ancestral Europe). The Southeast Appalachian Geosyncline bordered Gondwanaland opposite Ancestral North America (see Gondwanaland).

As discussed in Chapter 9, the Northern Appalachian Geosyncline probably formed as a result of the separation of Ancestral Europe from Ancestral North America about 675 million years ago (**6**). During the opening of the Iapetus Ocean between these two continents, geoclines would have bordered the ocean. However, when the Iapetus Ocean began to close, subduction zones must have bordered one or both sides of the ocean. There are three possible configurations for the subduction zones bordering the Iapetus Ocean during its closing (Fig. 10.7):

1. A subduction zone dipped westward under the Northwest Appalachian Geosyncline while the Northeast Appalachian Geosyncline was located on a passive margin.

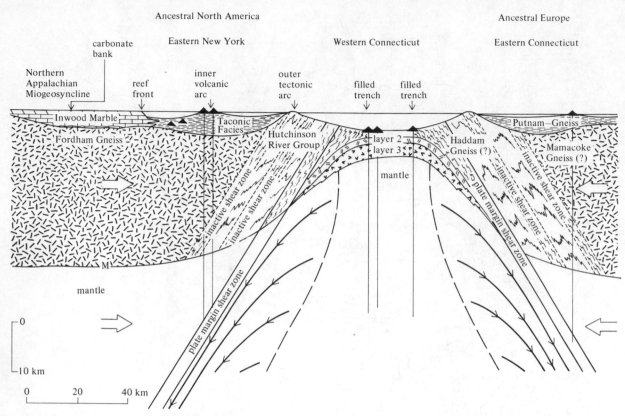

**FIGURE 10.6** Restored cross section across the North Appalachian Geosyncline from Ancestral North America to Ancestral Europe during the Cambrian. (After Seyfert and Leveson, Ref. 9).

2. A subduction zone dipped eastward under the Northeast Appalachian Geosyncline while the Northwest Appalachian Geosyncline was located on a passive margin.

3. There may have been two different subduction zones present, one dipping westward under the Northwest Appalachian Geosyncline, and one dipping eastward under the Northeast Appalachian Geosyncline.

### Northwest Appalachian Geosyncline

Thick sequences of sedimentary and volcanic rocks were deposited in shallow water in a belt extending from western Newfoundland through southern Quebec, eastern New York State, western Vermont, southeastern Pennsylvania, and Eastern West Virginia to Maryland (Fig. 10.4). Considerable controversy has developed over whether these deposits accumulated in a geosyncline or a geocline. Marshall Kay proposed a model for the northwestern Appalachian region in which island arcs bordered the eastern margin of North America during the early part of the early Paleozoic (10). In this case, the Appalachian region would have been a geosyncline at that time. On the other hand, Robert Dietz proposed a model in which a geocline bordered North America during the early part of the early Paleozoic (11). Which of these models is correct depends on whether or not a subduction zone underlay the northwestern Appalachian region at that time. Certainly,

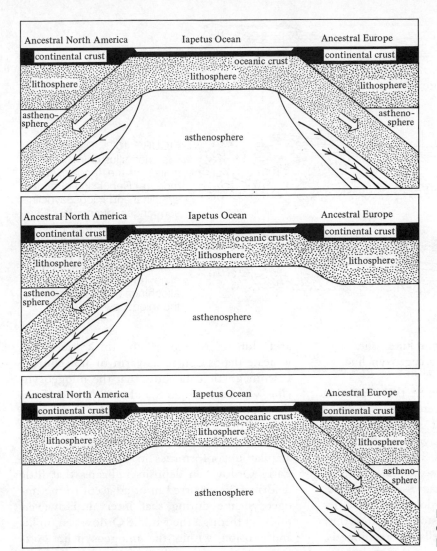

**FIGURE 10.7** Three possible cross sections across the Appalachian Geosyncline.

subduction would not have started until the Iapetus Ocean began to close, but it is not at all certain when this closure began. Deformation occurred in the northwestern Appalachian region during the Middle Ordovician, and therefore it is likely that closure began during or before this time. However, closure may have begun as early as the beginning of the Cambrian. There is not enough known at the present time to determine exactly when closure began or whether or not there was a subduction zone beneath the northwestern Appalachian region during the closure.

Within the northern Appalachian region, there are two contrasting sequences of early Paleozoic age. One consists of thick accumulations of limestone, dolostone, and sandstone with minor shale. This sequence was deposited in a miogeosyncline. The other sequence consists of thick accumulations of shale, gray-

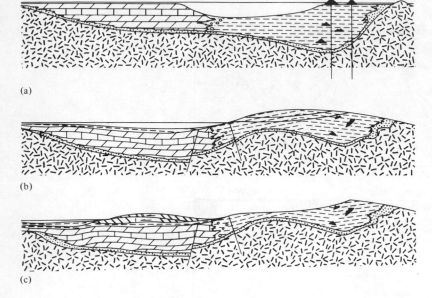

(a)

(b)

(c)

**FIGURE 10.8** Diagrammatic sketches illustrating the formation of the Taconic Allochthon: **(a)** Deposition of eugeosynclinal and miogeosynclinal sequences during the Cambrian. **(b)** Uplift within the eugeosyncline with clastic sediment deposited over the carbonates of the miogeosyncline. **(c)** Gravity sliding of the allochthon onto the deposits of the miogeosyncline.

wacke, volcanics, and minor limestone. This sequence was deposited in a eugeosyncline.

MIOGEOSYNCLINE.   At the beginning of the Cambrian, the seas were largely confined to the eugeosyncline, but gradually spread westward into the miogeosyncline during the Early Cambrian (Fig. 10.4). Sandstone was deposited in near-shore depositional environments, and as the sea transgressed westward, fossiliferous limestone and dolostone were deposited on top of the sandstone. Because they contain abundant fossil shells of brachiopods, pelecypods, and other groups, they belong to the *shelly facies.* The presence of oolites, algal stromatolites, and mud cracks indicate deposition in shallow, well-aerated water. John Rodgers has proposed that the deposits of the shelly facies were laid down on a very large carbonate bank, analogous to the Bahama Banks (**12**). In this case, the boundary between the miogeosyncline and the eugeosyncline was probably a reef front (Fig. 10.8).

Dolostones, which are abundant within the Cambrian through Early Ordovician deposits, are thought to be indicative of deposition in an arid climate. Moreover, the thick limestones indicate deposition in a warm or hot climate. Paleomagnetic data show that the miogeosyncline was between 25°S and 35°S latitude from the Cambrian through the Early Ordovician (Fig. 10.6).

The abundance of carbonates and absence of angular unconformities in Cambrian through Early Ordovician deposits indicates that little deformation occurred in or adjacent to the miogeosyncline during that interval. However, midway through the Middle Ordovician, uplift and erosion within the miogeosyncline was followed by deposition of clastic sediments from the east. The source of the clastics was a broad geanticline that developed close to the boundary between the miogeosyncline and the eugeosyncline.

Continued uplift of the geanticline resulted in the sliding of great masses of rock from the geanticline onto the deposits of the miogeosyncline (Fig. 10.8). The slide masses are collectively referred to as the Taconic Allochthon. E-an Zen has postulated that the deposits of the allochthon were originally deposited along the site of what is now the Green Mountain Anti-

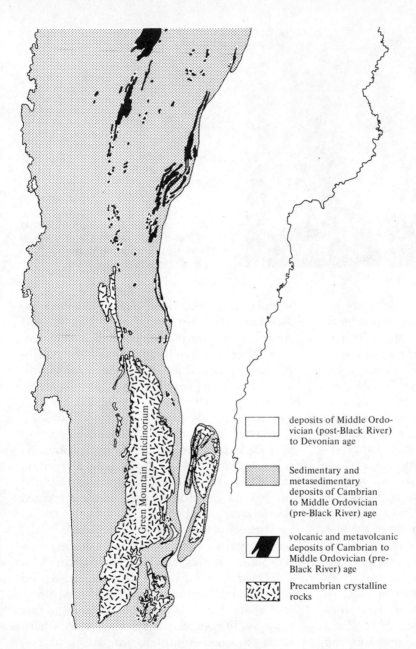

deposits of Middle Ordovician (post-Black River) to Devonian age

Sedimentary and metasedimentary deposits of Cambrian to Middle Ordovician (pre-Black River) age

volcanic and metavolcanic deposits of Cambrian to Middle Ordovician (pre-Black River) age

Precambrian crystalline rocks

**FIGURE 10.9** Geologic map of Vermont showing the distribution of volcanic rocks of Cambrian through early Middle Ordovician age. The concentration of volcanics indicates that this area was a geosyncline at the time. (Modified from the Centennial Geological Map of Vermont, published by the Vermont Geological Survey.)

clinorium, an area along which Precambrian rocks are exposed (**13**) (Fig. 10.9). Boulders, cobbles, and pebbles within the deposits beneath the allochthon can be matched with rocks in the allochthon (Fig. 10.10). Thus deposition of the beds beneath the allochthon was contemporaneous with its emplacement. Middle Or-

dovician fossils are found beneath, within, and on top of the allochthon, and therefore it must have been emplaced during the Middle Ordovician (**14**).

At about the same time, large ophiolite sheets were thrust across the shallow waters of the miogeosyncline in western Newfoundland

**FIGURE 10.10** Conglomerate beneath the Taconic Allochthon east of Albany, New York. The large blocks within this deposit are thought to have slid or have been eroded from the advancing thrust sheet (Photo by E-an Zen, U.S. Geological Survey.)

(15). As discussed in Chapter 4, ophiolite sequences are thought to represent segments of the ocean crust and possibly upper mantle which have been shoved upward and onto or under the continent during plate collisions. It is thought that when an ophiolite sheet has been emplaced more or less as a unit on top of shallow water deposits that *obduction,* rather than *subduction,* has taken place. Obduction occurs when a continent is thrust beneath a plate containing oceanic crust at its leading edge. Subduction, on the other hand, occurs where oceanic crust is thrust beneath a plate that contains a continent on its leading edge. From the presence of an ophiolite sequence on top of shallow water deposits on the west side of the Appalachian Geosyncline, it may be inferred that a subduction zone dipped eastward away from the Northwestern Appalachian Geosyncline prior to emplacement of the ophiolite. The ophiolite at Bay of Islands has been dated at 508 million years by the uranium-lead method (15).

The emplacement of the Taconic Allochthon and the ophiolite sheets indicates that a major orogenic episode occurred in the Appalachian Geosyncline during the Middle Ordovician. This orogeny, which will be referred to as the first phase of the Taconic orogeny, was the only

significant deformation to have affected the Northeastern Appalachian Geosyncline since its formation during the Middle or Late Proterozoic. After the Middle Ordovician, the Northeastern Appalachian Geosyncline was almost certainly a true geosyncline.

During the later part of the Middle Ordovician and the early part of the Late Ordovician, a geanticline contributed sediments eastward into the eugeosyncline and westward into the miogeosyncline. The Martinsburg Formation thickens and increases in grain size eastward toward the geanticline, which is the presumed source area.

The sediments and sedimentary rocks of the miogeosyncline were folded, faulted, and metamorphosed during the Late Ordovician. This deformation will be referred to as the second phase of the Taconic orogeny, and it occurred about 8 million years after the first phase of that orogeny. In eastern Pennsylvania, Martinsburg strata deformed during this orogeny are unconformably overlain by the Early Silurian Tuscarora Sandstone. Since the Martinsburg Formation is Middle to Late Ordovician in age, its deformation must have occurred between the Late Ordovician and Early Silurian.

In southeastern New York and northern New Jersey, Precambrian gneisses have been thrust over shale correlative with the Martinsburg. A study of gravity and magnetic anomalies in the vicinity of the thrust sheet indicates that the displacement on the thrust fault is many tens of miles (**16**). The involvement of basement rocks shows that the thrusting was not a result of gravity sliding, as was the case in the Taconic Allochthon. Deep-seated stresses must have been involved in this thrusting. Since the youngest rocks involved in the faulting are of Late Ordovician age, it is likely that the displacement occurred during the second phase of the Taconic orogeny.

The geanticline produced during the first phase of the Taconic orogeny increased in width and height during its second phase. During the uplift, a thick sequence of red shales, sandstones, and conglomerates were deposited as a *clastic wedge* west of the geanticline (Fig. 10.11). A clastic wedge is a wedge-shaped body of sedimentary rocks which is usually continental and increases in thickness toward its source. The Late Ordovician clastic wedge includes the Queenston Shale of western New York and southeastern Ontario and the Juniata Formation of central Pennsylvania.

Following the second phase of the Taconic orogeny, a thick sequence of conglomerate and sandstone was deposited in the geosyncline by rapidly moving streams and rivers. These strata include the Tuscarora Sandstone and the Shawngunk Conglomerate. These units are time-transgressive, and they vary in age from Early Silurian to Middle Silurian.

During the Late Silurian, red beds belonging to the Salina group were deposited west of the miogeosyncline in New York State. These deposits thicken eastward and are probably due to a minor orogenic episode, the Silinic disturbance (**17**), that affected the geanticline to the east. The Bloomsburg Formation of eastern Pennsylvania probably correlates with the Salina group. Folding and uplift occurred during the Late Silurian in western Newfoundland (**18**), and this deformation may have taken place at the same time as the Silinic disturbance. The

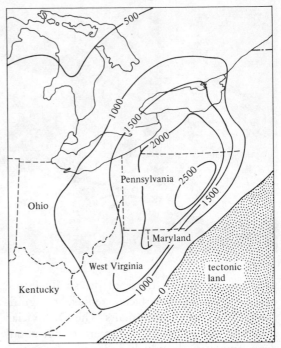

**FIGURE 10.11** Isopach map showing the variation in thickness of Upper Ordovician sediments in Pennsylvania and adjacent states. (After Kay, Ref. 10.)

Caledonian orogeny, a major episode of folding, metamorphism, and intrusion in Europe began near the end of the Late Silurian (**19**), and it too may correlate with the Silinic disturbance.

EUGEOSYNCLINE. The Cambrian to Middle Ordovician sedimentary rocks of the miogeosyncline grade eastward into eugeosynclinal sequences of shale, sandstone, and volcanics of the same age. The deposits of the eugeosyncline include the Taconic sequence of eastern New York and western Vermont and the East Vermont sequence of eastern Vermont. Graptolites are among the most abundant fossils in the eugeosynclinal deposits, and hence these deposits are referred to as the *graptolite facies*. Graptolites typically occur in black shales which probably were deposited below wave base in relatively deep, poorly aerated water. The trilobites in the graptolite facies seldom are

**FIGURE 10.12** Cow Head conglomerate in western Newfoundland. (Photo by E-an Zen, U.S. Geological Survey.)

the same species as those of the shelly facies; evidently these two facies represent significantly different environments. The Ordovician fossils of the graptolite facies are rather abundant and varied, which indicates deposition probably did not occur in waters of abyssal depths (**20**).

Large blocks of limestone, found in the Cambrian to Middle Ordovician deposits near the western margin of the eugeosyncline, may be the result of submarine landslides (Fig. 10.12). It is likely that these blocks were part of a reef complex bordering a large carbonate bank. Loosened by wave action, they may have slid down the reef front into the deeper waters of the eugeosyncline (**12**).

In eastern Vermont, eugeosynclinal deposits of the graptolite facies rest unconformably on Precambrian gneisses, granites, and schists (**13**). However, these deposits thicken rapidly eastward, and part of the sequence may have been deposited on oceanic crust. Some of the Cambrian to Middle Ordovician rocks of the eugeosyncline have no known basement and were probably deposited on oceanic crust. These deposits include the Dunnage Melange (**22**) and Campbellton sequence in central Newfoundland (**23**). Both the Dunnage and the Campbellton sequences were probably deposited between Cambrian and Middle Ordovician time. A melange is a chaotic mixture of various rock types. It is thought to be a deep ocean sediment that has been deformed along a subduction zone as the seafloor descended beneath a continent. The Campbellton sequence contains beds rich in manganese. Manganese-rich beds are common on the floors of modern deep oceans. Ophiolites associated with these deposits are probably slivers of ocean crust caught between plates as they collided. An ophiolite from central Newfoundland has been dated at 463 million years (Middle Ordovician) (**15**).

The first phase of the Taconic orogeny resulted in the uplift and deformation of the deposits of at least part if not all of the eugeosyncline. Following this deformation, a thick sequence of clastic sediments with a conglomerate at the base was deposited unconformably on the deformed deposits of the eugeosyncline (**13**).

During the second phase of the Taconic orogeny, the eugeosynclinal deposits were again folded and uplifted. However, the deformation was more intense and it was accompanied by strong metamorphism. Large *recumbent folds* (folds whose axial planes are sub-horizontal), which are similar in many respects to the *nappes* of the Alps, were transported westward

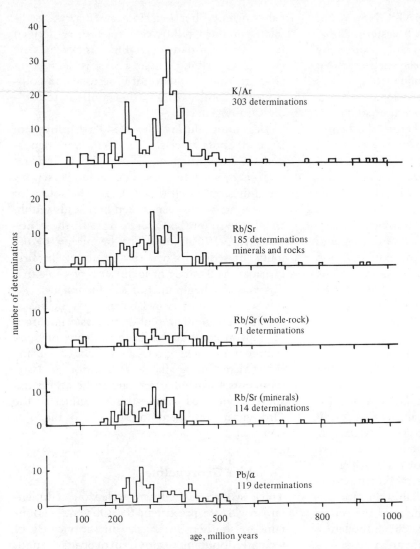

number of determinations

age, million years

K/Ar
303 determinations

Rb/Sr
185 determinations
minerals and rocks

Rb/Sr (whole-rock)
71 determinations

Rb/Sr (minerals)
114 determinations

Pb/*a*
119 determinations

**FIGURE 10.13** Frequency distribution of radiometric dates on igneous and metamorphic rocks from the Appalachian Geosyncline. Peaks of igneous and metamorphic activity occur during: Early, Middle, and Late Ordovician; Late Silurian; Middle and Late Devonian; Late Mississippian; Late Pennsylvanian; Middle Permian; and Late Triassic or Early Jurassic. (From Lyons and Faul, Ref. 27.)

toward the miogeosyncline (**24**). The highland that resulted from this deformation was rapidly eroded and during the Early Silurian, the sea transgressed across its borders. A basal conglomerate followed by a thick sequence of clastic sediments, volcanic rocks, and limestone was deposited in this sea.

In central Maine, southeastern Quebec, and central Newfoundland, an episode of folding, metamorphism, and intrusion occurred during the Late Silurian or Early Devonian (**25**). Pebbles of Silurian slates from Early Devonian deposits in central Maine indicate that the metamorphism in that area was relatively weak (**26**). The evidence for intrusive activity at this time is that there are granitic igneous rocks in the northern Appalachians that have been dated at 395 to 400 million years in age (**27**) (Fig. 10.13). This Late Silurian or Early Devonian orogenic activity may correlate with the Silinic disturbance and/or with the Caledonian orogeny of Europe.

CAUSE OF DEFORMATIONS. The orogenic activity in the northwestern Appalachian Geosyncline was probably caused by the collision of plates. It has been suggested that the first phase of the Taconic orogeny was caused by the beginning of closure of the Iapetus Ocean, but the evidence presented here suggests that closure began before the beginning of the Cambrian. A possibility is that it was caused by the collision between Ancestral Europe and Ancestral North America. The initial contact may have been only between one or more island arcs during the Middle Ordovician, but the intensity of the Late Ordovician deformation (second phase of the Taconic orogeny) indicates that Ancestral North America and Ancestral Europe were joined at that time. The Late Silurian orogenies may have been caused by continued compression between these two continents.

## Southwest Appalachian Geosyncline

The Southwest Appalachian Geosyncline (or Geocline) developed on the margin of Ancestral North America adjacent to Gondwanaland. Subsequently, these two continents were brought together as a result of plate collisions. The boundary between Ancestral North America and Gondwanaland is probably the Brevard Zone, a major zone of faulting near the center of the Appalachian Fold Belt (**28**). Subduction zones may have dipped westward under Ancestral North America, eastward under Gondwanaland, or both. The Brevard zone dips eastward under a block which was once part of Gondwanaland (**28**), and therefore it is likely that a subduction zone dipped eastward under Gondwanaland at the time of collision of Gondwanaland with Ancestral North America.

MIOGEOSYNCLINE. The Chilhowee Group consists of 2000 m (6000 ft) of sandstone, shale, and conglomerate, and contains Cambrian trilobites near its top. Studies of paleocurrent directions indicate that the sediments of the Chilhowee Group were derived from Precambrian crystalline rocks of the craton. Overlying these rocks is a 3000-m (9000-ft) thick sequence of limestone and dolostone that ranges from Early Cambrian to Middle Ordovician in age.

Beginning during the Middle Ordovician, large amounts of clastic sediment were deposited in the miogeosyncline. These clastics thicken toward the east and grade westward into limestones (**29**) (Fig. 10.14). The source of the sediments may have been highlands within an island arc on the outer margin of the continent (Fig. 10.15) (**30**). The beginning of the influx of sediment occurred at about the first phase of the Taconic orogeny. It is possible that this was the beginning of subduction beneath the Southwestern Appalachian Geosyncline. If an island arc were present on the eastern margin of the continent at this time the zone of thick sedimentary rocks would be a miogeosyncline rather than a miogeocline. Deposition of clastic sediments from what appears to be an eastern source continued during the remainder of the Ordovician and lasted until the end of the Silurian.

## Ouachita Geosyncline

Thick sequences of lower Paleozoic sedimentary rocks are present in the Ouachita Mountains of southern Arkansas. In the vicinity of Crystal Mountain, Ordovician deposits thought to have a shallow water origin are 1870 m (6120 ft) thick (**31**). The Ordovician lasted for 100 million years and therefore the average rate of sedimentation was 19 m (61 ft) per million years, more than the minimum necessary for it to qualify as a geosyncline. The geosyncline extends from west-central Alabama through southern Arkansas to southern Texas (Fig. 10.4).

Deposits in the geosyncline are dominantly sandstone, black shale, chert, and limestone. Therefore it would be either a miogeosyncline

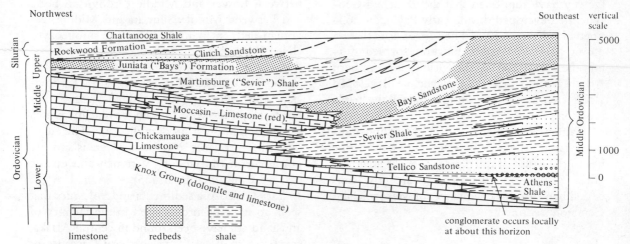

FIGURE 10.14  Restored cross section of the Ordovician and Silurian rocks in the Valley and Ridge Province of eastern Tennessee. The thickening of the clastic sediments eastward indicates that their source lay in that direction. The unconformity at the base of the Chattanooga Shale indicates a Devonian deformation affected this area. (After P. B. King, Ref 29. Reproduced with permission of The American Association of Petroleum Geologists.)

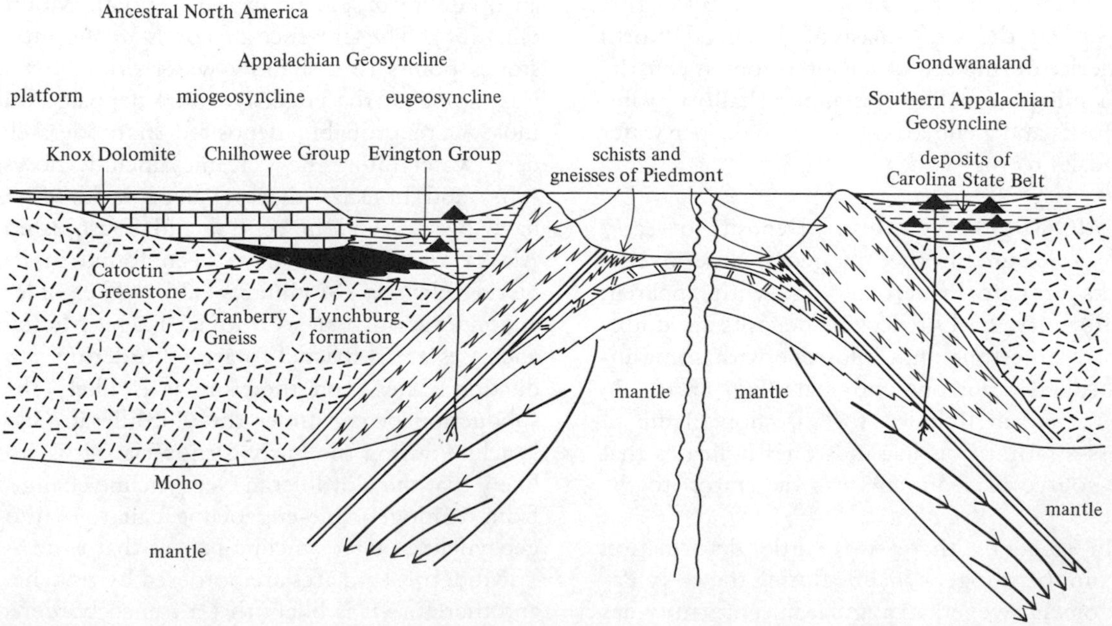

FIGURE 10.15  Restored cross section across the Southern Appalachian Geosyncline during the Cambrian. The section extends southeastward from eastern Kentucky through Virginia to eastern North Carolina.

or a miogeocline. Kevin Burke and John Dewey have suggested that the Ouachita Geosyncline developed during early Paleozoic continental rifting (**32**). They propose that a plume-generated triple junction formed during the Cambrian in the vicinity of southern Arkansas, and that the geosyncline developed along two active spreading arms of the triple junction.

Magnetic surveying and deep drilling indicate that the axis of the Ouachita Geosyncline lies to the south of the present Ouachita Mountains. The geosynclinal deposits are now concealed beneath thick sequences of Mesozoic and Cenozoic sediments.

## Cordilleran Geosyncline

The Cordilleran Geosyncline (or Geocline) extends from Alaska through Canada and the western United States into Mexico (Fig. 10.4). It joins the Franklin Geosyncline in the north and the Ouachita Geosyncline in the south. As discussed in Chapter 9, the Cordilleran Geosyncline may have developed as a result of rifting along the west coast of Ancestral North America during the Late Proterozoic. It consists of a miogeosyncline containing shallow water deposits and a eugeosyncline with deeper water deposits to the west.

MIOGEOSYNCLINE. Deposits of early Paleozoic age in the miogeosyncline overlie rocks of Late Proterozoic age with apparent conformity. Early Paleozoic deposits are dominantly limestone and dolostone with some interbedded sandstone and shale (Fig. 10.16). In the Canadian Rockies, the direction of dip of cross-stratification in sandstones indicates that the source of sediment was the craton to the east.

In general, there was little deformation within the miogeosyncline during the early Paleozoic. However, an angular unconformity has been reported from Middle Cambrian deposits in Montana (**33**), and disconformities occur between Cambrian and Lower Ordovician beds, between Lower and Middle Ordovician beds, and between Middle Silurian and Middle Devonian beds in many parts of the miogeosyncline (**34**).

EUGEOSYNCLINE. During the Ordovician and Silurian, the miogeosyncline was bordered on the west by a eugeosyncline which was underlain by a subduction zone. The eugeosyncline probably consisted of an island arc bordered on the west by a trench and on the east by a marginal basin (**35**).

Early Paleozoic sediments and volcanics that seem to have been deposited in the vicinity of an island arc have been found in central Alaska, southeastern Alaska, the northern Cascades of Washington State, the Klamath Mountains of northern California, and in the Sierra Nevada Mountains of east-central California (**36**). These deposits are dominantly shale, sandstone, and conglomerate with some interbedded limestone and volcanics. They rest on a crystalline basement in one area in southeastern Alaska and on an ophiolite basement in the Klamath Mountains (**36**). The presence of corals in the limestones points to a shallow water origin for at least some of the eugeosynclinal deposits, but most were probably deposited in moderately deep water (**36**). The volcanics include flows, tuffs, and breccias. Early Paleozoic andesites have been found in central and southeastern Alaska and rhyolites have been found in the Sierra Nevada Mountains (**36**). The oldest andesites are of Early Ordovician age. Modern andesites are restricted to areas bordering subduction zones, and therefore it is likely that subduction began under the Cordilleran Geosyncline during the Early Ordovician. It is also likely that the Cordilleran Geosyncline changed from a miogeocline-eugeocline pair to a miogeosyncline-eugeosyncline pair at that time.

Modern island arcs are bordered by trenches, and therefore it is likely that a trench bordered the early Paleozoic island arc in the Cordilleran region. W. Porter Irwin has divided the eugeo-

**FIGURE 10.16**   Breached anticline in Banff National Park, Canada. The cliff at the left is composed of Cambrian quartzite overlain by Cambrian limestone. The great thickness of these deposits indicates deposition in a miogeosyncline.

synclinal deposits of the Klamath Mountains into five elongated belts that more or less parallel the Pacific shoreline (**38**) (Fig. 10.17). Metamorphosed sedimentary and volcanic rocks in both the Central Metamorphic Belt and the Western Paleozoic and Triassic Belt have no known basement and were probably deposited either in a trench or in the deep ocean seaward of a trench. Metamorphosed graywackes and black shales in these two belts were probably deposited in a trench, while rhythmically bedded cherts and greenstones were probably deposited in the deep ocean (**39**) (Fig. 10.18). The rhythmically bedded cherts probably originated as radiolarian oozes like those found in the modern deep oceans. Chaotic mixtures of angular fragments in a shaley matrix in the

rocks of the Western Paleozoic and Triassic Belt are probably a melange formed by faulting along a subduction zone as the seafloor moved under a continent (**36, 37**) (Fig. 10.19). Glaucophane and lawsonite have been found in deposits of the Central Metamorphic Belt (**41**), and these minerals form only under conditions of high pressure and low temperature (**42**). Continental areas have a geothermal gradient too high for these minerals to form, but high pressures and low temperatures could occur along subduction zones as sediments and volcanics are carried rapidly beneath a continent. Radiometric dating of schists and amphibolites from the Central Metamorphic Belt indicates that metamorphism occurred during the Devonian and Carboniferous (**38**), and therefore

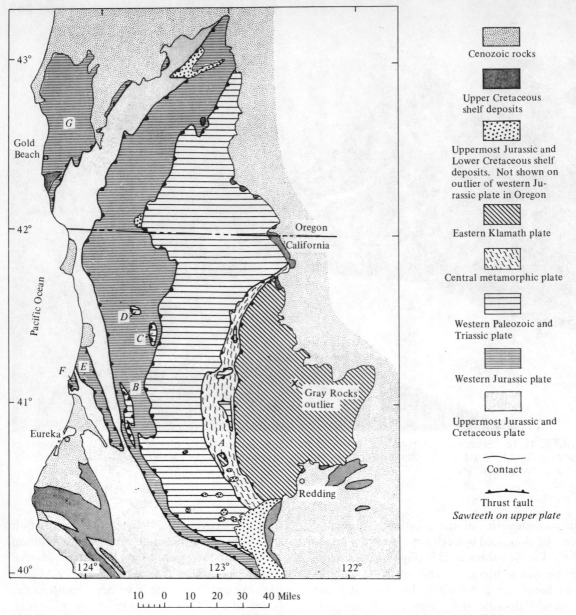

FIGURE 10.17 Geologic map of northern California and southern Oregon showing the distribution of Irwin's five belts and the thrust faults forming their boundaries. (From Irwin, Ref. 38.)

deposition of the parent rocks of the schists and amphibolites probably occurred during the early Paleozoic. Fossils in the Western Paleozoic and Triassic Belt show that deposition occurred from the Silurian or Devonian to the Jurassic (38).

A eugeosynclinal sequence of early Paleozoic age in western Nevada includes shale, chert, quartz sandstone, and minor volcanics. Some of the sedimentary rocks appear to have a western source, perhaps from an island arc (43). The composition of the volcanics and character of the sedimentary rocks suggests deposition on oceanic crust, probably in a marginal sea (43).

**Map legend:**

- Cenozoic rocks
- Upper Cretaceous shelf deposits
- Uppermost Jurassic and Lower Cretaceous shelf deposits. Not shown on outlier of western Jurassic plate in Oregon
- Eastern Klamath plate
- Central metamorphic plate
- Western Paleozoic and Triassic plate
- Western Jurassic plate
- Uppermost Jurassic and Cretaceous plate
- Contact
- Thrust fault
  *Sawteeth on upper plate*

(a)

(b)

**FIGURE 10.18** Rocks of the Western Paleozoic and Triassic Belt in Northern California: **(a)** Rhythmically bedded chert near Sawyers Bar, California. **(b)** Pillow greenstone near Cecilville, California.

## Franklin Geosyncline

The Franklin Geosyncline (or Geocline) extends from the Arctic Islands in northernmost Canada to northeastern Greenland (Fig. 10.4). It joins the Cordilleran Geosyncline on its western end and the East Greenland Geosyncline on its eastern end.

Thick Cambrian sequences are not exposed at the surface in the Franklin Geosyncline, but the Ordovician is up to 1200 m (4000 ft) thick, and the Silurian is 3000 m (10,000 ft) thick, one of the thickest sequences of that age in the world. Thus, the early Paleozoic deposits in this area are of geosynclinal thickness. Evaporites are abundant in Ordovician sequences, and limestone and dolostone are abundant in Silurian sequences. These deposits probably accumulated in a hot, arid climate. Paleomagnetic data place the Franklin Geosyncline between 10°N and 15°N of the equator during most of the early Paleozoic.

In the southern part of the geosyncline the deposits belong to the shelly facies, but the carbonates of this facies grade northward into black shales of the graptolite facies. Isolated carbonate bodies within these shales may have been reefs built up above the muddy sea bottom to form shoals and low islands (**44**) (Fig. 10.20). Volcanics are not found in the southern part of the area of the graptolite facies, but a eugeosynclinal sequence of graywacke, shale, sandstone, and volcanic rocks is exposed in northern Ellesmere Island and northwestern Axel Heiberg Island.

There was little deformation in the geosyncline during the early Paleozoic, but a mild deformation near the end of the Silurian created a geanticline known as the Boothia Arch (Fig. 10.20). The age of this deformation is approximately the same as that of the terminal phase of the Caledonian orogeny in Europe.

### East Greenland Geosyncline

The East Greenland Geosyncline (or Geocline) extends from northeastern Greenland southward along the coast to Scoresby Sound. It joins the Franklin Geosyncline on the north, and was connected to the southern end of the Northwestern Caledonian Geosyncline before Europe and North America broke apart during the early Tertiary.

A basal sandstone of Cambrian age was deposited disconformably on Late Proterozoic strata in many parts of the geosyncline. These sandstones are overlain by a 2000- to 3000-m (6000- to 9000-ft) thick sequence of carbonates. The absence of Late Cambrian fossils seems to indicate that there is a disconformity between Middle Cambrian and Early Ordovician beds. The early Paleozoic sequence in East Greenland

**FIGURE 10.19** Melange from the Klamath Mountains of northern California.

is remarkably similar in lithology and fossils to early Paleozoic miogeosynclinal deposits in Scotland, Spitsbergen, and eastern North America. At least nine invertebrate fossil species are common to all of these areas, and therefore it is likely that shallow water connections existed between them.

Intense deformation affected the East Greenland Geosyncline near the end of the Silurian. Great blocks were displaced westward along major thrust faults and the geosynclinal deposits were intensely folded. This deformation occurred at the same time as the terminal phase of the Caledonian orogeny in Europe, and it was probably caused by the joining of Ancestral North America and Ancestral Europe.

### Caledonian Geosyncline

The Caledonian Geosyncline (or Geocline) is a complex system of depositional troughs that may be divided into two separate geosynclines (Fig. 10.21). During the early Paleozoic, the Northwest Caledonian Geosyncline bordered Ancestral North America, and the Southeast Caledonian Geosyncline bordered Ancestral Europe. The Southeast Caledonian Geosyncline will be discussed under the section on Ancestral Europe.

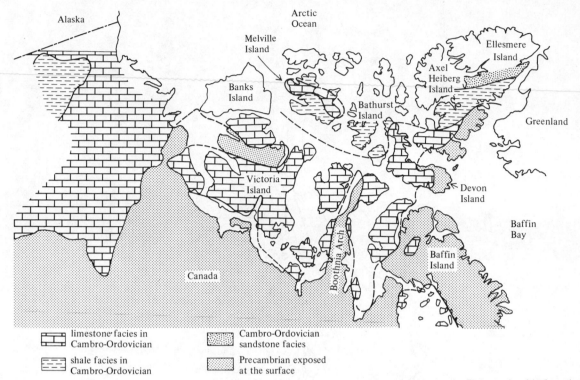

**FIGURE 10.20** Changes in facies within the Franklin Geosyncline during the early Paleozoic. Notice the carbonate bank on Melville Island and the coarsening of sediments toward the north end of Ellesmere Island. (After Douglas *et al.,* Ref. 44.)

Like the Appalachian Geosyncline, the Caledonian Geosyncline may have formed as a result of separation of Ancestral North America and Ancestral Europe during the Middle or Late Proterozoic (**6**). During the closing phase of the Iapetus Ocean, a subduction zone dipped either: (1) westward under the Northwest Caledonian Geosyncline (**45**); (2) eastward under the Southeast Caledonian Geosyncline (**46**); or (3) both westward under the Northwest Caledonian Geosyncline and eastward under the Southeast Caledonian Geosyncline (Fig. 10.22) (**47**).

### Northwest Caledonian Geosyncline

The Northwest Caledonian Geosyncline (or Geocline) extends northeastward from Ireland through Scotland, Norway, and western Sweden to Spitsbergen. Before separation of the continents, this geosyncline was connected to both the East Greenland Geosyncline and the Northwest Appalachian Geosyncline. The Northwest Caledonian Geosyncline contains both miogeosynclinal and eugeosynclinal deposits.

MIOGEOSYNCLINE. A 1300- to 2600-m (4000- to 8000-ft) thick sequence of Cambrian and Ordovician carbonates rests unconformably on Late Proterozoic deposits in northern Scotland and west Spitsbergen. As in east Greenland, a disconformity separates Middle Cambrian and Early Ordovician deposits (**48**). Silurian deposits are absent from the miogeosyncline, perhaps as a result of uplift in the

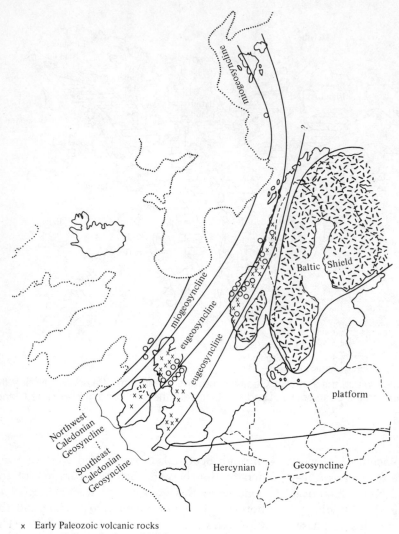

**FIGURE 10.21** Distribution of early Paleozoic volcanics, serpentinites, and related rocks within the Caledonian Geosyncline. The boundary between Ancestral North America and Ancestral Europe is inferred to have been along the belt of serpentinites and other ophiolitic rocks.

x  Early Paleozoic volcanic rocks
o  serpentinites and other ophiolitic rocks
·········  1000 m depth contour

geosyncline following Ordovician orogenic activity. Between the Late Silurian and the Middle Devonian, miogeosynclinal deposits were thrust northward several tens of miles along the Moine Thrust (**49**). This deformation probably occurred during the terminal phase of the Caledonian orogeny near the end of the Silurian.

EUGEOSYNCLINE.  A belt of Late Proterozoic and early Paleozoic schist, gneiss, amphibolite, and marble extends from northwestern Ireland through central Scotland to western Norway. These deposits include the 4000-m (13,000-ft) thick Dalradian sequence in the highlands of Scotland. A trilobite of Early or

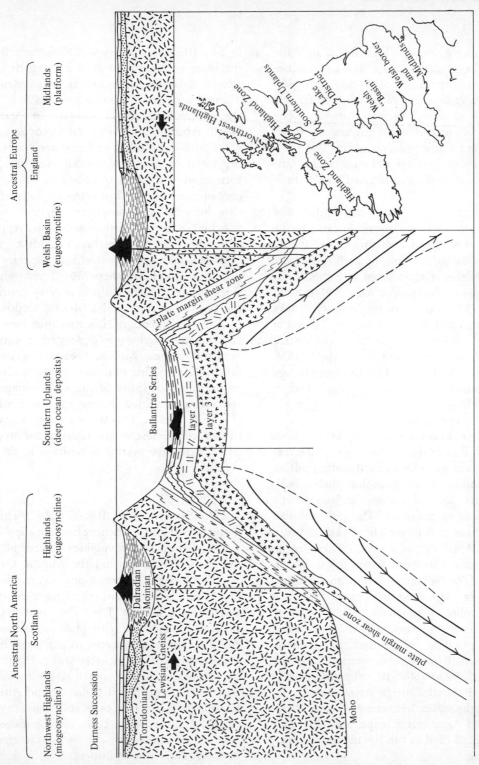

**FIGURE 10.22** Restored cross section across the Caledonian Geosyncline during the Early Ordovician. Insert shows the major geographic divisions of the United Kingdom and Ireland.

Middle Cambrian age has been found near the top of the Dalradian sequence, and a diamictite is present near the middle of the sequence (**50**). Thus, the Dalradian may range from Late Proterozoic (**51**) to early Paleozoic in age. The Upper Dalradian consists of graywacke, shale, limestone, and a few volcanic units. The presence of algal stromatolites and mud cracks indicates that it was deposited in relatively shallow water (**50**).

The Dalradian sequence was folded and metamorphosed at least three times prior to the deposition of Early Devonian strata. Potassium-argon dating of biotite and muscovite from the Dalradian shows that one episode of metamorphism occurred at least 450 million years ago (**52**). Furthermore, granitic rocks from Scotland and Sweden have been dated at 447 to 460 million years by the whole-rock rubidium-strontium isochron method (**53**). These dates indicate a Middle Ordovician age for the metamorphism and intrusion, which is the time of phase one of the Taconic orogeny in the northern Appalachians.

The oldest known sequence in southern Scotland is the Early Ordovician Ballantrae series. This unit includes mafic (basaltic) pillow lavas, volcanic breccias, graptolitic shales, radiolarian cherts, sheeted diabase, and ultramafic rocks, such as serpentinite. The pillow lavas, sheeted diabase, and ultramafic rocks are probably part of an ophiolite (**54**). There is no known granitic basement under the sedimentary rocks of the Ballantrae Series and it is likely that they were deposited on oceanic crust, that is, the ophiolite. Ordovician pillow lavas in western Norway are associated with red jasper and black phyllitic schist, and they contain structures which indicate deposition in water more than 2000 m (6000 ft) deep (**55**). These deposits and the Ballantrae series may be located along the suture between Ancestral North America and Ancestral Europe, and they may have been deposited in the Iapetus Ocean prior to its closing.

The Ballantrae series was deformed during the Early or Middle Ordovician, which would also be about the same time as phase one of the Taconic orogeny. This deformation may mark the time of the initial contact of Ancestral North America and Ancestral Europe.

Middle Ordovician sedimentary rocks were deposited in southern Scotland following deformation of the Ballantrae Series. These rocks contain a fauna which is remarkably similar to that in Middle Ordovician deposits of the Southwest Appalachian Geosyncline. Approximately 45 percent of the brachiopods are of the same species, and many of these species have not been found elsewhere (**56**). These relationships suggest that a shallow seaway connected the two areas during the Middle Ordovician.

A thick sequence of clastic sedimentary rocks and volcanic rocks was deposited in southern Scotland during Middle Ordovician to Late Silurian time. This sequence, along with most of the other deposits of the Caledonian Geosyncline was folded during the latest Silurian (**18**) (Fig. 10.23). This deformation, known as the terminal phase of the Caledonian orogeny, ended extensive marine deposition in the geosyncline.

CAUSE OF DEFORMATION. The first deformation and metamorphism in the Northwest Caledonian Geosyncline occurred between the Early Cambrian and the Middle Ordovician. Subsequent deformations occurred during the Middle Ordovician, the Late Ordovician, and the Late Silurian. These deformations were probably associated with plate collisions. The Middle Ordovician seems to mark the ending of deposition of sedimentary and volcanic rocks on oceanic crust, and therefore the initial contact of Ancestral North America and Ancestral Europe (possibly an arc-arc collision) may have occurred at this time. Later deformations may have been caused by continued compression between these two continents.

FIGURE 10.23 Folded Silurian graywackes and shales in southeastern Scotland. These rocks were folded during the Caledonian orogeny at the end of the Silurian. (NERC Copyright. Reproduced by permission of the Director, Institute of Geological Sciences, London.

## Continental Interior

The continental interior of Ancestral North America includes that part of the continent which lies between the geosynclines. It consists mainly of lowlands and plains except in the western region where Cenozoic folding and faulting has produced a number of mountain ranges in areas of relatively thin sedimentary cover.

At the beginning of the Cambrian, the seas were largely confined to the ocean basins and the geosynclines, but beginning during the Middle Cambrian, the seas transgressed onto the continental interior (Fig. 10.24). This transgression reached its maximum extent during the Late Cambrian (Fig. 10.7b). Basal sandstones of this transgression, such as the Potsdam Sandstone of northern New York, the Flathead Sandstone of Wyoming, and the Tapeats Sandstone in the Grand Canyon, were deposited in near-shore environments. These sandstones range in age from Middle Cambrian to Early Ordovician, a period covering about 75 million years. Minor regressions during the overall transgression produced a complex interfingering of rock units (Fig. 10.24).

As the seas moved toward the center of the continent, basal sandstones were covered by shale or carbonates. During the maximum transgression of the seas, much of the continental interior was inundated, and only higher

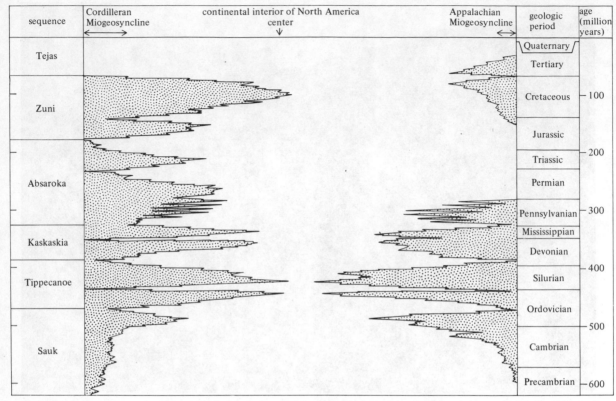

**FIGURE 10.24**  Transgressions and regressions of the seas across the continental interior of North America during the Phanerozoic. (Modified considerably after Sloss, Ref. 57.)

areas of the Canadian Shield and portions along the Transcontinental Arch remained above sea level (Fig. 10.4). Thus sedimentation was continuous over much of the continental interior between Cambrian and Ordovician time. The warm, shallow seas of the continental interior were favorable to the development of marine life, and fossils are abundant in early Paleozoic deposits. The seas retreated from the continental interior during the Early Ordovician (Fig. 10.7c).

In the Middle Ordovician, the seas again returned to the continental interior. A brief regression in mid-Late Ordovician time may have been associated with Late Ordovician glaciation in the southern continents. Another advance and withdrawal of the seas in latest Ordovician was followed by another trans-

gression, which reached its maximum in the Middle Silurian.

During times of crustal unrest in the Appalachian and Cordilleran geosynclines, clastic sediments were carried across the geosynclines onto the continental interior. For example, extensive marine shales, such as the Utica Shale in central New York, were deposited in the eastern part of the continental interior during and after the first phase of the Taconic orogeny (Middle Ordovician), and continental sandstones and shales were deposited in the eastern continental interior as highlands were uplifted during the second phase of the Taconic orogeny (Late Ordovician).

Dolostones of Early and Middle Silurian age form prominent escarpments in the northeastern and north-central United States. One of

FIGURE 10.25 Silurian marine and continental deposits at the Whirlpool below Niagara Falls, New York. (Photo by G. P. Morehead, Jr.)

these, the Lockport Dolostone, is the cap rock of Niagara Falls (Fig. 10.25). During the Late Silurian, extensive red beds and evaporites were deposited in several basins in the eastern continental interior (Fig. 10.26). The circulation within these basins had been restricted by the development of coral reefs and only limited amounts of seawater of normal salinity entered the basins. The presence of thick evaporites, red beds, and dolomites suggests that deposition occurred in an arid climate, and the abundant reefs show that the climate was rather warm. Paleomagnetic studies indicate that the northeastern and northcentral areas of the United States were at a latitude of about 20°S during the Silurian. A gradual withdrawal of the seas from the continent began during the Middle Silurian and continued to the end of the period.

## ANCESTRAL EUROPE

The geosynclines that encircled Ancestral Europe during the early Paleozoic included the Caledonian, Uralian, and Hercynian geosynclines, as well as the eastern half of the Northern Appalachian Geosyncline (Fig. 10.27). Nova Scotia, eastern Newfoundland, New Brunswick, and eastern New England were left on the North American side of the Atlantic Ocean when Europe and North America separated during the late Mesozoic.

### Southeast Caledonian Geosyncline

The Southeast Caledonian Geosyncline extends from southern Ireland and northwestern Eng-

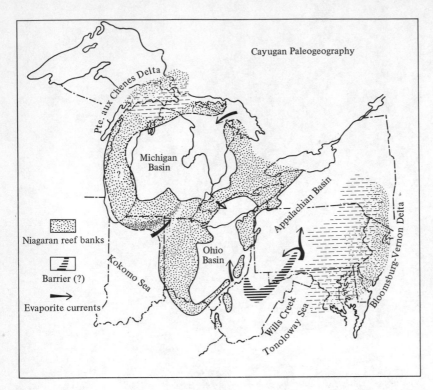

FIGURE 10.26 Paleographic map of north-central United States during the Cayugan epoch of the Silurian period. Evaporites were deposited in the enclosed basins during this time. (Courtesy of Briggs. Published by permission of The American Association of Petroleum Geologists, Ref. 58.)

land through western Norway to Spitsbergen (Fig. 10.27). The geosyncline (or geocline) contains as much as 10,000 m (30,000 ft) of shale, quartz sandstone, and graywacke, as well as minor amounts of limestone, conglomerate, and andesitic to basaltic volcanic rocks. These eugeosynclinal deposits grade eastward into relatively thin platform deposits. The Cambrian trilobite fauna of the Southern Caledonian Geosyncline is remarkably similar to that found in eastern Newfoundland, Nova Scotia, New Brunswick, and eastern Massachusetts (59). This similarity suggests that there was a relatively shallow seaway between these areas.

Unconformities, which are common in early Paleozoic sequences, were the result of minor episodes of folding and uplift within the geosyncline. In Wales, there are two unconformities in the Cambrian, three in the Ordovician, and at least three in the Silurian sequence (18). The presence of andesitic volcanics and unconform-

ities suggests that the Southeast Caledonian Geosyncline was a eugeosyncline during the early Paleozoic, and that a subduction zone dipped southeastward under it during that time (Fig. 10.22). The deformations culminated at the end of the Silurian with the terminal phase of the Caledonian orogeny, which ended marine sedimentation in the geosyncline (Fig. 10.28). As discussed earlier, the Middle Ordovician to Silurian deformations were probably due to a collision between Ancestral North America and Ancestral Europe.

## Northeast Appalachian Geosyncline

The Northeast Appalachian Geosyncline (or Geocline) extends from eastern Newfoundland through eastern New Brunswick and Nova Scotia to eastern New England (Fig. 10.27). The deposits in the geosyncline include tens of

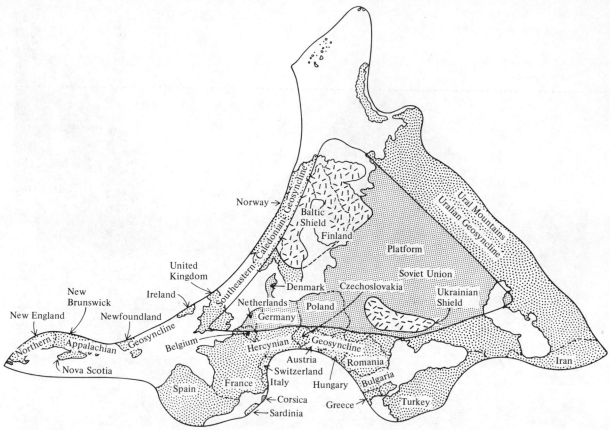

**FIGURE 10.27** Tectonic elements of Ancestral Europe during the early Paleozoic.

thousands of feet of deformed, metamorphosed clastics and volcanics, which grade eastward into mildly deformed limestone, shale, and sandstone. Thus, both the miogeosynclinal and eugeosynclinal deposits are present.

MIOGEOSYNCLINE. In eastern Massachusetts, Early Cambrian quartzite, red and green shales, and red limestones overlie granitic rocks that have been dated at between 580 and 610 million years (**60**). Sequences of similar rock types with similar fossils are also found in southeastern New Brunswick, Nova Scotia, and eastern Newfoundland (**59**). These deposits are between 170 and 3000 m (500 and 11,000 ft) thick.

Marine beds younger than Middle Cambrian have not been found along the Atlantic border of New England, which may have been above sea level during most of the Paleozoic. A relatively continuous sequence of marine strata ranging in age from Early Cambrian to Middle Ordovician occurs in eastern Newfoundland and southeastern New Brunswick. However, except for a narrow trough in western Nova Scotia, most of the miogeosyncline was probably emergent after the Late Ordovician (**61**).

EUGEOSYNCLINE. The eugeosyncline, extending from eastern Connecticut and western Rhode Island through eastern Massachusetts, central New Hampshire, eastern Maine,

**FIGURE 10.28** The Aberystwyth grits of Silurian age were tilted during the Caledonian orogeny at the end of the Silurian. Western coast of Wales, England. (NERC Copyright. Reproduced by permission of the Director, Institute of Geological Sciences, London.)

and central New Brunswick to central Newfoundland, contains more than 13,000 m (40,000 ft) of early Paleozoic deposits. These deposits have been subjected to deformation and metamorphism ranging from broad folding and slight recrystallization to intense folding and high-grade metamorphism. Before metamorphism, the eugeosynclinal deposits were graywackes, shales, siltstones, conglomerates, volcanic rocks, and limestones.

The absence of latest Ordovician and earliest Silurian deposits from much of the eugeosyncline and the presence of angular unconformities at the base of most Silurian deposits indicates that an episode of folding and erosion occurred near the end of the Ordovician (**62**). Furthermore, an absence of latest Silurian and earliest Devonian deposits in the eugeosyncline indicates that the early Paleozoic deposits were uplifted at about the same time as the terminal phase of the Caledonian orogeny (**61**). Silurian eugeosynclinal deposits presumably accumulated on continental crust because widespread subaerial erosion could occur only in areas underlain by crust thick enough to stand above sea level. If this is correct, Ancestral North America and Ancestral Europe were joined before the beginning of the Silurian.

## Hercynian Geosyncline

Extending eastward across southern Europe from Spain to the Caspian Sea, the Hercynian Geosyncline forms a link between the Caledonian and Uralian geosynclines (Fig. 10.27). Deposits of the Hercynian Geosyncline are dominantly clastics and volcanics with only minor limestone. There are no miogeosynclinal deposits. Volcanic activity was especially intense during the Late Cambrian and Early and Middle Ordovician.

In Sardinia, on the southernmost margin of Ancestral Europe, Ordovician shale rests with angular unconformity on Cambrian strata. In southern France, an angular unconformity has been reported between Middle and Lower Ordovician beds. These unconformities provide evidence that this region was tectonically active during the early Paleozoic. The cause of the deformation and volcanism may have been movement of oceanic plates under Ancestral Europe as it moved toward Gondwanaland. In this case, the geosyncline would be a eugeosyncline.

## Uralian Geosyncline

The Uralian Geosyncline is located on the eastern margin of Ancestral Europe and extends the length of the Ural Mountains (Fig. 10.27). The geosyncline consists of a eugeosyncline bordered by miogeosynclines both to the east and to the west. Thick sequences of limestone, dolostone, shale, and quartzite in the western miogeosyncline grade eastward into even thicker sequences of shale, arkosic sandstone, and felsic volcanics. Little is known about the eastern miogeosyncline, since it is largely covered by thick Cenozoic sediments.

Unconformities in the early Paleozoic sequences provide evidence of several episodes of folding, crustal uplift, and withdrawal of the seas from the geosyncline. Large mafic and ultramafic intrusions (ophiolites) that contain important deposits of chromite, asbestos, and platinum, occur on the east flank of the Urals. These intrusions were probably emplaced during an episode of folding in Late Ordovician or Early Silurian time (**63**). Folding also occurred in the geosyncline at the end of the Silurian. Early Paleozoic deformations may have been related to the joining of Ancestral Europe and Ancestral Siberia. Cambrian and Ordovician paleomagnetic poles are widely separated, but Silurian and later paleomagnetic poles agree rather well with each other (Fig. 7.12b). Ordovician deformation may have been caused by the movement of the seafloor under Ancestral Europe as the continents approached each other. The late Silurian phase of deformation may have resulted from the collision of Ancestral North America and Ancestral Europe.

## Continental Interior

The seas transgressed across the Russian Platform in the continental interior of Ancestral Europe at least twice and probably three times during the early Paleozoic. These and later transgressions appear to have been contemporaneous with those in the continental interior of Ancestral North America (**64**).

The lower Paleozoic strata of the continental interior include sandstone, shale, limestone, and blue clay. The lower Paleozoic rocks of the continental interior are quite fossiliferous. On the island of Gotland, in the Baltic Sea, limestones contain corals that are remarkably similar to those of North America. Evidently, shallow seas connected these areas during the Silurian. Several disconformities in the lower Paleozoic sequences provide evidence of transgressions and regressions of the seas into the interior of the continent.

## GONDWANALAND

Paleomagnetic poles have been recalculated for a Pangaea reconstruction of the continents

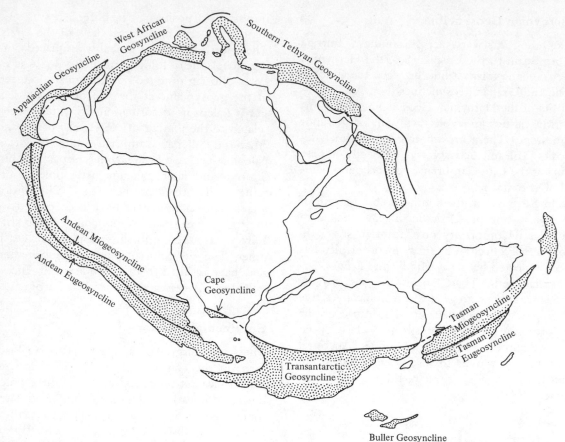

**FIGURE 10.29** Tectonic elements of Gondwanaland during the early Paleozoic.

(Fig. 9.4) and it is evident the Cambrian, Ordovician, Silurian, and Devonian poles of Gondwanaland do not agree with poles of other continents. Therefore, it is likely that Gondwanaland was separated from the other continents during the early Paleozoic (Fig. 10.3).

Geosynclines that formed a continuous belt around the continent (Fig. 10.29) included the Tasman Geosyncline in eastern Australia, the Buller Geosyncline in New Zealand, the Southern Tethyan Geosyncline extending from northern India to northern Africa, the West African Geosyncline, the southeastern Appalachian Geosyncline, the Andean Geosyncline,

the Cape Geosyncline in southern Africa, and Transantarctic Geosyncline. Deformation often occurred at the same time in several different geosynclines (Fig. 10.30).

**Tasman Geosyncline**

During the Cambrian, the Tasman Geosyncline was divided into a western miogeosyncline with a shallow water, shelly facies and an eastern eugeosyncline with a deeper water, graptolitic facies.

A thick sequence of Lower and Middle Cambrian carbonates in the miogeosyncline contains algal stromatolites, oolites, and intra-

Geosynclines

| Age | Transantarctic, Antarctica | Ref. | Tasman, Australia | Ref. | Buller, New Zealand | Ref. | Andean, South America | Ref. | West African | Ref. | Cape, South Africa | Ref. | Southern Tethyan, India | Ref. |
|---|---|---|---|---|---|---|---|---|---|---|---|---|---|---|
| Permian Upper | | | deformation and intrusion | 21 | | | deformation and uplift | 25 | | | deformation | | | |
| Permian Middle | | | | | | | | | | | ←---?--→ | | | |
| Permian Lower | intrusion | 28* | | | | | | | | | | | | |
| Carboniferous Upper | | | | | uplift and intrusion | 21 | intrusion | 26 | ←---?----- deformation ----?---→ | | ←?→ deformation | | | |
| Carboniferous Lower | | | intrusion and deformation | 21 | | | deformation | 25 | | | | | | |
| Devonian Upper | intrusion | 28 | deformation and intrusion | 21 | | | deformation | 25 | | | | | uplift | 23 |
| Devonian Middle | | | | | | | | | | | | | | |
| Devonian Lower | | | | | | | | | | | | | | |
| Silurian Upper | | | deformation and intrusion | 37 | | | deformation and uplift | 42 43 | uplift | 39 | | | | |
| Silurian Middle | | | | | | | | | | | | | | |
| Silurian Lower | | | | | | | | | | | | | | |
| Ordovician Upper | | | deformation and intrusion | 37 | | | deformation and uplift | 43 | | | | | | |
| Ordovician Middle | | | | | | | | | | | | | | |
| Ordovician Lower | | | ←---?—uplift and deformation→ | | | | | | | | | | | |
| Cambrian Upper | deformation and intrusion | 46 | | 37 | | | | | | | intrusion and deformation | 44 | | |
| Cambrian Middle | | | | | | | | | | | | | | |
| Cambrian Lower | | | | | | | | | | | | | | |

* References numbers for Late Paleozoic data refer to Chapter 11

FIGURE 10.30 Comparison of the history of deformation of Australia, New Zealand, Antarctica, South America, Africa, and India, during the Paleozoic.

formational breccias, all features of shallow water deposition. Sedimentation ceased during the Middle Cambrian, and the deposits were folded in Middle or Late Cambrian time (**65**).

During the early Paleozoic, geanticlines were uplifted within the eugeosyncline and furnished clastics to adjacent troughs (Fig. 10.30). Basaltic and andesitic volcanic rocks are common along the geanticlines, which may have been part of an island arc on the continental margin.

The deformation and volcanism may have been associated with movement of the seafloor under the eugeosyncline.

## Buller Geosyncline

The Buller Geosyncline, which underlies most of New Zealand, was the site of rapid deposition of clastic sediments and volcanic rocks throughout the early Paleozoic (**65**). Andesites and basalts in the northwestern part of the geosyncline may have been deposited in an island arc environment. The fauna in the associated sedimentary rocks is similar to that in the early Paleozoic deposits of Australia, and it is probable that New Zealand and Australia were relatively close during the early Paleozoic.

In the southeastern part of the geosyncline, a sequence of metamorphosed graywackes, shales, and basalts contains fossils as old as the Permian. These deposits are underlain by a great thickness of unfossiliferous rocks, including some that may have been deposited on ocean crust during the early Paleozoic.

## Southern Tethyan Geosyncline

The Southern Tethyan Geosyncline extends westward from the Himalayas across West Pakistan, Iran, Turkey, Greece, Yugoslavia, and Italy to North Africa (Fig. 10.29). From Cambrian through Devonian time, this geosyncline was continuous with the West African Geosyncline. Until deposition ceased in the Southern Tethyan Geosyncline in the early Tertiary, it was separated from the Northern Tethyan Ge-

osyncline by a deep ocean. The two geosynclines were brought together as India, Africa, and Arabia collided with Eurasia.

The lower Paleozoic deposits of the Southern Tethyan Geosyncline include between 1500 and 55,000 m (5000 and 17,000 ft) of sandstone, shale, limestone, and minor conglomerate. The scarcity of volcanic rocks here indicates that these sequences are miogeosynclinal in character.

The lack of deformation and volcanism in the geosyncline may indicate that there was no subduction zone near the geosyncline during the early Paleozoic, in which case, it may have been a miogeosyncline during the early Paleozoic.

## West African Geosyncline

A geosyncline extended from the western end of the Southern Tethyan Geosyncline in Morocco along the west coast of Africa to Portuguese Guinea (Fig. 10.29).

## Southeastern Appalachian Geosyncline

Before separation of the continents took place, the West African Geosyncline was continuous with the southeastern Appalachian Geosyncline (Fig. 10.29). The early Paleozoic sedimentary rocks in this geosyncline are largely covered by younger deposits, but they crop out in the Carolina Slate Belt of North Carolina and South Carolina, as well as in adjacent parts of Virginia and Georgia. It is probable that these deposits were originally laid down in a eugeosyncline. They are at least 10,000 m (30,000 ft) thick and include volcanic flows, breccias, and tuffs interbedded with sandstone and slate. Fossils are rare, but it is likely that a Cambrian trilobite found in a stream bed in central North Carolina was derived from this sequence (**67**). Radiometric dating of rhyolites from the Carolina Slate Belt indicates that volcanism occurred during the Cambrian (**68**). The felsic composi-

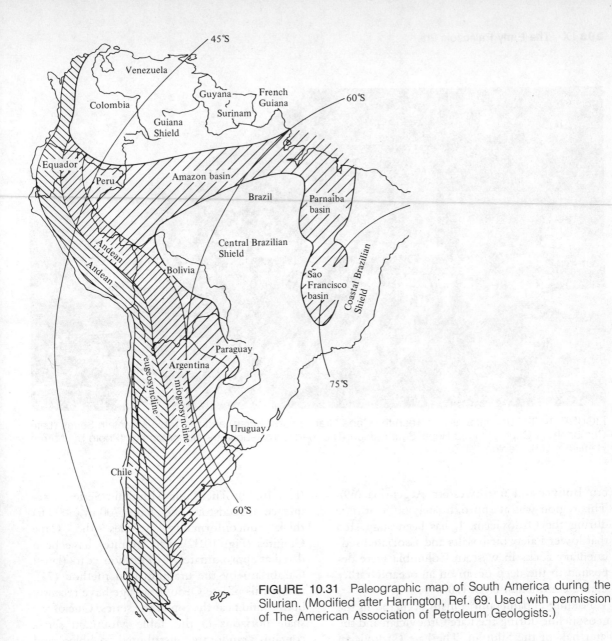

**FIGURE 10.31** Paleographic map of South America during the Silurian. (Modified after Harrington, Ref. 69. Used with permission of The American Association of Petroleum Geologists.)

tion of some of the volcanic rocks suggests that the eugeosyncline may have been part of an island arc (Fig. 10.15). In this case, a subduction zone would have dipped southeastward under Gondwanaland.

### Andean Geosyncline

The Andean Geosyncline occupied the western margin of South America more or less along the present trend of the Andean Mountains (Fig. 10.29). Although the early Paleozoic deposits in the eastern part of the geosyncline do not contain volcanic rocks, those to the west contain appreciable volcanics. The Andean Geosyncline consisted of both a miogeosyncline and eugeosyncline (Fig. 10.31).

The miogeosynclinal deposits are 1500 to 3000 m (5000 to 10,000 ft) thick and are dominantly clastics with only minor carbonates. Glacial deposits of Early Ordovician and Middle Silurian age have been reported from west-

FIGURE 10.32 Unconformity between the Cape Granite (Ordovician) and the Table Mountain Series (Late Cambrian) at Cape of Good Hope, South Africa. The road is just above the unconformity. (Photo by Warren Hamilton, U.S. Geological Survey.)

ern Bolivia and northwestern Argentina (**69**). This region was at approximately 60°S latitude during the Ordovician. It has been suggested that lower Paleozoic basalts and associated sedimentary rocks in western Columbia were deposited in the deep ocean on an oceanic rather than a continental crust (**70**).

Folding and metamorphism occurred in the geosyncline during the Late Ordovician and the later half of the Silurian. The Late Ordovician orogeny may have been related to the movement of an oceanic plate under an island arc on the margin of South America. From that time on, this area has been a miogeosyncline.

### Cape Geosyncline

The Cape Geosyncline, located on the southern tip of South Africa, is a small fragment of a geosyncline that extended from South America to Antarctica during the early Paleozoic

(Fig. 10.29). The Table Mountain Series, a sequence of sandstone and shale 1700 m (5000 ft) thick, unconformably overlies the Cape Granites (Fig. 10.32). The granites have been dated at approximately 510 million years (latest Cambrian) by the uranium-lead method (**71**), and fossils of late Ordovician age have recently been found near the top of the series. One of the shale horizons in the Table Mountain series contains randomly distributed pebbles and boulders, some having striated and faceted surfaces (**72**) (Fig. 10.33). The age of this horizon is the same as that of glacial deposits in northern Africa, and it is likely that at least part of the Table Mountain series has a glacial origin.

### Transantarctic Geosyncline

The Transantarctic Geosyncline parallels the Transantarctic Mountains from the Wedell Sea to Victoria Land (Fig. 10.29). During the Early

**FIGURE 10.33** Striated pebbles from the Table Mountain Series, South Africa. (Photo from DuToit, Ref. 72. Reproduced with permission of Oliver and Boyd, Edinburgh.)

and Middle Cambrian, clastics, limestones, and volcanics were deposited with apparent conformity on late Proterozoic deposits. The entire sequence was deformed, metamorphosed, and intruded by granites in Late Cambrian or Early Ordovician time between 450 and 520 million years ago (**73**) (Fig. 10.34). This deformation occurred at approximately the same time as the intrusion of the Cape granites in South Africa and the deformation of the Cambrian deposits in the southern part of the Tasman Geosyncline.

The axis of the Transantarctic Geosyncline shifted toward the Pacific Ocean during Ordovician and Silurian time. A thick sequence of clastics that are probably of early Paleozoic age unconformably overlies metamorphosed Cambrian deposits in northeast Victoria Land on the

Pacific end of the Transantarctic Geosyncline (Fig. 10.35).

## Continental Interior

Marine sedimentary rocks of early Paleozoic age are rather restricted in the continental interior of Gondwanaland. During the early Paleozoic, a large depositional basin was located in central Australia, and several smaller basins lay along the western and northern borders of the continent. Deposits in these basins include sandstone, shale, limestone, dolostone, gypsum, and anhydrite. In the Salt Range of northern Pakistan, a 400-m (1200-ft) thick sequence of Lower Cambrian clastics contains salt pseudomorphs and gypsum. The evaporites

FIGURE 10.34  Large anticline in early Paleozoic carbonate and clastic rocks in the Neptune Range, Pensacola Mountains, Antarctica. (Photo by J. R. Ege.)

FIGURE 10.35  Folded graywacke and siltstone of ordovician (?) age, northern Victoria Land, Antarctica. (Photo by Warren Hamilton, U.S. Geological Survey.)

indicate that deposition occurred under somewhat arid conditions. Evidently, both Australia and India were in an arid belt during the Early Paleozoic. Paleomagnetic data support this conclusion (Fig. 10.3).

Lower Paleozoic sedimentary rocks have been found in several basins along the west coast of Africa. In the Sahara Desert, shales and sandstones of Late Ordovician age are overlain by tillites that contain striated and faceted boulders and rest on grooved and striated pavements (Fig. 10.36). The tillites interfinger with marine deposits of Late Ordovician age near the northern margin of the continental interior (**74**). Late Ordovician glacial deposits cover a very large part of northern Africa and were probably deposited by continental glaciers (Fig. 10.3b). The direction of ice movement is from south to north, *away* from the present equator (**75**). Paleomagnetic data indicate that the South Pole was located in west central Africa during the Late Ordovician. This is approximately the region from which the Late Ordovician ice sheets appear to have radiated. The Ordovician glacial deposits all occur within 50° of the Late Ordovician South Pole (Fig. 10.3b).

Lower Paleozoic sedimentary rocks have been identified in the Amazon, Parnaiba, and Paraña basins (Fig. 10.33). The Iapo Formation of southern Brazil contains striated boulders, presumably of glacial origin. The exact age of these deposits is not known, but they underlie Early Devonian strata. They may be the same age as the Saharan glacial deposits (Late Ordovician).

The southeastern United States and southern Mexico were part of the continental interior of Gondwanaland during the Early Paleozoic. Wells in southeastern Alabama, southern Georgia, and northern Florida have penetrated relatively undisturbed shales of Ordovician and Silurian age (**76**). In central Florida, volcanic rocks, possibly of early Paleozoic age, overlie granitic and metamorphic rocks dated at between 480 and 530 million years (Late Cambrian to Early Ordovician) (**77**). Radiometric dates in this range are common elsewhere in Gondwanaland.

## ANCESTRAL SIBERIA

The geosynclines that bordered the continental interior of Ancestral Siberia include the Uralian Geosyncline in the west, the Angara Geosyncline in the south, the Northern Pacific Ocean Geosyncline in the east, and the Taimyr Geosyncline in the north (Fig. 10.37). The depositional and deformational histories of these geosynclines are summarized in Table 10.1.

Volcanics and associated sedimentary rocks in the Angara Geosyncline were probably deposited in an island arc environment on the margin of Ancestral Siberia. The present location of these geosynclinal deposits near the center of the Asian continent was probably the result of the joining of Ancestral Siberia and Ancestral China. The initial contact of these two plates may have occurred during the Silurian.

It is likely that the volcanic and associated sedimentary rocks of the Northern Pacific Ocean Geosyncline were probably also deposited in the vicinity of an island arc. Early Paleozoic deformation and volcanism were probably due to the westward movement of an oceanic plate under the continent. The folding and metamorphism that occurred within the Taimyr Geosyncline at the end of the Silurian may have been caused by the movement of an oceanic plate southward under the continent.

Flat-lying early Paleozoic sedimentary rocks are present in the continental interior of Ancestral Siberia (Fig. 10.38).

## ANCESTRAL CHINA

During the early Paleozoic, Ancestral China was bordered on the north by the Central Asia Geosyncline, on the southwest by the Northern Tethyan Geosyncline, on the south by the Indonesian Island Arc, and on the east by the Southern Pacific Ocean Geosyncline (Fig.

**FIGURE 10.36** Ordovician glacial features in the Sahara Desert: **(a)** tillite, **(b)** faceted boulder, **(c)** striated pavement, **(d)** *rouche moutonnée*. [**(a)**, **(c)**, and **(d)** From S. Beuf *et al.*, Ref. 74; **(b)** from R. W. Fairbridge.]

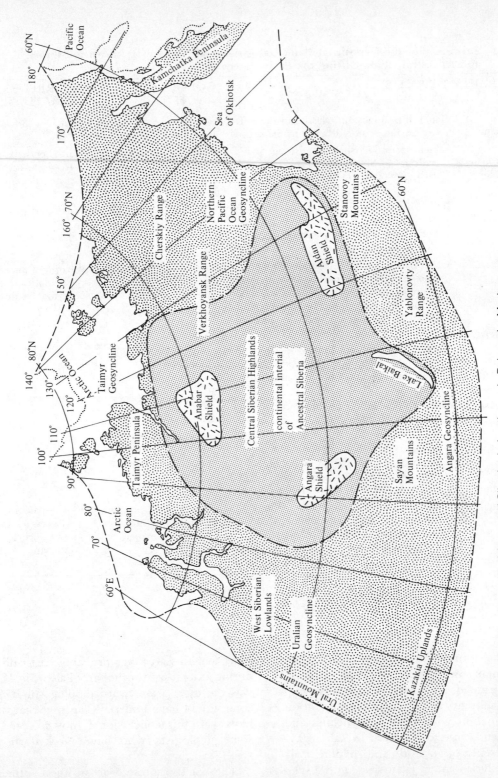

**FIGURE 10.37** Tectonic elements of Ancestral Siberia during the early Paleozoic. Modern latitudes and mountains are included for the purposes of location.

TABLE 10.1 **Summary of the Depositional and Deformational History of Ancestral Siberia and Ancestral China During the Early Paleozoic.**

| ANCESTRAL SIBERIA | DEPOSITS | OROGENIC ACTIVITY | INFERRED CLIMATE |
|---|---|---|---|
| Angara Geosyncline | Thick limestones and clastics in the north, mafic and felsic volcanics in the south | Folding at the end of the Silurian | Warm |
| Northern Pacific Ocean Geosyncline | Thick limestone, shale, sandstone, and volcanics—Silurian deposits are more than 10,000 ft (3000 m) thick | Uplift during the Ordovician; folding near the close of the Silurian | Warm, perhaps tropical—area was near paleoequator |
| Taimyr Geosyncline | Thick limestones, and also shale, sandstone, and felsic volcanics; sedimentary rocks dominantly marine; red beds and evaporites also present | Clastic wedge indicates deformation during the Late Silurian | Warm and dry |
| Continental Interior | Cambrian: sandstone, clay, dolostone, limestone, and evaporites—Ordovician: abundant limestone, clay, sandstone, evaporites and red beds—Silurian: argillaceous limestones, clays, and red beds | Local uplift during the Silurian | Warm and dry—paleolatitude 10°S during Cambrian |
| **ANCESTRAL CHINA** Southern Pacific Ocean Geosyncline | Very thick Cambrian limestones in southeast; felsic volcanics interbedded with Silurian shale and limestone in Japan | No major orogenic activity | Warm |
| Central Asia Geosyncline | Great thicknesses of clastics and volcanics; some limestones | Early Cambrian uplift and late Silurian folding and metamorphism | ? |
| Northern Tethyan Geosyncline | Conglomerate, sandstone, shale, limestone, and volcanics; sequence is 15,000 ft (5000 m) thick in Tibet | None | ? |
| Continental Interior | Cambrian shales grading upward into limestones—Early and Middle Ordovician limestones—Late Ordovician to Late Silurian shale, sandstone, and limestone | Middle Ordovician and Late Silurian uplift | Warm—paleolatitude between 30 and 60°N during Ordovician |

10.39). The continental interior of Ancestral China contains a number of small, isolated shields separated by platform deposits (Fig. 10.39). A summary of the history of deposition and deformation within the geosynclines and the continental interior is given in Table 10.1.

The presence of andesitic and rhyolitic volcanics of Silurian age in Japan (Fig. 11.31) suggests that this area was probably an island arc during at least part of the early Paleozoic. If this was the case, a westward dipping subduction zone would have underlain Japan at that time. However, this conclusion is not accepted by some geologists who have worked in this area (78).

The sea transgressed across the continental

**FIGURE 10.38** Cambrian deposits exposed at the "Pillars of the Lena" along the south bank of the Lena River in Siberia. (Photo by Allison R. Palmer.)

interior of Ancestral China during the Cambrian, and thick limestones were deposited. The sea retreated from this area during the Middle Ordovician, but later in the period, it invaded the southern part of the continental interior and remained until the end of the Silurian.

## EARLY PALEOZOIC LIFE

The base of the Cambrian is generally placed at the first appearance of fossils with hard parts. However, recent study of sequences near the boundary between the Cambrian and the Late Proterozoic have indicated the presence of skeletal remains in rocks that are generally considered to be Precambrian in age. Tubelike structures which may be related to the archeocyathids are found in southern Africa associated with fossils which have affinities to

the Late Proterozoic Ediacaran Fauna (**10**). Conical tubes composed of calcium phosphate and chitinous tubes have been found in the Soviet Union below Cambrian strata (**81**). It is not certain whether these deposits should be considered earliest Cambrian or latest Precambrian. If they are considered to be Precambrian, the base of the Cambrian can no longer be defined as the first appearance of fossils with hard parts. It appears likely that the Cambrian-Precambrian boundary will eventually be based on the particular fauna present, rather than at the first appearance of fossils with hard parts.

The Early Cambrian lasted 30 to 40 million years, and rocks deposited during the first 10 million years of the Early Cambrian do not contain trilobites (**81**). These rocks are designated as belonging to the Tommotian stage and

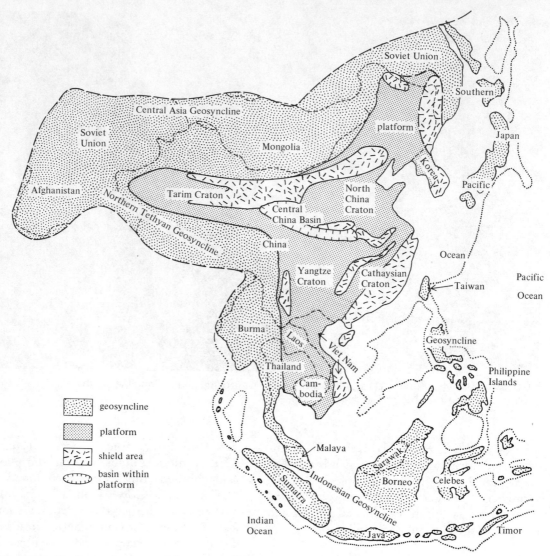

**FIGURE 10.39** Tectonic elements of Ancestral China during the early Paleozoic.

they contain gastropods, brachiopods, sponges, archeocyathids, and wormlike fossils. Trilobites, pelecypods, and echinoderms first appeared after the Tommotian stage. Nearly all the important invertebrate phyla have been found in Lower Cambrian strata. Of the important index forms, only bryozoans, cephalopods, conodonts, stony corals, echinoids, pelecypods, graptolites, and vertebrates are missing from the Early Cambrian fossil record. All of these groups appear by Middle Ordovician time.

## Origin of Calcareous Shells

The sudden appearance of invertebrates with calcareous shells at the beginning of the Cambrian remains one of the great mysteries of geology. How could such a varied, advanced, and abundant fauna evolve with almost no evidence of its development? As discussed in Chapter 9, impressions of a variety of advanced, soft-bodied invertebrates have been found in Late Proterozoic deposits in many parts of the world. Thus the lack of a record of the development of the Cambrian fauna is probably due to the absence of calcareous shells among Precambrian organisms. However, this raises the question of why so many groups developed calcareous shells in such a relatively short interval of time. A number of theories have been proposed to explain this phenomenon:

1. It has been suggested that the atmosphere may have been relatively low in oxygen during the Precambrian and earliest Paleozoic time. In this case, the ozone content of the atmosphere would also have been low. Ozone in the modern atmosphere helps to screen out harmful ultraviolet radiation. Calcareous shells may have provided protection from such radiation, and therefore selection may have operated in favor of organisms with calcareous shells in earliest Paleozoic time.

2. Some paleontologists have suggested that Precambrian seas were too acidic to allow invertebrates to extract carbonate for shell formation. The presence of abundant calcareous stromatolites in Precambrian deposits shows that this was probably not the case. . Furthermore, high acidity cannot account for the absence of siliceous skeletons in Precambrian rocks, since silica is readily precipitated in an acid medium. Since siliceous sponge spicules first appear in Cambrian deposits, it is unlikely that the chemistry of the seawater was a significant factor in the appearance of organisms with hard parts.

3. The appearance of calcareous shells may have been brought about as a result of significant changes in the chemistry, depth, and temperature of the oceans following reces-

sion of the Late Proterozoic glaciers. Decreased ocean temperatures during the ice age produced an increase in the solubility of calcium carbonate. Warming of the ocean water following the ice age then resulted in an increase in the calcium carbonate available for shell formation. At the end of the Late Proterozoic, the seas were quite restricted. As sea level rose with the melting of the glaciers, new environments developed into which expanding faunas could migrate. The transgression of the seas into the miogeosynclines during the Early Cambrian coincided approximately with the appearance of fossils with hard parts. With an increase in the area of the seas, there was less competition among marine forms, and this may have facilitated evolution and dispersal.

4. It is possible that shell development is related to the rise of predators. If Precambrian invertebrates were herbivores or scavengers, for example, hard skeletons would not have been advantageous. As predatory forms evolved, shells became essential for the survival of many invertebrates.

5. It has also been suggested that the development of hard parts is related to high tides associated with the close approach of the moon to the earth following lunar capture (**79**). Organisms with protective coverings may have had an evolutionary advantage in the strong currents which would have been generated by high tides.

## Protista and Monera

Stromatolites produced by blue-green algae are common in lower Paleozoic deposits (Fig. 10.40). They often formed reefs in shallow waters. In British Columbia, fossil green algae and red algae have been found in the Burgess Shale along with a rich invertebrate fauna.

Agglutinated foraminifera, those that built tests by cementing sand grains together, have been reported from strata of Early Cambrian age. Radiolarians have also been reported from

**FIGURE 10.40**  Algal stromatolites of Cambrian age from Saratoga, New York. (New York State Museum and Science Service, Albany.)

Cambrian deposits, but some paleontologists have questioned whether these fossils have been correctly identified.

### Archeocyathids and Sponges

The spongelike archeocyathids occur in Early and Middle Cambrian sequences throughout the world (Fig. 10.41). Since they are restricted to this time range, they are very useful for intercontinental correlations. Archeocyathids are the oldest known sessile organisms and were probably among the first to secrete a hard, calcareous shell or test.

Sponges first appear in Lower Cambrian sequences and are abundant enough in Silurian deposits to be useful as guide fossils. Complete sponges have been found in the Middle Cambrian Burgess shale.

### Coelenterates

Soft-bodied anemones were present in Cambrian seas. Early Paleozoic stony corals belong to the classes Tabulata and Rugosa. Tabulate corals first appeared during the Early Ordovician, and rugose corals first appeared during the Middle Ordovician (**81**). Stony corals were

**FIGURE 10.41** Restoration of the fauna from the Middle Cambrian Burgess Shale in British Columbia. The fauna includes: **(a)** archeocyathids. **(b)** Trilobites. **(c)** Sea cucumber. **(d)** An annelid worm. **(e)** Jellyfish. **(f)** Trilobitelike arthropods. (Courtesy of the Field Museum of Natural History, Chicago; prepared by George Marchland under the direction of I. G. Reimann.)

common during the Ordovician and Silurian. Ordovician corals were mainly colonial, tabulate corals that formed both horizontally bedded biostromes and occasionally moundlike bioherms (Fig. 10.42). True coral reefs are common in Silurian limestones and dolostones (Fig. 10.43). These reefs were a mixture of colonial corals, solitary corals, stromatoporoids, and a variety of other invertebrates such as gastropods, brachiopods, and trilobites (Fig. 10.44).

## Bryozoans and Brachiopods

Bryozoans are not known with certainty from Cambrian strata. They first appear in rocks of earliest Ordovician age (**81**) and they are abundant in the Ordovician and Silurian strata (Fig. 10.45). In shaly limestones of this age, they frequently comprise a significant percentage of the rock (Fig. 10.46).

Brachiopods are among the oldest fossils with hard parts. *Inarticulate brachiopods,* those in

**FIGURE 10.42** Reconstruction of the sea bottom as it might have appeared during the Middle Ordovician in Illinois. The fauna includes: **(a)** Straight-shelled nautiloid cephalopods. **(b)** Trilobites. **(c)** Colonial corals. **(d)** Solitary corals. **(e)** Gastropods. **(f)** Brachiopods. (Courtesy of the Field Museum of Natural History, Chicago.)

**FIGURE 10.43** Reconstruction of a coral reef as it might have appeared during the Middle Silurian in Illinois. The fauna includes: **(a)** Colonial tabulate coral *Favosites*. **(b)** Colonial tabulate coral *Halysites*. **(c)** Solitary corals. **(d)** Trilobites. **(e)** Brachiopods. **(f)** Cystoids. **(g)** A nautiloid cephalopod. (Courtesy of the Field Museum of Natural History, Chicago; prepared by George Marchland under the direction of I. G. Reimann.)

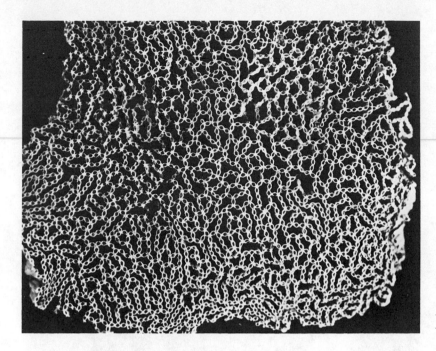

**FIGURE 10.44** The Silurian "Chain Coral" *Halysites* from Louisville, Kentucky. (Courtesy of the Smithsonian Institution.)

which the valves are not rigidly hinged to one another, first appear in the fossil record at the base of the Tommotian stage (**81**). *Articulate brachiopods,* those in which there is a hinge of interlocking teeth and sockets between valves, first appear just above the top of the Tommotian stage (**81**). All five orders of inarticulate brachiopods appear first in Lower Cambrian deposits. Of these, two became extinct in the Middle Cambrian, and one became extinct in the Middle Ordovician (**83**). Of the seven orders of articulate brachiopods, one first appeared in the Early Cambrian, one in the Middle Cambrian, one in the Lower Ordovician, two in the Middle Ordovician, and one in the Lower Silurian. No orders of articulate brachiopods became extinct during the early Paleozoic.

The diversity of the brachiopods (as indicated by the number of families present at any one time) increased from the beginning of the Cambrian and reached a peak during the Middle Ordovician. From the Middle Ordovician to the end of the Silurian, diversity declined

somewhat. The abundance of brachiopods in deposits of early Paleozoic age makes them very useful as guide fossils.

## Mollusks

Pelecypods first appear in rocks of Early Cambrian age (**82**) and are fairly common in Upper Ordovician sequences. Gastropods are found in Early Cambrian strata and are relatively abundant in Ordovician and Silurian sequences. Since some species of gastropods have relatively short time ranges, they may be useful as guide fossils.

The oldest fossils that have been classed as cephalopods are small conical shells in rocks of Early and Middle Cambrian age in Europe (**83**). However, it is not certain if these are cephalopods. Fossils unquestionably identified as cephalopods have been found in Late Cambrian deposits in North America and Asia. The shells of these cephalopods are 1 to 5 cm (0.5 to 2 in.) long. During the Early Ordovician, many new groups appeared and some developed shells up

**FIGURE 10.45** Middle Silurian (Niagaran) fossils from Western New York: **(a)** *Eucalyptocrinites caelatus*, crinoid calyx (X3.5). **(b)** *Rhynchotreta americana*, brachiopod (X1.5). **(c)** *Fardenian subplana*, brachiopod (X1.5). **(d)** *Heliolites*, sp., tabulate coral (X5). **(e)** *Trematopora tuberculosa*, (X5). **(f)** *Dalmanites limulurus*, trilobite (X2). **(g)** *Fenestella* (lower right), bryozoan (X3). **(h)** *Dictyonema retiforme*, a dendroid graptolite (X2). (Photographs courtesy of Carlton Brett.)

**FIGURE 10.46** Bryozoans from the Rochester shale, Niagara Falls, New York.

to 5 m (17 ft) long (Fig. 10.42a). The first coiled cephalopods appeared during the Ordovician, but the straight-shelled forms were dominant for the remainder of the early Paleozoic.

## Annelids and Similar Organisms

Carbon residues of segmented worms have been found in Middle Cambrian Burgess Shale (Fig. 10.47a–10.47c). Conodonts became abundant during the Silurian and are useful in intercontinental correlations because of their widespread occurrence and distinctive characteristics.

## Arthropods

Trilobites are the most abundant fossils in Cambrian sequences (Fig. 10.47d). Since they evolved rapidly and range widely, they are good index fossils for Cambrian and Ordovician sequences. They reached their climax in the Late Cambrian and declined with the rise of predators such as cephalopods and fish. Trilobites remained abundant throughout the Ordovician and Silurian, but their numbers were less than during the Cambrian. Some trilobites were blind burrowers; others crawled about on the sea bottom or were adapted to swimming or floating. They did not have a biting mouth, which restricted their diet to relatively soft material. They may have been scavengers feeding on decaying matter, or they may have fed on microorganisms.

Ostracods appeared first in the Cambrian and became very abundant during the Ordovician. These microscopic, bivalved crustaceans have proven very useful in subsurface correlations.

Eurypterids have been found in beds of Or-

(a)

(b)

(c)

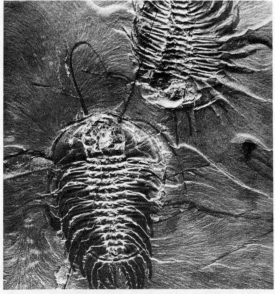

(d)

**FIGURE 10.47** Fossils from the Middle Cambrian Burgess shale from British Columbia: **(a)**–**(c)** annelids. **(d)** Trilobites belonging to the species *Olenoides serratus*. (Courtesy of Smithsonian Institution.)

dovician age in association with normal marine assemblages. During the Silurian they appear to have migrated into hypersaline environments. Specimens from Silurian deposits in the vicinity of Buffalo, New York, reach a maximum of 3 m (9 ft) long, but most species were just under 25 cm (10 in.) long. (Figs. 10.48 and 10.49).

It has been suggested that the earliest land animals were scorpionlike arthropods that first appeared during the Late Silurian (**85**). It is possible that these organisms lived in fresh or marine waters rather than on land. Undoubted

FIGURE 10.48 *Eurypterus remipes lacustris,* most common eurypterid in vicinity of Buffalo Museum of Science.)

land-dwelling animals do not appear in the fossil record until the Devonian.

## Echinoderms

The oldest known echinoderms are found in Early Cambrian strata in southern California. These include a primitive cystoid and a spindle-shaped form quite unlike later echinoderms. Crinoids first appeared during the Ordovician, and from then until the end of the Paleozoic, they were abundant in shallow epicontinental seas (Fig. 10.45). Starfish arose during the Ordovician.

## Graptolites

Graptolites made their first appearance in the Cambrian, and they became abundant during the Ordovician. They are common in many black shales where they are preserved as carbon residues (Fig. 10.45h). Since graptolites evolved rapidly and were quite widely dispersed in Ordovician and Silurian seas, they are useful in correlating both within and between the continents. Early Ordovician graptolites include *Dictyonema,* which was a colonial form having a large number of branches attached to a floating

FIGURE 10.49 Reconstruction of eurypterids that lived during the Late Silurian in the Buffalo area. (Courtesy of Buffalo Museum of Science; diorama prepared by George and Paul Marchland, under the direction of I. G. Reimann.)

**FIGURE 10.50** A restoration of the ostracoderm *Hemicyclaspis*. (Courtesy of the Field Museum of Natural History, Chicago.)

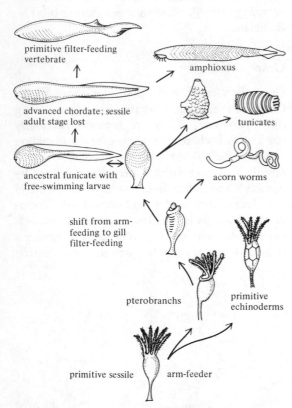

**FIGURE 10.51** Diagrammatic family tree suggesting the possible mode of evolution of the vertebrates. [Copyright © 1967 by American Association for the Advancement of Science (From Romer, Ref. 87.)]

structure. *Tetragraptus,* a four-branched graptolite, appeared somewhat later in the Early Ordovician, and *Didymograptus,* a two-branched form, is found in Middle Ordovician deposits.

The evolutionary trend toward simple forms continued into the Silurian with the appearance of *Monograptus,* which had but a single row of thecae.

## Vertebrates

The oldest fossil remains of vertebrates are fragments of fish from Upper Cambrian deposits of northeastern Wyoming (**86**). Fragments of fish are relatively abundant in the Middle Ordovician Harding Sandstone in the central Rockies. Like the modern lamprey, these fish had no jaws and belong to the class Agnatha. These bony-plated forms, which are known as ostracoderms, were seldom more than a few inches long (Fig. 10.50). They were probably filter feeders, since the head contained a set of gill pouches that could have been used for straining out small particles of food. Unlike many of the more primitive filter feeders, the ostracoderms had the advantage of being mobile and therefore were able to search out sources of food. Attempts to trace the evolution of the vertebrates from the invertebrates have led to the suggestion that the invertebrate ancestor may have been a filter feeder closely related to the echinoderms (**87**) (Fig. 10.51).

## Plants

Plant fossils resembling seaweed are found in many limestones of Cambrian and Ordovician age. These probably represent multicellular marine plants. There have been reports of land plants of Cambrian age, but it is not certain that these really did live on land; they might have been aquatic plants. The earliest fossils of vascular plants (plants having tissues which conduct water to the top of the plant) occur in rocks of Middle Ordovician age (**88**). These remains include spores and plant tissues, presumably from land plants. Spores that may have been derived from vascular land or semi-aquatic plants have been found in Early Silurian deposits from western New York (**87**). Well-preserved fossil land plants have been found in Late Silurian strata.

## EARLY PALEOZOIC CLIMATES

The widespread occurrence of thick limestones and coral reefs of early Paleozoic age has led some geologists to conclude that the entire earth had a warm and equable climate during this interval. However, this evaluation was made by comparing climatic indicators with modern latitudes, not paleolatitudes.

### Cambrian

In the Canadian Rockies, southern California, and southwestern Nevada, Cambrian deposits are dominantly carbonates of considerable thickness. These areas were very close to the Cambrian equator (Fig. 10.3a). The Cambrian and Ordovician rocks of the Appalachian miogeosyncline are also very thick and are dominantly carbonates. This region was between 20°S and 35°S during deposition of the carbonates. Relatively thick limestones were deposited at 75°S in Morocco and between 30°S and 50°S in China. Reefs or reeflike bodies formed by algae and archeocyathids generally occur within 30° of the Cambrian equator. However, reefs are found at 75°S in Morocco. The thick limestones, reefs, and evaporites at relatively high latitudes in China and Morocco indicate that the climate may have been warmer than aver-

age in the middle to high latitudes during the Cambrian. Red beds and evaporites are generally formed between 15° and 30° north or south of the paleoequator (Fig. 10.3a) and are an indication of extreme aridity during the Cambrian.

### Ordovician

As in the Cambrian, red beds and evaporites are generally found between 15° and 30° north or south of the Paleoequator (Fig. 10.3b). Ordovician reefs generally fall within 30° of the paleoequator. Thus the climate for most of the Ordovician was not abnormal with respect to temperature and humidity. Glacial deposits of Late Ordovician age occur within 50° of the Ordovician South Pole in Africa and South America. Those in the Saharan region are so extensive that they must have been deposited by continental rather than alpine glaciers (Fig. 10.3b).

During this time, the climate at middle and high latitudes in the southern hemisphere was colder than average, perhaps similar to that during the Pleistocene. Presumably the climate in the northern hemisphere would also have cooled somewhat at the time of the glaciation.

### Silurian

Coral reefs, evaporites, and red beds generally fall within 40° of the Silurian equator, whereas glacial deposits occur only within 25° of the Silurian South Pole (Fig. 10.3c). Thus climates during the Silurian probably resembled modern climates.

## SUMMARY

At the beginning of the Cambrian Era, the five ancestral continents were separated from each other by deep oceans. These oceans gradually became smaller throughout the early Paleozoic as the ancestral continents moved closer together. The epicontinental seas were largely confined to the outermost margins of the continents at the beginning of the Cambrian, but they transgressed toward the centers of the continents during the Cambrian. By the Late Cambrian, seas covered more than half of the area of the continents.

Sedimentation was most rapid in the geosynclines bordering the ancestral continents. The eugeosynclines received thick deposits of clastic sediments and volcanics, while thick limestones were deposited in most miogeosynclines. Sedimentation on the platforms was rather slow, and there was deposition of limestone, shale, or sandstone.

Extensive deformation at the end of the Cambrian resulted in the formation of mountains in much of the southern hemisphere and parts of the northern hemisphere as well. Deformation during the Middle and Late Ordovician in the Appalachian, Caledonian, and Uralian regions resulted in the uplift of elongated ridges within the geosynclines and deposition of clastic wedges both in the miogeosyncline and on the platform. Near the close of the Silurian, the deposits of geosynclines were deformed and metamorphosed. The early Paleozoic deformations may have been caused by the progressive joining of Ancestral North America, Europe, Siberia, and China.

The Cambrian marks the first appearance of fossils with calcareous and siliceous skeletons. The trilobites were the most important invertebrates in the Cambrian, whereas the Ordovician saw the rise of the brachiopods, graptolites, bryozoans, crinoids, and mollusks. The first fish appeared during the Cambrian, but did not become an important part of the marine fauna until the late Paleozoic. The first land plants and the first land animals are found in Ordovician or Cambrian rocks and in Silurian or Devonian rocks, respectively.

The climate during the Cambrian was probably warmer than average. The Early and Middle Ordovician were warm, but during the Late Ordovician, the climate cooled, especially in the middle and high latitudes of the southern hemisphere. At that time, extensive glaciation occurred in Africa and possibly in South America as well. The climate remained cool in the middle and high latitudes of the southern hemisphere during the Silurian.

## REFERENCES CITED

1. J. C. Briden, W. A. Morris, and J. D. A. Piper, 1973, *Geophys. J. Roy. Ast. Soc.,* v. 34, p. 107.
2. R. W. Fairbridge, 1970, *Geotimes,* v. 15, no. 6, p. 18.
3. W. S. McKerrow and A. M. Ziegler, 1972, *Nature Phys. Sci.,* v. 240, p. 92; J. D. Keppie, 1977, *Province of Nova Scotia Dept. of Mines Pap. 77-3;* J. C. Briden, G. E. Drewry, and A. G. Smith, 1974, *J. Geol.,* v. 82, p. 555.
4. M. W. McElhinny, 1973, *in* D. H. Tarling and S. K. Runcorn, eds., *Implications of Continental Drift to the Earth Sciences:* Academic Press, New York, p. 77; E. A. Hailwood and D. H. Tarling, 1973, *in* D. H. Tarling and S. K. Runcorn, eds., *Implications of Continental Drift to the Earth Sciences:* Academic Press, New York, p. 37; J. F. Dewey and W. S. F. Kidd, 1974, *Geology,* v. 2, p. 543; G. R. Keller and S. E. Cebull, 1973, *Geol. Soc. Amer. Bull.,* v. 84, p. 1659; W. S. McKerrow and A. M. Ziegler, 1972, *Proc. 24th Int. Geol. Cong.,* section 6, p. 4.
5. L. L. Sloss, E. C. Dapples, and W. C. Krumbein, 1960, *Lithofacies Maps:* Wiley, New York.
6. J. T. Wilson, 1966, *Nature,* v. 211, p. 676.
7. D. F. Strong, W. L. Dickson, C. F. O'Driscoll, B. F. Kean, and R. K. Stevens, 1974, *Nature,* v. 248, p. 37; J. M. Bird and J. F. Dewey, 1970, *Geol. Soc. Amer. Bull.,* v. 81, p. 1031; J. F. Dewey and J. M. Bird, 1971, *J. Geophys. Res.,* v. 76, p. 3179; M. J. Kennedy, 1973, *Geol. Soc. Amer. Northeastern Sec. Eighth A. Mtg. Abstr.,* v. 5, p. 183; W. R. Church and R. K. Stevens, 1971, *J. Geophys. Res.,* v. 76, p. 1460.
8. M. Kay, 1972, *Proc. 24th Int. Geol. Cong.,* section 3, p. 122; R. D. Hatcher, 1972, *Geol. Soc. Amer. Bull.,* v. 83, p. 2735.
9. C. K. Seyfert and D. J. Leveson, 1969, *Geol. Bull.,* v. 3, p. 33.
10. G. M. Kay, 1951, *North American Geosynclines:*

Geol. Soc. Amer. Mem. 48; R. S. Dietz, 1972, *Sci. Amer.,* v. 226, p. 30.

11. R. S. Dietz, 1972, *Sci. Amer.,* v. 226, p. 30.

12. J. Rodgers, 1968, *in E-an* Zen *et al.,* eds., *Studies of Appalachian Geology: Northern and Maritime:* Wiley, New York, p. 141.

13. *E-an* Zen, 1967, *Geol. Soc. Amer. Spec. Pap. 97.*

14. J. M. Bird, 1949, *in* M. Kay, ed., *Amer. Assoc. Petrol. Geol. Mem. 12,* p. 630.

15. J. M. Mattinson, 1975, *Geology,* v. 3, p. 181; P. A. Brown, 1976, *Nature,* v. 264, p. 712; J. J. Peterson, P. J. Fox, and E. Schreiber, 1974, *Nature,* v. 247, p. 194.

16. Y. W. Isachsen, 1964, *Trans. N.Y. Acad. Sci.,* v. 26, p. 812.

17. A. J. Boncot, 1962, *in* K. Coc, ed., *Some Aspects of the Variscan Fold Belt:* Manchester University Press, Manchester, England, p. 155.

18. A. J. Eardley, 1962, *Structural Geology of North America:* Harper and Row, New York, p. 205.

19. G. M. Bennison and A. E. Wright, 1969, *The Geological History of the British Isles:* St. Martin's Press, New York, p. 121.

20. R. B. Neuman, 1968, *in E-an* Zen *et al,* eds., *Studies of Appalachian Geology: Northern and Maritime:* Wiley, New York, p. 35.

21. J. T. Wilson, 1966, *Nature,* v. 211, p. 676.

22. M. Kay, 1972, *Proc. 24th Int. Geol. Cong.,* section 3, p. 122.

23. M. Kay, 1975, *Geol. Soc. Amer. Bull.,* v. 86, p. 105.

24. J. L. Rosenfield, 1960, *Geol. Soc. Amer. Bull.,* v. 71, p. 1960.

25. B. A. Hall, 1966, *in* D. W. Caldwell, ed., *New England Intercollegiate Geological Conference Guidebook,* p. 42.

26. R. S. Naylor and A. J. Boucot, 1965, *Amer. J. Sci.,* v. 263, p. 153.

27. J. B. Lyons and H. Faul, 1970, *in E-an* Zen *et al,* eds., *Studies in Appalachian Geology: Northern and Maritime:* Wiley, New York, p. 305.

28. J. Rodgers, 1970, *The Tectonics of the Appalachians:* Wiley, New York.

29. P. B. King, 1950, *Amer. Assoc. Petrol. Geol. Bull.,* v. 34, p. 635.

30. R. D. Hatcher, 1972, *Geol. Soc. Amer. Bull.,* v. 83, p. 2735.

31. D. K. Davies and E. A. Williamson, 1976, *Geology,* v. 4, p. 340.

32. K. Burke and J. F. Dewey, 1973, *J. Geol.,* v. 81, p. 406.

33. J. Gilluly, 1965, *Geol. Soc. Amer. Spec. Pap. 80.*

34. J. R. Patterson and T. P. Storey, 1960, *Proc. 21st Int. Geol. Cong.,* p. 150; W. H. Twehhofel *et al.,* 1954, *Geol. Soc. Amer. Bull.,* v. 65, p. 247.

35. M. Churkin, 1974, *Geology,* v. 2, p. 339.

36. J. T. Dutro, W. P. Bresgé, M. A. Lanphere, and H. N. Reiser, 1976, *Amer. Assoc. Petrol. Geol. Bull.,* v. 60, p. 952; M. Churkin and G. D. Eberlein, 1977, *Geol. Soc. Amer. Bull.,* v. 88, p. 769.

37. V. E. McMath, 1966, *in* E. H. Bailey, ed., *Geology of Northern California:* Calif. Div. Mines Geol. Bull. 190, San Francisco, p. 178.

38. W. P. Irwin, 1966, *in* E. H. Bailey, ed., *Geology of Northern California:* Calif. Div. Mines Geol. Bull. 190, San Francisco, p. 19.

39. C. K. Seyfert, 1974, *in* D. F. R. McGeary, ed., *Geological Guide to the Klamath Mountains:* Department of Geology, California State University, Sacramento, p. 69.

40. D. P. Cox, 1973, *Geol. Soc. Amer. Bull.,* v. 84, p. 1423.

41. P. E. Hotz, 1973, *J. Res. U.S. Geol. Survey,* v. 1, p. 53.

42. F. J. Turner and J. Verhoogen, 1960, *Igneous and Metamorphic Petrology:* McGraw-Hill, New York.

43. J. J. W. Rogers *et al.,* 1974, *Geol. Soc. Amer. Bull.,* v. 85, p. 1913.

44. R. J. W. Douglas *et al.,* 1963, *Geology and Petroleum Potentialities of Northern Canada:* Geological Survey of Canada Pap. 63–31.

45. W. R. Church and R. A. Gayer, 1973, *Geol. Mag.,* v. 110, p. 497.

46. G. H. Gale and D. Roberts, 1972, *Nature Phys. Sci.,* v. 238, p. 60; D. F. Strong, W. L. Dickson, C. F. O'Driscoll, and B. F. Kean, 1973, *Reconnaissance Geochemistry of Eastern Newfoundland Granitoid Rocks:* Newfoundland Dept. Mines and Energy; J. G. Fitton and D. J. Hughes, 1970, *Earth Planet. Sci. Lett.,* v. 8, p. 223; R. S. Thorpe, 1972, *Geol. Soc. Amer. Bull.,* v. 83, p. 3663.

47. J. M. Bird and J. F. Dewey, 1970, *Geol. Soc. Amer. Bull.,* v. 81, p. 1031; F. Moseley, 1977, *Geol. Soc. Amer. Bull.,* v. 88, p. 764.

48. A. Hallam, 1958, *Geol. Mag.,* v. 95, p. 71; R. A. Fortey and D. L. Bruton, 1973, *Geol. Soc. Amer. Bull.,* v. 84, p. 2227.

**49.** T. N. George, 1965, *in* G. Y. Craig, ed., *The Geology of Scotland:* Archon, Hamden, Conn., p. 1.

**50.** M. R. W. Johnson, 1965, *in* G. Y. Craig, ed., *The Geology of Scotland:* Archon, Hamden, Conn., p. 115.

**51.** A. Kröner, 1977, *J. Geol.,* v. 85, p. 289.

**52.** K. Bell, 1968, *Geol. Soc. Amer. Bull.,* v. 79, p. 1167.

**53.** R. J. Pankhurst, 1974, *Geol. Soc. Amer. Bull.,* v. 85, p. 345; D. G. Gee and M. R. Wilson, 1974, *Amer. J. Sci.,* v. 274, p. 1.

**54.** W. R. Church and R. A. Gayer, 1973, *Geol. Mag.,* v. 110, p. 497.

**55.** H. Furnes, 1973, *Geology,* v. 1, p. 27.

**56.** E. K. Walton, 1965, *in* G. Y. Craig, ed., *The Geology of Scotland:* Archon, Hamden, Conn., p. 161.

**57.** L. L. Sloss, 1963, *Geol. Soc. Amer. Bull.,* v. 74, p. 93.

**58.** H. L. Alling and L. I. Briggs, Jr., 1961, *Amer. Assoc. Petrol. Geol. Bull.,* v. 45, p. 515.

**59.** J. W. Skehan, 1969, *in* M. Kay, ed., *Amer. Assoc. Petrol. Geol. Mem. 12,* p. 793.

**60.** W. D. McCarthy, 1969, *in* M. Kay, ed., *Amer. Assoc. Petrol. Geol. Mem. 12,* p. 115.

**61.** W. B. N. Berry, 1968, *in* E-an Zen *et al.,* eds., *Appalachian Geology: Northern and Maritime:* Wiley, New York, p. 23; A. J. Boucot, 1969, *in* M. Kay, ed., *Amer. Assoc. Petrol. Geol. Mem. 12,* p. 477.

**62.** L. Pavlides, A. J. Boucot, and W. B. Skidmore, 1968, *in* E-an Zen *et al.,* eds., *Studies in Appalachian Geology: Northern and Maritime:* Wiley, New York, p. 61.

**63.** D. V. Nalivkin, 1960, *The Geology of the U.S.S.R.:* Pergamon Press, New York.

**64.** L. L. Sloss, 1972, *Proc. 24th Int. Geol. Cong.,* section 6, p. 24.

**65.** D. A. Brown, K. S. W. Campbell, and K. A. W. Crook, 1968, *The Geological Evolution of Australia and New Zealand:* Pergamon Press, New York.

**66.** A. Gansser, 1964, *Geology of the Himalayas:* Wiley, New York.

**67.** H. W. Sundelius, 1970, *in* G. W. Fisher *et al.,* eds., *Studies in Appalachian Geology: Central and Southern:* Wiley, New York, p. 351.

**68.** F. A. Hills, personal communication, 1970.

**69.** H. J. Harrington, 1962, *Amer. Assoc. Petrol. Geol. Bull.,* v. 46, p. 1173.

**70.** H. Burgl, 1967, *Tectonophysics,* v. 4, p. 429.

**71.** H. L. Allsopp and P. Kolbe, 1965, *Geochim. Cosmochim. Acta,* v. 29, p. 1115.

**72.** A. L. Du Toit, 1954, *Geology of South Africa:* Oliver and Boyd, London.

**73.** E. Picciotto and A. Coppez, 1963, *in* R. J. Adie, ed., *Antarctic Geology:* North-Holland, Amsterdam, p. 563.

**74.** S. Beuf *et al.,* 1966, *Rev. Inst. Franc. Pétrole,* v. 21, p. 363.

**75.** F. Arbey, 1968, *C. R. Acad. Sci. Paris,* v. 266, Ser. D, p. 76.

**76.** P. L. Applin, 1951, *U.S. Geol. Survey Circular 91.*

**77.** C. Milton and R. Grasty, 1969, *Amer. Assoc. Petrol. Geol. Bull.,* v. 53, p. 2483.

**78.** Tushio Kimura, 1974, *in* C. A. Burk and C. L. Drake, eds., *The Geology of Continental Margins:* Springer-Verlag, New York, p. 817.

**79.** D. L. Lamar and P. M. Merifield, 1967, *Geol. Soc. Amer. Bull.,* v. 78, p. 1359.

**80.** G. J. B. Germs, 1972, *Amer. J. Sci.,* v. 272, p. 752.

**81.** S. M. Stanley, 1976, *Amer. J. Sci.,* v. 276, p. 56; J. D. McLeod, 1978, *Science,* v. 200, p. 771.

**82.** J. Pojeta, B. Runnegar, and J. Kriz, 1973, *Science,* v. 180, p. 866.

**83.** M. J. S. Rudwick, 1970, *Living and Fossil Brachiopods:* Hutchinson, London.

**84.** C. Teichert, 1967, *in* Teichert and E. L. Yochelson, eds., *Essays in Paleontology and Stratigraphy:* Special Publ. 2, Department of Geology, University of Kansas, p. 162.

**85.** L. Stormer, 1977, *Science,* v. 197, p. 1362.

**86.** J. E. Repetski, 1978, *Science,* v. 200, p. 529.

**87.** A. S. Romer, 1967, *Science,* v. 158, p. 1629.

**88.** J. Gray and A. J. Boucot, 1971, *Science,* v. 173, p. 918.

# The Late Paleozoic Era

# 11

THE LATE PALEOZOIC was a time of great tectonic activity. During this time, many of the world's geosynclines, including the Appalachian, Ouachita, Hercynian, Uralian, and Angara geosynclines, were gradually transformed into lofty mountain ranges. Collision of major crustal plates and uplift of the continents caused a general but irregular withdrawal of the seas.

The emergence of the amphibians and subsequent development of the reptiles during the late Paleozoic were accompanied by widespread colonization of the land by plants such as horsetails, tree ferns, seed ferns, and true conifers. During the Carboniferous, extensive coal-forming swamps developed under tropical conditions along the paleoequator. At the same time the area within 45° of the Carboniferous South Pole was extensively glaciated. This glaciation may also have been related to late Paleozoic orogenic activity.

## LATE PALEOZOIC GEOGRAPHY

The relative positions of the continents during the late Paleozoic may be inferred by comparing polar wandering curves, determining radiometric ages of granitic and metamorphic rocks in orogenic belts, and correlating folded sedimentary rocks. Comparison of the recalculated apparent polar wandering curves suggests that all the continents were in close proximity during the Carboniferous and Permian.

The last major orogeny in the Caledonian and Northern Appalachian geosynclines occurred during the Middle Devonian, at which time the joining of Ancestral North America and Ancestral Europe was probably completed. Since deformation within the Uralian Geosyncline did not cease until the Late Permian, the joining of Ancestral Europe and Ancestral Siberia was probably not completed until the end of the Paleozoic. Radiometric dating of orogenic activity in the Angara Geosyncline indicates that the last major metamorphism and intrusion occurred during the Late Permian. Thus Ancestral China and Ancestral Siberia were finally joined by the end of the Paleozoic. Deformation in the Appalachian and Ouachita geosynclines did not cease until the Middle Permian, when Ancestral North America and Gondwanaland were welded together. The proposed sequence of joining of the continents appears in Fig. 11.1.

## ANCESTRAL NORTH AMERICA

Throughout most of the late Paleozoic, Ancestral North America was almost completely encircled by geosynclines (Fig. 11.2). Limestones and shales were deposited in the geosynclines

(a)

**FIGURE 11.1** Proposed reconstruction of the continents showing paleoclimatic indicators: (a) Devonian. (b) Carboniferous—the alpine glacial deposits in western South America were laid down during the Early Carboniferous. (c) Permian—the alpine glacial deposits in India, Australia, and South America were laid down during the Late Permian.

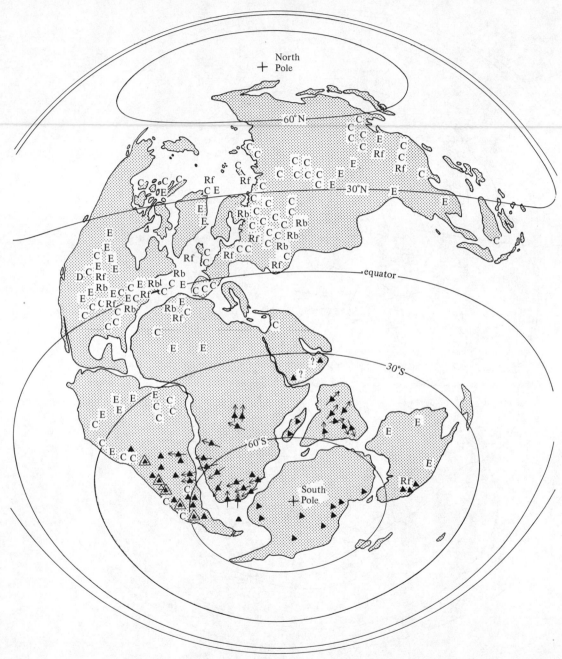

(b)

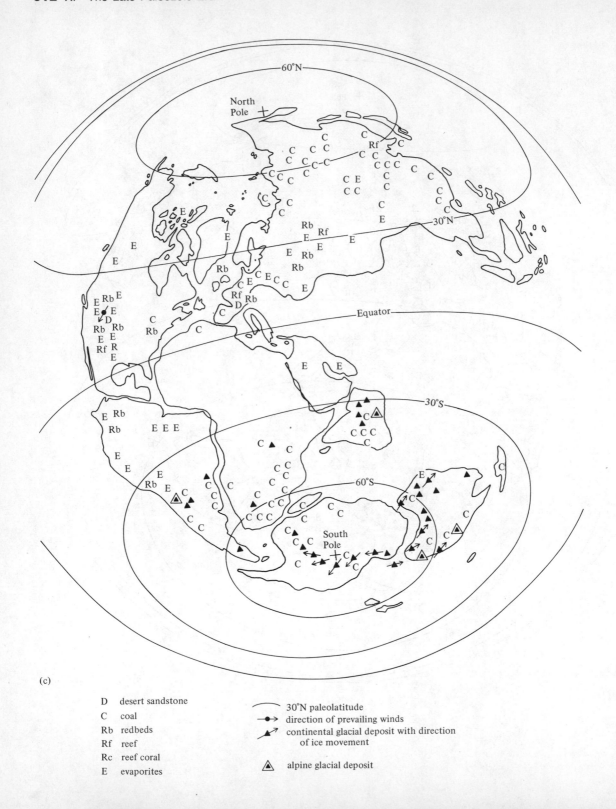

(c)

D    desert sandstone
C    coal
Rb   redbeds
Rf   reef
Rc   reef coral
E    evaporites

—— 30°N paleolatitude
•—► direction of prevailing winds
▲—► continental glacial deposit with direction
      of ice movement

⟁   alpine glacial deposit

and on the adjacent platform during times of little tectonic activity, and coarse clastics were deposited in these areas during orogenic episodes.

## Northern Appalachian Geosyncline

At the beginning of the late Paleozoic, the Northern Appalachian Geosyncline consisted of a miogeosyncline and a eugeosyncline separated by a geanticline. Deposition ended in the eugeosyncline at the beginning of the Middle Devonian and in the miogeosyncline during the Early Permian.

MIOGEOSYNCLINE. Late Paleozoic miogeosynclinal deposits are almost completely restricted to the southern part of the geosyncline. In southeastern New York, eastern Pennsylvania, and Maryland, Devonian beds overlie the Silurian with apparent conformity. The Lower Devonian sequence is thin and consists mainly of carbonates with minor shales. Evidently, the geanticline to the east was rather low at this time.

The Middle Devonian Acadian orogeny produced a mountain range that extended from eastern Pennsylvania northward to Newfoundland and from the Hudson Valley eastward to the Atlantic Ocean (Fig. 11.2b). Sediments derived from the mountains were deposited in a large, subaerial alluvial plain (the Catskill Delta) which was built out into the miogeosyncline (Fig. 11.3). Continental red beds were deposited on the flood plains of rivers that crossed the alluvial plain. These arkosic sandstones, shales, and conglomerates were probably deposited in an arid or semiarid climate. According to paleomagnetic data, this region was located at approximately 20°S during the Devonian. The Middle and Late Devonian red beds are part of an eastward thickening clastic wedge that reaches a maximum thickness of more than 3000 m (10,000 ft) (Figs. 11.4 and 11.5). The contact between marine and continental facies migrated westward as sediments spread outward from the base of the mountains. By the end of the Devonian, continental red beds were deposited on the platform adjacent to the geosyncline.

Deposition of great thicknesses of sandstone and conglomerate continued during the Carboniferous and Early Permian. Some limestone and coal are intercalated with these clastic sediments. The direction of dip of cross-beds in sandstones and the change in pebble size in conglomerates indicate that the clastics were derived from a mountain range to the east (Fig. 11.6). The mountain range extended from central Mexico through the southeastern United States and into southern Europe (Fig. 11.2). The deposition of great thicknesses of coarse clastics throughout the late Paleozoic required repeated uplift of the mountain range. There is no evidence that the Carboniferous uplifts were accompanied by folding, but radiometric dating of crystalline rocks from the core of the Appalachian Fold Belt indicates that granites were intruded during the Carboniferous (Fig. 10.13 and 11.7).

During the Pennsylvanian, extensive swamps and marshes developed within the miogeosyncline. The deposition and decay of the lush vegetation in these areas produced economically important coal deposits. Although these coal-forming environments were probably restricted to a relatively narrow belt along the coastline, coal deposits cover a considerable area because the seas repeatedly inundated low-lying coastal regions.

By the beginning of the Permian, the seas had withdrawn completely from the geosyncline, and the continental red beds of the Dunkard series were deposited in West Virginia, Pennsylvania, and Ohio (Fig. 11.2c). The Alleghenian orogeny, an episode of intensive folding, uplift, and intrusion which probably occurred during the Middle Permian, terminated deposition in the geosyncline and converted it into a mountain range.

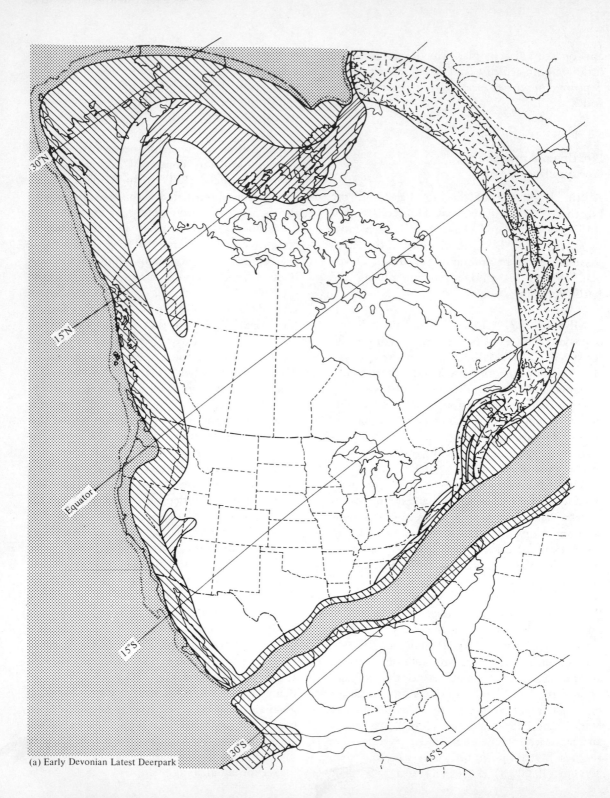

(a) Early Devonian Latest Deerpark

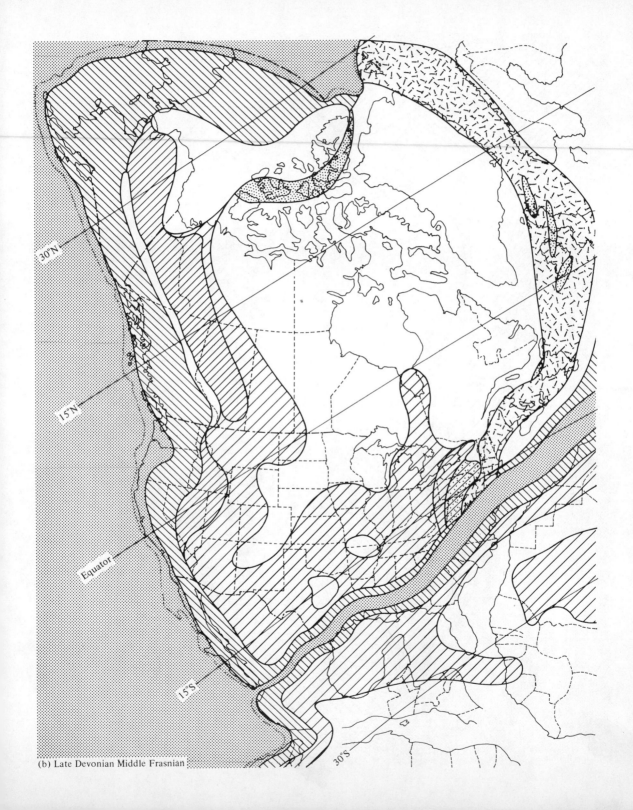

(b) Late Devonian Middle Frasnian

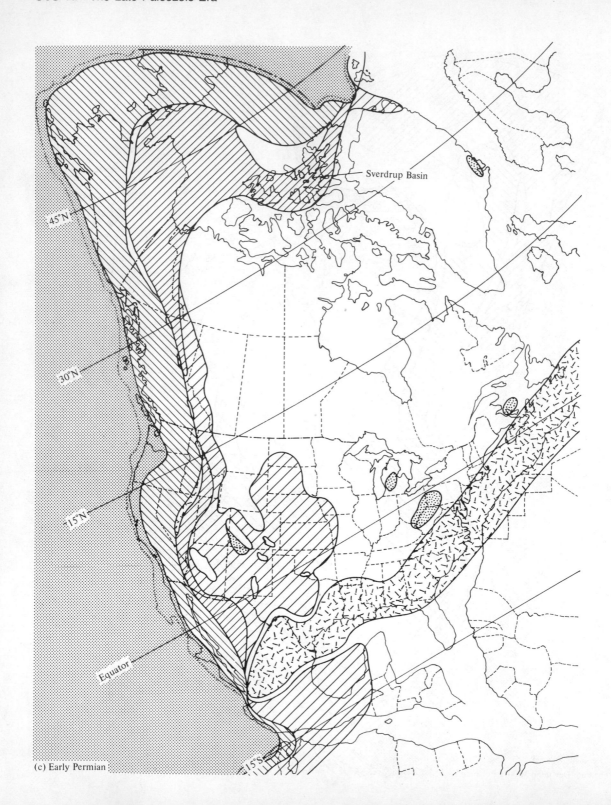

Sverdrup Basin

45°N

30°N

15°N

Equator

15°S

(c) Early Permian

| | | | | | |
|---|---|---|---|---|---|
| ▨ | marine miogeosynclinal deposits | generally shallow water | ⊠ | geanticline or island arc | mountains |
| ▧ | marine eugeosynclinal deposits | generally moderate depth water | ☐ | land which is undergoing erosion | |
| ▨ | marine platform deposits | shallow water | ▦ | deep ocean | |
| ▨ | terrestrial deposits | land | ★ | volcanics | |

**FIGURE 11.2** Paleogeographic maps for Ancestral North America: **(a)** Early Devonian, latest Deerpark. **(b)** Late Devonian, middle Frasnian. **(c)** Early Permian.

EUGEOSYNCLINE. At the beginning of the late Paleozoic, a eugeosyncline extended from southern New England through the Maritime Provinces of Canada to central Newfoundland. A very thick sequence of clastic sediments and volcanic rocks was deposited in the eugeosyncline during the Early Devonian. The sediments were derived mainly from geanticlines within and adjacent to the eugeosyncline (Fig. 11.2a). The rhyolitic composition of many of the volcanic rocks suggests that they were deposited on continental rather than oceanic crust.

The eugeosynclinal deposits were folded, metamorphosed, and intruded during the Acadian orogeny. In central Maine, slightly deformed sandstones and conglomerates of probable Middle Devonian age rest with angular unconformity on rocks as young as Early Devonian (5). This relationship indicates that the principal Acadian deformation in Maine occurred between Early and Middle Devonian time. In central New England, sediments and volcanics were metamorphosed into schists, quarzites, and amphibolites. Early students of New England geology believed that these rocks

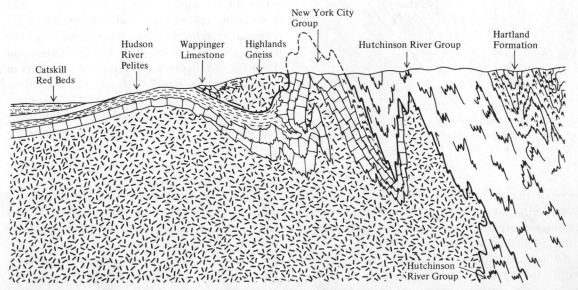

**FIGURE 11.3** Restored cross section across southeastern New York and southwestern Connecticut during the Late Devonian, following the Acadian orogeny. (After Seyfert and Leveson, Ref. 1.)

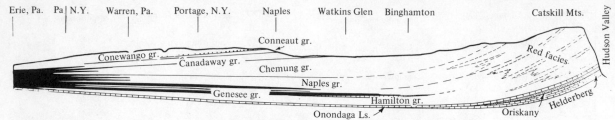

**FIGURE 11.4** Cross section from the Catskill Mountains to Erie, Pennsylvania, showing the clastic wedge of sediments deposited in and adjacent to the Catskill delta during the Middle and Late Devonian. (From A. J. Eardley, Ref. 2.)

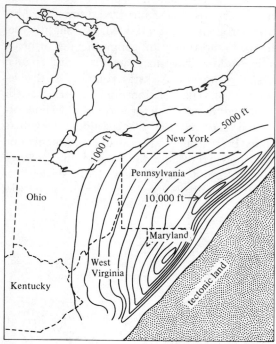

**FIGURE 11.5** Isopach map of Middle and Upper Devonian strata in Pennsylvania and adjacent states. (After Kay, Ref. 3.)

were of Precambrian age. However, some of these units have been traced northward into less-metamorphosed, sedimentary rocks containing Paleozoic fossils.

Laboratory studies indicate that the minerals within the intensely metamorphosed Paleozoic rocks of the eugeosyncline were formed under very high temperatures and pressures. Silli-

manite, a common mineral in many of these rocks, forms under conditions of regional metamorphism at a minimum temperature of about 500°C (900°F) and a pressure equivalent to 15,000 m (30,000 ft) of rock (**6**). Such conditions might reasonably be expected in the core of a mountain range. Since rocks containing sillimanite are now exposed at the earth's surface, a considerable amount of rock must have been eroded from the mountains. The weathering of this rock produced the clastics that were deposited in the miogeosyncline during the Middle and Late Devonian.

Although the Acadian orogeny ended deposition in the eugeosyncline, both marine and nonmarine sedimentary and volcanic rocks accumulated in local basins along the trend of the eugeosyncline. For example, Carboniferous and Permian sediments were deposited in basins in Massachusetts, Rhode Island, the Maritime provinces, and Newfoundland. They contain diamictites of possible glacial origin, such as the Squantum "tillite." The Squantum contains angular, striated pebbles, is poorly sorted, lacks stratification, and sand grains from it show surface textures characteristic of glacial origin (**7**). An associated unit, the Cambridge argillite, contains what appears to be dropstones and varves. However, striated pebbles can be produced in mudflows as well. Furthermore, the Squantum does not rest on a striated basement, and glacial textures have been observed on modern Florida beach sands so that it could be of nonglacial origin (**8**). The Squantum has

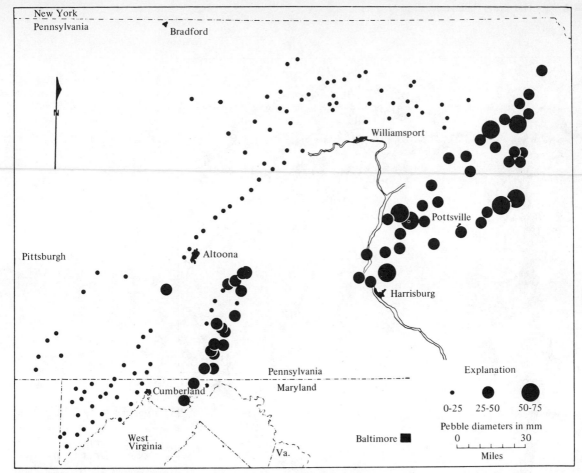

**FIGURE 11.6** Distribution in the size of pebbles in the Pocono Formation of Early Carboniferous age in Pennsylvania. The westward decrease in the size of the pebbles indicates that they were derived from an eastern source. The large size of some of the pebbles indicates that the source had a considerable relief. (After Pelletier, Ref. 4.)

also been interpreted as having been deposited by subaqueous mudflows (**9**). Angular unconformities in these sequences and axial plane foliation in Late Devonian and/or Carboniferous deposits (**10**) indicate that the deformation continued during the Carboniferous and Permian. In Massachusetts and Rhode Island, sediments of Pennsylvanian age were folded and metamorphosed prior to the deposition of nearby Upper Triassic sediments. Potassium-argon dating of granitic and metamorphic rocks in western Connecticut indicates that an episode of intrusion and metamorphism occurred in Middle Permian time, approximately 250 million years ago (**11**). The metamorphism of the

Carboniferous rocks in Massachusetts and Rhode Island and the folding of the deposits of the miogeosyncline during the Alleghenian orogeny may have occurred at this time.

### Southern Appalachian Geosyncline

Late Paleozoic clastics of the Southern Appalachian Geosyncline interfinger with carbonates to the west. Pebbles in conglomerate include granitic and metamorphic rocks derived from a crystalline highland to the east. Marine and continental clastics of Devonian age, which are very thick in the northern part of the geosyn-

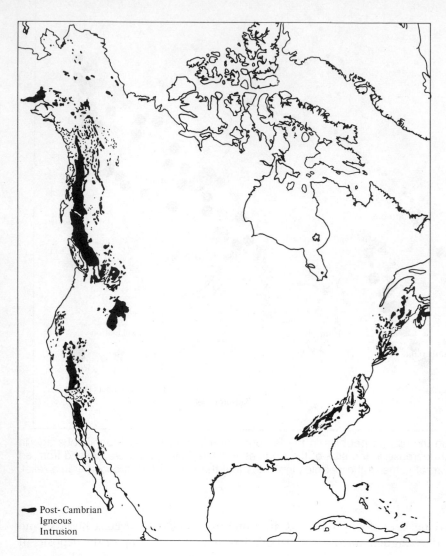

FIGURE 11.7     Granitic rocks (in black) in the Appalachian and Cordilleran regions. Most of the granites of the Appalachian were emplaced during the late paleozoic; most of those of the Cordilleran region were emplaced during the Mesozoic. (Data from the Geologic Map of North America, published by the U.S. Geological Survey.)

— Post-Cambrian
Igneous
Intrusion

cline, thin rapidly southward (Fig. 11.5). These clastics were derived from highlands uplifted during the Acadian orogeny (**12**). Miogeosynclinal deposits in eastern Tennessee were tilted and eroded during the Middle or Late Devonian. The overlying Chattanooga Shale is black, has few fossils, and was probably deposited in a reducing environment. The shale contains small percentages of uranium and constitutes a major reserve of low-grade uranium ore. It is approximately 150 m (500 ft) thick in the miogeosyn-

cline, but thins to a few tens of feet on the platform to the west.

Late Paleozoic uplift to the east of the geosyncline resulted in the deposition of clastic wedges during the Mississippian, Pennsylvanian, and Permian. The Pennsylvanian clastic wedge is especially widespread and reaches a maximum of 3000 m (10,000 ft) in northern Alabama. The culminating deformation in the geosyncline was the Alleghenian orogeny (Figs. 11.8 and 11.9). Deformation was so intense that

**FIGURE 11.8** Folds in sedimentary rocks at Laurel Run Dam, Pennsylvania. These folds were formed during the Alleghenian orogeny. (Photo by Rodger T. Faill.)

the geosyncline was converted into a great mountain range. The Appalachian Mountains are the deeply eroded remains of this range.

Much of the folding in the Southern Appalachian Geosyncline appears to die out with depth (Fig. 11.9) and probably is related to large-scale thrust faults (**8**). The toe of the thrust was crumpled and then faulted in sequence from west to east so that the oldest faults lie to the west. The thrust faults could have moved on a gentle slope downhill away from the core of the mountain belt (**13**).

## Ouachita Geosyncline

From Early Devonian until Late Mississippian time, the rate of sedimentation was very slow in the area that was to become the Ouachita Geosyncline. However, toward the end of the Late Mississippian, a great flood of clastic sediments was deposited. These deposits, which include the Stanley Shale and the Jackfork Sandstone, are part of a 8000 m (25,000 ft) thick sequence whose age ranges from Late Mississippian to Early Pennsylvanian. Small fragments of granitic and metamorphic rocks in these deposits and a southward thickening of the sequence indicate that the source was a crystalline highland that lay to the south. Radiometric dating of basement rocks from the Gulf coastal plain shows that these rocks were metamorphosed during the late Paleozoic and may therefore have been part of the crystalline terrain from which the clastics were derived.

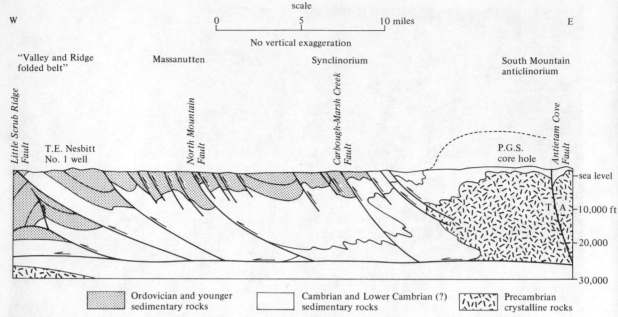

FIGURE 11.9 Generalized cross section from the Valley and Ridge Province to the Blue Ridge Province in Pennsylvania. Note that the angle of dip of these thrust faults decreases at depth. (From Root, Ref. 17.)

During the Middle Pennsylvanian, folding and faulting of the Ouachita Geosyncline produced a mountain range that extended along the southern border of Ancestral North America (Fig. 11.10). Folding also occurred during the late Pennsylvanian, but since Permian strata are relatively undisturbed, deformation had ended by that time. Subsequent erosion has largely leveled these mountains, leaving only the Ouachita Mountains in southwestern Arkansas and southeastern Oklahoma, and the Marathon Mountains of southwestern Texas. All deposition in the Ouachita Geosyncline had ceased by the end of the Permian.

## Cause of Deformation in the Appalachian and Ouachita Geosynclines

Deformations of Late Devonian, Carboniferous, and Permian age in the Appalachian and Ouachita geosynclines were probably caused by a collision of continental plates. Paleomagnetic evidence indicates that Ancestral North America and Gondwanaland were separated during the early Paleozoic but were joined from the Carboniferous to the Triassic. Thus the beginning of intense deformation in the Appalachian and Ouachita geosynclines coincides approximately with the initial contact of the continents. This contact may have occurred at about the same time as deposition of the Chattanooga Shale (uppermost Devonian and lowermost Mississippian). The suture between Ancestral North America and Gondwanaland is probably located along an eastward dipping zone of intense shearing and faulting known as the Brevard Zone (Fig. 11.11) (**14**). This zone was probably a subduction zone on the northwestern margin of Gondwanaland prior to collision with Ancestral North America (**15**). During the Middle Carboniferous, the zone of contact may have extended southward to the Ouachita Geosyncline. During closure between Gondwana-

FIGURE 11.10 Oblique aerial view of the Pena Blanca Hills, near Marathon in southwestern Texas. The white beds are cherts of Silurian and/or Devonian age. The outcrop pattern is a result of the erosion of plunging folds. A faint line of vegetation on the second hill in the foreground marks a thrust fault. (Photo by Earle F. McBride.)

land and Ancestral North America, the direction of dip of the subduction zone in the Ouachita region was probably toward the south (**16**).

### Cordilleran Geosyncline

The late Paleozoic deposits of the eastern Cordilleran Geosyncline consist largely of carbonates and clastics, but deposits to the west were dominantly clastics and volcanics (Fig. 11.12).

MIOGEOSYNCLINE. The complete absence of earliest Devonian marine deposits from the miogeosyncline suggests that the seas had withdrawn from the miogeosyncline at that time. However, the seas returned before the end of the Early Devonian. Devonian miogeosynclinal deposits are principally carbonates. Large coral reefs within this sequence indicate deposition in a warm climate. Paleomagnetic data place the miogeosyncline close to the equator throughout the Devonian (Fig. 11.2b).

At the end of the Devonian, geosynclinal deposits were deformed along a linear zone in central Nevada known as the Antler Orogenic Belt. Eugeosynclinal deposits were folded and thrust over miogeosynclinal deposits in a manner similar to that of the Taconic Allochthon in the Appalachians. The displacement along the Roberts Mountain Thrust, which separates the two sequences, has been estimated at more than

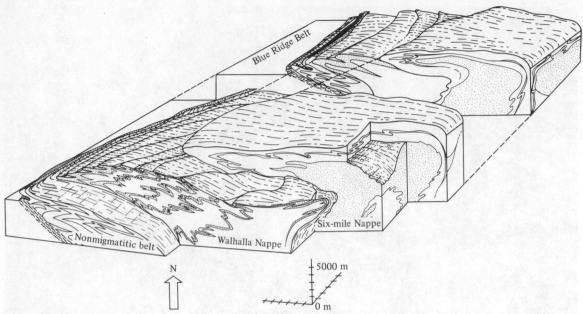

**FIGURE 11.11** Schematic cross section of Piedmont region in northwest South Carolina. The Brevard zone is the layer with short parallel lines located just above the N. (From V. S. Griffin, Jr., Ref. 14.)

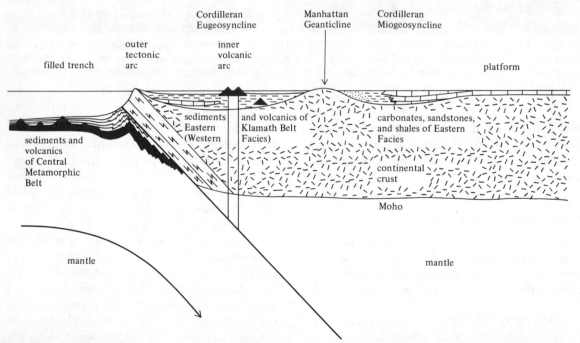

**FIGURE 11.12** Restored cross section across the Cordilleran Geosyncline during the Early Carboniferous.

129 km (77 mi) (**18**). Since the faulting does not involve basement rocks, it may have resulted from the sliding of giant slices from uplifted highlands to the west of the miogeosyncline (**18**). However, it has also been suggested that the Roberts Mountain Thrust is an underthrust formed as miogeosynclinal deposits were shoved under eugeosynclinal deposits to the west (**10**). The movement along the fault took place in only 2 to 5 million years, which would give a rate of movement of about 2.5 to 6.5 cm/year (**19**). An angular unconformity at the base of Carboniferous limestones in northern Alaska may have been formed during the same orogenic episode.

Throughout most of the Carboniferous and Permian, the Manhattan Geanticline was an island that shed sediments eastward into the miogeosyncline and westward into the eugeosyncline (Fig. 11.12). The Malay Peninsula in the Indonesian Island Arc may be a modern analog of the Manhattan Geanticline. This peninsula is bordered by volcanic islands on the west and south, and the deposits in the vicinity of the volcanic islands are eugeosynclinal in character, whereas those on the opposite side of the peninsula are miogeosynclinal.

In central Nevada, the Mississippian Tonka Conglomerate is 800 m (2500 ft) thick and grades eastward into thinner shale beds. Deposits in this clastic wedge were derived from the Manhattan Geanticline to the west. The presence of conglomerates throughout the Mississippian section implies continued uplift of the geanticline. Intermittent uplift continued during the Pennsylvanian, and occasionally deformation extended into the miogeosyncline (Fig. 11.13b). By Early Permian time, the geanticline had been eroded almost to sea level, and limestone was deposited in the western part of the miogeosyncline. In contrast to the abundant clastics in the miogeosyncline in Nevada, the deposits in the Canadian segment of the miogeosyncline are dominantly limestones and dolostones. These carbonates are exposed in

towering cliffs in Banff and Jasper National Parks in the Canadian Rockies (Fig. 11.14).

Toward the end of the Paleozoic, large faults displaced eugeosynclinal deposits over miogeosynclinal deposits. These faults, which include the Golconda Thrust in Nevada, do not involve basement rocks, and they may be gravity thrusts that slid from uplifted areas to the west of the miogeosyncline.

EUGEOSYNCLINE. A very thick sequence of clastic sediments and volcanic rocks was deposited in the eugeosyncline to the west of the Manhattan Geanticline. Volcanic and tectonic islands in the eugeosyncline also contributed sediments. This area was probably part of an island arc during the Late Paleozoic (**19**). Episodes of deformation, uplift, and erosion occurred within the eugeosyncline during the Late Devonian or Early Mississippian, during the Pennsylvanian, and at least twice in the Permian. Radiometric dating indicates that these deformations were accompanied by emplacement of granitic rocks.

In the western part of the eugeosyncline, an intensely deformed sequence of rhythmically bedded cherts, clastics, and volcanics contains late Paleozoic fossils (Fig. 11.15). These deposits include the Western Paleozoic and Triassic Belt in the Klamath Mountains of northern California and the western part of the Calaveras Formation in the Sierra Nevada Mountains of eastern California. Radiometric dates on rocks from the Central Metamorphic Belt in the Klamath Mountains (Fig. 10.18) indicate that metamorphism and deformation occurred during the Devonian and Carboniferous (**20**). The late Paleozoic volcanic and orogenic activity may have been due to the movement of an oceanic plate under the continent during the late Paleozoic.

### Franklin Geosyncline

From the Early to the Middle Devonian, carbonates and minor shales were deposited in the

(a)

(b)

**FIGURE 11.13** Late Paleozoic deposits of the Cordilleran Geosyncline. **(a)** A tilted angular unconformity exposed along Interstate 80 west of Elko, Nevada. The more gently dipping beds are limestones and siltstones of latest Pennsylvanian and Permian age; the underlying steeply dipping beds are Late Mississippian conglomerates. This area is close to the Manhattan Geanticline and shows a record of three phases of deformation. The first is recorded in the thick conglomerates, which require a nearby uplift. The angular unconformity was produced and the unconformity was tilted during the second and third phases of deformation, respectively; **(b)** Gently dipping carbonates of Carboniferous age unconformably overlying steeply dipping rocks of early Paleozoic or Late Proterozoic age. (Photo by J. T. Dutro, Jr., U.S. Geological Survey.)

**FIGURE 11.14** Mount Rundle in Banff National Park in the Canadian Rockies, showing late Paleozoic sedimentary rocks. The resistant unit capping the mountain is the limestone-dolostone Rundle Formation of Early Carboniferous age. The underlying unit is the less resistant Banff Formation, of limestone, argillaceous limestone, and calcareous shale also of Early Carboniferous age. The lowest cliffs are composed of massive limestone and dolostones of the Palliser Formation, which is Devonian. (Photo by T. L. Tanton, courtesy of the Geological Survey of Canada, Ottawa.)

**FIGURE 11.15** Aerial view of the Klamath Mountains in northern California. The rocks in the foreground are metasedimentary and metavolcanic and belong to the Western Paleozoic and Triassic belt. The white cliff in the middle is composed of marble. Mt. Shasta, a relatively young volcano, appears on the skyline. (Photo courtesy of W. P. Irwin, U.S. Geological Survey.)

FIGURE 11.16 Middle and Upper Devonian shale and limestones at 18 Mile Creek in western New York. There is a disconformity between Middle and Upper Devonian strata just below the highest overhang. (Courtesy of the Buffalo Museum of Science; from E. J. Buehler and I. H. Tesmer, Ref. 22.)

Franklin Geosyncline. Coral reefs are abundant in Middle Devonian deposits, even though this area is now located between 75°N and 80°N latitude. Paleomagnetic measurements show that this area was at approximately 15°N latitude during Devonian time (Fig. 11.2).

Marine sedimentation ended in the Middle Devonian when a geanticline was uplifted on the northern border of the geosyncline (21). Continental clastics spread southward from this uplift into the miogeosyncline. These deposits reach a thickness of 3000 m (10,000 ft) and include coalbearing sandstones and shales. In the latest Devonian or earliest Mississippian, the deposits of the geosyncline were strongly folded. This deformation, the Ellesmerian orogeny, ended sedimentation in the Franklin Geosyncline. Subsequently, the Sverdrup Basin developed approximately along the former boundary between the miogeosyncline and the eugeosyncline.

## Sverdrup Basin

The seas transgressed over the folded rocks of the Franklin Geosyncline during the Carboniferous and deposited a thick sequence of sediments in the Sverdrup Basin (Fig. 11.2c). Thick deposits of carbonates and evaporites interbedded with clastics indicate that this region had an arid climate during the late Paleozoic. Paleomagnetic data show that the Sverdrup Basin probably lay between 25°N and 50°N latitude during the Carboniferous (Fig. 11.2). The Permian deposits of the Sverdrup Basin appear to be conformable with those of Triassic age, so that, at least in this region, the Paleozoic era was not terminated by orogeny.

## Continental Interior

The continental interior of Ancestral North America was periodically inundated by shallow seas during the late Paleozoic. Following withdrawal in the Late Silurian, the seas returned to cover much of the continent and a sequence of limestone and shale was deposited (Fig. 11.16).

An extensive evaporite basin developed east of the Cordilleran Geosyncline in western Canada during the Middle Devonian. The basin, which was partially enclosed by reefs and geanticlines, contains halite, gypsum, anhydrite, and some of the largest deposits of sylvite in the world (23, 24). Sylvite (potassium chloride)

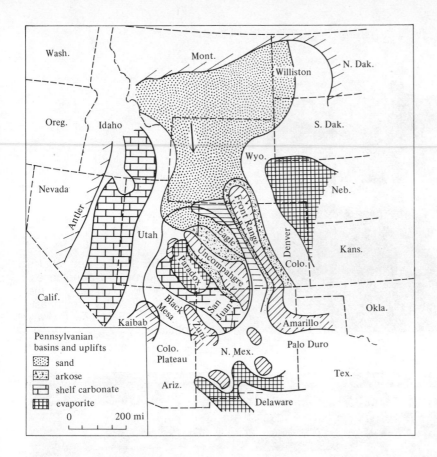

**FIGURE 11.17** Late Paleozoic uplifts and basins in Colorado, Oklahoma, and Texas. (From Peterson and Hite, Ref. 25. Used with the permission of American Association of petroleum Geologists.)

requires almost complete evaporation of seawater before it precipitates. Thus this region must have been very arid during the Middle Devonian. However, according to paleomagnetic data, it was located very near the equator during the Middle Devonian.

During the Late Devonian and Early Mississippian, a very extensive black shale was deposited over much of the interior of the continent. This shale is continuous with the Chattanooga Shale and was probably derived from highlands to the east. The seas became shallower during the late Early Mississippian, and limestones were deposited over most of the southern and western continental interior. These include the Madison Limestone of Wyoming and Montana and the equivalent Redwall Limestone of the Grand Canyon region.

Throughout most of the Paleozoic, there was very little deformation in the interior of the continent. However, folding and uplift during the late Mississippian and Early Pennsylvanian formed the Arbuckle, Wichita, and Amarillo mountains in northern Texas and southern Oklahoma and the Ancestral Rockies in Colorado, New Mexico, and eastern Utah (Fig. 11.17). Great thicknesses of clastic sediments were eroded from these uplifts and deposited in the adjacent basins. The Central Paradox Basin, located west of the Uncompahgre Uplift, received up to 4000 m (13,000 ft) of red and gray sandstones, conglomerates, siltstones, salt, and gypsum during the Pennsylvanian and Permian (Figs. 11.18 and 11.19).

Highlands to the east and south increased in width and height during the Pennsylvanian, and

FIGURE 11.18 Red shales overlain by red sandstones, both of Permian age, in the Monument Valley, southeastern Utah. (Santa Fe Railway photo.)

FIGURE 11.19 Well-bedded Late Carboniferous limestones, shales, and sandstones in southeastern Utah. These clastic sediments and those in Figure 11.18 were eroded from the Ancestral Rockies. (Courtesy of the U.S. Department of the Interior, Bureau of Reclamation.)

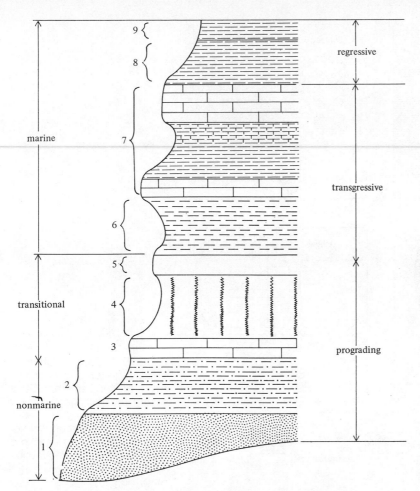

FIGURE 11.20 Idealized stratigraphic section showing a Late Pennsylvanian cyclothem. The strata depicted are: (1) nonmarine sandstone. (2) Nonmarine sandy shales. (3) Brackishwater limestone. (4) Underclay. (5) Coal. (6) Marine, shallow water shale. (7) Interbedded shales and limestones (marine). (8) and (9) Marine shales formed during a regression (shallowing) of the sea. Rarely are all layers found at any one locality. (modified from Weller, Ref. 26.)

clastic sediments spread to the eastern, central, and southern part of the continental interior. Alternating rapid transgressions and gradual regressions of the sea across a very flat craton coupled with progradation of continental clastics and development of coastal marshes produced a series of cyclic deposits, known as cyclothems, in this region. A typical cyclothem consists of nine different rock types each representing a successive depositional environment and facies at any one locale during sea level changes (Fig. 11.20). Such changes in sea level may have been the result of: (1) glaciation in the southern continents; (2) variations in the rate of sediment deposition; (3) variations in the rate of

subsidence; or (4) changes in the location of major rivers.

The sea withdrew from the eastern part of the continental interior in the Permian. During the Middle Permian, thick accumulations of salt were laid down in coastal lagoons in Kansas and Oklahoma. The uplift of the Marathon Mountains in southern Texas created the Delaware Basin in western Texas and southeastern New Mexico. The deeper parts of this basin were bordered by reefs during the Middle and Late Permian (Fig. 11.21). One of these, El Capitan Reef, is comprised mainly of the remains of calcareous algae, sponges, bryozoans, pelecypods, brachiopods, and foraminiferans

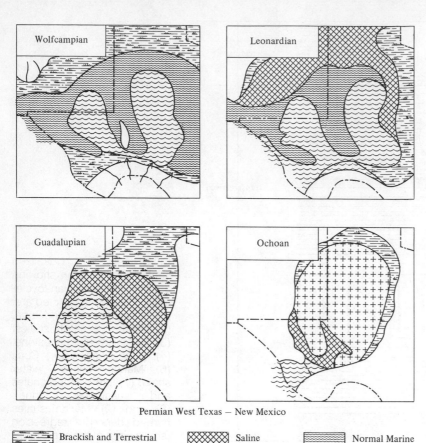

Permian West Texas — New Mexico

| | | | | | |
|---|---|---|---|---|---|
| ≡≡≡ | Brackish and Terrestrial | ⧓⧓⧓ | Saline | ▤▤▤ | Normal Marine |
| ∿∿∿ | Euxinic | ++++ | Supersaline | ☐ | Interval Eroded |

(a)

FIGURE 11.21(a)   Changes in paleogeography in the vicinity of the Delaware Basin during the four stages of the Permian period. (After Sloss, Ref. 27.). (b) Cross section through the Permian rocks of the Guadalupe Mountains, west Texas. (In Matthews, after King, Ref. 27.)

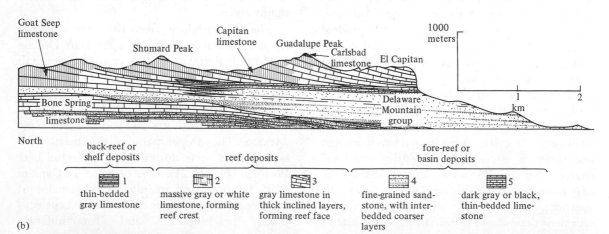

(b)

**FIGURE 11.22** Oblique aerial view of El Capitan and Guadalupe Peak at the southern end of the Guadalupe Mountains in western Texas. (Photo by P. B. King, U.S. Geological Survey.)

(Fig. 11.22). Broken fragments of fossils and limestone are found as much as 600 m (1800 ft) below the top of the reef. Limestone and dolostone were deposited in shallow lagoons behind the reefs, and gypsum, anhydrite, and shales were deposited in the deeper parts of the basin. Groundwater percolating through the limestones has carved vast networks of caves, including Carlsbad Caverns, one of the largest and most beautiful caves in the United States. Toward the end of the Middle Permian, the waters of the Delaware Basin became highly saline and thick layers of halite and sylvite were deposited in what was presumably a very arid climate. Paleomagnetic studies indicate that the Delaware Basin was at a latitude of about 15°N during the Middle and Late Permian.

## ANCESTRAL EUROPE

Ancestral Europe was bordered by the Hercynian Geosyncline on the south and the Uralian Geosyncline on the east during most of the late Paleozoic. By the end of the Paleozoic, orogenic movements had converted these geosynclines into mountains. This orogenic activity was probably associated with plate collisions. Deformation in the Uralian Geosyncline seems to have been associated with the joining of Ancestral Europe and Ancestral Siberia, and deformation in the Hercynian Geosyncline may have been caused by the joining of Ancestral Europe and Gondwanaland.

### Hercynian Geosyncline

Marine eugeosynclinal deposits of the Hercynian Geosyncline grade northward into marine and continental platform deposits with no intervening miogeosynclinal deposits. The Devonian and Carboniferous deposits are largely shale and sandstone, with minor limestone and volcanic rocks. Early Devonian sediments were derived mainly from the Caledonian Mountains to the northwest, but as these mountains were worn down, an increasing amount of sediment was derived from highlands to the south.

At the end of the Early Devonian, red beds were deposited in the Hercynian Geosyncline in the Ardennes region of northern France (**28**). Thick limestones containing coral reefs were deposited in the same region and in the Harz Mountains of Germany during the latter part of the Middle Devonian and early part of the Late Devonian (**28, 29**). These observations suggest that shallow water or nonmarine conditions existed during the Devonian in the Hercynian region, and they conflict with a proposal that the Devonian deposits of the Harz region were deposited in deep water on oceanic crust (**29**).

Gondwanaland and Ancestral Europe were moving toward each other during the Devonian as the ocean between them closed. A subduction zone may have dipped northward under Ancestral Europe at this time (**30**), or it may have dipped southward under Gondwanaland (**29**). Changes in composition of volcanic rocks in France seem to indicate that the subduction zone dipped northward (**31**). In this case, there would have been an island arc on the southern border of Ancestral Europe during the Devonian. Local episodes of folding during the Devonian in the Hercynian Geosyncline may have been related to subduction along this margin.

The Hercynian Geosyncline was not subjected to strong folding until the end of the Devonian. Deformation at that time was largely restricted to the southern portion of the geosyncline, and it may have been caused by the initial contact (perhaps an arc-arc or arc-continent collision) between Ancestral Europe and Gondwanaland. Toward the end of the Early Carboniferous, the entire geosyncline was intensely folded, metamorphosed, and intruded by granites (Fig. 11.22). Whole-rock rubidium-strontium isochron dating indicates that granitic rocks were emplaced 350 to 330 million years ago (Late Devonian and Early Carbonif-

erous) and 290 to 280 million years ago (Late Carboniferous) (**32**).

A few local episodes of folding occurred during the Devonian, but the geosyncline was not subjected to strong folding until the end of the Devonian. Deformation was largely restricted to the southern portion of the geosyncline, but toward the end of the Early Carboniferous, the entire geosyncline was intensely folded, faulted, metamorphosed, and intruded by granites during the Variscan orogeny. Extensive strike-slip faults developed in southern Europe as a result of compression between Gondwanaland and Ancestral Europe (**33**). These resulted, in part, from the adjustment of small blocks caught in the collision zone in a manner similar to the behavior of small blocks near the contact of the African and Eurasian plates at the present time.

The Early Carboniferous deformation, which was the main phase of the Hercynian orogeny, ended marine sedimentation in the geosyncline and produced a mountain range extending across the southern part of Ancestral Europe. The Ore Mountains, on the border between Germany and Czechoslovakia, are remnants of this ancient range. During the Late Carboniferous, coal swamps developed in intermontane basins north of the new mountain range. Subsequent episodes of folding occurred toward the end of the Late Carboniferous and between the Early and Middle Permian. All episodes of deformation correlate with deformations in the Southern Appalachian and Ouachita geosynclines and may have been caused by the collision of Ancestral Europe with Gondwanaland.

### Uralian Geosyncline

Late Paleozoic sedimentary and volcanic rocks attain geosynclinal thickness in the Ural Mountains in easternmost Ancestral Europe. A thick sequence of limestone, coal, and clastic sedimentary rocks in the western part of the geosyncline grades eastward into an even thicker sequence of clastics, lavas, and tuffs of the same age. Accordingly, the deposits in the west are miogeosynclinal in character, whereas those in the east are eugeosynclinal. The climate of this region was probably warm and moist. According to paleomagnetic data, the Uralian Geosyncline was near the equator during the Carboniferous (Fig. 11.1b).

The uplift of a geanticline along the trend of the eugeosyncline late in the Early Carboniferous ended sedimentation and resulted in deposition of clastic wedges in the eastern part of the miogeosyncline. In the Permian, an increasing amount of arkosic sediment was deposited in the miogeosyncline. These clastics grade westward into limestones and evaporites (Fig. 11.23). The salt basin of the Upper Kama region of the Soviet Union is said to be the largest in the world (**35**). Large quantities of sylvite in these evaporites were probably deposited under conditions of extreme aridity. Paleomagnetic studies indicate that the basins were located at 25°N latitude, the same latitude occupied by the Sahara Desert today. The entire geosyncline was transformed into a mountain range during intense deformation in Middle Permian time. The deformation was presumably the result of compression between the plate containing Ancestral Siberia and the plate containing Ancestral Europe.

## Caledonian Mountains

Following the final phase of the Caledonian orogeny, continental sediments accumulated between and adjacent to the new mountain ranges. These deposits, which are collectively known as the Old Red Sandstone, are largely deltaic and lacustrine in origin. Although similar in character, only the Middle and Upper Old Red Sandstone are the time-stratigraphic equivalents of the Devonian "Catskill" red beds of New York and Pennsylvania. The Lower Old Red Sandstone reaches a maximum thickness of 13,000 m (40,000 ft) and rests unconformably on folded early Paleozoic strata (Fig. 11.24). In Scotland, the Upper Old Red Sandstone of Late Devonian age rests unconformably on mildly deformed Lower Old Red Sandstone and on Middle Devonian granitic rocks (Fig. 11.25). The deformation indicated by this unconformity occurred at approximately the same time as the Acadian orogeny in the northern Appalachian Geosyncline. However, the deformation was much less intense in Scotland than it was in New England.

The Carboniferous deposits of the Caledonian region are rather thin and include extensive coal-bearing clastics interbedded with lavas. Movement along large-scale strike-slip faults was contemporaneous with deposition of these sequences. One such fault, the Great Glen Fault in northern Scotland, has a left-lateral displacement of approximately 100 km (65 mi).

## Continental Interior

Sediments derived from the erosion of the Caledonian Mountains were deposited over much of the interior of Ancestral Europe in Devonian time. Continental sequences grade eastward and southward into marine deposits. As the mountains were leveled, finer sediments were deposited and the sea transgressed onto the platform. By the early Carboniferous, limestone covered much of the continental interior.

Uplift of the Hercynian Mountains on the southern border of the platform during the late Early Carboniferous resulted in the deposition of clastic wedges on the adjacent platform. By mid-Late Carboniferous time, the sea had withdrawn from all but the eastern continental interior. The sea became even more restricted in the Permian, and a large basin of evaporation formed in central Europe. The evaporites, which include thick deposits of halite and sylvite, formed in an arid region that lay, according to paleomagnetic data, 10°N of the equator during the Permian.

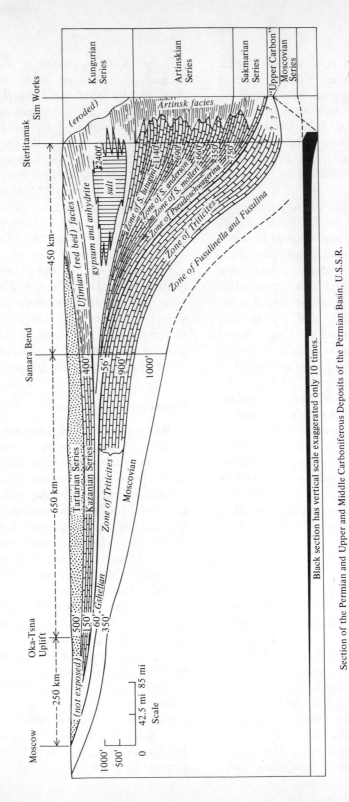

FIGURE 11.23  Restored cross section across the western part of the Uralian Geosyncline during the Late Permian. (After Dunbar, Ref. 34. Used with the permission of American Association of Petroleum Geologists.)

Section of the Permian and Upper and Middle Carboniferous Deposits of the Permian Basin, U.S.S.R.

FIGURE 11.24 Angular unconformity between vertical Silurian beds and gently dipping deposits of the Old Red Sandstone. Near Cockburnspath on the coast of southeastern Scotland. (NERC Copyright. Reproduced by permission of the Director, Institute of Geological Sciences, London.)

## GONDWANALAND

Thick sequences of sedimentary and volcanic rocks were deposited in the geosynclines which encircled Gondwanaland during the late Paleozoic (Fig. 10.29). The eugeosynclinal sequences contain andesitic volcanic rocks and were deformed several times during the late Paleozoic. It is likely that the continent was bordered by island arcs and trenches, under which oceanic plates were moving. Upper Paleozoic sequences of the continental interiors of South America, Africa, Antarctica, India, and Australia are remarkably similar. A typical sequence includes Devonian shales and sandstones overlain unconformably by tillites of Carboniferous and/or Permian age, which in turn are followed by coal-bearing shales of Permian age (Fig. 7.7). The similarity of these units provides one of the strongest arguments that the southern continents were part of a single landmass during the late Paleozoic.

## Tasman Geosyncline

Situated on the east coast of Australia, the Tasman Geosyncline consists of a western miogeosyncline and an eastern eugeosyncline. Miogeosynclinal sandstones, limestones, and shales contain fossils indicating deposition in shallow water. Coral reefs and evaporites of Devonian age were deposited at a latitude of approximately 35°S according to paleomagnetic data (Fig. 11.1a). It is interesting to note that at the present time this region is at approximately the same latitude and has extensive living coral reefs nearby.

Eugeosynclinal deposits include graywacke, shale, and felsic to mafic volcanics, with only minor limestone. Early to Middle Devonian terrestrial volcanics are surrounded by marine deposits of the same age (**36**), suggesting deposition in a volcanic island arc. Toward the end of the Middle Devonian, an episode of intense deformation affected the entire geosyncline,

**FIGURE 11.25** Angular unconformity between the moderately dipping beds of the Lower Old Red Sandstone and the gently dipping beds of the Upper Old Red Sandstone, in the central part of eastern Scotland. (NERC Copyright. Reproduced by permission of the Director, Institute of Geological Sciences, London.)

with granites intruding the eastern part of the geosyncline. This deformation resulted in an eastward shift of the axis of the geosyncline.

From the Late Devonian until the end of the Permian, the eugeosyncline was restricted to the easternmost border of Australia. During the Late Devonian and Early Carboniferous, it received both marine and continental sediments. At the end of the Early Carboniferous, a very intense deformation converted much of the geosyncline into a mountain range. This orogeny was contemporaneous with the main phase of the Hercynian orogeny in Europe and was accompanied by numerous granitic intrusions. Tillites and fluvioglacial sediments were deposited in the eugeosyncline during the Late Carboniferous and Permian. Some of these deposits may have resulted from alpine glaciers which

originated in nearby highlands. A final deformation at the end of the Permian ended sedimentation in the Tasman Geosyncline.

## Buller and New Zealand Geosynclines

During the Early and Middle Devonian, quartz sandstone, shale, limestone, and volcanic rocks were deposited in the Buller Geosyncline of western New Zealand. The limestone contains a rich fauna of brachiopods, corals, and trilobites, which indicates deposition in relatively shallow water (**36**). The deposits of the geosyncline were uplifted and intruded during the Late Devonian and Carboniferous. Sedimentation did not resume until the Permian when clastics, limestone, and volcanics were deposited. At the same time, thick deposits of graywacke, silt-stone, and mafic volcanics accumulated in the New Zealand Geosyncline east of the Buller Geosyncline. Ultramafic intrusives are common. It has been suggested that the Buller Geosyncline formed on a sialic (continental) basement, and the New Zealand Geosyncline formed on a simatic (oceanic) basement (**37**).

## Timor

A relatively thin sequence of Permian sedimentary and volcanic rocks occurs on the island of Timor. The volcanics include basalts and related rocks having an oceanic affinity. These deposits may have accumulated at bathyal to abyssal depths on oceanic crust. Possibly they were deposited seaward of the margin of Australia and were subsequently carried into the Java Trench.

## Southern Tethyan Geosyncline

Upper Paleozoic deposits in the Himalayas are miogeosynclinal in character. Although Devonian fossils have not been found in the Himalayas, an unfossiliferous quartzite, which occurs between Silurian and Lower Carboniferous beds, may be of Devonian age. Thick limestones of Lower Carboniferous age grade upward into marine clastics, which in turn are disconformably overlain by continental deposits, including tillites and basaltic lavas of Carboniferous and Permian age (**38**). Since the volcanics do not include andesites and are associated with thick limestones, the sequence is unlike that of a typical eugeosyncline. Permian continental beds containing a *Glossopteris* flora are interbedded with fossiliferous marine strata.

Thick upper Paleozoic deposits are also found in the Southern Alps of northern Italy. Paleomagnetic measurements indicate that the location of the Permian pole as determined from the Southern Alps is quite different from the Permian pole determined from northern and central Europe (**38**). However, the Permian pole as determined from the Southern Alps is very similar to the Permian pole as determined from Africa. Therefore, it is likely that the Southern Alps and most of Italy were part of Gondwanaland during the Permian.

## West African Geosyncline

In Mauritania, Devonian rocks of the West African Geosyncline are intensely folded. Since Mesozoic rocks in this region are relatively undisturbed, the Devonian sequence was presumably deformed during the late Paleozoic (**39**).

## Southern Appalachian Geosyncline

In the late Paleozoic, the western half of the Southern Appalachian Geosyncline bordered Gondwanaland and linked the Andean and West African geosynclines (Fig. 10.29). Either the late Paleozoic sequences of the geosyncline are covered by coastal plain deposits of Mesozoic and Cenozoic age, or they are so strongly metamorphosed that fossils have not been preserved. Radiometric dates in the Piedmont province of Virginia, North Carolina, South Carolina, and Georgia generally fall into the range of 350 to 250 million years, that is, latest Devonian to Middle Permian. Thus some of the parent sediments may be of late Paleozoic age. Deposition in the southern Appalachian Geosyncline ended by the close of the Paleozoic.

## Andean Geosyncline

Deposits in the Andean Geosyncline record a long history of glaciation during the late Paleozoic. In western Argentina, Early Devonian conglomerates may represent outwash deposits from local alpine glaciation (**40**). Tillites and fluvioglacial deposits of Early Carboniferous age which occur in southern Bolivia and western Argentina also may have been deposited by alpine glaciers (**41**). However, Late Carbonifer-

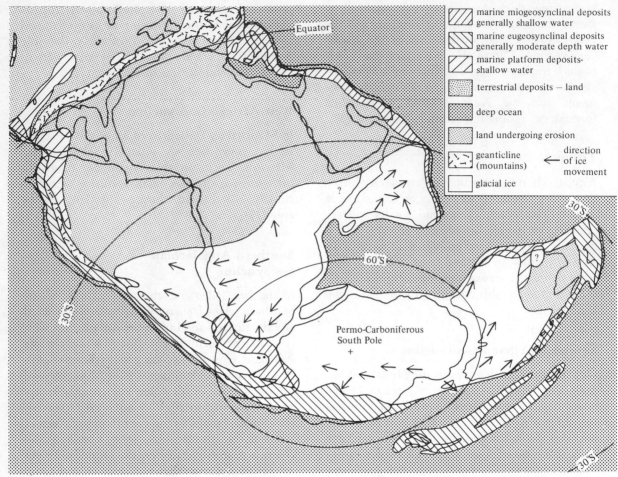

**FIGURE 11.26** Paleogeographic map of Gondwanaland during the earliest Permian, showing the inferred distribution of ice sheets.

ous glacial deposits in the eastern part of the Andean Geosyncline were almost certainly deposited by a very large continental ice sheet (Fig. 11.26). Tillites containing faceted pebbles grade westward into marine deposits, which in Bolivia reach a thickness of 2000 m (6500 ft). Glacial deposits of Early Permian age, which occur in southern Bolivia and western Argentina, may have been deposited by alpine glaciers. Somewhat younger limestones in Bolivia indicate that warmer climates succeeded the glacial episode in the Andean Geosyncline (Fig. 11.27).

Repeated deformation during the late Paleozoic produced highlands that became the source of many alpine glaciers. An episode of folding and faulting affected the southern part of the geosyncline during the Late Devonian (**40**). Before separation of the continents, this fold belt was continuous with a Late Devonian fold belt on the southern tip of Africa. Near the end of the Early Carboniferous, the deposits of the geosyncline were again folded. Potassium-argon dates on granitic rocks indicate that an intrusive episode occurred during the Late Carboniferous (**42**). Deformation and uplift during the Middle and Late Permian resulted in the widespread deposition of continental red beds in the geosyncline (**43**).

## Cape Geosyncline

In the Cape Geosyncline marine clastics of Early Devonian age are succeeded by quartzites and shales of Middle and perhaps Late Devonian age. These units were folded prior to the deposition of the overlying Dwyka Series. A thick sequence of tillites and fluvioglacial deposits in the lower Dwyka Series is overlain by the Ecca Series of lowermost Permian age. Apparently the glaciation occurred during Carboniferous and possibly earliest Permian time.

The Ecca Series is comprised of marine and coal-bearing continental deposits containing a *Glossopteris* flora. Since the fossil wood within the continental deposits exhibits seasonal growth rings, the coals were probably deposited in a moist, temperate climate (Fig. 11.28). Paleomagnetic data indicate that this area was at approximately 60°S latitude during the Permian. The source of the Ecca Series was south of the Cape region. Since separation of the continents occurred after the Permian, the source may have been in Antarctica. Folding during the Permian in the southern part of the Cape Geosyncline ended sedimentation in the geosyncline.

## Antarctic Geosyncline

Thick sequences of late Paleozoic rocks have been found in the Ellsworth Mountains, the Pensacola Mountains, and the Horlick Mountains of Antarctica. The deposits in these ranges were once part of a major geosyncline that extended from the Wedell Sea to the Ross Sea (**44**). In the Horlick Mountains, Early Devonian marine sandstones and shales rest unconformably on granitic basement and are overlain by the Buckeye tillite (**45**). The tillite is 300 m (900 ft) thick, and shales near the top of the tillite contain spores that are probably of

**FIGURE 11.27** Tight folding in thin bedded limestones of the Copacabana Formation of Permian age in western Peru. (Photo by C. R. Peterson, La Oroya, Peru.)

Permian age. However, the lower part of the unit may be Carboniferous (**46**). Elsewhere, correlative tillites rest unconformably on grooved and striated basement. Shales, sandstones, and coals that overlie the tillites have yielded a *Glossopteris* flora of Late Permian age (Fig. 11.28).

No evidence of late Paleozoic deformation has been found in the Antarctic Geosyncline, although radiometric ages of granitic rocks indicate that intrusive episodes occurred 350 million years ago (Late Devonian) and 280 million years ago (earliest Permian) (**44**).

## Continental Interior

Thick accumulations of sediment were laid down during the late Paleozoic in basins adja-

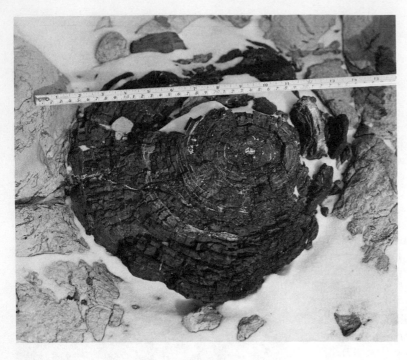

**FIGURE 11.28** Fossil wood showing growth rings from Permian deposits in Antarctica. (Photo by W. E. Long.)

cent to geosynclinal belts. For example, nearly 3000 m (10,000 ft) of Devonian strata were deposited in eastern Brazil, and 3000 m of Permian sediments were deposited in northwestern Australia. The rate of sedimentation of these deposits is equal to that of many geosynclines, but the basins are not significantly elongated.

Evidence of glaciation is found in late Paleozoic deposits in all the southern continents and in India as well (Fig. 11.29). During the Middle Devonian, glaciers formed on the northern edge of the Central Brazilian Shield and moved into both the Amazon and Parnaiba basins, where they deposited tillites containing striated and faceted pebbles (**40**).

The most extensive of the late Paleozoic glaciations occurred during the latest Carboniferous and earliest Permian. The areas glaciated include Australia, South Africa, Malagasy, South America, Falkland Islands, Antarctica, India, Pakistan, and possibly Siberia (**48**). Most of the features associated with Pleistocene glaciers have also been found in these glacial deposits:

1. Poorly sorted tillites contain striated and faceted pebbles and rest on grooved, striated, and polished basement (Fig. 11.29).
2. In some areas, the separation of several tillites by marine deposits suggests alternating glacial and interglacial episodes (**50**).
3. Banded shales associated with tillites may represent varves that were deposited in proglacial lakes (**51**).
4. Linear bodies of sand interbedded with tillites may be buried eskers.

The glacial deposits are preserved in a series of unconnected basins. Although it has been suggested that glaciation was confined to these basins (**52**), studies of the orientation of glacial striations and *roches moutonnées* indicate that a single large ice sheet originated in two principal centers, one in southwestern Africa and one in eastern Antarctica (Fig. 11.26). Both centers are located near the Permo-Carboniferous South Pole as determined from paleomagnetic studies. The wide distribution of the glacial deposits and the radial pattern of the ice movement

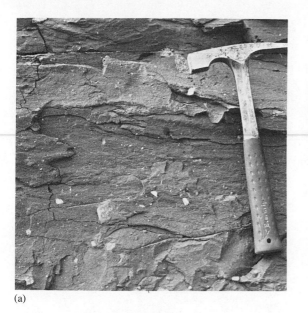

(a)

(b)

(c)

(d)

FIGURE 11.29 Permo-Carboniferous glacial features: **(a)** Pebbly tillite from Natal, southern Africa. **(b)** Striated facet on cobble in bouldery tillite, Wynyard, Tasmania, Australia. **(c)** Glacially polished basalt, Kimberly, South Africa (Bushmen etching on the left). **(d)** Exhumed *roche moutonnée* of Precambrian basalt, Kimberly, South Africa (ice movement was toward upper left). (Courtesy of Warren Hamilton, U.S. Geological Survey, from Hamilton and Krinsley, Ref. 47.)

indicate continental rather than alpine glacia-
tion. Although local alpine glaciers probably
formed in upland areas at this time, most of the
drift may be attributed to continental glaciers.
This glaciation extended to within 30° of the
Permo-Carboniferous equator in northern India
(Fig. 11.26). By comparison, Pleistocene ice
sheets reached to within 40° of the equator.
Thus the extent of glaciation in the Permian and
Carboniferous time is comparable to that of the
Pleistocene.

Glacial deposits are often overlain by continental and marine deposits of Early Permian
age. The continental deposits and sometimes
the glacial deposits themselves commonly contain coal with a *Glossopteris* flora. In Antarctica
some of these coal deposits are located within 5°
of the Permian South Pole (Fig. 11.30). Presumably these deposits formed under a cool
climate in swamps that developed on the un-
even, glaciated terrain.

## ANCESTRAL SIBERIA

During the Devonian and Early Carboniferous,
geosynclines bordered Ancestral Siberia in essentially the same locations as those of the early
Paleozoic (Fig. 10.37). The Uralian and Angara
geosynclines were converted into mountain
ranges toward the end of the late Paleozoic.
Deformation in the Angara Geosyncline may
have been caused by plate collisions involving
the joining of Ancestral Siberia and Ancestral
China. By the end of the late Paleozoic, Ancestral Siberia was firmly connected to the other
continents that comprise Pangaea.

### Angara Geosyncline

A sequence of continental red beds and minor
volcanic rocks was deposited in the northern
part of the Angara Geosyncline after an intense
episode of deformation at the close of the Silurian. This sequence, which ranges in age from
Devonian to Early Carboniferous, grades
southward into marine limestones and shales. It
was folded and uplifted near the end of the
Early Carboniferous, and subsequent deposits
are almost entirely continental. Deformation
during the Permian ended deposition in the
Angara Geosyncline.

### Northern Pacific Ocean Geosyncline

The upper Paleozoic deposits of the Northern
Pacific Ocean Geosyncline are almost entirely
marine and consist of limestone, shale, sandstone, and minor volcanics. The Verkhoyansk
Complex, in the northern part of the geosyncline, ranges from Permian to Middle Triassic
and reaches a thickness of more than 10,000 m
(30,000 ft).

### Taimyr Geosyncline

A sequence of limestones and shales ranging
from Middle Devonian to Early Carboniferous
rests unconformably on Silurian deposits in the
Taimyr Geosyncline. Folding near the close of
the Early Carboniferous may have been caused
by movement of an oceanic plate under the
geosyncline. According to paleomagnetic data,
Late Carboniferous coal-bearing beds were deposited at approximately 45°N. Permian strata
have not been reported from the Taimyr Geosyncline, and Mesozoic deposits are not of geosynclinal thickness. Evidently, deposition in the
geosyncline had ceased by the end of the Carboniferous.

### Continental Interior

Devonian red beds crop out in the western part
of the Siberian Platform. Since they grade eastward into limestones, it is probable that they
were derived from highlands in or adjacent to
the Uralian Geosyncline. Carboniferous and
Lower Permian deposits are mainly limestone
with some interbedded clastics and coals. The
Angara flora found associated with Permian
coals is similar to the *Glossopteris* flora. How-

**FIGURE 11.30** Late Paleozoic sedimentary rocks on the face of Mt. Weaver, Queen Maud Mountains, Antarctica. Coals containing a *Glossopteris* flora are present near the top of the mountain. (Photo by George A. Doumani, in Ref. 49.)

ever, paleomagnetic studies indicate that the Angara flora grew at approximately 35°N latitude, whereas the *Glossopteris* flora grew at higher paleolatitudes. Very thick volcanic rocks and minor continental sediments were deposited over most of the western part of the Siberian platform during the Late Permian.

## ANCESTRAL CHINA

The arrangement of geosynclines during the late Paleozoic was essentially the same as that of the early Paleozoic (Fig. 10.39). However, when Ancestral China was joined to Ancestral Siberia

as a result of plate collisions, the Central Asia Geosyncline was converted into a mountain range.

### Southern Pacific Ocean Geosyncline

Deposits in the Southern Pacific Ocean Geosyncline have been thoroughly studied in Japan, where thick layers of clastics and felsic to mafic volcanics were deposited during the late Paleozoic (Fig. 11.31). The clastics are almost entirely marine, but the presence of several unconformities in late Paleozoic sequences provides evi-

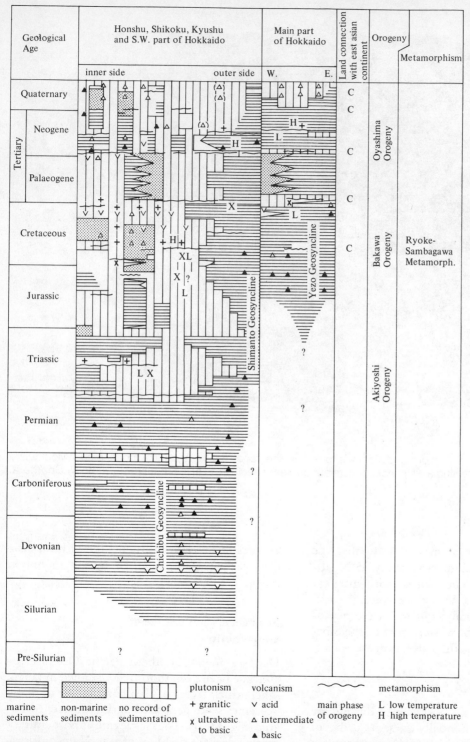

**FIGURE 11.31** A schematic illustration of the geologic history of Japan. (From Matsushita, Ref. 53.)

dence of local emergence. Uranium-lead dates on sphene and zircon indicate that an episode of regional metamorphism occurred in northwestern Japan in the Middle Permian, approximately 240 million years ago (**54**). This metamorphism was accompanied by uplift and erosion of a granitic basement. Granitic pebbles in Permian deposits were probably derived from this terrain.

Japan is now separated from the mainland of Asia by the Sea of Japan, which is more than 2000 m (6300 ft) deep in most parts. Three theories have been proposed for the origin of this sea:

1. If Japan was at one time connected with the Asian mainland, a small block of continental crust may have subsided to form the Sea of Japan. However, the subsidence of such a block would have produced a large negative gravity anomaly, which would be easily detected by seismic refraction profiling. Geophysical surveys in the Sea of Japan have yet to detect a layer with a thickness and density comparable to that of continental crust.
2. Japan may have drifted southeastward away from Asia (**55**). In this case, Korea would have been located along the trend of structures in Japan prior to separation from the mainland. However, Japan is not geologically comparable to Korea. Whereas Korea was largely above sea level during the Paleozoic, Japan received a great thickness of sediment during that time.
3. The Sea of Japan may have formed behind an island arc in essentially the same position it occupies today. In this case, Japan would be a relatively young continental block composed of volcanics, sediments derived from the weathering of the volcanics, and granites formed by the partial melting of the sediments and volcanics.

## Central Asia Geosyncline

The late Paleozoic was a time of great tectonic activity in the Central Asia Geosyncline. Erosion of uplifts within and adjacent to the geosyncline resulted in the deposition of an average of 5000 m (15,000 ft) of late Paleozoic clastic sediments and volcanic rocks (Fig. 11.32).

## Northern Tethyan Geosyncline

Geanticlinal uplifts within the Northern Tethyan Geosyncline furnished great thicknesses of clastic sediment during the Devonian and Early Carboniferous. More than 5000 m (15,000 ft) of Devonian clastics were deposited in the central part of the geosyncline (**55**). After tectonic activity subsided during the Late Carboniferous and Permian, limestone and coal were deposited in the geosyncline (Fig. 11.32).

## Indonesian Geosyncline

The oldest known deposits in the Indonesian Geosyncline are Devonian marine clastics, which crop out in western Borneo. Carboniferous deposits in Sumatra contain gneiss and schist pebbles, which indicate that a pre-Carboniferous basement is present in the region. Carboniferous and Permian deposits are dominantly clastics (probably eroded from nearby geanticlines) and volcanics (**56**).

## Continental Interior

Limestones were deposited over much of the southern part of the continental interior during the Devonian and Early Carboniferous, but a fluctuating withdrawal of the seas during the Late Carboniferous and Permian resulted in the deposition of coal-bearing clastics in many areas (Fig. 11.32). Both the Carboniferous and Permian coals contain floras similar to those of Carboniferous and Permian coal deposits in North America and Europe. However, according to paleomagnetic studies, the continental interior of Ancestral China was at approximately 45°N latitude, whereas the Carboniferous coals of North America and Europe were deposited within 10° of the equator.

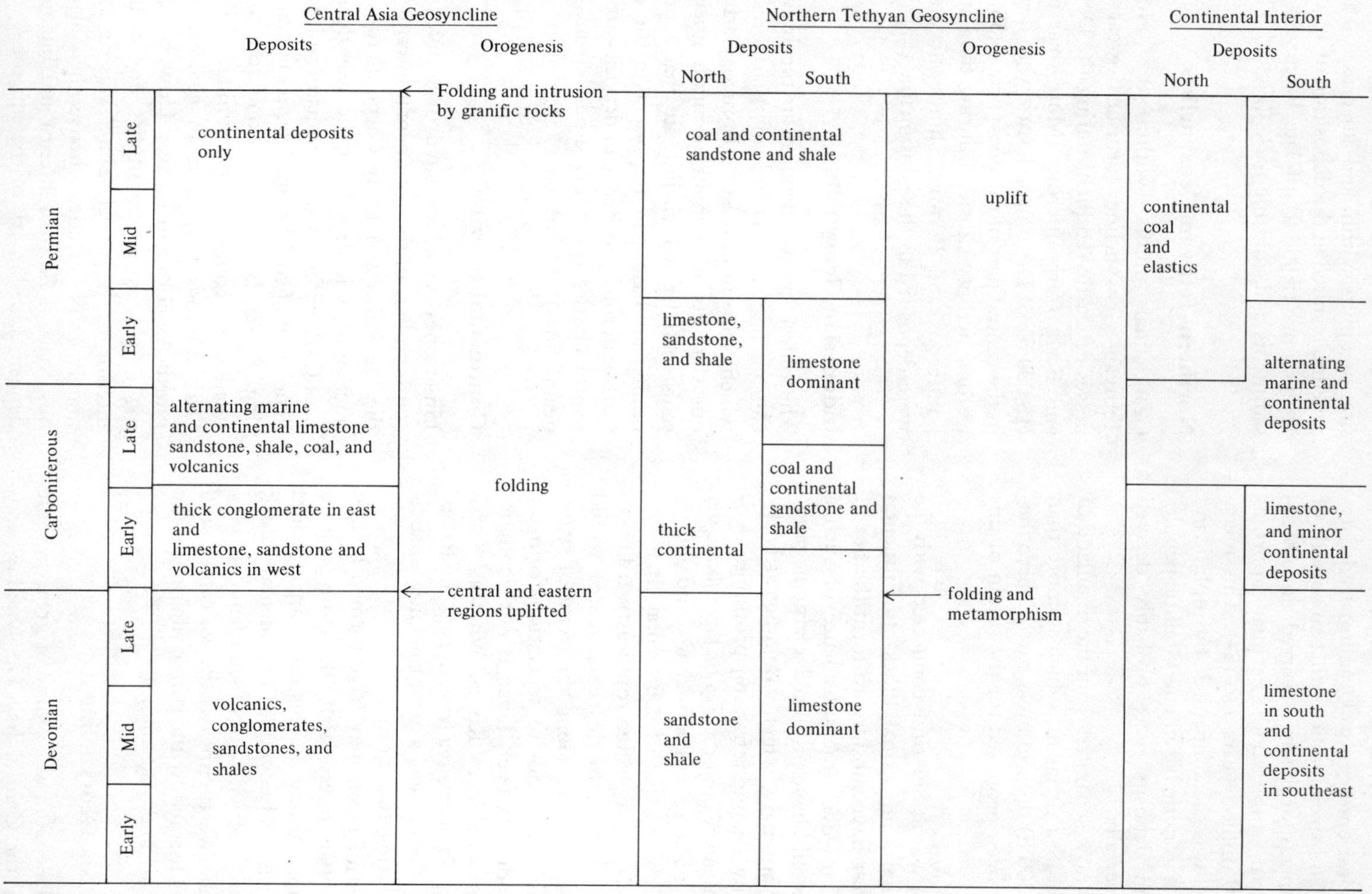

FIGURE 11.32 A schematic illustration of the late Paleozoic geologic history of Ancestral China.

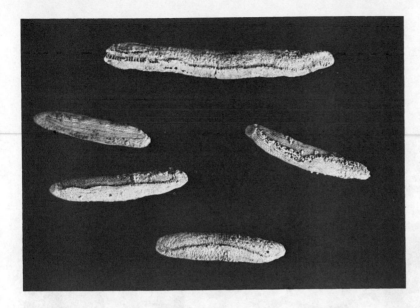

**FIGURE 11.33** *Parafusulina,* fusulinid foraminiferans from Permian deposits near Marathon, Texas. (Courtesy of Smithsonian Institution.)

## LATE PALEOZOIC LIFE

Early in the late Paleozoic, rapid evolutionary changes occurred among invertebrates, vertebrates, and plants. All groups, but particularly the invertebrates, were responding to pressures that resulted from the merging of continents. As geosynclines were uplifted to form mountains, many marine environments were eliminated.

The gradual decline of the trilobites, which began in the Ordovician, continued into the late Paleozoic. Corals and stromatoporoids replaced algae as the principal reef-forming organisms, and the graptolites were well on their way to extinction. The spiny, productid brachiopods and coiled, ammonoid cephalopods exhibited unusual morphologic development during the late Paleozoic. By the end of the Permian, most invertebrate groups had diminished considerably in importance and many became extinct.

### Protists

Foraminifera with calcareous tests occur in rocks as old as Devonian. One group, the endothyroids, were so abundant in Mississip-pian seas that some limestones of that age are composed almost entirely of their shells. The endothyroids gave rise to the fusulinids in the Pennsylvanian (Fig. 11.33). Fusulinids are large, spindle-shaped forms which diversified sufficiently to provide important index fossils for Pennsylvanian and Permian sequences.

### Sponges

Large colonies of delicate glass sponges flourished during the Devonian, particularly in western New York (Fig. 11.34). Calcareous sponges were important reef builders in the area of present-day Texas and New Mexico. The rise in importance of sponges as reef builders paralleled the decline of the rugose and tabulate corals in the Permian.

### Coelenterates

Tabulate corals, such as *Favosites,* and rugose corals, such as *Heliophyllum,* were abundant in the warm seas of the Devonian (Fig. 11.35). However, the rugose corals declined during the Carboniferous and had become extinct by the end of the Permian. Coral reefs contained

**FIGURE 11.34** Restoration of part of a Devonian coral reef in western New York: **(a)** Crinoids. **(b)** Glass sponges. **(c)** Cephalopods. **(d)** Corals. **(e)** Seaweed. **(f)** A gastropod. (Courtesy of the Buffalo Museum of Science.)

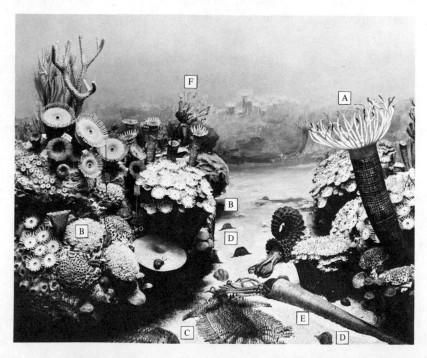

**FIGURE 11.35** Restoration of the sea bottom in Michigan during the Middle Devonian. Organisms include: **(a)** Solitary corals. **(b)** Colonial corals. **(c)** Trilobites. **(d)** Brachiopods. **(e)** Nautiloid cephalopods. **(f)** Crinoids. Some of the genera are *Heliophyllum,* the large solitary corals at the upper left; *Hexagonaria,* the solitary corals at the bottom left; *Atrypa,* the brachiopods at the bottom right; and *Stropheodonta,* the brachiopod at the bottom near the center. (Courtesy of the Field Museum of Natural History, Chicago.)

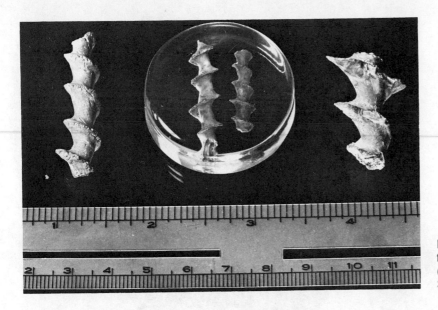

FIGURE 11.36 *Archimedes,* the screw-shaped bryozoa. (courtesy of Wards Natural Science Establishment.)

abundant colonial as well as rugose corals during the Devonian. *Pleurodictium,* a small colonial coral, is an excellent index fossil, since certain species are widespread and occupy rather short time spans.

## Bryozoans

Delicately branching and lacy bryozoans are very common in late Paleozoic sedimentary rocks. *Fenestella,* whose name means little windows, was attached to a screw-shaped support, *Archimedes* (Fig. 11.36). Because of its distinctive shape and widespread occurrence, *Archimedes* is an excellent guide fossil in Mississippian strata. Like the calcareous sponges, bryozoans were important reef-builders during the Permian.

## Brachiopods

Devonian marine invertebrates were dominated by the long-ranging, spiriferid brachiopods (Fig. 11.37). These forms characterized the benthonic shelf and deltaic environments. As the sea level fluctuated, the spirifers and their asso-

ciated faunas migrated within their respective sedimentary facies. With time, the faunas experienced significant evolutionary changes which provided the basis for zonation of the marine Devonian. This zonation, which was developed by H. S. Williams, stands as an early example of the use of facies faunas and population statistics in biostratigraphy.

The productids evolved from the strophomenid brachiopods. They were the dominant brachiopods in the Carboniferous and Permian, but became extinct by the end of the Paleozoic. Productids are characterized by large size and long spines. Some species were extraordinarily large—for example, *Gigantella gigantea,* which grew up to 30 cm (12 in.) or more. Late Permian limestones in the Glass Mountains of western Texas contain large members of productids along with unusual oysterlike, conical, and horn-shaped brachiopods (Fig. 11.38).

## Mollusks

Pelecypods were generally far less abundant than brachiopods during most of the late Paleozoic. In the Devonian, some groups of pelecy-

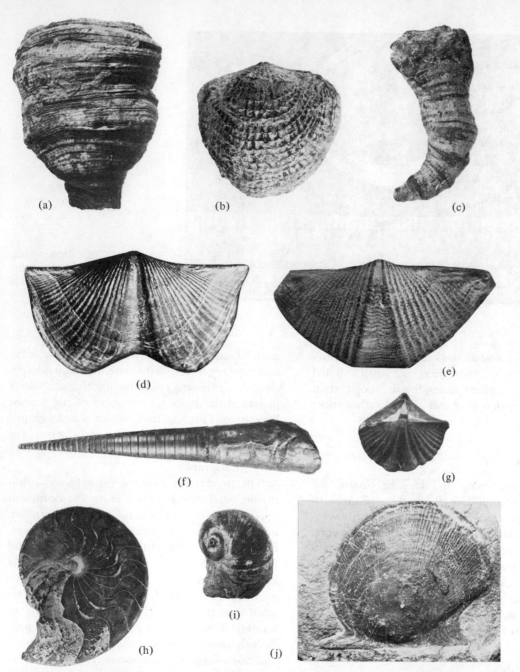

**FIGURE 11.37** Middle Devonian (Hamilton) fossils from western New York: **(a)** *Heliophyllum halli,* a rugose coral (X1). **(b)** *Spinatrypa spinosa,* brachial view of a brachiopod (X1.5). **(c)** *Amplexiphyllum hamiltoniae,* a rugose coral (X2). **(d)** *Spinocyrtia granulosa,* pedicle view of a brachiopod (X1). **(e)** *Mucrospirifer mucronatus,* pedicle view of a brachiopod (X1.5). **(f)** *Michelinoceras aldenense,* a nautiloid cephalopod (X1). **(g)** *Cyrtina hamiltonensis,* brachial view of a brachiopod (X2). **(h)** *Tornoceras uniangulare,* an ammonoid cephalopod (X1). **(i)** *Naticonema limeata,* a gastropod (X1). **(j)** *Pseudaviculopectan princeps,* a pelecypod (X1). (Photo courtesy of Carlton Brett.)

**FIGURE 11.38** Restoration of part of a Permian coral reef in western Texas: **(a)** Spiny productid brachiopods.
**(b)** Leptodid brachiopods. **(c)** Siliceous sponges. **(d)** Nautiloid cephalopods. **(e)** Calcareous sponges.
**(f)** Gastropods. **(g)** Rugose corals. (Courtesy of the Smithsonian Institution.)

pods moved from marine into freshwater environments, where they are still prevalent. Ammonoids are thought to have evolved from straight-shelled nautiloids. They appeared first in the Early Devonian and progressed rapidly from straight shells to loosely coiled and then tightly coiled shells in the later part of the Early Devonian (57).

The ammonoids became important index fossils in late Paleozoic sequences, mainly owing to the development of distinctive sutures (Fig. 11.37). Devonian and Early Carboniferous ammonoids had relatively simple, goniatite su-

tures. Ceratite and ammonite sutures developed independently in several families during the Late Carboniferous and Permian (57). Many groups of cephalopods became extinct at the end of the Paleozoic, and only one family with complex sutures survived into the Triassic.

### Arthropods

The gradual decline of the trilobites, which began in the Ordovician, continued during the late Paleozoic. In some Devonian sequences, however, trilobites such as *Phacops rana,* are

FIGURE 11.39 The trilobite *Phacops rana* from the Middle Devonian of western New York. (Photograph courtesy of Carlton Brett.)

abundant (Fig. 11.39). They became scarce during the Carboniferous and were extinct by the end of the Paleozoic.

Late Paleozoic ostracodes are distinctive and are well represented. Although a number of ostracode families became extinct before the end of the Permian, enough survived to give rise to a great number of new forms in the Mesozoic.

By early Devonian time, many arthropods had developed structures, such as lungs, pseudotracheae, and preoral cavities, which would enable them to exist on land (**58**). Arachnids, which are arthropods with jointed legs and pincerlike appendages in the head region, have been discovered in beds of late Early Devonian age in western Germany (**59**). These spiderlike fossils are the oldest known nonscorpion arachnids and may be the oldest terrestrial animals. The Rhynie Chert in Scotland contains very small (0.3–3.5 mm) nonscorpion arachnids of Middle or possibly late Early Devonian age.

A great variety of insects developed during the Pennsylvanian, and some reached enormous size (Fig. 11.40). Tropical forests and swamps in North America and Europe contained cockroaches 10 cm (4 in.) long and dragonflies with 75 cm (30 in.) wingspreads.

## Echinoderms

The shallow epicontinental seas of the platform were increasingly dominated by crinoids, blastoids, and starfish. Crinoids were so numerous that many limestones are comprised almost entirely of crinoid fragments. Occasionally, large numbers of complete crinoids are preserved in limestones or shales that were deposited in very quiet waters (Figs. 11.41 and 11.42). Blastoids reached their climax during the Mississippian but died out in most areas by the end of the Pennsylvanian. In Indonesia, they persisted until the end of the Permian.

## Graptolites

By the end of the Silurian, the number of graptolite genera had declined drastically. Only one group lived on into the Devonian and Carboniferous. Therefore, late Paleozoic graptolites are not useful guide fossils.

## Vertebrates

FISH. During the Devonian, fish were abundant and well diversified—the most advanced animals on the evolutionary scale. Consequently, the Devonian has been termed the Age of Fishes. Indeed, one of the major events in the history of the vertebrates was the development of jaws in the placoderms during the Late Silurian. Jaws are thought to have developed through ossification of the front set of V-shaped gill arches of the placoderms. The gill arches, which are useful to aid in pumping

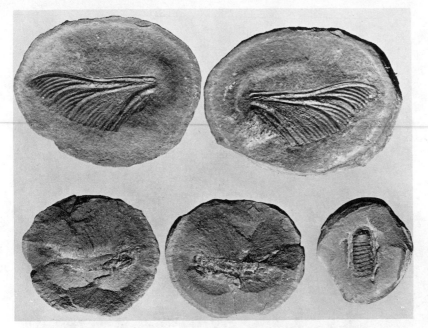

**FIGURE 11.40** Insects in concretions from Late Carboniferous beds at Mazon Creek, Illinois. (Courtesy of the Field Museum of Natural History, Chicago.)

**FIGURE 11.41** Fossil crinoids of Early Carboniferous age from Le Grand, Iowa. (Courtesy of the Buffalo Museum of Science.)

**FIGURE 11.42**  Restoration of a midcontinent seafloor during Early Carboniferous. (Courtesy of the Smithsonian Institution.)

water through the gill openings, developed hinges which allowed for opening and closing. Subsequently, teeth developed, and were used in biting and chewing. Released from the necessity of filter feeding, the placoderms could then compete with eurypterids and other predators. The placoderms evolved rapidly during the Devonian, and some became predators of an impressive size. *Dinichthys* was 10 m (30 ft) long.

The Osteichthyes (bony fish), the Chondrichthyes (sharks, rays, and skates), and the first freshwater fish are thought to have evolved from the placoderms. The placoderms and the agnaths declined in importance after the Devonian with the rise of more advanced fish. The Chondrichthyes, which have cartilaginous skeletons, have persisted with few changes into recent times (Fig. 11.43). The Osteichthyes arose from placoderms during the Devonian, but the oldest fossil remains have come from Late Devonian shales. These are the dominant fish today and include ray-finned and lobe-finned fish. Most of the modern bony fish descended from ray-finned forms.

Amphibians are thought to have evolved from lobe-finned fish. The history of their evolution is interesting because of its importance in the development of the higher animals and because the evolutionary advances can easily be traced to the demands of the environment.

Arkosic red beds of late Paleozoic age frequently contain the remains of lobe-finned fish. As discussed in Chapter 4, it is believed that

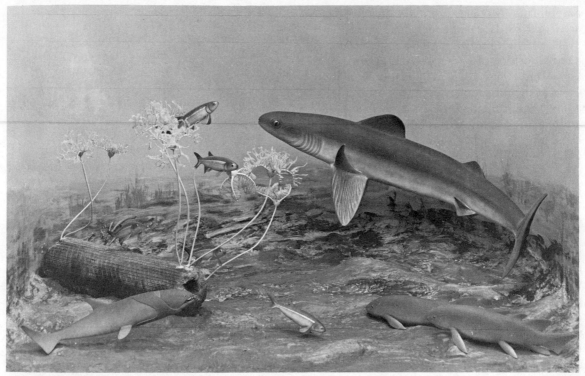

**FIGURE 11.43** Upper Devonian fish and associated fauna of western New York. Crinoids in the foreground are attached to a fragment of a Devonian tree. The small fish belong to the genus *Rhadinichthys*. Other genera include *Coccostes* at the bottom left; *Ctenecanthus,* the large sharklike form; and the lungfish, *Dipterus,* at the bottom right. (Courtesy of the Buffalo Museum of Science.)

such deposits were laid down under arid conditions. Only fish adapted to these conditions could survive. For this reason, most late Paleozoic freshwater fish had lungs and most saltwater forms did not. One order of lobe-finned fish could survive prolonged drought by burrowing deep in the mud and slowing their metabolism in a manner similar to hibernation. Lungfish in South America and Africa still burrow in the mud during dry spells.

The crossopterygians, another order of lobe-finned fish, coped with the problem of drought in another way. They had strong fins with which they could push themselves from one pool to another. The crossopterygians are anatomically similar to the amphibians. They had a well-ossified skeleton and their skull

bones can be matched, bone for bone, with those of amphibians and the higher vertebrates as well. Moreover, the teeth of crossopterygians show a marked similarity to the teeth of the early amphibians. For these reasons, it is almost certain that crossopterygians gave rise to the amphibians.

AMPHIBIANS. Amphibians are a group of tetrapods (four-footed vertebrates) that live either on land or in the water and lay their eggs in water or in a damp place. They hatch out at a very immature stage (larva stage) and generally live in water and breathe by means of gills during the early stage of their life. When approaching maturity, a metamorphosis takes place and lungs replace gills in land-dwelling amphibians. Modern examples of amphibians are frogs, toads, salamanders, and newts.

The earliest known amphibians are found in a red sandstone of Late Devonian age in eastern Greenland. This area was at a latitude of about 15°N at the time, and therefore it probably had a hot, dry climate (Fig. 11.2b). The earliest amphibians are known as labyrinthodonts because their teeth have intricately infolded enamel which resembles a labyrinth (**61**) (Fig.11.44). Before the evolution of the amphibians, there were only insects, scorpionlike arthropods, spiders, and plants in terrestrial environments.

The amphibians flourished during the Carboniferous (Fig. 11.45), and before the end of the Mississippian, the labyrinthodonts gave rise to the first reptiles. These reptiles belong to a group known as cotylosaurs. The amphibians declined during the Permian, probably as a result of competition from the more advanced reptiles. Labyrinthodonts finally became extinct at the end of the Triassic.

REPTILES. The most significant difference between the reptiles and the amphibians is that the reptiles lay their eggs on land and those eggs have a tough shell which serves not only as a protection against predators, but it also protects the egg from drying out. The reptilian egg is called an amniote egg, and it contains a yolk to nourish the growing embryo and a water-filled membrane (the amnion) surrounding the embryo. The egg of a chicken is an example of this type of egg. The development of the amniote egg is a major factor in the rise of the reptiles to dominance since it allowed the reptiles to overcome their dependence on water for reproduction. The eggs of amphibians are fertilized externally in water, whereas those of reptiles are fertilized internally.

The oldest known reptile, a cotylosaur known as *Romeriscus,* was found in rocks of Early Carboniferous (Early Pennsylvanian) age in Nova Scotia. At that time, Nova Scotia was connected to Greenland, the presumed birth-

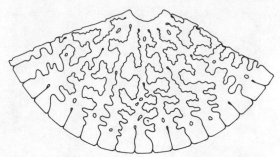

FIGURE 11.44 Diagrammatic cross-section of part of a labyrinthodont tooth. The sinuous lines are complex infolds of the enamel of the tooth. (After Colbert, Ref. 60.)

place of the amphibians. The first reptiles probably walked with the sprawling gait which characterized their amphibian ancestors and which characterize modern lizards and crocodiles.

The cotylosaurs are the "stem" reptiles from which the other reptile groups evolved. During the Late Pennsylvanian, the cotylosaurs gave rise to the pelycosarus and the ancestors of the thecodonts (which in turn gave rise to the dinosaurs, crocodiles, birds, etc.), the ichthyosaurs, the plesiosaurs, the mosasaurs, and the rhynchocephalians (**62**). The cotylosaurs reached their zenith during the Early Permian, but declined thereafter and became extinct during the Late Triassic.

By the Early Permian, the pelycosaurs had become the dominant land animals. They had splayed-out legs like the cotylosaurs and early amphibians. The differentiation of teeth began in the carnivorous pelycosaurs with the enlargement of the canines. Among the pelycosaurs were a group of large, finback reptiles that probably used their fins for regulation of body temperature (Fig. 11.46). Blood circulating through the fins would be warmed when the fins were oriented perpendicular to the sun's rays, and it would be cooled when oriented parallel to the sun's rays. During the Permian, pelycosaurs spread from North America to Eurasia and Gondwanaland. At that time, all of

FIGURE 11.45 Skeleton of a large amphibian, *Megacephalus*. (Courtesy of the Field Museum of Natural History, Chicago.)

FIGURE 11.46 Restoration of a group of late Paleozoic reptiles. The five large finback reptiles are carnivores belonging to the genus *Dimetrodon*. On the left are a group of small pelycosaurs, and the small amphibian *Diplocaulis* can be seen in the water in the bottom right. (Courtesy of the Field Museum of Natural History, Chicago.)

the continents were joined into a single land-mass (Pangaea) and this would have greatly facilitated migration of these reptiles. From the distribution of the fossil remains of pelycosaurs, it may be inferred that their range was restricted to tropical and subtropical parts of the Permian globe (**63**). This is to be expected, since large cold-blooded reptiles would have difficulty living in cool or cold climates.

Pelycosaurs declined during the Late Permian and became extinct before the end of that period, probably because of competition from more advanced reptiles. These included the therapsids which evolved from the pelycosaurs during the Early Permian and quickly became the dominant reptiles. This dominance lasted from the Late Permian to the Middle Triassic. There is some evidence that the therapsids were warm-blooded and may have had fur (**63**). This would not be too surprising because mammals are thought to have evolved from the therapsids.

A cold-blooded animal or *ectotherm* must use an external heat source (the sun) to raise its body temperature above that of the surrounding air. A warm-blooded animal or *endotherm* creates its own body heat during the process of metabolism. The following criteria may be used to determine whether an extinct animal was warm-blooded or cold-blooded:

1. The geographic range of a cold-blooded animal would be expected to be restricted to the warmer latitudes (**63**). In a cold or temperate zone, a cold-blooded animal must be small enough to find a safe place for hibernation during the winter.
2. The bones of modern cold-blooded animals differ substantially from those of warm-blooded animals (**63**). The bones of ectotherms have a low density of blood vessels. Furthermore, when an ectotherm hibernates, its bones develop growth rings in the outer layers. Modern warm-blooded animals have bones that are rich in blood vessels and almost never show growth rings.
3. The predator/prey ratio (by weight) is higher when the predator is an ectotherm than when it is an endotherm (**63**). This is because an endotherm must take in much more food than an ectotherm in order to have enough energy to create its own heat and sustain physical activity. Since an endotherm must eat more, a given body of prey can support fewer endotherms than ectotherms. The predator/prey ratio is not dependent on the size of the predator or of the prey. For a mammal or a bird (both of which are endotherms), the predator/prey ratio is typically about .01 to .03, whereas the ratio for ectotherms (for example, a crocodile) typically ranges from .33 to .40, more than 10 times as much. The predator/prey ratio of extinct animals can be determined by quantitative measurements of fossil assemblages.
4. Present day cold-blooded animals alternate long periods of rest (often with their bellies on the ground) with short bursts of activity (**64**). It is only the warm-blooded animals which can spend long periods of time

standing on their feet or which can sustain long periods of activity. Modern cold-blooded animals just do not have the energy. Long-legged modern animals which appear to be built for running are invariably warm-blooded. It would be strange for an animal to appear to be built for running and not have the energy to do so.

Other advancements possessed by the therapsids included a second bony roof of the mouth which allowed the animal to breathe and chew at the same time. Not only were their teeth differentiated, but they were rooted in the jaw bones which allowed for a more efficient consumption of their prey. Knees were turned forward and their stride lengthened. The resulting increase in speed was a distinct advantage in escaping predators and in running down prey.

The therapsids became the dominant land form during the Late Permian. In contrast to the pelycosaurs, the therapsids ventured into higher latitudes and cooler climates. Furry coats would be almost a necessity if the therapsids were warm-blooded because the heat loss for an animal without insulation in a cool or cold climate would be so great that it would require the animal to consume enormous quantities of food. The therapsids declined during the Triassic and became extinct in the Jurassic.

## Plants

The Early Devonian plant record is sparse, possibly because the colonization of the land by plants had just begun. Early Devonian plants included primitive varieties or mosses and vascular plants. By the Middle Devonian, the expansion of terrestrial plants was well underway. Forests of tree ferns, such as *Eospermatopteris,* scale trees, such as *Protolepidodendron,* scouring rushes, and seed ferns were abundant at that time.

Recent discoveries indicate a great antiquity of seed-bearing plants. Primitive pinelike coni-

**FIGURE 11.47** Restoration of a Pennsylvanian coal swamp in the midcontinent. Genera of the plants include: **(a)** *Sigillaria.* **(b)** *Lepidodendron.* **(c)** *Cordaites.* Also notice the large cockroach on the tree at the left. (Courtesy of the Field Museum of Natural History, Chicago.)

fers arose during the Late Devonian. In the Carboniferous, scouring rushes, such as *Cala-mites,* achieved tree size, and the scale tree *Lepidodendron* and the tree fern *Sigillaria* grew to heights exceeding 30 m (100 ft) (Figs. 11.47 and 11.48). The more advanced seed-bearing ferns and gymnosperms had achieved dominance over the seedless trees by Late Carboniferous time. Prominent among the gymnosperms were the cordaites, true conifers, ginkgoes, and cycads.

### Extinctions

A number of invertebrate groups became extinct near the end of the Paleozoic. These include trilobites, eurypterids, blastoids, and ru-

gose corals. The number of families of foraminiferans and bryozoans declined markedly during the Permian. Among the vertebrates, the agnaths became extinct in the Middle Permian. Amphibians, cotylosaurs, and several groups of plants declined throughout the Permian.

The declines in both fauna and flora are too irregular to be assigned to any single cause. One contributing factor may have been the gradual uplift of the land and the resulting withdrawal of the seas, which occurred during the late Paleozoic. The uplifts were associated with deformations such as the Hercynian orogeny between the Early and Late Carboniferous and the Alleghenian orogeny in the Permian. The Permian marked the end of marine deposition in

(a)

(b)

**FIGURE 11.48**  Late Carboniferous fossilized land plants: **(a)** Part of the trunk of the lycopod *Lepidodendron,* showing leaf scars. **(b)** Leaves of the horsetail rush, *Annularia,* from a concretion found at Mazon Creek, Illinois. (Courtesy of the Field Museum of Natural History, Chicago.)

the Appalachian, Uralian, Angara, and Hercynian geosynclines. Furthermore, the seas rapidly withdrew from the continental interiors and portions of the geosynclines bordering the Pacific Ocean and the Tethyan Sea at that time (**65, 66**) (Fig. 11.49). The uplifts drained many of the coal-forming swamps in both the northern and southern hemisphere. This alone may have resulted in the decline of swamp-dwelling plants and animals. The uplift of the lands also re-

sulted in a drastic reduction in the area of the shallow seas, which greatly increased competition among their inhabitants. Such competition might account for the decline in the foraminiferans, bryozoans, and brachiopods. For the trilobites it may have been the final blow to their survival.

The principal cause for the decline of the amphibians was probably the rise of the reptiles, which could prey on all but the deepest

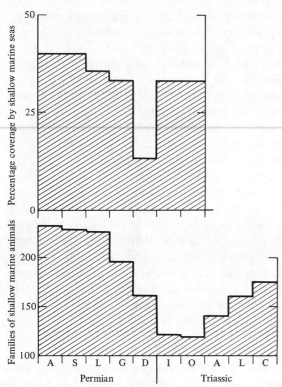

**FIGURE 11.49** Summary of changes in families of shallow marine invertebrates and in the area of shallow marine seas for the Permian and Triassic. (Reprinted from T. J. M. Schopf, Ref. 66, by permission of The University of Chicago Press.)

diving amphibians. Furthermore, the development of the amniote egg insured that a greater percentage of young reptiles survived.

Changes in the salinity of the oceans near the end of the Permian has been suggested as a cause of marine extinctions at the end of the Paleozoic. Calvin Stevens has suggested that Late Permian seas became somewhat brackish as a result of extensive deposition of halite in marginal basins during the Permian (**67**). On the other hand, R. L. Bowen has suggested that the Late Permian seas were hypersaline and that this was a cause of extinctions (**68**). The hypersalinity might have been caused by warming of the seas following the Permo-Carboniferous

glaciation. The end of glaciation could result in the elimination of cold water ecological niches. Furthermore, warmer waters would have reduced the amount of nutrients in the sea (**68**). Both of these factors might cause marine extinctions.

## LATE PALEOZOIC CLIMATE

On a late Paleozoic reconstruction of the continents, climatic zones are aligned approximately parallel to paleolatitudes in a pattern similar to that of today (Fig. 11.1). In general, tropical coals, reefs, red beds, evaporites, and dune sands are located within 40° of the paleoequator and glacial deposits occur within 40° of the south paleopole. The absence of glacial deposits of late Paleozoic age in the northern hemisphere is probably due to the location of the north paleopole over an open ocean (Fig. 11.1).

### Devonian

Glacial deposits are not common in Devonian sequences, but they have been reported from several areas in South America and one in Africa (Fig. 11.1a). The limited extent of these deposits suggests that they were a result of alpine glaciation. All Devonian glacial deposits occur within 40° of the south paleopole.

Devonian coals are found on Bear Island north of Norway, in the European part of the Soviet Union, and in the Arctic Islands of northern Canada. All these areas fall within 15° of the paleoequator. Coral and other organic reefs were widespread, and most were formed within 30° of the Devonian equator. However, in northwestern India, well-developed reefs are found at a paleolatitude of approximately 40°S. Red beds and evaporites generally fall within 30° of the paleoequator, but some have been found as far from the equator as 40°. From paleoclimatic indicators, it may be inferred that on the whole the Devonian climate was somewhat like that of today.

## Carboniferous

The Carboniferous was a time of gradual cooling in the vicinity of the south paleopole. There are three distinctly different glacial horizons in the Carboniferous sequences of Bolivia. The first and possibly the second are of Early Carboniferous age (**42**). Early Carboniferous glacial deposits have also been reported from Argentina (**40**). The association of these deposits with highlands suggests that they may have been the result of alpine glaciation. The lower sections of the Dwyka Tillite of central and southern Africa may also be of Early Carboniferous age (**50**).

The climate began to cool rapidly toward the end of the Early Carboniferous, and just prior to the Permian the southern continents were in the grip of a major ice age. Permian and Carboniferous glacial deposits have been found in all the southern continents and India. All these deposits occur within 60° of the south paleopole (Fig. 11.1b).

Much of North America and Europe experienced tropical, subtropical, or arid conditions during the Carboniferous, since these continents were situated near the paleoequator. Tropical coals formed within 15° of the Carboniferous equator and reefs, evaporites, and red beds generally fall within 40° of this equator (Fig. 11.2b). In Japan, however, small patch reefs are found at a paleolatitude of 65°N. This suggests that the area in the vicinity of the north paleopole was considerably warmer than that in the vicinity of the south paleopole during the Carboniferous. The waters near Japan may have been warmed by currents from the south similar to the modern Kuroshio Current.

## Permian

The climate was cool at the beginning of the Permian, but it gradually became warmer during the Middle and Late Permian. Continental glaciers formed during the Early Permian in Australia, Antarctica, and possibly Africa. In northeastern Australia, erratic blocks weighing up to several tons occur in Middle and Upper Permian marine deposits and may have been deposited by icebergs originating from alpine glaciers (**36**). Upper Permian glacial deposits are also present in the Himalayas (**38**) and in northwestern Argentina (**40**). These deposits are not extensive and were probably deposited by alpine glaciers. Overall, a general warming trend occurred during the Permian.

Very large Permian evaporite deposits occur in the United States and in the Soviet Union between 15° and 30° north of the paleoequator in what was presumably an arid belt. Minor gypsum and salt of Permian age in western Australia indicate local arid conditions. Permian faunas from the Canadian Arctic are thought to have developed in cool water (**69**), although this area was at a paleolatitude of about 45°N. Possibly the northern hemisphere was cooled as a result of the glaciation in the southern hemisphere. Tropical coals of Permian age are found within 15° of the paleoequator, but coals containing a *Glossopteris* flora occur within 5° of the Permian South Pole in Antarctica. They may have been deposited in an environment similar to that of the modern Arctic muskeg swamps.

## SUMMARY

At the beginning of the late Paleozoic there were only two discrete landmasses. These continents were separated by a rather narrow ocean during most of the Devonian, but by the end of the Devonian this ocean was largely closed. Several transgressions and regressions of the seas across the continents took place during the late Paleozoic. At the beginning of the Devonian, the seas were generally restricted to the geosynclines and the adjacent platform. Transgressions occurred during the Middle and Late Devonian and during the Early and Middle Pennsylvanian. The seas began an irregular withdrawal during the latest Pennsylvanian and by the end of the Permian they had largely withdrawn from the continents.

The late Paleozoic was a time of tectonic activity in most geosynclines, especially those located along the junctions between ancestral continents. Uplifts furnished large volumes of clastic sediment to the geosynclines. By the end of the Paleozoic, deformations had ended deposition within the Appalachian, Ouachita, Franklin, Hercynian, Uralian, Angara, Taimyr, Central Asia, and West African geosynclines and had converted them into mountains. These deformations were probably related to collisions of continental plates associated with the completion of joining of the continents into a single landmass, Pangaea. Deformations in the geosynclines on the outer margins of Pangaea were probably associated with movements of oceanic plates under the continent.

The land was largely barren of animal and plant life at the beginning of the late Paleozoic, but insects, vertebrates, and plants rapidly colonized the lands. Dispersal into new and unoccupied environments allowed rapid diversification of certain forms until most of the available habitats were filled. Fish with strong fins gave rise to amphibians, and reptiles evolved from the amphibians during the Carboniferous. Land plants evolved rapidly from a few simple varieties in the earliest Devonian to a rich flora including giant tree ferns, scouring rushes, and conifers.

Among the invertebrates, major trends include the evolution of the ammonoid cephalopods from the nautiloids, the development of the productid brachiopods, and the extinction of many invertebrate groups near the end of the Paleozoic. The groups that became extinct included the rugose corals, two orders of bryozoans, many molluscan families, many groups of stalked echinoderms, several groups of sponges, and the trilobites.

The climate was rather cool during the Devonian, probably similar to that of today. Rapid cooling began in the Late Carboniferous, and glaciation was widespread in the southern hemisphere during the latest Carboniferous and Early Permian. The climate warmed somewhat during the Early Permian, but probably remained cool throughout the period.

## REFERENCES CITED

1. C. K. Seyfert and D. J. Leveson, 1969, *Geological Bulletin:* v. 3, Queens College Press.
2. A. J. Eardley, 1962, *Structural Geology of North America:* Harper and Row, New York.
3. M. Kay, 1951, *North American Geosynclines:* Geol. Soc. Amer. Mem. 48.
4. B. R. Pelletier, 1958, *Geol. Soc. Amer. Bull.,* v. 69, p. 1033.
5. J. Rodgers, 1970, *The Tectonics of the Appalachians:* Wiley, New York.
6. A. Hietanen, 1967, *J. Geol.,* v. 75, p. 187.
7. J. A. Rehmer and J. C. Hepburn, 1974, *Geology,* v. 2, p. 413.
8. C. J. Stuart, 1975, *Geology,* v. 3, p. 153; H. W. Baker and R. H. Dott, Jr., 1975, *Geology,* v. 3, p. 153.
9. R. H. Dott, 1961, *Geol. Soc. Amer. Bull.,* v. 72, p. 1289.
10. A. A. Ruitenberg, D. V. Venugopal, and P. S. Giles, 1973, *Geol. Soc. Amer. Bull.,* v. 84, p. 3029.
11. G. S. Clark and J. L. Kulp, 1968, *Amer. J. Sci.,* v. 266, p. 865.
12. L. E. Long, J. L. Kulp, and F. D. Eckelmann, 1959, *Amer. J. Sci.,* v. 257, p. 585.
13. R. C. Milici, 1975, *Geol. Soc. Amer. Bull.,* v. 86, p. 1316.
14. V. S. Griffin, 1974, *Geol. Soc. Amer. Bull.,* v. 85, p. 1123.
15. J. S. Watkins, 1971, *Geol. Soc. Amer. Abstr. Programs,* v. 3, p. 357.
16. J. Wickham, D. Roeder, and G. Briggs, 1976, *Geology,* v. 4, p. 173.
17. S. I. Root, 1970, *Geol. Soc. Amer. Bull.,* v. 81, p. 815.
18. R. J. Roberts, 1964, *U.S. Geol. Survey Prof. Pap. 459-A;* M. Churkin, Jr., 1974, *Geology,* v. 2, p. 339.
19. J. G. Johnson, 1975, *Geology,* v. 3, p. 219.
20. W. P. Irwin, 1966, *in* E. H. Bailey, ed., *Geology of Northern California:* Calif. Div. of Mines and Geol. Bull. 190, p. 19.
21. M. Churkin, Jr. 1969, *Science,* v. 165, p. 549.
22. E. J. Buehler and I. H. Tesmer, 1963, *Buffalo Soc. Nat. Sci. Bull.,* v. 21, no. 3.

23. L. D. Grayston, D. F. Sherwin, and J. F. Allan, 1964, *in Geological History of Western Canada:* Alberta Soc. of Petrol. Geol., Calgary, p. 49.

24. A. M. Klingspor, 1969, *Amer. Assoc. Petrol. Geol. Bull.,* v. 53, p. 927.

25. J. A. Peterson and R. J. Hite, 1969, *Amer. Assoc. Petrol. Geol. Bull.,* v. 53, p. 884.

26. J. M. Weller, 1960, *Stratigraphic Principles and Practice:* Harper and Row, New York; J. M. Weller, 1957, *Geol. Soc. Amer. Mem. 67,* v. 2.

27. L. L. Sloss, 1953, *J. Sediment. Petrol.,* v. 23, p. 156. R. K. Matthews, 1974, *Dynamic Stratigraphy:* Prentice-Hall, Engelwood Cliffs; P. B. King, 1948, *U.S. Geological Survey Professional Paper* 215.

28. M. G. Rutten, 1969, *The Geology of Western Europe:* Elsevier, New York.

29. T. A. Anderson, 1975, *Geol. Soc. Amer. Bull.,* v. 86, p. 77.

30. G. C. Amstutz, R. A. Zimmerman, and E. H. Schot, 1971, *in Sedimentology of Parts of Central Europe:* Verlag Waldemar Kramer, Frankfurt am Main, p. 253.

31. C. Boyer, 1974, *Volcanismes acides Paléozoiques dans le massif Armoricain* (Ph.D. thesis): Université de Paris-sud, Paris, France, p. 381.

32. J. Hamet and C. J. Allègre, 1976, *Geol. Soc. Amer. Bull.,* v. 87, p. 1429.

33. F. Arthaud and P. Matte, 1977, *Geol. Soc. Amer. Bull.,* v. 88, p. 1305.

34. C. O. Dunbar, 1940, *Amer. Assoc. Petrol. Geol. Bull.,* v. 24, p. 237.

35. D. V. Nalivin, 1960, *The Geology of the U.S.S.R.* (transl. S. I. Tomkeieff): Pergamon Press, New York.

36. D. A. Brown, K. S. W. Campbell, and K. A. W. Crook, 1968, *The Geological Evolution of Australia and New Zealand:* Pergamon Press, New York.

37. C. A. Landis and D. S. Coombs, 1967, *Tectonophysics,* v. 4, p. 501.

38. J. D. A. Zijderweld *et al.,* 1970, *Tectonophysics,* v. 10, p. 639.

39. J. Sougy, 1962, *Geol. Soc. Amer. Bull.,* v. 73, p. 871.

40. H. J. Harrington, 1962, *Amer. Assoc. Petrol. Geol. Bull.,* v. 46, p. 1773.

41. L. A. Frakes and J. C. Crowell, 1969, *Geol. Soc. Amer. Bull.,* v. 80, p. 1007.

42. C. Schubert, 1969, *Geol. Soc. Amer. Bull.,* v. 80, p. 443.

43. J. W. Stewart, J. F. Evernden, and N. J. Snelling, 1974, *Geol. Soc. Amer. Bull.,* v. 85, p. 1107; J. Helwig, 1973, *Geol. Soc. Amer. Bull.,* v. 84, p. 1493.

44. W. Hamilton, 1967, *Tectonophysics,* v. 4, p. 555.

45. W. E. Long, 1963, *in* R. J. Adie, ed., *Antarctic Geology:* North-Holland, Amsterdam, p. 352.

46. W. E. Long, 1965, *in* J. B. Hadley, ed., *Geology and Paleontology of the Antarctic:* Antarctic Research Series, v. 6, American Geophysical Union, p. 71; L. A. Frakes, J. L. Matthews, and J. C. Crowell, 1971, *Geol. Soc. Amer. Bull.,* v. 82, p. 1581.

47. W. Hamilton and D. Krinsley, 1967, *Geol. Soc. Amer. Bull.,* v. 78, p. 783.

48. J. Steiner and E. Grillmair, 1973, *Geol. Soc. Amer. Bull.,* v. 84, p. 1003.

49. G. A. Doumani and V. H. Minshew, 1965, *in* J. B. Hadley, ed., *Geology and Paleontology of the Antarctic:* Antarctic Research Series, v. 6, American Geophysical Union, p. 127.

50. L. A. Frakes and J. C. Crowell, 1970, *Geol. Soc. Amer. Bull.,* v. 81, p. 2261.

51. L. A. Frakes, P. M. de Figueiredo, and V. Fulfaro, 1968, *J. Sediment. Petrol.,* v. 38, p. 5.

52. A. A. Meyerhoff and C. Teichert, 1971, *J. Geol.,* v. 79, p. 285.

53. S. Matsushita, 1963, *in* F. Takai *et al.,* eds., *Geology of Japan:* University of California Press, Berkeley.

54. K. Ishizaka and M. Yamaguchi, 1969, *Earth Planet. Sci. Lett.,* v. 6, p. 179.

55. S. W. Carey, 1958, *Continental Drift: Symposium:* Geology Department, University of Tasmania, Hobart, p. 177.

56. J. H. F. Umbgrove, 1938, *Amer. Assoc. Petrol. Geol. Bull.,* v. 22, p. 1.

57. C. Teichert, 1967, *in* C. Teichert and E. L. Yochelson, eds., *Essays in Paleontology and Stratigraphy:* Special Publ. No. 2, Department of Geology, University of Kansas, p. 162.

58. L. Stormer, 1977, *Science,* v. 197, p. 1362.

59. L. Stormer, 1969, *Science,* v. 164, p. 1276.

60. E. H. Colbert, 1961, *Evolution of the Vertebrates:* Science Editions, New York.

61. D. Baird and R. L. Carroll, 1967, *Science,* v. 157, p. 56.

62. A. S. Romer, 1966, *Vertebrate Paleontology:* University of Chicago Press, Chicago.

63. R. T. Bakker, 1975, *Sci. Amer.,* v. 232, no. 4, p. 58.

**64.** A. J. Desmond, 1976, *The Hot-Blooded Dinosaurs: A Revolution in Paleontology:* Dial Press, New York.

**65.** J. G. Johnson, 1974, *Geology,* v. 2, p. 479.

**66.** T. J. M. Schopf, 1974, *J. Geol.,* v. 82, p. 131.

**67.** R. J. Lantzy, M. F. Dacey, F. T. Mackenzie, 1977, *Geology,* v. 5, p. 724; C. H. Stevens, 1977, *Geol. Soc. Amer. Bull.,* v. 88, p. 133.

**68.** R. L. Bowen, 1975, *Geol. Soc. Amer. Abstr. Programs,* v. 7, p. 1005.

**69.** J. B. Waterhouse, 1967, *Nature,* v. 216, p. 47.

# The Mesozoic Era 12

THE SUPERCONTINENT of Pangaea, which was created by the joining of the ancestral continents during the Paleozoic, broke apart during the Mesozoic. The rate of separation of the newly formed continents was very slow, averaging a few centimeters per year. As the continents moved apart, the rifts between them deepened and widened. At an early stage in the separation, these rifts probably resembled the modern Red Sea trough. Rifting was caused by the development of rising convection currents beneath Pangaea.

The Atlantic and Indian oceans widened throughout the Mesozoic, while the Pacific Ocean decreased in area. Movement of the seafloor under the Pacific margins of the continents resulted in extensive deformation, volcanism, plutonism, and metamorphism in the geosynclines bordering the Pacific Ocean. As a result of this tectonic activity, thick layers of clastic sediment were deposited in these geosynclines.

The widespread extinctions within many plant and animal groups at the end of the Paleozoic were followed by the development of a great variety of new forms during the Mesozoic. Evolution was rapid in a number of groups, but changes were most dramatic among the reptiles. Dinosaurs, pterosaurs, plesiosaurs, mososaurs, and ichthyosaurs occupied nearly all of the available land, sea, and air habitats. Dinosaurs were so widely dispersed and so abundant that the Mesozoic is known as the Age of Dinosaurs. The great expansion in

dinosaurs and other terrestrial faunas was matched by an equally important expansion in the Mesozoic floras. These expansions may have resulted from the increase in land area during the early Mesozoic.

## MESOZOIC GEOGRAPHY

The most significant change in geography during the Mesozoic was the breakup of Pangaea. The beginning of separation of the continents which comprised Pangaea may be dated by various paleomagnetic, structural, geochronological, and paleontological techniques (see Chapter 7). These data indicate that the breakup of Pangaea did not occur all at once, but occurred in five different stages (1) (Table 12.1 and Fig. 12.1). The continents that began to separate during the Mesozoic are: North America from Eurasia and Gondwanaland; Antarctica, Australia, and India from Africa; South America from Africa; Greenland from North America; Alaska from northern Canada; Spain and Portugal from France; Greenland from Eurasia; India from Australia and Antarctica; Malagasy from Africa; and India from the Seychelles Islands.

### Separation of North America and Eurasia from Gondwanaland

A plate containing North America and Eurasia began to separate from a plate containing Africa, South America, India, Antarctica, and

TABLE 12.1A **Dates of the Beginning of Continental Separations**

Late Triassic—Karnian: about 202 million years ago
  1. Begin separation North America and Eurasia from Gondwanaland
  2. Begin separation Africa from India, Australia, and Antarctica

Late Jurassic—Kimmeridgian: about 147 million years ago—Anomaly M-22
  1. Begin separation Africa from South America
  2. Begin separation Antarctica and Australia from India
  3. Begin separation Malagasy from Africa
  4. Begin separation of Greenland from North America
  5. Begin rotation of Spain and Portugal away from Europe
  6. Begin rotation of Alaska away from Northern Canada

Late Cretaceous—Maestrichtian: about 70 million years ago—Anomaly 30
  1. Begin separation of Greenland from Europe
  2. Begin separation of India from the Seychelles Islands

Early Tertiary—latest Paleocene: about 54 million years ago—Anomaly 22
  1. Begin separation of Australia from Antarctica

Late Tertiary—Middle Miocene: about 15 million years ago
  1. Begin separation of Arabia from Africa
  2. Begin separation of Baja, California, from Mexico
  3. Begin separation of Sardinia and Corsica from Europe

TABLE 12.1B **Dates of Changes in the Direction of Seafloor Spreading**

Late Jurassic—Kimmeridgian: about 147 million years ago—Anomaly M-22
  1. Change in the direction of seafloor spreading in the North Atlantic

Middle Cretaceous—Albian-Cenomanian boundary: about 100 million years ago
  1. Change in the direction of seafloor spreading in the North Atlantic

Late Cretaceous—Maestrichtian: about 70 million years ago
  1. Change in the direction of separation of India and Antarctica
  2. Change in the direction of separation of Africa and South America

Early Tertiary—Latest Paleocene: about 53 million years ago
  1. Change in the direction of seafloor spreading in Pacific Ocean
  2. Temporary halt to seafloor spreading on SW Indian Ocean Ridge

Early Tertiary—Late Eocene: about 42 million years ago
  1. Change in the direction of seafloor spreading in Pacific Ocean
  2. Shift in the location of the Mid-Atlantic Ridge north of Iceland to the west
  3. Change in the direction of seafloor spreading on Pacific-Antarctic Ridge
  4. Restart seafloor spreading on SW Indian Ocean Ridge

Middle Tertiary—Oligocene-Miocene boundary: about 26 million years ago
  1. Beginning of spreading on the Galapagos Ridge
  2. Change in the direction of seafloor spreading in the western Pacific
  3. Temporary halt to seafloor spreading on SW Indian Ocean Ridge

Late Tertiary—Middle Miocene: about 15 million years ago
  1. Shift in the location of the Mid-Atlantic Ridge north of Iceland to the west
  2. Change in the direction of seafloor spreading in the North Atlantic
  3. Beginning of spreading on the Chile Ridge
  4. Southern segment of the fossil Galapagos Rise jumped to the East Pacific Rise
  5. Restart seafloor spreading on SW Indian Ocean Ridge

Late Tertiary—Pliocene: about 2 to 4 million years ago
  1. Change in the direction of seafloor spreading on the Juan de Fuca and Gorda ridges
  2. Restart seafloor spreading in the Red Sea
  3. Restart seafloor spreading in the Gulf of California

(a)

**FIGURE 12.1** Proposed reconstruction of the continents showing paleoclimatic indicators: **(a)** Triassic. **(b)** Jurassic. **(c)** Cretaceous. (Considerably modified after Phillips and Forsyth, Ref. 1.)

(b)

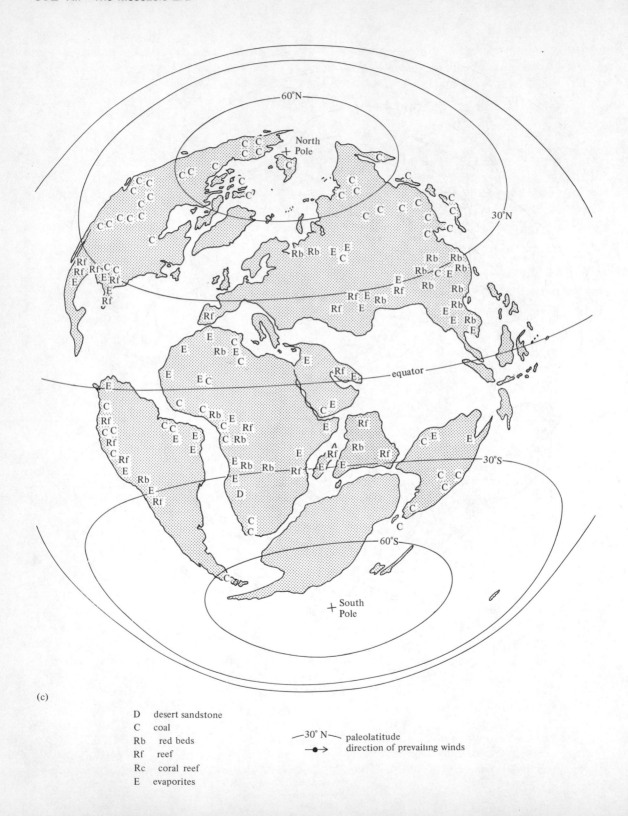

(c)

D  desert sandstone
C  coal
Rb  red beds
Rf  reef
Rc  coral reef
E  evaporites

—30° N— paleolatitude
•→ direction of prevailing winds

Australia sometime between the beginning of the Late Triassic and the end of the Early Jurassic. This conclusion was made on the basis of the following observations:

1. A series of fault basins bounded by normal faults developed along the east coast of North America and the northwest coast of Africa during the Late Triassic. The oldest sedimentary rocks in these basins are of upper Carnian age, about 202 million years old (**2**).
2. Mafic (basaltic) sills, dikes, and flows (flood basalts) were emplaced contemporaneously with the deposition of sedimentary rocks in the Triassic fault basins in North America, and basaltic dikes were intruded in Liberia and Morocco in Africa (**3**). Potassium-argon and argon 40-argon 39 dating of these igneous rocks indicate that they were probably emplaced between 192 and 202 million years ago (Late Triassic to Early Jurassic) (**4**).
3. Alkalic (rich in sodium and/or potassium) intrusive and extrusive igneous rocks were emplaced along a north-northwesterly trending zone near the New Hampshire-Vermont border during the Mesozoic (Fig. 12.2). Radiometric dates on these rocks indicate that the alkalic igneous activity began about 190 million years ago (**5**).
4. The oldest known sediments deposited on oceanic crust between Africa and North America are of Late Jurassic age (Callovian or Oxfordian stage), deposited between 151 and 162 million years ago. These sediments were recovered by the *Glomar Challenger* from a site just northeast of the Bahama Islands (**6**). Oxfordian deposits (about 155 million years old) were recovered from another site about 530 km (318 mi) from the continental margin east of North Carolina (**7**). Based on the distance between a nearby magnetic anomaly dated at 153 million years and another dated at 107 million

years, the rate of seafloor spreading between 153 and 107 million years ago was 1.1 cm/year. At this rate, 48 million years would have elapsed between the time of separation of the continents and the deposition of the 155 million-year-old deposits 530 km from the continental margin (*distance = rate × time*). This would place the time of the beginning of separation of North America and Africa at 203 million years ago.
5. The oldest known deep ocean deposits between North America and South America are Late Jurassic evaporites that were recovered by the *Glomar Challenger* from a site near the middle of the Gulf of Mexico (**8**). Evidently, North and South America began to separate well before the Late Jurassic.
6. The oldest magnetic anomaly in the Atlantic Ocean between North America and Africa (M–25) has been dated at about 153 million years (**7**). This anomaly is 535 km (321 mi) from the continental margin, and if the rate of seafloor spreading averaged 1.1 cm/year, North America and Africa would have begun to separate about 202 million years ago. However, the correlative anomaly on the other side of the Mid-Atlantic Ridge is significantly closer to the African continental margin (**9**) (Fig. 12.3). Its distance varies from 270 to 400 km (162 to 240 mi). If the rate of seafloor spreading averaged 1.1 cm/year, separation of North America and Africa would have begun between 178 and 193 million years ago. The reason for the discrepency between the eastern and western sides of the Atlantic is not clear, but it is possible that sometime between 153 and about 200 million years ago, the spreading ridge suddenly shifted several hundred kilometers to the east. Such a *ridge jump* occurred twice during the Tertiary, north of Iceland (see Chapter 13). In this case, the separation of North America and Africa would have begun about 194

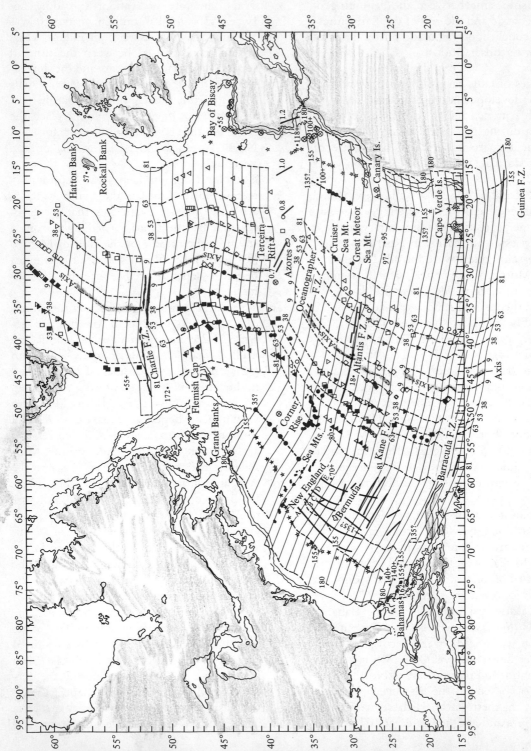

**FIGURE 12.3** Correlation of magnetic anomalies on either side of the Mid-Atlantic Ridge in the North Atlantic Ocean Basin. Numbers next to anomalies are the approximate ages in millions of years of those anomalies (From Pitman and Talwani, Ref. 9.)

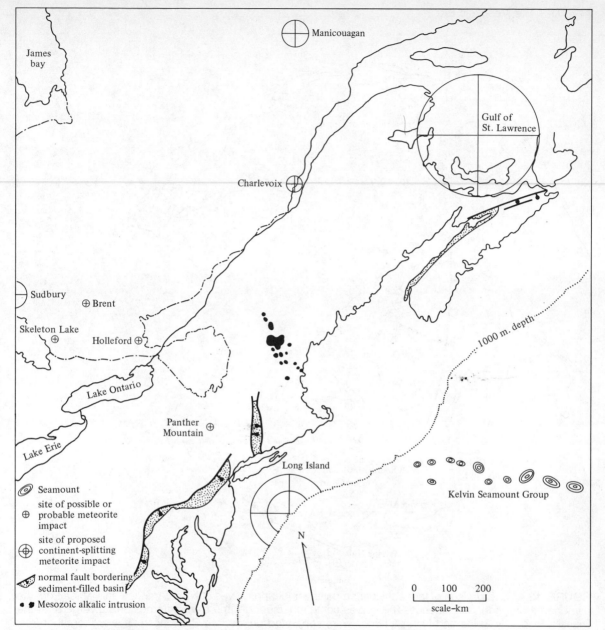

Manicouagan

James
bay

Gulf of
St. Lawrence

Charlevoix ⊕

Sudbury ⊕
⊕ Brent

Skeleton Lake
⊕
Holleford ⊕

1000 m. depth

Lake Ontario

Panther ⊕
Mountain

Long Island

Lake Erie

Kelvin Seamount Group

⊘ Seamount
⊕ site of possible or
   probable meteorite
   impact
⊕ site of proposed
   continent-splitting
   meteorite impact
▨ normal fault bordering
   sediment-filled basin
• ⦁ Mesozoic alkalic intrusion

N

0   100   200
scale–km

**FIGURE 12.2** Map showing the relationship between meteorite impact sites, alkalic intrusives, and normal faults.

million years ago (averaging the results from either side of the Atlantic).

7. The recalculated apparent polar wander curves for Africa and North America agree reasonably well for the Carboniferous through the Triassic, but they diverge thereafter (Fig. 12.4). This indicates that the two continents began to separate sometime between the Middle Triassic and Middle Jurassic (between 210 and 165 million years ago).

8. The oldest known miogeoclinal sedimentary rocks deposited between Africa and North America following separation of

Cambrian Af

Cambrian
NA

Carboniferous Af ⊕ / Ordovician NA
Silurian-Devonian NA
Carboniferous
NA

Permian Af ⊕

Permian NA

Cretaceous NA

Jurassic
NA

Triassic NA
Cretaceous Af ⊕⊕↗ ⊕ Triassic Af

Jurassic Af

**FIGURE 12.4** Comparison of the recalculated polar wandering curves of North America and Africa from the Cambrian through the Cretaceous. The curves come together near the Carboniferous and diverge after the Triassic. Evidently Africa and North America were separated prior to the Carboniferous, were together from the Carboniferous to the Triassic, and became separated after the mid-Triassic.

those continents are Early Jurassic evaporites (about 182 million years old), which were recovered by deep drilling on the Nova Scotia Shelf (**10**). Upper Jurassic sedimentary rocks have been recovered by deep drilling on the coast of Mauritania, Africa, and Late Jurassic (Oxfordian) limestones (about 155 million years old) have

been dredged from the base of the continental slope off northwest Africa (**11**).

9. The oldest known miogeoclinal sedimentary rocks deposited between North America and South America following their separation are evaporites of Jurassic or Late Triassic age which are found in salt domes in southern Louisiana (**12**). If the salt is of

Late Triassic age, separation of North America from Gondwanaland began between 192 and 205 million years ago.

10. In southern Europe, Early Jurassic (Pleinsbachian to Toarcian) deposits show evidence of having been deposited in progressively deepening water. John Dewey proposes that the separation of North America from Eurasia began during the Pleinsbachian (about 180 million years ago) and that the deepening water was due to plastic stretching and thinning of the crust during separation (**13**). However, continental separation in the Red Sea area has resulted in uplift, not subsidence, of the continental margins. It is possible that the increase in water depth was caused by the weight of an adjacent thick wedge of miogeoclinal deposits. Such a wedge might carry nearby sediments to great depths as the wedge sank. Therefore, separation of North America from Eurasia may have begun before (but not after) the Pleinsbachian.

Based on the evidence presented above, we favor the Late Triassic (about 202 million years ago) for the beginning of separation of North America from Gondwanaland. This date is in agreement with that suggested by E. D. Schneider and G. L. Johnson (**14**), but other geologists favor an early Jurassic age (about 180 million years ago) for the onset of separation of these continents (**9, 13**). The Late Triassic was also a time of increased orogenic activity in the circum-Pacific region, in contrast to the Early Jurassic, which was tectonically rather quiet. It is likely that the beginning of separation of the continents would result in folding, faulting, and intrusion on the leading edges of the continents.

## Separation of Antarctica, Australia, and India from Africa and Malagasy

Separation of a plate containing Antarctica, Australia, and India from a plate containing Africa and Malagasy (Madagascar) probably began sometime between the Late Triassic and Early Jurassic (205 to 160 million years ago).

1. Grabens, which are an indication of tension, developed on the western margin of southern Africa during the Late Triassic or Jurassic (**14**).
2. Extrusion of basaltic lavas, such as the Karroo Dolerites, Kaoko Basalts, and Stormberg Lavas, began in southern Africa about 200 million years ago (Late Triassic) (**15**).
3. The oldest known sediments deposited on oceanic crust between Africa and Antarctica are of Early Cretaceous (Aptian) age, about 109 million years old (**15**). The oldest known sediments deposited on oceanic crust between Africa and Australia are of Early Cretaceous (Barremian) age, about 116 million years old (**16**).
4. The oldest known magnetic anomaly between Australia and Africa is M-22, dated at 147 million years old (**17**).
5. The oldest known sediments deposited on continental crust following the separation of the continents are of Middle Jurassic (Oxfordian or Callovian) age, about 157 million years old (**18**), but nonmarine clastics on the southeast coast of Africa may be of Late Triassic age (**17**).
6. The last episode of compressive folding in southern Africa ended between 235 and 200 million years ago. Thus, continental separation must have begun before this time.

Considering the evidence presented above, Antarctica, Australia, and India probably began to separate from Africa during the Late Triassic, about 200 million years ago.

## Separation of South America from Africa

A number of lines of evidence indicate that the breakup of Gondwanaland occurred after its separation from North America and Eurasia. The separation of South America from Africa began sometime between the Late Jurassic and the earliest Cretaceous (between 130 and 150 million years ago).

1. Widespread extrusion and intrusion of mafic volcanic rocks occurred in Gondwanaland during the Jurassic. Radiometric dates on the Serra Geral Basalts from South America range from 161 to 111 million years (**19**). There appears to have been a peak in igneous activity about 147 million years ago (Late Jurassic, Kimmeridgian) (**20**).
2. Widespread alkalic igneous activity occurred in South America and Africa during the Late Jurassic and Early Cretaceous. The oldest radiometric dates on these rocks is 159 million years, and a peak of alkalic igneous activity began about 138 million years ago (latest Jurassic, Purbeckian) (**20**).
3. Intrusion of coastal diabases in South America began about 137 million years ago (latest Jurassic, Purbeckian) (**20**).
4. The oldest known sediments deposited on oceanic crust between South America and Africa are of Early Cretaceous (Aptian) age, about 112 million years old (**21**). A plot of the oldest sediments against their distance from the Mid-Atlantic Ridge indicates that the rate of seafloor spreading has been constant for the past 70 million years (Fig. 7.26b). If this rate has been constant since the beginning of separation of South America and Africa, the separation would have begun about 150 million years ago (Fig. 7.26b).
5. Magnetic anomalies present near the continental margin of Africa have been used to date the time of beginning of separation of South America and Africa. One estimate places their age at 165 million years (Early Jurassic) (**22**), while another places their age at 127 million years (Early Cretaceous) (**23**). The oldest magnetic anomaly in the South

Atlantic may be M-22 which has been dated at 147 million years (Late Jurassic, Kimmeridgian). This anomaly is located very close to the continental margin, and dates the beginning of separation of South America and Africa at about 147 million years ago.
6. The oldest known miogeoclinal sediments deposited between Africa and South America following the separation of these continents are of Late Jurassic age (**24**).
7. The recalculated apparent polar wander paths of South America and Africa begin to diverge from each other between the Late Triassic and the Early Cretaceous (**25**).

Although several geologists favor an Early Cretaceous age for the beginning of separation of South America and Africa (**22**, **26**), a Late Jurassic (Kimmeridgian) age seems more likely. This was a time of greatly increased tectonic activity in the circum-Pacific region. This tectonic activity, known as the Nevadan orogeny, may have been caused by a change in the rate and perhaps the direction of subduction of the seafloor under the Pacific borders of the continents when Africa and South America began to separate. There was little tectonic activity in the circum-Pacific region during the Early Cretaceous.

## Separation of Greenland from North America

The separation of a plate containing Europe and Greenland from a plate containing North America began sometime between the Late Jurassic and the Early Cretaceous.

1. The oldest sediments recovered from the deep ocean between Greenland and North America are of Eocene age. These deposits were cored by the *Glomar Challenger* from a site near the crest of an extinct spreading ridge located midway between North America and Greenland. Evidently, these

two continents began to separate well before the Eocene (which lasted from 53.5 to 37.5 million years ago).

2. Marine sediments of Late Cretaceous (Cenomanian) age, 94 to 100 million years old, are present on an isolated knoll north of Newfoundland which was probably once part of North America before sinking to great depths following separation of North America from Greenland (**27**). Thus, separation of Greenland from North America probably began more than 94 million years ago.

3. The oldest identified magnetic anomaly in the Labrador Sea between Greenland and North America is anomaly 24, dated at 60 million years old (early Paleocene) (**28**). This anomaly is 160 to 190 km (96 to 110 mi) from the continental margins of Greenland and North America. It is difficult to date the time of separation of Greenland and North America because there is little to indicate what the rate of seafloor spreading would have been in this area before 60 million years ago.

4. The oldest identified magnetic anomaly in the North Atlantic between North America and Europe is number 34, dated at 77 million years (Late Cretaceous, Santonian) (**29**). The distance between this anomaly and the North American continental margin is approximately equal to the distance between anomaly 31 and anomalies dated at about 142 million years not far to the south (Fig. 12.3). Therefore, the beginning of separation of North America and Europe (and consequently the beginning of separation of North America and Greenland) may have begun about 142 million years ago (Late Jurassic, Portlandian).

5. Marine and nonmarine sedimentary rocks of Early Cretaceous age are present on Disko Island in southwestern Greenland. The oldest known nonmarine deposits here are of early Cretaceous age (about 115 million

years old) (**30**). If these deposits were laid down after separation of North America and Greenland, the separation must have begun more than 115 million years ago.

6. Mafic dikes which approximately parallel the shoreline of southwestern Greenland have been radiometrically dated at 138 to 162 million years (Late Jurassic) (**31**). These dikes may have been emplaced during the initial separation of North America and Greenland.

All things considered, North America probably began to separate from Greenland and Europe about 147 million years ago (Late Jurassic, Kimmeridgian). However, it should be pointed out that some geologists favor a somewhat younger date (**32**).

### Rotation of Alaska

On the basis of a change in the direction of the Alaska Range in southern Alaska, S. Warren Carey proposed that Alaska rotated away from the rest of the continent of North America (**33**). This rotation began sometime between the Late Triassic and the Late Cretaceous, and it resulted in the formation of the Amerasian Basin between Alaska and Siberia.

1. Mafic (basaltic) dikes were emplaced in the Brooks Range of Alaska during the Early Jurassic and in the Sverdrup Basin in northern Canada during the Early Triassic (**34**). However, it is not certain that these dikes were associated with the beginning of the rotation of Alaska.

2. Sediments in the Sverdrup Basin were derived in part from a northerly landmass (Alaska?) during the Paleozoic and Triassic, but there is no indication of a northerly landmass after the Triassic (**34**). This suggests a post-Triassic age for the beginning of rotation of Alaska.

3. The oldest known sediments in the Amerasian Basin are of Late Cretaceous (Mae-

strichtian) age, more than 65 million years old (**35**). These deposits are 230 km (129 mi) from the continental margin, and therefore the rotation of Alaska began well before 65 million years ago.

The rotation of Alaska probably began during the Late Jurassic (Kimmeridgian), about 147 million years ago, at the same time as the beginning of separation of Greenland and North America (**36**). However, it may have begun during the Early Cretaceous (**36**). Paleomagnetic study of Mesozoic deposits in Alaska might provide a more definite time for the beginning of rotation.

## Rotation of Spain and Portugal

Paleomagnetic data and the fit of North America against Europe indicate that the northern border of Spain was in contact with the southwestern border of France before the breakup of Pangaea (**33**) (Fig. 12.1a). Spain and Portugal rotated about a spreading pole in northern France sometime between the Late Jurassic and the Late Cretaceous (**37**). This rotation resulted in the opening of the Bay of Biscay (Fig. 12.5).

1. Permian and Triassic paleomagnetic poles for Spain and Portugal do not agree with paleomagnetic poles of the same age from the rest of Europe (**37**). However, they agree rather well if the poles are rotated 35° in a counterclockwise direction about a pole in northern France (**37**). Late Cretaceous paleomagnetic poles from Spain agree with those from the rest of Europe (**31**). Therefore, rotation of Spain and Portugal took place between the Triassic and Late Cretaceous.
2. The oldest known sediments in the Bay of Biscay are of Late Cretaceous (Maestrichtian) age (**38**).
3. Normal faulting occurred during the Late Jurassic (Kimmeridgian) on the northern border of the Bay of Biscay, and this suggests that the initiation of the Bay of Biscay

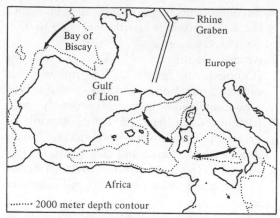

**FIGURE 12.5** Map of southwestern Europe showing the rotations that were suggested by S. W. Carey. (Redrawn from S. W. Carey, Ref. 33.)

occurred at about the same time (**39**). However, it is possible that the actual spreading in the Bay of Biscay began at a later date (**39**).

The rotation of the Iberian Peninsula may have begun during the Late Jurassic (Kimmeridgian) about 147 million years ago (**40**). It is likely that the rotation of Spain and Portugal was rather slow, as might be expected from the fact that this area was rather close to its spreading pole. Even though the distance traveled by Spain was only about 350 km (210 mi) at the most, the angular rotation about the spreading pole was rather large (35°). A typical rate of rotation during the Mesozoic and Cenozoic for continents bordering the Atlantic was about .5° per million years (**9**). At this rate, 70 million years would have elapsed between the beginning and end of rotation of the Iberian Peninsula. This would place the time of the end of rotation at about 65 million years (latest Cretaceous), which is consistent with the paleomagnetic data indicating that rotation ended by the Late Cretaceous.

Before the beginning of breakup of Pangaea, the boundary between Greenland and North America was aligned with the boundary between Spain and southwestern France

(Fig. 12.1a). It is possible that the rotation of the Iberian Peninsula and the opening of the Labrador Sea (between Greenland and North America) began at the same time.

Faulting occurred during the Late Jurassic (Kimmeridgian) in Scotland and uplift and erosion occurred in East Greenland during the Late Jurassic (**41**). Widespread uplift and erosion took place in Europe sometime between middle Kimmeridgian time and the early part of the Cretaceous. All of this faulting, uplift, and erosion may have been related to the beginning of separation of Greenland and North America and to the beginning of rotation of the Iberian Peninsula.

Rockall Bank is a largely submerged continental block located west of Britain, and it is thought that rifting occurred between this block and Britain during the Late Jurassic or Early Cretaceous (**41**). The rift may have been an aulacogen formed during the separation of Greenland from North America and the rotation of the Iberian Peninsula. In this case there would have been a triple junction (presumably underlain by a mantle plume) at the southern end of this aulacogen where it joined a spreading ridge (**42**).

## Separation of Greenland from Eurasia

The separation of Greenland from Eurasia began sometime between the Late Cretaceous and the latest Paleocene.

1. Between 67 and 50 million years ago, there was a great deal of igneous activity in the North Atlantic region (**43**). This activity included eruption of flood basalts and intrusion of mafic dikes. Although much of this activity occurred during the latest Paleocene, it began during the Late Cretaceous or Early Paleocene.
2. The oldest known sediments from the deep ocean between Greenland and Europe are of Early Eocene age (49 to 53.5 million years old). Deposits of this age have been recovered from two sites, the Loften Basin and the Voring Plateau (**44**). The site in the Loften Basin is about 130 km from the continental margin. The average rate of seafloor spreading in the North Atlantic between Greenland and Norway has been about 1.0 cm/year. At this rate, the separation of Greenland and Europe began between 62 and 66.5 million years ago.
3. The oldest identified magnetic anomaly in the North Atlantic between Greenland and Europe is 24 anomaly dated at about 60 million years old (**26**). This anomaly is 35 to 120 km from the edge of the continent, and at a rate of seafloor spreading of 1.0 cm/year, separation of Greenland and Eurasia began between 63.5 and 72 million years ago.
4. The oldest miogeoclinal sedimentary rocks on the Atlantic border of Greenland or Europe are of latest Paleocene age (**45, 27**). This sets a minimum date on the separation of Greenland and Europe of 53.5 million years.

From the above, it is evident that the beginning of separation of Greenland and Eurasia probably began during the Late Cretaceous, perhaps at the beginning of the Maestrichtian about 70 million years ago. This age is close to that estimated by Manik Talwani and Olav Eldholm of 60 to 63 million years for the beginning of separation of these continents (**45**). It is interesting to note that the Larimide orogeny also began at the beginning of the Maestrichtian (**46**). This orogeny may have been caused by a change in the rate and/or direction of seafloor spreading as Greenland and North America began to separate.

## Separation of India from Australia and Antarctica

India began to separate from a plate which included Australia and Antarctica during the Early Cretaceous or Late Jurassic, sometime between 125 and 155 million years ago.

1. Widespread intrusion and extrusion of mafic (basaltic) igneous rocks occurred in Australia and India during the Jurassic. This activity includes the intrusion of the Tasmanian dolerites in southern Australia (radiometrically dated at 143 to 167 million years old) (**39**) and the Ferrar dolerites of Antarctica (radiometrically dated at 147 to 170 million years) (**47**). Potassium-argon dates on sanadine (a potassium feldspar) average about 150 million years, while potassium-argon dates on the Ferrar dolerites have a peak at about 151 million years (**47**). Furthermore, a whole-rock rubidium-strontium isochron date of 151 million years was obtained in the Ferrar dolerites (**47**). These dates suggest a Late Jurassic (Kimmeridgian) age for the igneous activity.

2. The oldest known sediments recovered from the deep ocean floor between India and Australia are of Early Cretaceous (Barremian) age, about 116 million years old (**16**).

3. The oldest identified magnetic anomaly in the Indian Ocean is M-22 dated at 147 million years (**17**). If this identification is correct, the beginning of separation of India from Australia and Antarctica must have begun during or before the Late Jurassic (Kimmeridgian).

4. The oldest miogeoclinal deposits along those coastlines of India and Antarctica which were in contact before they separated are Cenomanian (Cretaceous) sedimentary rocks on the southeastern margin of India (**40**).

5. In Africa and India, the oldest marine miogeoclinal deposits on the new coastlines created when these two landmasses separated are sedimentary rocks of Early Cretaceous (Neocomian) age (about 130 million years old) (**40**).

6. Graben formation off the south coast of western Australia may have begun during the Late Jurassic (**48**). This graben may have been an aulacogen connected with the formation of a plume at the southwestern corner of Australia. This plume may have been instrumental in the initiation of the separation of India from Australia and Africa.

India probably began to separate from Australia and Antarctica during the Late Jurassic (Kimmeridgian), about 147 million years ago. As discussed earlier, this was the time of widespread orogenic activity in the circum-Pacific area. However, some geologists favor a time of the beginning of separation of India from Australia at between 120 and 135 million years ago (**49**, **17**). The younger dates are based on what may be misidentified anomalies.

## Separation of Malagasy (Madagascar) from Africa

Paleomagnetic data indicate that during the interval from the late Carboniferous to the Middle Jurassic, Malagasy (Madagascar) was in contact with the east coast of Africa adjacent to Kenya and Tanzania (Fig. 12.1a) (**50**). Deep-sea drilling by the *Glomar Challenger* indicates that Malagasy was not in that position during the Late Cretaceous (Turonian) about 90 million years ago (**50**). Therefore, separation of Malagasy and Africa must have begun between the Middle Jurassic, 172 million years ago, and the Late Cretaceous (Turonian), about 90 million years ago. Igneous activity began along the central section of the west coast of Africa (adjacent to the position thought to have been occupied by Malagasy before breakup of the continents) at about the Jurassic/Cretaceous boundary (**51**). This activity may have begun at about the same time as the beginning of separation of Malagasy and Africa. The separation may have begun during the Late Jurassic (Kimmeridgian) when India began to separate from Australia and Antarctica.

## Separation of India from the Seychelles Islands

The Seychelles Islands are a group of islands located in the western part of the Indian Ocean between India and Malagasy. They are the smallest islands which are underlain by continental crust of Precambrian age. Separation of India from the Seychelles began sometime between the beginning of the Late Cretaceous and the early Tertiary (early Paleocene) between 61 and 100 million years ago.

1. Widespread intrusion of mafic dikes and extrusion of mafic flood basalts occurred in India and the Seychelles during the early Tertiary between 60 and 65 million years ago (**52**). These deposits are about 300 km (180 mi) from the continental margin of India.
2. The oldest known sediments deposited on oceanic crust between India and the Seychelles are of early Paleocene age (**53**), deposited between about 61 and 65 million years ago.
3. The oldest known magnetic anomaly in the Indian Ocean between India and the Seychelles is M-32, which has been dated at about 70 million years (**52**).

From the evidence presented above, it is likely that India and the Seychelles began to separate about 70 million years ago during the Late Cretaceous (Maestrichtian).

## Cause of Breakup of the Continents

The separation of the continents probably occurred in five stages (see Table 12.1) which are summarized below.

1. Late Triassic, Karnian: about 202 million years ago
2. Late Jurassic, Kimmeridgian: about 147 million years ago
3. Late Cretaceous, Maestrichtian: about 70 million years ago
4. Early Tertiary, latest Paleocene: about 54 million years ago
5. Late Tertiary, middle Miocene: about 15 million years ago

Changes in the direction of seafloor spreading occurred in eight stages (see Table 12.1) which are summarized below.

1. Late Jurassic, Kimmeridgian: about 147 million years ago
2. Middle Cretaceous, Albian/Cenomanian boundary: about 100 million years ago
3. Late Cretaceous, Maestrichtian: about 70 million years ago
4. Early Tertiary, latest Paleocene: about 53 million years ago
5. Early Tertiary, Late Eocene: about 42 million years ago
6. Middle Tertiary, Oligocene/Miocene boundary: about 26 million years ago
7. Late Tertiary, Middle Miocene: about 15 million years ago
8. Late Tertiary, Pliocene: about 2 to 4 million years ago

Five of the eight times of change in the direction of seafloor spreading also correspond to times when the continents began to separate. Dates of increased tectonic activity (folding, faulting, uplift, etc.) during the Mesozoic and Cenozoic are summarized in Table 12.2. All of the dates of the beginning of separation of the continents and all of the dates of changes in the direction of seafloor spreading correspond to times of increased tectonic activity. It is likely that the reason for this is that all of these activities are caused by changes in the pattern of convection within the earth.

In Chapter 6, we suggested that the impact of large meteorites might cause mantle plumes and that these plumes might break apart plates and cause continents to separate. The principal reason for suggesting this idea is that there appears to be a correspondence between the times of impact of large meteorites and both the beginning of separation of continents and changes in the direction of seafloor spreading.

Manicouagan, a 65-km (39-mi) diameter circular structure located in southeastern Quebec (Fig. 12.6), contains most of the features associ-

TABLE 12.2  **Mesozoic and Cenozoic Tectonic Activity and Correlative Impact Epochs**

| OROGENY | DATE | TECTONIC ACTIVITY | CORRELATIVE IMPACT EPOCH |
|---|---|---|---|
| #1 Cascadian Orogeny | Pleistocene to Present<br><br>0–2 Million Years Ago | 1. Begin second stage of opening of Gulf of Southern California<br>2. Begin second stage of opening of Red Sea<br>3. Change in the direction of seafloor spreading on the Juan de Fuca and Gorda ridges<br>4. Folding and faulting in California, Oregon, Washington, and Japan<br>5. Increase in volcanism in Indonesia, Japan, Cascades, and Central America<br>6. Widespread unconformity in western North America<br>7. Uplift and faulting in Kenya<br>8. Increase in sedimentation rate in deep oceans<br>9. Alkalic volcanism at Lucite Hills and Pleasant Valley | Impact Epoch #1 |
| #3 Andean Orogeny | Middle Miocene<br><br>About 15 Million Years Ago | 1. Begin separation of Arabia from Africa<br>2. Begin separation of Baja California from Mexico<br>3. Begin separation of Sardenia and Carsica from Europe<br>4. Begin spreading on the Chile Ridge<br>5. Westward shift of Mid-Atlantic Ridge north of Iceland<br>6. Southern segment of fossil Galapagos Rise jumped to the East Pacific Rise<br>7. Begin spreading on Chile Ridge<br>8. Restart spreading on SW Indian Ocean Ridge<br>9. Intense folding and faulting in the Andes Mountains<br>10. Folding in Washington, Oregon, and Spitsbergen<br>11. Increase in volcanism Oregon, Great Basin, New Mexico, Central America, Cascades, Azores, Canary Islands, Hawaiian Ridge, southern California, and Iceland<br>12. Flood basalts in Columbia River Plateau<br>13. Increase in explosive volcanism near Marianas Islands<br>14. Intrusions of granitic rocks in Great Basin and Japan<br>15. Thrust faulting in California<br>16. Uplift and erosion in North Carolina and Virginia in the Coastal Plain<br>17. Begin uplift and normal faulting in Basin-Range Province<br>18. Intrusion, faulting, and volcanism in Aleutians<br>19. Metamorphism in the Alps | Impact Epoch #2 |
| #7 Unnamed Orogeny | Middle Late Eocene | 1. Change in the direction of seafloor spreading in the Pacific Ocean<br>2. Westward shift in the Mid-Atlantic Ridge north of Iceland | Impact Epoch #3 |

TABLE 12.2   (*Continued*)

| OROGENY | DATE | TECTONIC ACTIVITY | CORRELATIVE IMPACT EPOCH |
|---|---|---|---|
| | About 42 Million Years Ago | 3. Restart spreading on SW Indian Ocean Ridge<br>4. Seafloor spreading ends in Labrador Sea<br>5. Intrusion of granitic rocks in Black Hills and Idaho<br>6. Uplift and erosion in western North America<br>7. Uplift and faulting in Kenya | |
| #10 Larimide Orogeny | Between Maestrichtian and Campanian<br><br>About 70 Million Years Ago | 1. Beginning of separation of Greenland from Europe<br>2. Change in the direction of seafloor spreading in the South Atlantic<br>3. Change in the spreading rate in the North Atlantic<br>4. Increase in volcanism in Emperor Seamount Chain, Marshal-Ellice Island Chain, and southern Alaska<br>5. Increase in intrusion of granitic rocks central California, southern Alaska, Oregon, Washington, Idaho, and St. Mathews Island<br>6. Increase in intrusion of alkalic igneous rocks in Africa and South America<br>7. Folding and Faulting in Rocky Mountains, Andes Mountains, and Oman Mountains, Saudi Arabia<br>8. Uplift and erosion on the Gulf Coastal Plain and in the Sacramento Valley of California | Impact Epoch #4 |
| #12 Santa Lucian Orogeny | Between Albian and Cenomanian<br><br>About 100 Million Years Ago | 1. Uplift and erosion in southern Oregon, Sevier Orogenic Belt (Utah), Gulf and Atlantic Coastal Plains, central California<br>2. Increase in intrusion of granitic rocks in Baja, California, Sierra Nevada Mountains, Boulder Batholith, Japan, Oregon, Washington, and southeastern China<br>3. Increase in intrusion of alkalic rocks in southwestern Arkansas and southeastern Canada<br>4. Increase in volcanism in Andes Mountains (Peru) | Impact Epoch #5 |
| #15 Unnamed Orogeny | Between Cretaceous and Jurassic<br><br>About 136 Million Years Ago | 1. Uplift and erosion in Sevier Orogenic Belt (Utah) and in the Gulf Coastal Plain<br>2. Increase in intrusion of granitic rocks in California | Impact Epoch #6 |
| #16 Nevadan Orogeny | During the Kimmeridgian<br><br>About 159 (?) Million Years Ago | 1. Beginning of separation of India from Australia and Antarctica<br>2. Beginning of separation of India from Africa<br>3. Beginning of separation of Madagaskar from Africa<br>4. Beginning of separation of Africa from South America | Impact Epoch #7 |

(*Table continued*)

TABLE 12.2    (*Continued*)

| OROGENY DATE | TECTONIC ACTIVITY | CORRELATIVE IMPACT EPOCH |
|---|---|---|
| | 5. Possible beginning of separation of Greenland from North America, Spain and Portugal from Europe, and Alaska from northern Canada<br>6. Intense folding in Cordilleran Geosyncline and in the Andes Mountains<br>7. Metamorphism in Yugoslavia and in the Cordilleran Geosyncline<br>8. Increase in intrusion of granitic rocks in California, Oregon, Washington, Idaho, and Alaska<br>9. Increase in intrusion of alkalic rocks in central New England<br>10. Uplift and erosion in the western interior of North America<br>11. Increase in volcanism in South America, Africa, and Antarctica | |
| #17 Unnamed Orogeny<br><br>During Sinemurian<br><br>About 185 Million Years Ago | 1. Beginning of intrusion of alkalic igneous rocks in New England<br>2. Folding in Sierra Nevada Mountains and southwest Nevada<br>3. Increase in volcanism in southern Africa | Impact Epoch #8 |
| #18 Palisades Disturbance<br><br>Between Middle and Late Triassic<br><br>About 205 Million Years Ago | 1. Beginning of separation of Africa from North America<br>2. Intrusion of Palisades Sill, New Jersey (?)<br>3. Beginning of normal faulting in New Jersey, North Carolina, and central Connecticut<br>4. Increase in intrusion of granitic rocks in western North America<br>5. Uplift and erosion in southwestern Nevada and elsewhere in the western interior of North America, and in the Sierra Nevadas and Canadian Rockies | Impact Epoch #9 |

ated with intense shock such as would be produced by the impact of a large meteorite (see Chapter 4). Most geologists who have studied Manicouagan have concluded that it was formed by the impact of a large meteorite (**54**). Igneous rocks in the structure are thought to have been produced by the melting of surface rocks during meteorite impact. Whole-rock potassium-argon dates on these rocks average about 203 million years (Late Triassic, Karnian) and paleomagnetic measurements on these

rocks also indicate that they were formed during the Late Triassic (**55**).

There are several other probable impact structures (astroblems) which were also probably formed during the Late Triassic (Table 12.3). Manicouagan and these smaller impact structures may have been produced during a period of greatly increased rate of impacts of large meteorites. There have been a number of such times in the past (Table 12.3). Times of increased rate of meteorite impacts

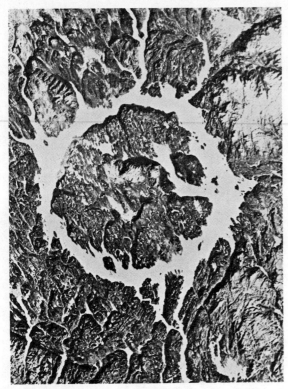

**FIGURE 12.6** Satellite view of Manicouagan, a probable meteorite impact structure in southern Quebec. The lake (covered with snow) is about 65 km in diameter. (From NASA.)

will be referred to as *impact epochs*. Nine impact epochs have been identified in the last 210 million years.

Epoch #1: 0–2.5 million years ago
Epoch #2: 15 million years ago
Epoch #3: about 42 million years ago
Epoch #4: about 70 million years ago
Epoch #5: about 100 million years ago
Epoch #6: about 135 million years ago
Epoch #7: about 160 million years ago
Epoch #8: about 180 million years ago
Epoch #9: about 205 million years ago

Five of the nine impact epochs correspond to times during which continents began to separate. Only the latest Paleocene (54 million years ago) time of separation does not correspond to an impact epoch. All but one of the impact epochs correspond to a time of change in the direction of seafloor spreading and this one time (Late Triassic, 203 million years ago) was a time of the beginning of seafloor spreading in the Atlantic Ocean. Six of the eight times of changes in the direction of seafloor spreading correspond to impact epochs. Therefore it is evident that there is a close relationship between times of the beginning of separation of the continents, changes in the direction of seafloor spreading, increased tectonic activity, and the impact of large meteorites. It is possible that the impact of large meteorites acts as a trigger for all of these other activities.

The impact structures which we see are not the ones which actually caused the separation of continents or changes in the direction of seafloor spreading. Impacts which cause continents to separate (*continent-splitting impacts*) are caused by even larger meteorites, or perhaps meteorites comparable in size to the ones producing the largest observed impact structures, but which hit in areas with high heat flow (such as a recently active orogenic belt). It would seem that plumes would be more easily developed in an area with a high heat flow than in an area of average or below-average heat flow. Craters formed by continent-splitting impacts would not be easily observed because they would be covered by thick sediments in miogeoclines on the margins of the continents.

One of the proposed sites of a continent-splitting impact is the Atlantic margin of North America southeast of New York City (Fig. 12.2). A chain of seamounts terminates here (the Kelvin Seamount Group) and these are thought to be the trace of a hot spot generated by a mantle plume (**42**). The Long Island four-arm junction (with one more arm than a triple junction) is thought to be the site of a mantle plume during the Late Triassic (**42**). It has been proposed that this area was the site of a continent-splitting impact during the Late Tri-

TABLE 12.3 **Impact Epochs and Associated Craters**

| IMPACT EPOCH | STRUCTURE | AGE MILLION YEARS | DIAMETER KM |
|---|---|---|---|
| #1—2.5 Million Years to Present | 1. Barringer Crater | 0.01–0.04 | 1.2 |
| | 2. Sithlemenkat | 0.02 | 12.4 |
| | 3. Lonar Crater | 0.05 | 1.8 |
| | 4. Kofels | 0.01–0.10 | 5.0 |
| | 5. Wilkes Land (?) 70S 120°W | 0.69 *Antarctica* | 240. |
| | 6. Ashanti | 1.0 ± .2 | 10.5 |
| | 7. Talemzane | 1.0 | 1.8 |
| | 8. Pretoria Salt Pan 26S 28E | 1.0 | 1.1 |
| | 9. New Quebec 62N 75W Que | 2.0 ± 1.0 | 3.0 |
| | 10. Tenoumer | 2.5 ± 0.5 | 1.8 |
| #2—About 15 Million Years Ago | 1. Reis | 14.8 | 22. |
| | 2. Steinheim | 15.1 | 3.5 |
| | 3. Pfahldorf | 15.0 (?) | 2.5 |
| | 4. Sornhul | 15.0 (?) | 1.5 |
| | 5. Mandelgrund | 15.0 (?) | 1.0 |
| | 6. Mendord | 15.0 (?) | 2.5 |
| | 7. Willenhofen (3) | 15.0 (?) | 2.0 |
| | 8. Hemauer Pulk (14) | 15.0 (?) | 2.0 |
| | 9. Wipfelsfurt | 15.0 (?) | 0.9 |
| | 10. Sausthal (2) | 15.0 (?) | 1.0 |
| | 11. Haughton | 15.0 (?) | 17. |
| #3—About 42 Million Years Ago | 1. Lake Mistastin 51N 74W Que | 35–40 | 20 |
| | 2. Lake Wanapitei | 37 ± 2 | 10 |
| | 3. Popigay | 40–45 | 100 |
| | 4. El'gytkhyn | 40 | 23 |
| #4—About 70 Million Years Ago | 1. Kamensk 56N 62E | 65 | 25 |
| | 2. Gusev 55N 22½E LITHUANIA | 65 | 3 |
| | 3. Boltyshsk | 70 (?) | 25 |
| | 4. Minsk 54N 28E | 70 | (?) |
| | 5. Eagle Butte | 74–78 | 10 |
| | 6. Dumas | 74–78 | 2 |
| #5—About 100 Million Years Ago | 1. Mien | 92 ± 6 | 6 |
| | 2. Steen River | 95 ± 7 | 13.5 |
| #6—About 135 Million Years Ago | 1. Gosses Bluff | 130 ± 6 | 50 |
| | 2. Bolytsh | 140 (?) | 25 |
| #7—About 160 Million Years Ago | 1. Rochechart | 154–173 | 15 |
| #8—About 180 Million Years Ago | 1. Puchezh-Katunksk 57N 43E | 183 ± 3 *N.Gorkiy* | 80 |
| #9—About 205 Million Years Ago | 1. Manicouagan 51N 68W Que | 202–210 | 65 |
| | 2. Lake St. Martin 45.5N 66W N-B | 225 ± 40 *N.St John* | 24 |
| | 3. Red Wing Creek | Triassic (?) | 8 |
| | 4. Hartney | Triassic (?) | 7 |
| | 5. Viewfield | Triassic (?) | 2 |

TABLE 12.3 (*Continued*)

| IMPACT EPOCH | STRUCTURE | AGE MILLION YEARS | DIAMETER KM |
|---|---|---|---|
| #10—About 360 Million Years Ago | 1. Flynn Creek<br>2. Kaluga  *54N 36E  SW of morgan*<br>3. Charlevoix  *45N 85W Michigan*<br>4. Lake La Moinerie | Late Dev.<br>L–M Dev.<br>372–321<br>350 ± 50 | 3.2<br>15<br>37<br>11 |
| #11—About 450 Million Years Ago | 1. Brent  *46N 79W Ont*<br>2. Carswell<br>3. Skeleton Lake | 450 ± 30<br>485 ± 50<br>450 ± 30 | 4<br>38<br>3.5 |

assic (**56**). Two aulacogens, the Newark Basin and the Connecticut Valley Basin, and swarms of diabase dikes radiate away from the site of the proposed impact. It is interesting to note that the Connecticut Valley Basin heads almost due north directly toward Manicouagan (Fig. 12.2). Thus there is not only a temporal but a spatial connection as well between a mantle plume and a probable impact structure.

In addition to the continent-splitting impact near Long Island, there may have been continent-splitting impacts during the Late Triassic at several other sites along the Atlantic margin (Fig. 12.7). One of these sites, near the Bahama Islands, is a center of radiation of several very large swarms of diabase dikes (**57**) (Fig. 12.7). Furthermore, Dale Krause has proposed that Freetown, in Sierra Leone, Africa, was the site of the impact of a large meteorite (**58**). Sierra Leone was directly opposite the Bahama Islands prior to separation of Africa and North America.

Other possible sites of continent-splitting impacts are Disko Island in southwestern Greenland (during the Late Jurassic), at the ends of the aseismic ridge on which Iceland is located (during the Late Cretaceous), at the ends of the Rio Grande and Walvis ridges in the South Atlantic (during the Late Jurassic), at the south end of the Red Sea (during the middle Miocene), and the Gulf of Lion in southern France (during the middle Miocene). Some of these proposed impact sites will be discussed later in this chapter and in Chapter 13.

A note of caution should be added here. The idea that the impact of large meteorites caused breakup of the continents is a *hypothesis,* and it should not be regarded as an established fact. It is an idea that must be further tested before it can even be considered a theory. In part, this testing might involve radiometric dating of meteorite impact structures and the comparison of these dates with the times of the beginning of continental separations.

## NORTH AMERICA

During the Early and Middle Triassic, North America was still part of Pangaea. The only areas that received thick marine deposits during this interval were the Cordilleran and Chukota geosynclines and the Brooks and Sverdrup basins (Fig. 12.8). During the Late Triassic, a series of fault basins developed along the east coast of North America. Subsequently, the East Coast and the Gulf Coast geoclines formed along the Atlantic and Gulf margins of North America (Fig. 12.8). In the Cordilleran Geosyncline, an episode of folding during the Late Jurassic resulted in the development of a new depositional trough, the Coast Range Geosyncline.

### Triassic Fault Basins

Deposits of Early and Middle Triassic age are not known on the east coast of North America. However, up to 8000 m (25,000 ft.) of Late Triassic sedimentary and volcanic rocks were

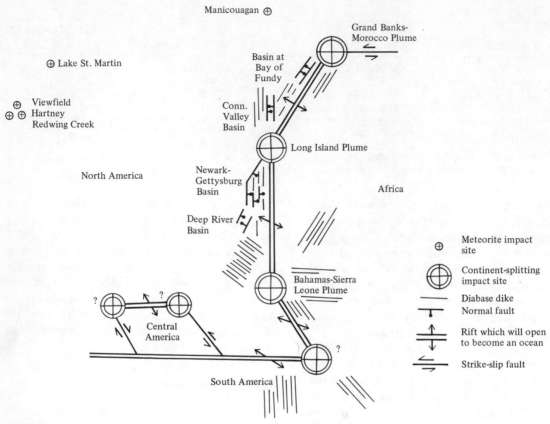

**FIGURE 12.7** Proposed sites of continent-splitting impacts during or soon after the Late Triassic, before breakup of the continents.

deposited in fault basins paralleling the Atlantic margin of North America (Fig. 12.2). The basins are bounded by normal faults with large vertical displacements.

The Newark Basin extends from southeastern New York through northern New Jersey to southeastern Pennsylvania (Fig. 12.2). The Newark Group, which was deposited in this basin, includes arkosic red beds, black shales, and volcanics. Sandstones and shales grade westward into conglomerates derived from Precambrian igneous and metamorphic rocks to the west (Fig. 12.9). The black shales of the Newark Group contain fossil branchiopods. These small bivalved crustaceans resemble

modern forms found in freshwater lakes. Several varieties of freshwater fish are also present in these deposits. This evidence, along with the lack of marine fossils, indicates that the black shales were probably deposited in a freshwater lake. The red clastics, which contain abundant dinosaur tracks, mud cracks, and raindrop imprints, were presumably deposited on flood plains and subaerial deltas of the streams and rivers that drained into such lakes. Their red color and arkosic composition suggests that they were laid down in an arid region (Fig. 12.1a).

Three basalt flows in the Newark Group form the prominent ridges of the Watchung

**FIGURE 12.8** Geosynclines bordering North America during the late Mesozoic.

**FIGURE 12.9** Late Triassic conglomerate of the Newark Series in southeastern New York. (Courtesy of Charles Regan.)

Mountains in central New Jersey. Diabase intrusives in the Newark Group include the Palisades Sill in New Jersey and New York and the Gettysburg Sill of Pennsylvania. The Palisades Sill is up to 300 m (1000 ft) thick and forms a high cliff along the west bank of the Hudson River opposite Manhattan Island (Fig. 12.10). Potassium-argon dates indicate that this sill was intruded during the Late Triassic, between 202 and 190 million years ago (**4**).

The Connecticut Valley Basin contains a sequence quite similar to that of the Newark Basin, that is, red clastics and black shales interbedded with three volcanic flows and intruded by a diabase sill. However, the Connecticut Valley Basin is bordered on the east by a normal fault, whereas the deposits of the Newark Basin are bordered on the west by a normal fault. The deposits of the Connecticut Valley Basin dip 5° to 15° to the east, and those of the Newark

Basin dip 5° to 15° to the west. Several authors have suggested that the Newark and Connecticut Valley basins were originally connected and later separated as a result of uplift and erosion of the deposits between the basins. However, paleomagnetic data indicate that the lava flows in the Connecticut Valley Basin do not correlate with those of the Newark Basin. Moreover, the dip of cross-bedding and potassium-argon ages of clastics in the basins suggest that some sediment was transported into the Newark Basin from the east and into the Connecticut Valley Basin from the west (**58**). The general trend of the Newark Basin is northeasterly, while the Connecticut Valley Basin trends north-south (Fig. 12.5).

The tilting of the deposits within the basins may have been caused by differential subsidence during deposition. Since most of the sediments were derived from uplifted blocks bordering

**FIGURE 12.10** The Palisades along the Hudson River, west of New York City, is part of a concordant diabase intrusion, the Palisades Sill. The top of the sill was removed by erosion prior to the Cretaceous period. Notice the prominent columnar jointing. (Photo by Horace Gilmore, courtesy of the Palisades Interstate Park Commission.)

the basins, the deposits tend to thicken toward these blocks. Therefore, the maximum subsidence occurred along these margins. The youngest beds have a lower angle of dip than the older beds in the Newark Basin, perhaps owing to a smaller amount of subsidence of the younger beds (Fig. 12.11).

Other Triassic fault basins are located in North Carolina, Virginia, and Nova Scotia, and subsurface fault basins have been discovered in northern Florida, southeastern Alabama, and southwestern Georgia (Fig. 12.2).

The faulting and volcanism associated with the Triassic basins probably resulted from stretching of the earth's crust along the entire east coast of North America. While the east coast was being pulled apart, the west coast was being compressed and folded as an oceanic plate

moved beneath the continent. Both conditions were probably related to the separation of plates. Burke and Dewey have proposed that a mantle plume was located at the junction of the Newark and Connecticut Valley basins, which they believe are aulacogens (**42**). This may be true of other Triassic fault basins of eastern North America.

### Gulf Coast Geocline

The Gulf Coast Geocline borders the Gulf of Mexico and joins the southern end of the East Coast Geocline in Florida (Fig. 12.8). Neither geocline contains significant volcanic rocks, nor has either one been affected by compressional deformation. The lack of volcanic and tectonic

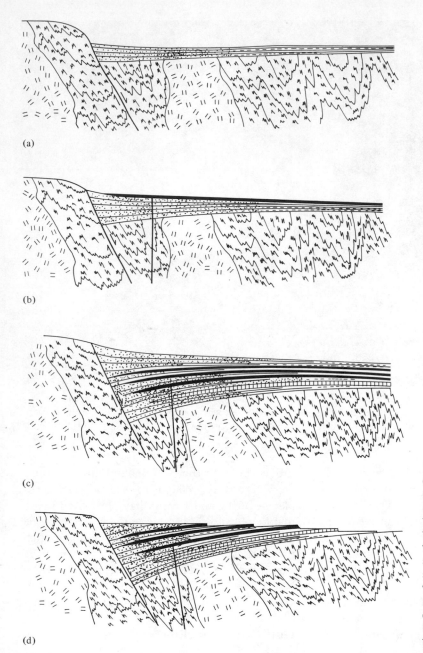

(a)

(b)

(c)

(d)

**FIGURE 12.11** Suggested sequence of events during the formation of the Newark Basin: **(a)** Normal faulting with deposition of clastic sediments within the basin. **(b)** Continued faulting with extrusion of lavas and deposition of clastic sediments (subsidence is greatest near the fault, which results in a tilting of the beds during deposition. **(c)** End of movement on the fault. **(d)** Uplift and erosion of the deposits of the basin.

activity indicates that an oceanic plate has not moved under these geosynclines since their inception.

A maximum of 20,000 m (60,000 ft) of Mesozoic and Cenozoic sediment has been depos-

ited in the Gulf Coast Geocline on a basement of pre-Mesozoic rocks. In addition, as much as 10,000 m (30,000 ft) of sediment has been deposited on oceanic crust in the deeper parts of the Gulf of Mexico. The oldest known sedi-

TABLE 12.4  **Generalized Late Mesozoic Stratigraphy of the Texas-Gulf Coast and the Mid-East Coast Geoclines.**

| | TEXAS-GULF COAST | MID-EAST COAST |
|---|---|---|
| **Upper Cretaceous** | Navarro Group<br>Taylor Marl<br>Austin Chalk<br>Eagle Ford Shale | Monmouth Group<br>Matawan Group<br>Magothy Group<br><br>Raritan Formation |
| **Lower Cretaceous** | Washita Group<br>Fredericksburg Group<br>Trinity Group | Patapsco Formation<br><br>Arundel Formation<br>Patuxtent Formation |
| **Upper Jurassic** | Cotton Valley Group<br>Smackover Formation<br>Eagle Mills Formation | |

ments in the geocline are salt deposits between 600 and 1300 m (2000 and 5000 ft) thick. These deposits have been penetrated in deep wells in the northern part of the geosyncline, and numerous salt domes rise from them in the central and southern parts of the geocline. The salt bed may be as old as Late Triassic (**12**). Since the Gulf of Mexico probably formed after the separation of South America and Africa from North America, deposition of salt must also have postdated the separation.

In the northern part of the geocline, Middle Jurassic salt beds are overlain by a sequence of limestone, sandstone, shale, and salt of Late Jurassic age. Cretaceous clastics in this region grade southward into chalks and reef-bearing limestones.

### East Coast Geocline

The East Coast Geocline, which extends along the continental margin from Newfoundland to Florida, received thousands of feet of sediments during the Mesozoic. Seismic profiling has revealed two wedge-shaped sedimentary bodies, one beneath the continental shelf and the other at the base of the continental slope (**60**) (Fig. 6.16). The inner miogeocline contains a maxi-

mum of 5200 m (17,000 ft) of Mesozoic and Cenozoic sediments; the outer eugeocline has up to 6000 m (20,000 ft) of sediments.

The oldest sediments in the East Coast Geocline are from the Early Cretaceous (Table 12.4). However, a deep well at Cape Hatteras, North Carolina, penetrated several hundred feet of red beds below Early Cretaceous deposits. These red beds may be of Jurassic age (**11**). A well in the Bahama Islands went through 5000 m (16,000 ft) of limestone, the oldest of which was of Early Cretaceous age (**61**). The well did not reach the basement, and older sedimentary rocks may lie beneath the limestone.

Cretaceous deposits are generally limestone or chalk in the south, limey shales in the central parts of the geocline, and unconsolidated sands and clays in New Jersey and New York. This may be due to the northward decrease in seawater temperature, since warmer waters favor precipitation of calcium carbonate.

During the Cretaceous, the seas began a gradual regression from the geocline. Spore and pollen zonation of Cretaceous sediments indicates that the coastal plain contains a number of interfingering deltaic wedges, and that each can be related to a specific river system (**62**). The sediments in the oldest of these deltas were deposited by the Ancestral Potomac drainage during the Early Cretaceous. The sediments deposited by the Ancestral Delaware River are of late Early Cretaceous age, and the sediments in the delta of the Ancestral Schuylkill River in central New Jersey are of early and middle Late Cretaceous age. Finally, the sediments of the Ancestral Hudson River drainage are of Late Cretaceous age.

## The Greater Antilles

The islands of the Greater Antilles include Cuba, Puerto Rico, Jamaica, Hispaniola, the Isle of Pines, and the Virgin Islands. The oldest known rocks in the Greater Antilles are deformed shales, slates, schists, and marbles that crop out in Cuba. These rocks, known as the San Cayetano series, range in age from Early Jurassic to Middle Jurassic. The San Cayetano series is unconformably overlain by less deformed and less metamorphosed limestones of Early Cretaceous (Aptian) age. The absence of Late Jurassic and earliest Cretaceous deposits indicates that folding and metamorphism occurred during latest Jurassic or earliest Cretaceous time.

The Bermeja Complex of Puerto Rico, which contains amphibolites, slates, and serpentinites, is unconformably overlain by deposits of Cretaceous age and may be of Jurassic age.

Cretaceous deposits are widely exposed in the Greater Antilles. In most areas, these deposits include considerable clastic sedimentary rocks, volcanic flows, and tuffs. In Puerto Rico, the lavas are almost entirely andesitic and basaltic [63] and the Cretaceous section contains numerous angular unconformities. This suggests deposition in an unstable environment, probably an island arc. Most of the clastic sediments were probably derived from the weathering of volcanic and tectonic islands located near the site of deposition. The Greater Antilles are part of an inactive island arc that is concave toward the south. Active island arcs are concave in the direction of movement of an oceanic plate relative to the island arc. Thus, when this island arc was active, the oceanic plate probably moved in a southerly direction beneath South America.

Radiometric dates from Cuba point to an Early to Middle Cretaceous thermal event [65]. Near the end of the Cretaceous, an episode of folding, metamorphism, and intrusion affected the Greater Antillean region and may have been caused by a change in the direction and/or rate of seafloor spreading.

## Sverdrup Basin, Brooks Basin, and Chukota Basin

Thick sequences of clastic sediments were deposited during the Mesozoic in the Sverdrup Basin in northern Canada, the Brooks Basin of northern Alaska, and the Chukota Basin of northeastern Siberia (Fig. 12.2). Facies changes in the Sverdrup Basin indicate easterly and southerly sediment sources, possibly the Canadian Shield. More than 3000 m (10,000 ft) of siltstone and sandstone accumulated in the Sverdrup Basin during the Triassic. These deposits grade upward into coal-bearing Jurassic clastics. Explosive volcanism during the Early Cretaceous produced beds of volcanic breccia interbedded with continental sandstones. During the Late Cretaceous, sandstone, shale, red beds, and coals were deposited in the basin. The rate of deposition began to increase during the Early Triassic (225 million years ago) and Early Cretaceous (124 million years ago) [66]. The Early Cretaceous increase may have been related to the beginning of the separation of Alaska from northern Canada as Alaska began to rotate in a counter-clockwise direction during the Late Jurassic or earliest Cretaceous.

## Cordilleran Geosyncline

The Cordilleran Geosyncline was divided into a miogeosyncline and a eugeosyncline during the

Mesozoic. Throughout the era, movement of the Pacific Ocean plate under the margin of the continent generated volcanic magmas and resulted in at least two episodes of deformation and metamorphism in the geosyncline. Accretion of sedimentary and volcanic rocks caused a westward growth of the continent (**67**).

MIOGEOSYNCLINE.  The seas had withdrawn from the geosyncline at the beginning of the Mesozoic, but they returned early in the Triassic. Marine limestones and sandstones of Early Triassic age grade eastward into a thick sequence of continental clastics. Middle Triassic beds are not present in many parts of the miogeosyncline, presumably because of upwarping and erosion. In Late Triassic time the Mesocordilleran Geanticline was uplifted along a zone located approximately at the boundary between the miogeosyncline and the eugeosyncline (Fig. 12.12a). In southwestern Nevada and southeastern California, folding and thrust faulting occurred along the geanticline, and sediments eroded from the resulting uplift spread eastward into the miogeosyncline (**68**). This orogenic activity may have been associated with the beginning of movement of North America westward relative to Africa.

The Mesocordilleran Geanticline remained above sea level during the Jurassic and furnished clastic sediments for the adjacent miogeosyncline. During the Early Jurassic, dune sandstones were deposited near the seacoast in the southeastern part of the miogeosyncline (Fig. 12.13). The sea spread southward during the Middle Jurassic, and by the Late Jurassic covered much of the miogeosyncline (Fig. 12.12b). The Middle and Late Jurassic sequences include limestone, shale, and sandstone.

Deposition in the Cordilleran Geosyncline ended during the Late Jurassic (Kimmeridgian) as a result of folding and uplift along the entire length of the geosyncline. This orogenic episode broadened the Mesocordilleran Geanticline into a mountain range extending from Central America to Alaska and from California to central Utah. A great thickness of clastic sediment was deposited on both the eastern and western flanks of the mountain range. Numerous unconformities in the eastern sequences record almost continuous uplift of the mountain range during the Cretaceous. Six distinct phases of deformation have been noted, and the entire orogenic episode has been referred to as the Nevadan-Laramide orogeny (**69**).

EUGEOSYNCLINE.  The sedimentary deposits of the eugeosyncline are dominantly shale, sandstone, and conglomerate. Volcanic rocks are locally abundant, but limestones are generally minor. During the Mesozoic, the eugeosyncline may have resembled the modern Indonesian Island Arc. Clastic sediments were eroded from the Mesocordilleran Geanticline on the eastern margin of the eugeosyncline and from volcanic and tectonic islands within the eugeosyncline. As much as 8000 m (25,000 ft) of Triassic sediments eroded from these highlands were deposited in the eugeosyncline (Fig. 12.14).

Folding, faulting, and uplift occurred during the Late Triassic in central Nevada, the northern Sierras, central Oregon, and northwestern Washington. Radiometric ages of basement rocks indicate that the first widespread intrusion of granites occurred at that time (**70**).

In the eastern part of the eugeosyncline, up to 5000 m (15,000 ft) of Jurassic sediments were deposited on the western margin of the continent. Upper Jurassic slates, graywackes, cherts, and volcanics, such as the Galice Formation of southern Oregon, the Western Jurassic Belt of northern California, and the Mariposa Formation in the foothills of the Sierra Nevadas, may have been deposited on ocean crust.

The deposits of the eugeosyncline were intensely deformed during the Nevadan orogeny. Radiometric dating has shown that numerous granites were emplaced at that time (**71**). The eugeosynclinal deposits were again folded and intruded during Middle Cretaceous

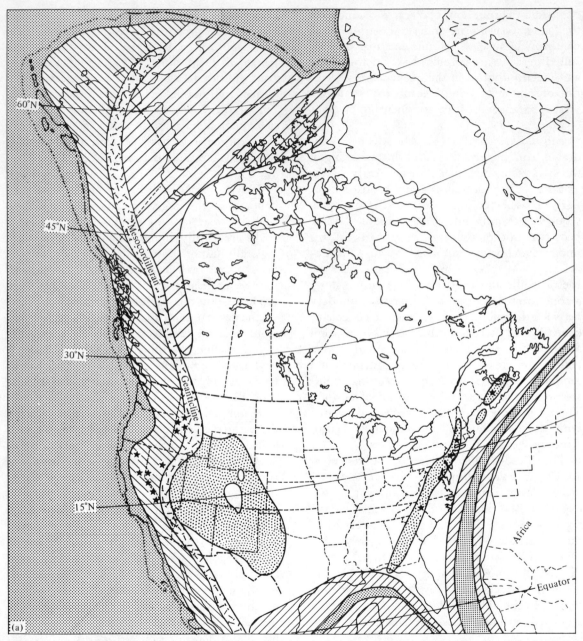

**FIGURE 12.12** Paleogeographic map of North America during: **(a)** Late Triassic. **(b)** Late Jurassic, Oxfordian age. **(c)** Late Jurassic, Portlandian age.

marine miogeosynclinal deposits     generally shallow water     marine eugeosynclinal deposits     generally moderate depth water

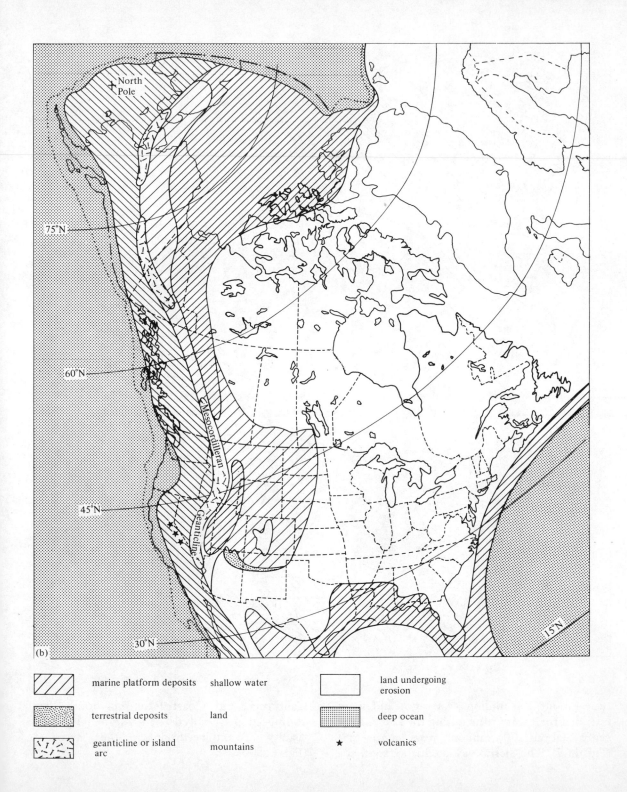

North Pole

75°N

60°N

45°N

30°N

15°N

Mesocordilleran Geanticline

(b)

| | marine platform deposits | shallow water | | land undergoing erosion |
| | terrestrial deposits | land | | deep ocean |
| | geanticline or island arc | mountains | ★ | volcanics |

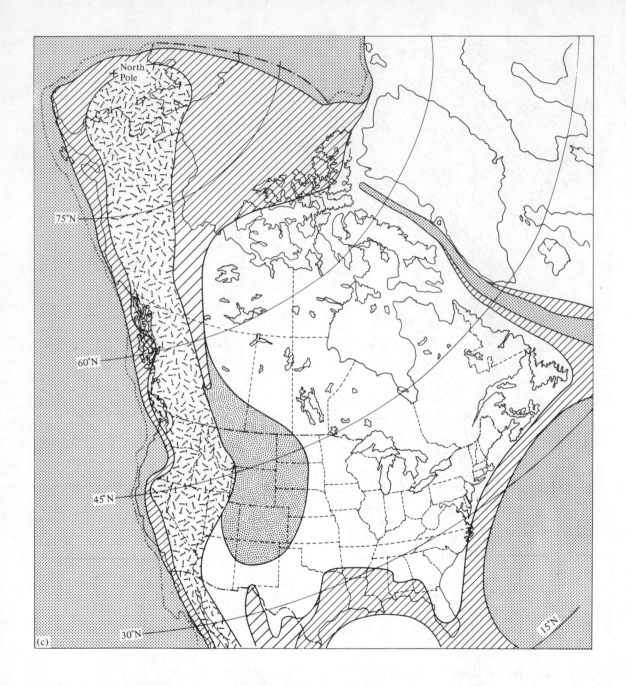

time (about 100 million years ago) and were intruded for a third time during the Late Cretaceous (about 70 million years ago) (**65**) (Fig. 11.7). The Sierra Nevada, Idaho, southern California, and Coast Range batholiths are composed of hundreds of separate intrusions that were emplaced during the Mesozoic (Fig. 12.15).

**FIGURE 12.13** Cross-bedding in dune sandstones of the Navajo sandstone of Jurassic age in Checkerboard Mesa, Zion National Park, Utah. (Union Pacific Railroad photo.)

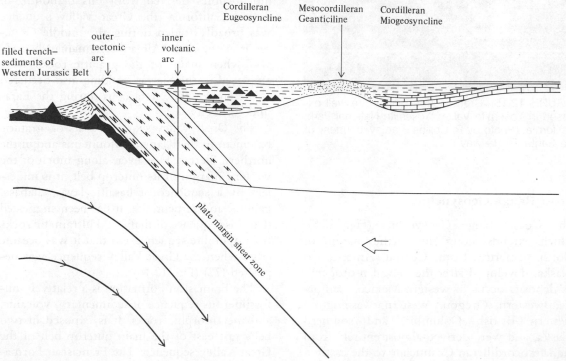

**FIGURE 12.14** Restored cross section across the Cordilleran Geosyncline during the Late Jurassic, Oxfordian age.

FIGURE 12.15 Cretaceous granites are well exposed in Yosemite Valley, Yosemite National Park, California. (Photo by R. Collins Bradley, courtesy of the Santa Fe Railway.)

## Coast Range Geosyncline

The Coast Range Geosyncline (Fig. 12.8), which extends along the Pacific margin of North America from Central America to Alaska, developed after the Nevadan orogeny. Its deposits occur in western Mexico, California, western Oregon, western Washington, western British Columbia, and southern Alaska, and were derived almost entirely from the Mesocordilleran Geanticline to the east.

In California, the Coast Range Geosyncline contains two contrasting sequences, the Great Valley sequence and the Franciscan Formation. The Great Valley sequence is comprised largely of sandstone and shale. Volcanic rocks and chert are found only near the base of the unit (**72**). The sequence is between 8000 and 16,000 m (25,000 and 50,000 ft) thick and is exposed in two belts, one extending from northern California to the southern end of the Sacramento Valley, and another extending from Point Arena to Santa Barbara. Deposition of the Great Valley sequence occurred between the Late Jurassic (Kimmeridgian) and the Late Cretaceous (Maestrichtian) (**73**).

Sandstones in the Great Valley sequence are generally poorly sorted, especially in the lower part of the section. An increase in arkosic sandstones in the upper part of the section is thought to be due to a progressive unroofing of the Sierra Nevada Batholith to the east. Conglomerates in the Great Valley sequence contain granitic pebbles derived from this batholith. In central California, the Great Valley sequence was broadly folded during the middle Cretaceous at the time (Albian/Cenomanian boundary) when many of the granitic rocks were emplaced in the Sierras. Folding also occurred at the end of the Cretaceous during the Laramide orogeny.

The Great Valley sequence rests on granitic basement in the Klamath Mountains and in the northern Sierras. However, along much of the western margin of its outcrop belt, it is underlain by a sequence of basaltic lavas, diabase, gabbro, and serpentinite. It has been suggested that this sequence of mafic and ultramafic rocks is an ophiolite sequence and that it was oceanic crust when the Great Valley sequence was deposited (**72**) (Fig. 12.16).

The Franciscan Formation is a relatively unfossiliferous sequence of sedimentary, volcanic, and metamorphic rocks. It is exposed in two belts just east of the main outcrop belt of the Great Valley sequence. The Franciscan Formation was deposited between Late Jurassic (Portlandian) and early Tertiary (Eocene) (**74**, **75**),

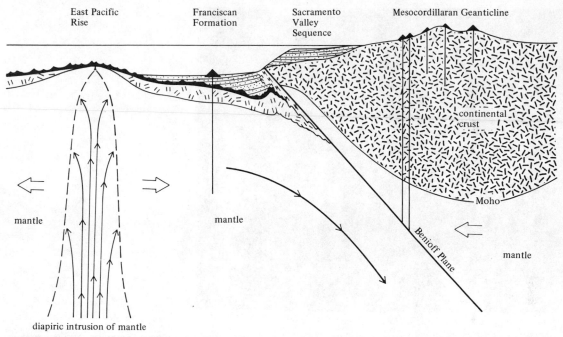

**FIGURE 12.16** Restored cross section across the Coast Range Geosyncline during the Early Cretaceous.

nearly contemporaneously with the Great Valley sequence. However, its lithology is quite different.

Although dominantly graywacke, the Franciscan Formation also contains shale, conglomerate, rhythmically bedded chert, greenstone, and limestone (Fig. 12.17). The graywackes are quite feldspathic and contain fragments of volcanic rocks, shale, and chert. Much of the clastic sediment in the Franciscan Formation was probably derived from the Mesocordilleran Geanticline to the east. Potassium feldspar, which is present in the younger Franciscan beds, may have been derived from the granitic intrusions of the geanticline.

The Franciscan Formation is intensely deformed and has been strongly metamorphosed in the vicinity of the fault contact with rocks to the east. The lithologic character and degree of deformation of the Franciscan Formation suggest that it was deposited in deep water on

oceanic crust. The rhythmically bedded cherts, which often contain fossil Radiolaria, may represent abyssal radiolarian oozes.

The continuing deformation in the Mesocordilleran Geanticline during the Late Jurassic and Cretaceous indicates that an oceanic plate was moving under North America. It is assumed that this movement was caused by westward drifting of North America relative to Europe and eastward spreading of the seafloor away from the Pacific Rise. Dating of magnetic anomalies in the Atlantic and Pacific oceans suggest that seafloor spreading has been active since at least the end of the Jurassic. Underflow of the oceanic plate may have caused folding and metamorphism of the Franciscan Formation and may have resulted in accretion of the sedimentary and volcanic rocks onto the North American continent.

Radiometric dating has shown that schists just west of the eastern boundary of the Fran-

(a)

(b)

(c)

FIGURE 12.17 Deposits of the Franciscan Formation: (a) Folded cherts north of San Francisco. (b) Pillow lavas north of San Francisco—the shape and orientation indicates that the lava is right-side up and dips steeply toward the right. (c) Thick-bedded graywacke near Crescent City, California. [(a) and (b) Photos by Mary Hill]

ciscan Formation were metamorphosed during the Late Jurassic; yet they grade into Franciscan rocks containing Early Cretaceous fossils (**76**). Evidently metamorphism of the Franciscan Formation was nearly contemporaneous with its deposition. The metamorphism produced high-pressure, low-temperature minerals such as glaucophane, lawsonite, jadite, and aragonite (**72**). High-temperature, low-pressure conditions would have been produced if Franciscan deposits had been carried beneath the continent so rapidly that they failed to reach the temperature of the adjacent rocks.

It has been suggested that during the upper Cretaceous about 100 million years ago fragments of a continent that originated in tropical latitudes were carried with the Pacific seafloor into the North American continent between Vancouver Island and south-central Alaska (**73**). The allochthonous rocks, composed of thoeleiitic flows and inner platform carbonates, are of Triassic age and were emplaced against dissimilar Triassic strata of North America. The pieces of this hypothetical continent, named Wrangellia after a major fragment, the Wrangell Mountains of Alaska, may include the Hell's Canyon region of Washington, Idaho, and Oregon.

## Western Continental Interior

Red beds, such as those in the Spearfish Formation of South Dakota and the Moenkopi Formation of Arizona, were deposited over much of the continental interior during the Early Triassic. They were probably deposited on flood plains of slowly moving rivers, which built a vast alluvial plain sloping gently toward the sea. These clastic deposits grade westward into limestones and thus indicate that the source of the clastics must have been to the east (Fig. 12.18).

Uplift of the Mesocordilleran Geanticline and adjacent areas during the Late Triassic caused a major change in sedimentation pattern in the western continental interior. Streams and rivers carried sand, gravel, and silt northward

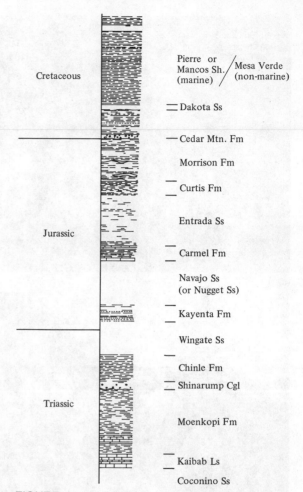

**FIGURE 12.18** Mesozoic rock units of the Colorado Plateau.

and northwestward from uplifts in Arizona toward central Nevada. During times of flood, logs were carried onto sand and gravel bars, where they were subsequently buried. Over the years, the wood was replaced by colorful chalcedony and agate. In the Petrified Forest National Monument in Arizona, these petrified logs have been weathered out of the weakly cemented sandstones and conglomerates of the Chinle Formation (Fig. 12.19a).

The Wingate Sandstone of uppermost Triassic age in southern Utah and northern Ari-

(a)

(b)

**FIGURE 12.19** Mesozoic deposits in the western United States. **(a)** Fossilized logs at Petrified Forest National Monument. (Photo by N. H. Darton, U.S. Geological Survey.) **(b)** Natural stone arches at Arches National Monument, southeastern Utah, are composed of Jurassic sandstone.

zona displays large cross-beds of aeolian origin. Paleomagnetic studies indicate that the southwestern United States was between 15° and 20°N latitude during the Triassic, in a climatic zone that was arid or semiarid. Dune sandstones were also widely deposited in the continental interior during the Early Jurassic (Fig. 12.19b). An example of this type of deposition, the Navajo Sandstone, forms much of the spectacular, erosional scenery of the canyon lands in Utah.

In the Middle Jurassic, an arm of the sea (the Sundance Sea) advanced from the north and covered much of the continental interior. Deposits of this transgression include gypsum, limestone, sandstone, and shale. The Mesocordilleran Geanticline was uplifted during the Late Jurassic, and the seas withdrew from the western continental interior. Streams and rivers deposited thick sequences of clastic sediments in a basin that extended from Montana to northern New Mexico. These deposits include the

**FIGURE 12.20** Dinosaur bones, exposed by a careful removal of the enclosing rock, at Dinosaur National Monument, northeastern Utah. The enclosing rock is of Late Jurassic age. (Photo by M. W. Williams, courtesy of the U.S. Department of Interior, National Park Service.)

Morrison Formation, which contains the remains of many large dinosaurs (Fig. 12.20). As the Mesocordilleran Geanticline continued to grow wider and higher during the Cretaceous, shale, sandstone, and conglomerate (for example, the Mesa Verde Formation) were deposited east of the geanticline (Fig. 12.21). The clastics interfingered eastward with marine shales (the Pierre and Mancos shales) that were deposited in the epicontinental sea.

The Laramide orogeny, which began near the end of the Cretaceous, produced a series of uplifts in the western continental interior (**77**) (Fig. 12.22 and 12.23). In the Medicine Bow Mountains, the earliest locally derived conglomerate is of Late Cretaceous (earliest Maestrichtian) age (**46**). Presumably this time marks the beginning of the Larimide orogeny. Cretaceous deposits are unconformably overlain by beds of Paleocene age, which are in turn un-

conformably overlain by undeformed Lower Eocene sediments (**46**). Uplifts of the western continental interior may have resulted from displacements along reverse faults. The faulting may have been caused by a combination of compression and shearing of a large segment of western North America.

## Southern Continental Interior

During most of the Mesozoic, the seas were confined to the Gulf Coast and East Coast geosynclines. During the Early Cretaceous, however, they spread northward and westward onto the adjacent platform. By the beginning of the Late Cretaceous, this sea was continuous with the sea in the western continental interior. Deposits of this transgression include sandstone, shale, and chalk.

0–2800 ft thick

\>2800 ft thick (sedimentation rate, >40 ft/m.y.)

FIGURE 12.21 Map of North America showing the thickness of Cretaceous deposits.

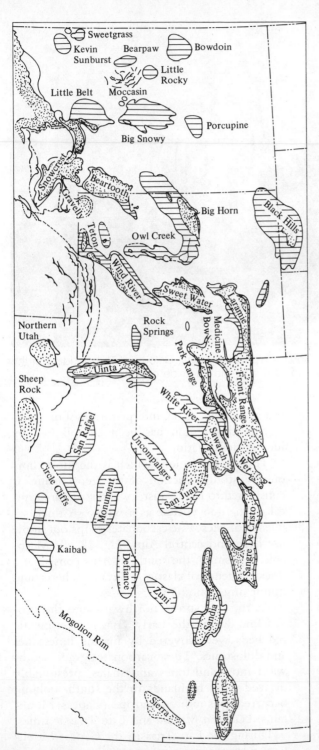

**FIGURE 12.22** Uplifts in the Rocky Mountains and adjacent regions, formed during the Late Cretaceous and early Tertiary. Stippled areas are exposed Precambrian rocks and the horizontal ruling indicates Paleozoic sedimentary rock. (From Eardley, Ref. 6.)

**FIGURE 12.23** Sandstones in the Garden of the Gods, Colorado Springs, Colorado. The bedding in the sandstones is approximately vertical as a result of folding during the Larimide orogeny.

## EURASIA

Following the general emergence of the continents at the end of the Paleozoic, the seas again transgressed across Eurasia in the early Mesozoic. Thick sequences of sedimentary and volcanic rocks were deposited in the northern Tethyan Geosyncline the Indonesian Island Arc and the Pacific Ocean Geosyncline. The seas were restricted to the geosynclines during most of the Triassic, but much of the continent was inundated during the Jurassic and Cretaceous.

### Northern Tethyan Geosyncline

The Northern Tethyan Geosyncline developed after the final deformation of the Hercynian Geosyncline. Its axis parallels that of the Hercynian Geosyncline but lies several hundred miles to the south. The Northern Tethyan Geosyncline extends from southern Europe, through the Middle East, and the Himalayas to

Indonesia (Fig. 12.1). Interpretation of the geology in this region has been complicated by intense deformation.

The Alpine region, which is the most thoroughly studied section of the geosyncline, is considered to have a history and structure typical of the geosyncline as a whole. A thick sequence of Mesozoic carbonates crops out in the northern and central Alps (Fig. 12.24); Mesozoic deposits in the southern Alps consist of a thick sequence of clastics, volcanics, cherts, and minor limestones (**78**).

As the seas spread northward into the geosyncline during the Early Triassic, continental red beds were covered by marine limestones and dolostones. These carbonates grade southward into sandstones and shales, presumably derived from highlands to the south. Folding occurred in southern Europe (Caucas Mountains) between Middle and Late Triassic times.

During the Late Triassic, the Caucas Mountains were the site of normal faulting and mafic

**FIGURE 12.24** Upper Jurassic limestone at Sisteron, France. These beds were folded during the early Tertiary.

volcanism (**79**), similar to that of the Newark Series. This may have occurred during the initial separation of Eurasia and Africa. Before the breakup of Pangaea, Italy and part of Yugoslavia and Greece were probably part of Europe, but when Africa and Eurasia began to separate, a rift developed between the Eurasian Plate and the Austroalpine-Adriatic Plate (Italy and western Yugoslavia and Greece) (**79, 80**). As discussed earlier in this chapter, the separation of Eurasia and Africa occurred sometime between the Late Triassic and Early Jurassic (202 to 180 million years ago), and a Late Triassic date is favored. During the Early and Middle Jurassic, deep-water deposits (such as radiolarian cherts and pelagic limestones) and breccias were deposited on the outermost margins of the newly formed ocean (Penninic Ocean), and shallow water limestones were deposited in the inner parts of the continental margins (Fig. 12.25) (**79, 80**).

While relatively shallow-water deposits accumulated on the southern margin of the European continent, radiolarian cherts, limestones, and shales were being deposited in a deep trough to the south. Some geologists have suggested that the cherts and limestones represent lithified oozes that were deposited on oceanic crust at a depth of several kilometers (**81**). The limestones contain only pelagic microfossils, and little detrital sediment is present in the cherts. Evidently these deposits were laid down far from their continental sources. Many of the fossil radiolarians in the cherts are similar to forms that live today only at depths below 4000 m (12,000 ft). The concentration of silica is high in such areas, and therefore an ample source of silica is available for radiolarian tests. A further indication of a deep-water depositional environment for the cherts is their association with manganese nodules, which are also found in the deeper parts of modern oceans. There is no known granitic basement, but these deposits are associated with ophiolites, a suite of greenish igneous rocks including pillow greenstone, serpentinites, and gabbros. The ophiolites, radiolarian cherts, and associated deposits are found in a belt extending from

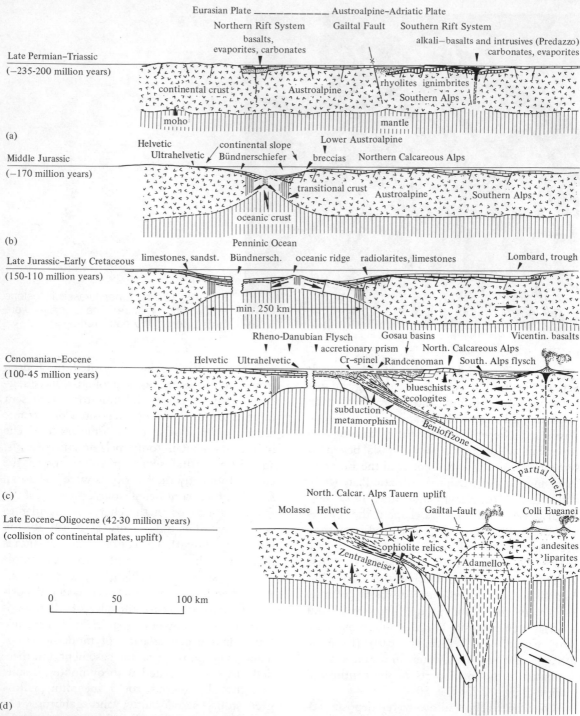

**FIGURE 12.25** Restored cross sections across southern Europe showing the opening and closing of a small ocean basin during the Mesozoic and early Tertiary. (From Dietrich, Ref. 80.)

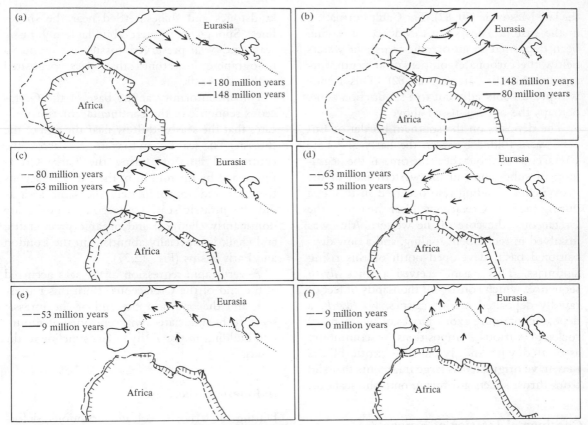

**FIGURE 12.26** Successive positions of Africa relative to Europe at 180, 148, 80, 63, 53, and 9 million years ago. The arrows show the simplest relative-motion paths for successive intervals and are segments of small circles concentric about the spreading pole for the interval indicated on the diagram. (From Dewey *et al.*, Ref. 79.)

northern Italy through southwestern Yugo-slavia, Albania, southwestern Greece, Cyprus, southeastern Turkey, and the Middle East to the Himalayas. It is likely that the ophiolites are part of a basement of oceanic crust on which the radiolarian cherts and associated deposits were laid down. The oceanic crust may have developed as the Eurasian and Austroalpine-Adriatic plates separated.

Folding occurred during the Late Jurassic (Kimmeridgian) in the Caucas Mountains and in northern Turkey (**79**). This folding was probably related to a change in the relative motions of Africa and Eurasia from moving apart (rifting) from the Late Triassic until the middle of the Late Jurassic to moving together (compression) in the eastern Tethyan region after this time (**79**) (Fig. 12.26).

During the Cretaceous, limestones and sandstones were deposited on the southern margin of Europe, while thick clastics and volcanics were deposited on the northern border of Italy. Apparently a southward dipping subduction zone had developed beneath Italy by the beginning of the Late Cretaceous (**80**). High-pressure, low-temperature rocks (such as glaucophane schists and eclogites) and melanges developed along this subduction zone during the late Mesozoic (**80, 82**). The Gault Formation in the eastern Alps shows evidence of having been deposited in an oceanic trench in waters below the carbonate compensation depth during the Middle Cretaceous (**82**). This trench was presumably adjacent to a subduction zone, perhaps the one discussed above.

The deposits on the southern border of Europe were folded between the Early and Late Cretaceous at about the same time as the beginning of subduction in the western part of the geosyncline. Folding also occurred in this area during the Late Cretaceous. Near the end of the Cretaceous, the core of the western Alps was involved in recumbent folding, and a new depositional basin developed north of this rising landmass. This basin received a thick *flysch* sequence, which consists of thousands of feet of rapidly deposited marine clastics. *Wildflysch,* a deposit containing exotic blocks of sedimentary rocks of various compositions, is commonly associated with the flysch. The exotic blocks may have originated as large fragments that slid from thrust sheets advancing onto the seafloor.

## Continental Interior of Europe

Relatively thick Mesozoic deposits are found in the Germanic, London, and Paris basins, and in a basin in Spain. These basins developed between broad uplifts that formed during the Hercynian orogeny. The continental interior was above sea level during most of the Triassic, but short transgressions occurred during the Middle Triassic and near the end of the Late Triassic. The Middle Triassic Muschelkalk Sea accounts for the marine interval in this three-part geologic period. Triassic continental deposits are mostly conglomerates and sandstones. Marine deposits include siltstone, evaporites, and limestone, which formed in environments that ranged from shallow lagoons to an open shallow sea.

Shallow seas covered much of Europe during the Jurassic. Limestone and marls were deposited near the centers of basins, and

sandstones and shales settled near the shorelines. Sponge reefs were abundant in Jurassic seas, and well-preserved fossils are found in lithographic limestones that were deposited within the reefs.

An unconformity at the base of the Cretaceous sequence in the continental interior indicates that the seas withdrew near the end of the Jurassic. The seas began to transgress across the continental interior during the Early Cretaceous, and by the middle of the Late Cretaceous they covered approximately the same area as did the Jurassic seas. Cretaceous deposits are dominantly chemical and organic precipitates, and chalk is especially abundant in the London and Paris basins (Fig. 12.27).

A very rapid regression of the seas occurred at the end of the Cretaceous. Both this regression and the one near the end of the Jurassic seem to be associated with episodes of mountain building in the Tethyan Geosyncline to the south.

## Indonesian Island Arc

During the Triassic, volcanic eruptions, deformation, and intrusion occurred in a linear belt along the axis of the Indonesian Geosyncline. The zone of volcanic and tectonic activity shifted progressively southward during the Mesozoic (**83**). Facies relationships indicate that there were islands within a eugeosyncline. Thus Indonesia was probably an active island arc during the Mesozoic, just as it is today.

In southwestern Borneo, folded Permian deposits are unconformably overlain by sedimentary and volcanic rocks of Late Triassic age. Andesitic and basaltic volcanics occur along a belt extending from Cambodia to southwestern Borneo. Near the end of the Triassic, the deposits of the geosyncline were folded, intruded, and uplifted along a belt extending from the Celebes Islands through central Borneo to the Malay Peninsula. The resulting geanticline persisted as a positive feature throughout much of the Mesozoic and was a source of clastic sedi-

**FIGURE 12.27** Chalk cliffs of Cretaceous age at Beer Cove, Devon, England. (NERC Copyright. Reproduced by permission of the Director, Institute of Geological Sciences, London Sw7.)

ment for the adjacent troughs. Geanticlines were also uplifted in the geosyncline during the Jurassic and Cretaceous. At the end of the Cretaceous, folding in the southern part of the geosyncline created a geanticline that extended from Java and Sumatra to Flores.

Lower Cretaceous deposits in Sumatra include radiolarian cherts, limestones, tuffs, and ophiolites. This sequence may have been deposited in the deep sea far from land (**83**). At the end of the Cretaceous, the deposits of the geosyncline were folded and uplifted, perhaps also becoming accreted to the continent. The cause of the Mesozoic deformation, volcanism, and intrusion in the Indonesian Island Arc may have been the movement of an oceanic plate under Asia as Australia moved away from Africa and toward Asia. The Late Triassic defor-

mation and volcanism in the Indonesian Island Arc may have been associated with the initial separation of the continents.

## Pacific Ocean Geosyncline

The Pacific Ocean Geosyncline extends from eastern Siberia through Japan, Taiwan, and the Philippines to the Indonesian Geosyncline (Fig. 7.9). Thick sequences of sedimentary and volcanic rocks were deposited during the Mesozoic in Japan, eastern Siberia, Taiwan, and the Philippines. The Mesozoic geosynclinal deposits of Japan have been divided into near-shore and deep-water facies. The near-shore facies is composed largely of coarse continental and marine clastics and volcanics; the deep-water facies contains cherts, limestones, and fine-

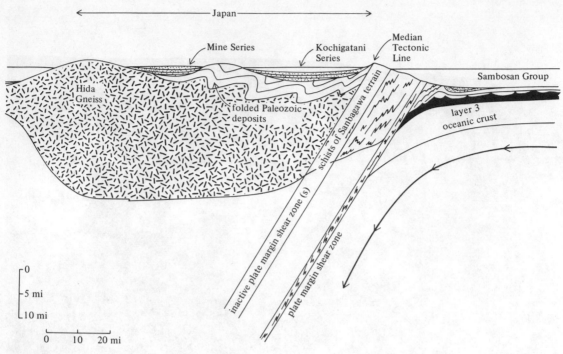

**FIGURE 12.28** Restored cross section across Japan during the Late Triassic.

grained clastics. These two facies are now separated by a major fault zone, the Median Tectonic Line. This line may mark a former subduction zone separating sedimentary and volcanic rocks deposited in deep water (on oceanic crust) from those deposited in relatively shallow water in an island arc environment. Numerous unconformities within the nearshore facies indicate that Japan was the site of repeated uplift and folding throughout the Mesozoic (Fig. 11.31). Potassium-argon ages of granites indicate that the deformations were accompanied by intrusion of granitic rocks (**84**).

Mesozoic deposits southeast of the Median Tectonic Line include the Sambosan Group, a rather thin sequence of bathyal cherts and fine clastics with only minor limestone. Fossils are not common in this sequence, but enough have been found to indicate that the sequence was deposited more or less continuously from the Permian through the Jurassic. The lithologic character of the Sambosan Group suggests deposition in the deep ocean. The Shimatogawa Group ranges from Triassic or Jurassic to Miocene in age. The Mesozoic deposits in this sequence include shale, sandstone, radiolarian chert, pillow greenstone, and limestone. These rocks have been intensely folded and are remarkably similar in lithology and structure to the Franciscan Formation in California. Both the Shimatogawa Group and the Sambosan Group may have been deposited in the deep ocean and incorporated into the continent as an oceanic plate moved under the continent (Fig. 12.28).

## GONDWANALAND

Geosynclines bordered Gondwanaland during the early part of the Mesozoic (Fig. 12.1). Following the breakup of Gondwanaland, geoclines formed on the Atlantic and Indian ocean borders of South America, Africa, Australia, Antarctica, and India. There were several episodes of widespread volcanism in the continental interior during the Mesozoic. This volcanic activity may have been the result of crustal tension as the continents separated.

### New Zealand Geosyncline

At the beginning of the Mesozoic, New Zealand and the adjacent Lord Howe rise may have been in contact with Australia (**85**). Two distinct facies have been recognized in the Mesozoic of the New Zealand Geosyncline, a relatively shallow-water "marginal facies" and a deep-water "axial facies." These facies are separated by a major fault zone which, as in Japan, is called the Median Tectonic Line. The marginal facies contains clastics, volcanic flows, and volcanic tuffs ranging in age from Early Triassic to Late Jurassic. The lack of Late Jurassic (Oxfordian) deposits in the marginal facies indicates that during this interval the geosyncline was uplifted and the seas withdrew. The deformation culminated during the Late Jurassic or Early Cretaceous (Rangitata orogeny), when the deposits of the marginal facies were folded. Intrusives into the marginal facies have been dated as Late Jurassic and Middle Cretaceous.

The axial facies of the New Zealand Geosyncline contains a thick layer of relatively unfossiliferous graywacke with only minor amounts of basalt, chert, and limestone. The deposits of the axial facies are more intensely deformed than those of the marginal facies (Fig. 12.29). The lithologies suggest that these rocks were deposited on oceanic crust (**86**). Ultramafic intrusions, which are common in this sequence, may be fragments of the upper mantle which underlay this facies during its deposition.

High-pressure, low-temperature minerals such as glaucophane and lawsonite also occur in the axial facies. It is possible that these minerals were produced when deep ocean deposits were carried beneath New Zealand during underflow

**FIGURE 12.29** Folded graywackes of the axial facies in the Southern Alps of New Zealand (Photo by D. L. Homer, courtesy of the New Zealand Geological Survey.)

of an oceanic plate. The final deformation of the axial facies occurred during the late Jurassic or Early Cretaceous during the Rangitata orogeny. This orogeny may have been caused by a separation of Australia from New Zealand. Subsequently, sedimentation shifted to the New Zealand East Coast Geosyncline east of the New Zealand Geosyncline.

### New Zealand East Coast Geosyncline

Sedimentary deposits in the New Zealand East Coast Geosyncline are dominantly marine clastics, such as mudstone, conglomerate, and sandstone. Throughout the Cretaceous, the geosyncline was bordered on the east by a chain of volcanic islands. Evidently then, as now, this region was an island arc. The seas withdrew during the Late Cretaceous, and coal-bearing continental sediments were laid down.

## Timor

The island of Timor is located between Australia and Indonesia. Few unconformities occur within a sequence that ranges in age from Permian to Late Cretaceous. The Mesozoic deposits in Timor include limestone, clastics, and cherts that have been intruded by ultramafic rocks such as serpentinites. Some of the sedimentary units are very fossiliferous. Over 900 species of marine invertebrates have been found in Triassic beds. The Mesozoic sequence is very thin, which suggests a slow rate of deposition in deep water away from highlands. However, the abundance of fossils is generally associated with deposition in shallow water. Some of the deposits contain a mixture of species of different ages, which suggests reworking of the sediments. Possibly these deposits were originally laid down in shallow water and were redeposited in deep water by submarine currents. The limestones in the sequence may have been derived from erosion of reefs on the outermost margin of Australia. The presence of manganese nodules in Cretaceous deposits of Timor suggests a deep-sea depositional environment (**87**). At the end of the Cretaceous, the deposits of Timor were folded and then accreted to the Asian continent, probably as a result of compression between Asia and Australia.

## Southern Tethyan Geosyncline

The Southern Tethyan Geosyncline extends along the outer margin of Gondwanaland from northern India through the Middle East and southern Europe to northern Africa (Fig. 7.19). In the early Mesozoic, a wide, deep ocean separated this trough from the Northern Tethyan Geosyncline on the Eurasian border. As Gondwanaland separated from North America, the Tethys Sea narrowed, and deformation occurred along the margin of Gondwanaland.

During the Triassic and Jurassic, carbonates were deposited in the Southern Tethyan Geosyncline (Fig. 12.30). However, thick shales were also deposited in the Himalayan region in Late Triassic and Late Jurassic time.

The Cretaceous was a time of increasing deformation. Thick sequences of Cretaceous clastics are found in the Himalayas, the Middle East, and southern Europe. In Iran and Iraq, Late Cretaceous conglomerates contain pebbles of radiolarian cherts and igneous rocks. Since these deposits grade southward into limestones, the source must have been to the north, possibly an island arc. The eastern part of the Zagros segment of the geosyncline in Iran was deformed and the western part of the basin subsided during the Late Cretaceous due to compression between the Arabian plate and Eurasia (**88**). In Italy, Albania, western Yugoslavia, and western Greece, unconformities in Cretaceous sequences provide evidence of deformation and uplift during that interval (**81**).

## Andean Geosyncline

During the Mesozoic, carbonates and clastics were deposited in the Andean Miogeosyncline, and clastics and volcanics were laid down in the eugeosyncline to the west. The boundary between these two troughs lies more or less along the Central Ranges of the Andes Mountains. In some areas a geanticline separated the miogeosyncline and eugeosyncline. The presence of mafic and intermediate volcanic rocks of early Cretaceous age on the western border of early Cretaceous sediments, all trending north-south along Tierra del Fuego, indicates that an island arc and marginal basin were present at that time (**89**).

MIOGEOSYNCLINE.   Seas occupied only a small part of the Andean Miogeosyncline during the Triassic and Jurassic. Continental red beds were deposited in some areas, but else-

**FIGURE 12.30** The Rock of Gibraltar is composed of limestone of Jurassic age. Gibraltar is located at the junction between the Northern and Southern Tethyan geosynclines. (Photograph by *The Times,* London.)

where uplift and erosion prevailed (Fig. 12.31). In Cretaceous time, the seas covered much of the northern and central regions of the miogeosyncline. Cretaceous deposits are largely carbonates and fine clastics, but they grade into sandstones to the east.

EUGEOSYNCLINE. Volcanic activity was especially intense in the eugeosyncline during the Late Triassic, Late Jurassic, and Late Cretaceous. Graywackes interbedded with the volcanics may have been deposited by turbidity currents, and deposits in the western part of the eugeosyncline may have been laid down on oceanic crust (**90**).

DEFORMATION. Widespread faulting and plutonism occurred in the Andean Geosyncline during the Late Triassic (**90**). It is quite possible that the deformation was caused by the movement of an oceanic plate under the geo-

syncline during the initial stages of separation of Gondwanaland and North America. A Late Jurassic deformation may have been related to an increase in the rate of movement of the plate under the continent as South America began to separate from Africa. The principal metamorphism and deformation in the geosyncline began during the late Early Cretaceous and culminated in the Late Cretaceous and early Tertiary (**91**). Radiometric dating has shown that numerous granite bodies were emplaced during this interval (**90**) (Fig. 12.32). Deformation near the end of the Cretaceous resulted in a withdrawal of the seas from much of the geosyncline. The seas did not return to the geosyncline during the remainder of the Mesozoic.

**Antarctic Geosyncline**

Mesozoic sedimentary and volcanic rocks of geosynclinal thickness are found in peaks rising

**FIGURE 12.31** Banded sediments of the Tonel Formation of Jurassic age in the Andes Mountains of northern Chile. (Photo by R. J. Dingman, U.S. Geological Survey.)

above the continental ice sheet on the Pacific margin of Antarctica. These sequences were folded, uplifted, and eroded several times during the Mesozoic. An especially intense deformation occurred during the Middle Cretaceous. Radiometric dating of granitic intrusions shows that granites were emplaced at that time (**92**).

### Continental Interior of Gondwanaland

The continental interior of Gondwanaland was subjected to widespread volcanic activity during the Jurassic and Cretaceous (Fig. 12.33). This volcanism was apparently related to the breakup of the southern continents. The most

widespread activity occurred during the Late Jurassic.

Hot spots at the ends of the Rio Grande and Walvis ridges are thought to have been underlain by mantle plumes (**42, 93**). Adjacent areas were the site of considerable mafic and alkalic igneous activity during the Late Jurassic and Early Cretaceous (**20**) (Fig. 13.23). Aulacogens originating at hot spots were active in Africa during the Mesozoic. (Fig. 13.23).

### Ocean Basins

An intensive effort has been made to recover pre-Tertiary sediments from the deeper parts of the ocean basins. Cretaceous sediments are

South America

**FIGURE 12.32** Mesozoic and Mesozoic(?) granitic rocks in South America. (Data from *Carte Geologique de l'Amerique du Sud* prepared by the Commission de la Carte Geologique du Monde, published by the Geological Society of America.)

widespread, but no samples older than Middle Jurassic have been obtained, even in areas where holes have been drilled to the basement. It appears, then, that the ocean floors are relatively young—a conclusion that is consistent with the concept of seafloor spreading (Fig. 7.22).

ATLANTIC OCEAN. Mesozoic deposits from the deeper parts of the Atlantic Ocean basin include clays, foraminiferal and radiolarian oozes, turbidites, and cherts. Cores taken in water depths greater than 4000 m (12,000 ft) generally have no calcareous sediments in their upper levels, since the solubility of calcium carbonate is greatly increased at these depths. However, calcium carbonate is commonly present near the bottom of Mesozoic sequences in regions which are now below 4000 meters deep. It may be that these areas were originally less than 4000 meters deep and have subsided as the seafloor moved away from the Mid-Atlantic Ridge.

Seafloor spreading probably began in the North Atlantic during the Late Triassic about 202 million years ago when North America and Eurasia began to separate from Gondwanaland. The Mid-Atlantic Ridge was the spreading center during separation. Studies of magnetic anomalies in the North Atlantic indicate that the direction of seafloor spreading changed during the Late Jurassic about 147 million years ago, possibly at the same time as the beginning of separation of Greenland and North America and the beginning of rotation of the Iberian peninsula (**9**) (Fig. 12.3). The direction of seafloor spreading changed a second time about 70 million years ago, probably at the same time as the beginning of separation of Greenland and Eurasia (**9**) (Fig. 12.3).

Seafloor spreading probably began in the South Atlantic during the Late Jurassic about 147 million years ago when South America began to separate from Africa. The southern section of the Mid-Atlantic Ridge was the spreading center during separation. A 20° bend

FIGURE 12.33 Sill of Ferrar dolerite (dark band) intruding sedimentary rocks of the Beacon group in South Victoria Land, Antarctica (U.S. Navy photo, courtesy of the U.S. Geological Survey.)

in the Rio Grande Ridge and a 40° bend in the Walvis Ridge occurred during the Late Cretaceous about 70 million years ago. This suggests that the direction of seafloor spreading relative to the underlying mantle plume changed at that time.

GULF OF MEXICO AND CARIBBEAN SEA. Seafloor spreading probably began in the Gulf of Mexico during the Late Triassic, about 202 million years ago, when North America began to separate from Gondwanaland. Seafloor spreading may have ended there during the Late Jurassic. The presence of Late Jurassic evaporites in a salt dome near the center of the Gulf of Mexico indicates that this area was the site of deposition of deep-water evaporites during at least part of the Mesozoic.

The oldest known sedimentary rocks deposited on oceanic crust in the Caribbean Sea are of Late Cretaceous (Coniacian) age, about 85 million years old (94). It is likely that the Caribbean was formed by seafloor spreading between the Late Jurassic and the Late Cretaceous as South America moved eastward relative to North America (95).

INDIAN OCEAN. Seafloor spreading probably began in the Indian Ocean during the Late Triassic about 202 million years ago when Australia, India, and Antarctica began to separate from Africa and Malagasy. The southwestern branch of the Mid-Indian Ocean Ridge was the center of spreading. Little is known about the Mesozoic spreading history of this ridge.

The initial separation of India from Australia and Antarctica took place during the Late Jurassic, about 147 million years ago. The spreading center was a now extinct ridge located west of Australia. Spreading ceased on this ridge about 100 million years ago and began about that time on a ridge to the northeast (Fig. 13.19). Seafloor spreading on the northwest section of the Indian Ocean Ridge probably began about 70 million years ago with the separation of India from the Seychelles.

Magnetic anomalies and deep sea drilling indicate that the seafloor northwest of Australia is of Late Jurassic (Oxfordian) age, 155 million years old (96). This date is older than the date of the beginning of separation of India from Australia. Perhaps this seafloor was formed during the separation of Australia, Antarctica, and India from Africa and Madagascar.

PACIFIC OCEAN. In general the lithology of the Mesozoic sediments from the Pacific Ocean is the same as that of the sediments from the Atlantic Ocean. The age of the oldest sediments in Pacific Ocean cores increases away from the East Pacific Rise. The oldest known sediments are of Late Jurassic age and were recovered by the *Glomar Challenger* approximately 8000 km (4800 mi) from the crest of the East Pacific Rise. Cretaceous sediments and volcanic rocks have been dredged and cored from the western and central Pacific, but none have been found within about 5000 km (3000 mi) of the rise.

## MESOZOIC LIFE

Many invertebrate groups that prospered during the Mesozoic then declined or became extinct at the end of the Cretaceous. For example, the ammonoids diversified and became very abundant during the Jurassic and Cretaceous, but by the end of the Cretaceous all were extinct.

The reptiles evolved rapidly during the early Mesozoic and adapted to most of the available environmental niches. Some developed the ability to fly, others became adapted to an aquatic environment. Some ran on two legs, and others lumbered about on four. Among them were vicious predators, armored herbivores, and scavengers. In spite of their success, however, almost all the reptile groups had become extinct by the end of the Cretaceous.

The plants also underwent dramatic changes during the Mesozoic. Of greatest importance was the evolution and rise to dominance of the flowering plants, the angiosperms. These advanced forms appeared during the Jurassic, and by the Late Cretaceous had replaced the gymnosperms in abundance in most areas.

### Protists

Foraminifera proliferated in the Jurassic and Cretaceous seas. They were dominantly microscopic varieties, and their tests make up a significant portion of Cretaceous chalk deposits in

northern Europe and the Gulf Coastal region. Foraminiferal oozes were deposited in many parts of the deep oceans during the Cretaceous.

Two types of one-celled, planktonic plants, the diatoms and the coccoliths, appeared for the first time in the Mesozoic fossil record. Diatoms, which have ornate, siliceous tests, are abundant in diatomaceous earth and in many cherts. Coccoliths secrete rings and groups of plates, each of which is a single calcite crystal (Fig. 12.34). Very fine-grained coccolith debris is a major component of many Cretaceous chalks.

### Sponges and Coelenterates

Siliceous sponges occur in Mesozoic deposits that were laid down in moderately deep water. The chert nodules in the Cretaceous chalks of northern Europe may have been formed by the replacement of chalk by silica from siliceous sponge spicules. Calcareous sponges grew in muddy waters in shallow, near-shore environments during the Mesozoic.

Mesozoic coelenterates include both jellyfish and hexacorals. Hexacorals, which succeeded tabulate and rugose corals when these forms became extinct, have a sixfold symmetry—their septa are arranged in multiples of six. During the Late Triassic, hexacorals were widely distributed in a narrow equatorial belt. In the Late Jurassic, however, coral reefs had spread throughout Europe and Africa in a wide latitudinal belt. Corals were also abundant in the shallow, epicontinental seas during the Cretaceous.

### Bryozoans and Brachiopods

The two dominant orders of Paleozoic bryozoans are not found in Mesozoic deposits. Instead, Mesozoic bryozoans are represented by two new orders. Bryozoans were abundant in Cretaceous seas and have remained so since that time.

The number of brachiopod species in early

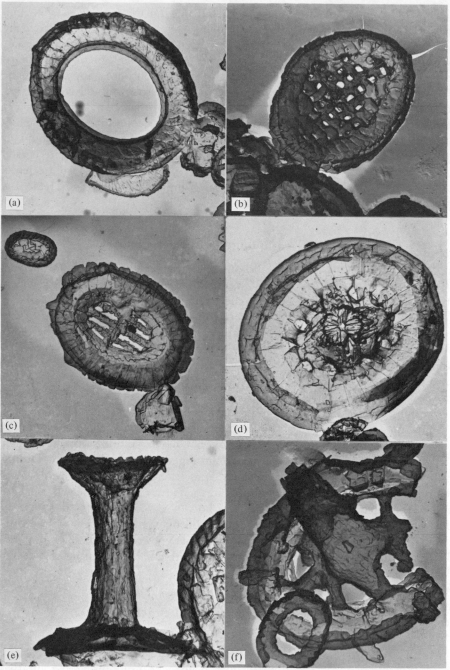

**FIGURE 12.34** Replica electronmicrographs of Early Cretaceous, late Albian, coccoliths recovered by the *Glomar Challenger* from site number 5 in the Atlantic Ocean, east of the Bahama Islands. Magnification approximately X10,000. **(a)** *Apertapetra gronosa.* **(b)** *Arkhangelskiella erratica.* **(c)** *Costacentrum horticum.* **(d)** *Cretarhabdus crenulatus.* **(e)** *Cretarhabdus decorus.* **(f)** *Cretarhabdus descussatus.* (Photo by David Burkry, U.S. Geological Survey, courtesy of the National Science Foundation.)

Mesozoic deposits was greatly reduced from that in late Paleozoic sequences, and it continued to decline throughout the Mesozoic. The simple, unspecialized forms have been the most persistent. *Lingula,* for example, has changed very little between the Paleozoic and the present. The terebratulids, which comprise most of the brachiopods living today, became numerous for the first time during the Mesozoic. They are important in correlating marine sequences, especially in the Tethyan region.

## Mollusks

The pelecypods, which expanded and diversified during the late Paleozoic, replaced the brachiopods as the dominant group in the near-shore, benthonic environment. They had already become well established in brackish and fresh waters as well. Important shallow-water marine pelecypods included oysterlike forms, such as *Exogyra* and *Gryphaea,* and forms closely related to the modern scallop. The pelecypod *Inoceramus,* which was abundant in Cretaceous seas (Fig. 12.35), grew to be as large as 1 m (3 ft) long—about the same size as the modern giant clam. Many Cretaceous chalks contain tiny prisms of calcite derived from the shells of this form. Some pelecypods had an external morphology similar to that of horn corals, and like horn corals, they attached themselves permanently to the sea bottom.

The ammonoids were close to extinction at the end of the Paleozoic. Only two families of ceratites persisted into the Triassic, and one of these did not survive beyond the Early Triassic. The remaining ammonoids dispersed and diversified so rapidly during the Triassic that this group has provided excellent index fossils. The ammonoids declined significantly in the Late Triassic and by the beginning of the Jurassic, only one family remained. However, a number of new ammonoid groups evolved from this stock during the Jurassic, and these spread throughout the oceans of the world. As in the Triassic, the rapid evolution and widespread occurrence of the Jurassic ammonoids makes them ideal guide fossils for intercontinental correlations. The 58 ammonoid zones that have been established for Jurassic marine deposits permit very accurate age determinations.

Most ammonoids are bilaterally symmetrical (Fig. 12.36), but quite a few of the Mesozoic forms diverged from this pattern (Fig. 12.35). As early as the Triassic, some ammonoids developed a spirally coiled shell, much like that of gastropods. Other forms developed straight shells or assumed a variety of shapes (Fig. 12.35). A few species grew to very large sizes. One specimen found near Graybull, Wyoming, measures more than 1.6 m (5 ft) across. Belemnoids, a group of squidlike cephalopods, were abundant in Cretaceous seas, and their cigar-shaped internal skeletons are common marine fossils (Fig. 12.37). By the end of the Cretaceous, all of the ammonoids and belemnoids were extinct.

## Arthropods

With the extinction of the trilobites at the end of the Paleozoic, the only important arthropods in the Mesozoic were crustaceans and insects. Mesozoic crustaceans include ostracods, branchiopods, barnacles, and lobsters; Mesozoic insects include flies, moths, lice, bees, beetles, and ants. The insects achieved a proficiency which has enabled them to persist relatively unchanged to modern times.

## Echinoderms

The Mesozoic echinoderms are represented mainly by crinoids and echinoids, and both are important guide fossils. Only one family of crinoids survived the mass extinctions at the end of the Paleozoic. This family gave rise to free-swimming, rootless forms which probably lived in communities much like modern crinoids. One species of rooted crinoid had a calyx more than 0.6 m (2 ft) across and a stem 16 m (50 ft) long.

**FIGURE 12.35** Upper Cretaceous fossils from the Ripley Formation in Tennessee: **(a)** *Nostoceras* sp., an aberrant ammonoid cephalopod (X1.1). **(b)** *Trigonia eufaulensis*, interiors of pelecypod valves showing original nacreous luster (X1). **(c)** *Trigonia eufaulensis*, exterior of a right valve (X1). **(d)** *Inoceramus sagensis*, typical large pelecypod (X1). **(e)** *Anchura*, gastropod (X2). **(f)** *Turritella vertebroides,* high, spired gastropod *Baculites* sp., straight shelled ammonoid, and other mollusks (X1.3). (Photo courtesy of Carlton Brett.)

**FIGURE 12.36** Restoration of the sea bottom during the Late Cretaceous showing ammonoids that are tightly coiled, loosely coiled (in weeds), and have straight shells. (Courtesy of the Field Museum of Natural History, Chicago.)

**FIGURE 12.37** Restoration of the Late Jurassic sea bottom in South Dakota with the squidlike belemnoid *Pachyteuthis* swimming above a bank of oysterlike pelecypods. (Courtesy of the Field Museum of Natural History, Chicago.)

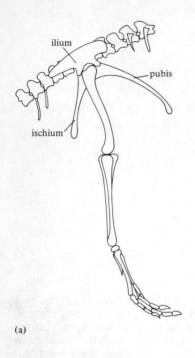

(a)

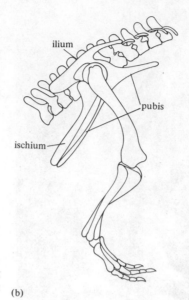

(b)

**FIGURE 12.38** Differences in the pelvis of the two types of dinosaurs: **(a)** Saurischian. **(b)** Ornithischian. (From Romer, Ref. 46.)

## Fish and Amphibians

Mesozoic fish are dominated by the ray-finned fish, the actinopterygians. The early Mesozoic forms had heavy scales and skeletons composed mostly of cartilage; later forms developed small scales and bony skeletons, which are characteristic of modern fish. The lungfish and lobe-finned fish were of little importance during the Mesozoic.

The amphibians continued to decline during the Mesozoic and the labyrinthodonts and steg-ocephalians became extinct by the end of the Triassic. Frogs appeared in the Jurassic and sala-manders in the Cretaceous. These small am-phibians are useful in ecological studies but are of little stratigraphic value.

## Terrestrial Reptiles

Early Triassic terrestrial reptiles include the cotylosaurs, therapsids, rhyncocephalians, eo-suchians, turtles, and thecodonts. The theco-donts are a group of Triassic reptiles which evolved from the eosuchians during the Late Permian or Early Triassic (**97**), and are the most primitive of the archosaurs or ruling reptiles. Descendents of the thecodonts included the dinosaurs, phytosaurs, birds, and crocodiles. Thecodonts and their descendents dominated the land for 160 million years. The earliest thecodonts were small, lizardlike carnivores. Among them were the first vertebrates able to walk on two legs. Many of the thecodonts had long, heavy tails which were probably used for balance while walking or running on two legs. Studies of their bones indicate that early theco-donts were probably cold-blooded, and that most of the later terrestrial thecodonts and their descendents were probably warm-blooded (**98**).

DINOSAURS. The dinosaurs are the most famous of the descendents of the thecodonts. The word *dinosaur* comes from the Greek words *deinos* (terrible) and *sauros* (lizard). Dino-saurs have been separated into two groups, saurischians (reptile-hipped) and ornithischians (bird-hipped). The difference between these

**FIGURE 12.39** Reconstruction of a confrontation between the largest carnivorous dinosaurs, *Tyrannosaurus rex* (right) and the horned dinosaur *Triceratops* (left.) (Courtesy of the Field Museum of Natural History, Chicago.)

two groups is in the orientation of the pubic bone. The pubic bone of the saurischians extends forward and down, while in the ornithischians one branch of the pubic bone points forward and up and the other branch points backward and down (Fig. 12.38). According to Robert Bakker, all of the dinosaur bones which have been studied show characteristics indicating that dinosaurs were warm-blooded (**99**). Furthermore, predator/prey ratios are generally 0.01 to 0.03, which is closer to that of warm-blooded animals than that of cold-blooded animals. At one time, dinosaurs were pictured as slow-moving animals similar to lizards or crocodiles. However, a number of paleontologists now view many dinosaurs as agile animals able to run for extended periods (**100**). Such a life style would be consistent with the viewpoint that dinosaurs were warm-blooded.

The dinosaurs show an evolutionary trend toward giantism, a trend common to many rapidly evolving groups. Some dinosaurs were truly monsters, reaching lengths of 15 m (50 ft) or more and were equally tall. However, there were also a few small dinosaurs; one was the size of a small dog. Big or small, however, most dinosaurs had disproportionately tiny brains for

their size. *Stegosaurus,* which was the size of a large elephant, had a brain the size of that of a kitten.

SAURISCHIANS.   The saurischians have been divided into two groups, the theropods and the sauropods. The theropods typically had five toes, but the fifth toe was greatly reduced in size and the first toe was off at an angle to one side. The theropods, which are closest anatomically to their thecodont ancestors, appear first in the fossil record during the Late Triassic. Theropods were bipedal predators, with long, strong hind legs and greatly shortened forelegs. The forelimbs of theropods were typically terminated by sharp claws. Among the theropods were large predators such as *Allosaurus, Gorgosaurus,* and *Tyranosaurus. Tyranosaurus rex,* the largest predator that ever lived, reached a length of 15 m (50 ft) stood 6 m (20 ft) tall and had teeth the size of railroad spikes (Fig. 12.39).

The ornithopods were a type of theropod that lived during the Late Jurassic and Cretaceous periods. They resembled the ostrich in size and form and are known as the ostrich dinosaurs. They had long arms that ended in large claws perhaps used for digging. Study of muscle scars on the bones of the ornithopods and study of the bones themselves indicate that ornithopods were strong runners (**101**). This characteristic suggests that these animals were warm-blooded. Ornithopods had exceptionally keen vision, and they had unusually large brains for a dinosaur. Their intelligence was probably comparable to that of large modern birds. Weak jaw muscles suggests a diet of soft-bodied animals, grubs, or eggs (**101**).

Sauropods, the other group of reptile-hipped dinosaurs, evolved from the theropods. They appeared first in the fossil record in the Early Jurassic, but did not become abundant until the Late Jurassic. Sauropods reverted to walking on four legs, and were typically herbivorous in contrast to the carnivorous nature of their theropod ancestors. The feet of sauropods were large and had five toes. Among the sauropods were the largest land animals that ever lived. Typically they had great, bulky bodies, elephantlike legs, long tails, heads that were tiny in proportion to their bodies, and long necks. The long necks would have allowed these sauropods to browse in the tops of trees. Their jaws were weak and were occupied by peglike teeth about the diameter of a lead pencil. The teeth were not adapted for chewing, and therefore food was probably swallowed whole. Food was ground up in the stomach by means of gizzard stones (as in the stomachs of modern chickens). Clusters of smooth, polished gastroliths are the fossil remains of these gizzard stones. It is somewhat of a mystery how the large sauropods could have consumed enough plants to sustain their great bulk, especially if they were warm-blooded. Perhaps the elimination of chewing helped.

One interesting sauropod is the well-known *Apatasaurus* (formerly known as *Brontosaurus*). Its neck was almost 6 m (20 ft) long, and its head was smaller than many of the vertebrae in its neck (Fig. 12.40). Each foot covered about a square yard! It weighed 27 to 36 metric tons (30 to 40 tons) (as much as the combined weight of 7 to 10 large modern elephants) and was 15 m (50 ft) long. Several other sauropods shared the same general body type. One of these, *Diplodocus,* grew to a length of 26 m (80 ft) and weighed more than 45 metric tons (50 tons). *Diplodocus* lived during the Jurassic Period. Another, *Brachiosaurus,* stood 13 m (40 ft) high and had nostrils at the top of its head. A recent find, unofficially known as "Supersaurus," may have stood 15 m (50 ft) tall and weighed 90 metric tons (100 tons), the equivalent of 15 large elephants!

The earliest researchers on dinosaurs could not believe that the large sauropods could live on dry land. Consequently, they postulated that the large sauropods walked on the bottoms of lakes or swamps, raising their heads above the water only to breathe. An aquatic environment would help to take the weight off their feet, but there is considerable evidence that the large sauropods were dry land animals. For one thing, the nostrils, rib cage, feet, knees, and neck of these sauropods are quite different from that of large semiaquatic mammals, such as the hippopotamus, but these structures do resemble those of large land mammals, such as the elephant. The structure of the rib cage of the large sauropods suggests that the lungs would collapse if the animal were largely submerged in water.

ORNITHISCHIANS. The ornithischians are generally believed to have evolved separately from the saurischians (**97**), but Robert Bakker has presented evidence that the ornithischians evolved instead from an early saurischian dinosaur (**99**).

Almost all ornithischians were herbivorous and had horny beaklike or bill-like structures in the place of front teeth. Their back teeth were crowded closely together forming an efficient grinding surface for chewing plants. Many ornithischians were bipedal (Fig. 12.41), as were the stock from which they evolved, but quite a few reverted to walking on all four feet.

There are three types of ornithischians dinosaurs: ornithopods, ankylosaurs, and ceratopsians. The ornithopods were the least specialized and most primitive of the three. *Ornithopod* means "bird feet," and they are so called be-

**FIGURE 12.40**  Reconstruction of the skeleton of *Apatosaurus* (*Brontosaurus*) *excelus*. (Courtesy of the Field Museum of Natural History, Chicago.)

**FIGURE 12.41**  Part of the exposed footprint horizon at Dinosaur State Park in central Connecticut. White crosses on the photo are 10-ft grid marks. The fine chalk lines represent an attempt to trace the path of individual animals. (Courtesy of the Connecticut Geological and Natural History Survery.)

**FIGURE 12.42** Restoration of a Late Cretaceous landscape with several kinds of dinosaurs. On the right are the duckbilled dinosaurs, *Trachodon,* in the center foreground is the armored *Ankylosaurus,* at the left in the water is *Corythosaurus,* and in the background at the left are the crested duckbills, *Parasaurolophus.* (Courtesy of the Field Museum of Natural History, Chicago.)

cause their feet resembled that of a bird. The best known ornithopods are the hadrosaurs and the trachodons, commonly known as duck-billed dinosaurs (Fig. 12.42). These large orni-thopods had a bony bill and webbed feet. They were probably semiaquatic, and, when on land, were able to walk on either two or four legs. Some developed bizarre crested skulls that en-closed elongated nasal passages. These nasal passages may have ben used to amplify their sense of smell (**100**), a distinct advantage in a world filled with predators such as *Tyrano-saurus.* Alternatively, the elongated nasal pas-sages may have been used to warm air coming into the lungs.

The Troodonts, another group of ornitho-pods, developed very thick skulls. *Pachycepha-losaurus* had a domed head with a 23 cm (9 in.) thick solid mass of bone above its brain.

The stegosaurs are a group of armored dino-saurs that lived during the Jurassic. They had a spiked tail and double row of bony plates ex-tending down their backs (Fig. 12.43). The

bony plates may have been used for the regula-tion of heat (**102**), perhaps in a manner similar to that of the fin-backed pelycosaurs (see Chap-ter 11). The stegosaur's spiked tail was doubt-lessly used for protection against the large Ju-rassic predators.

The ankylosaurs first appeared during the Cretaceous period. They were squat, heavily armored quadrupeds that had spikes on their bodies and a clublike tail containing spikes or bony knobs (Fig. 12.42).

The ceratopsians appeared during the Middle Cretaceous and they flourished, while other groups declined, until the end of the period. These dinosaurs were characterized by a plate-like extension of the skull that covered the back of the neck. Many had sharp horns, presumably for protection against predators. A typical cera-topsian was *Triceratops* (Figs. 12.39 and 12.44). These dinosaurs were 6 to 9 m (20 to 30 ft) long and stood 2.5 m (8 ft) high at the shoulder. Their neck and leg muscles were strong enough to make powerful lunges with their sharp horns

FIGURE 12.43 Reconstruction of a Jurassic landscape in the western part of the United States with two stegosaurs. (From a painting by Charles R. Knight, courtesy of the Field Museum of Natural History, Chicago.)

FIGURE 12.44 Reconstruction of the skeleton of the horned dinosaur *Triceratops prorsus* from the Upper Cretaceous Lance Formation in Niobrara County, Wyoming. (Courtesy of the Smithsonian Institution.)

at attacking predators. They roamed the North American continent in large numbers during the Late Cretaceous.

Toward the end of the Cretaceous, the numbers and variety of dinosaurs began to decline. *Triceratops* and some of the hadrosaurs were among the last dinosaurs. Of these two groups, the hadrosaurs became extinct first. Studies of eggs belonging to the remaining *Triceratops* give some indications of the conditions at the end of the Cretaceous. The egg shells of *Triceratops* show a progressive decrease in thickness in younger and younger strata (**100**). The youngest eggs found had shells so thin that there could not have been enough calcium in them for proper development of the bones of the embryos (**100**). The decrease in shell thickness is presumably a result of stresses in an increasingly hostile environment. The causes of these stresses are discussed later in this chapter.

PHYTOSAURS AND CROCODILES. The phytosaurs were cold-blooded quadrupedal thecodonts. They were very similar in appearance to crocodiles except that their nostrils were placed quite close to their eyes, while the nostrils of crocodiles are at the end of their noses. However, studies of phytosaur skeletons indicate that they were not the ancestors of the crocodiles. Phytosaurs appear first in the fossil record in the Late Triassic, and they became extinct before the end of that period (**97**).

The crocodiles evolved from the thecodonts and appear first during the Late Triassic or Early Jurassic (**103**). They were the only archosaur to survive the extinctions at the end of the Mesozoic (**97**). The crocodiles and the phytosaurs are excellent examples of parallel evolution. Apparently, this body shape is very well adapted to the environment in which these animals lived. Even today, crocodiles are the dominant animal of tropical swamps. Crocodiles are slow-moving, cold-blooded animals that spend much of their lives sprawled on their stomachs, basking in the sun. Modern crocodiles are

smaller than Mesozoic forms, which during the Cretaceous reached lengths of 15 m (50 ft).

RHYNCHOCEPHALIANS. The rhynchocephalians are a primitive, quadrupedal reptile group that evolved from the eosuchians during the Late Permian or Early Triassic (**97**). They are characterized by the presence of an overhanging beaklike structure on the upper jaw and teeth fused to the edges of their jaw rather than set into sockets. Most rhynchocephalians lived during the Triassic and they reached their climax during the middle of that period. All rhynchocephalians are now extinct except for the tuatara, which still survives on a seacoast of New Zealand. Triassic rhynchocephalians were up to 2 m (6 ft) long and most had short, flat skulls.

LIZARDS, SNAKES, AND TURTLES. Lizards evolved from the eosuchians during the Middle or Late Triassic (**97**). All modern lizards are cold-blooded, as were probably their ancestors. Most lizards, especially the larger varieties, move slowly except for brief spurts of rapid movement. They spend much of their lives sprawled on the ground. The forerunners of the skinks appeared first in the Late Jurassic, and the forerunners of the iguanids appeared first during the Jurassic or Cretaceous. The marine lizards first appeared in the Late Cretaceous and they will be discussed in the following section. Snakes appeared first in the Cretaceous. They probably evolved from the lizards.

The Chelonians (turtles) appeared first during the Triassic, and they have changed little since that time (**97**). Turtles probably evolved from the cotylosaurs. The armor plate of a typical turtle is composed of horny scutes, which developed from reptile scales, and bony plates beneath the scutes. Marine turtles evolved from terrestrial turtles during the Cretaceous, and they will be discussed in the next section.

FIGURE 12.45 Skeleton of an ichthyosaur. (Courtesy of the Field Museum of Natural History, Chicago.)

FIGURE 12.46 Restoration of a Jurassic seascape with plesiosaurs on the left and fishlike ichthyosaurs on the right. (From a painting by Charles R. Knight, courtesy of the Field Museum of Natural History, Chicago.)

## Marine Reptiles

When fierce competition for food developed among the terrestrial reptiles, some became adapted to a marine environment. Their bodies became streamlined and their legs were modified into flippers. The marine reptiles included mosasaurs, ichthyosaurs, plesiosaurs, and turtles.

The mosasaurs are lizards that became adapted to an aquatic life. They first appeared during the Late Cretaceous and rapidly spread throughout the seas of the world. Mosasaurs were up to 9 m (30 ft) long and had a long head,

short neck, long slim body, and a long tail. They had large, sharp teeth and probably fed primarily on fish. Their tails were laterally flattened for swimming. Paddlelike feet were probably used for steering. The mosasaurs became extinct before the end of the Cretaceous.

The ichthyosaurs are the most fishlike of the marine reptiles (Figs. 12.45 and 12.46). They first appear in the Early Triassic fossil record, but they probably evolved from the cotylosaurs during the Late Paleozoic. Ichthyosaurs swam like fish by undulations of the body and tail, and they bore a strong resemblance to modern porpoises. Those with long, pointed snouts probably ate fish, while those with short snouts probably ate shellfish. The presence of small ichthyosaur skeletons within larger ichthyosaur skeletons suggests that these marine reptiles gave birth to live young. Eggs were probably hatched within the mother's body, rather than outside the body as in most reptiles (**97**). The ichthyosaurs reached their peak during the Late Triassic or Early Jurassic and became extinct well before the end of the Cretaceous.

The plesiosaurs (Fig. 12.52) probably evolved from the lizardlike protorosaurs during the Permian, although they do not appear in the fossil record until the latest Triassic (**97**). They are characterized by small heads, long necks, and large bodies. The plesiosaurs probably paddled along the surface of the water or just beneath it by means of oarlike limbs (**103**). In this respect, they were probably just like the modern marine turtles. Some plesiosaurs had sharp teeth and probably ate fish, while others had large, flattened teeth and probably ate shellfish. Plesiosaurs were up to 15 m (50 ft) long, but most were 3 to 6 m (10 to 20 ft) long. They reached their peak during the Late Triassic and Early Jurassic, and became extinct before the end of the Cretaceous.

Marine turtles had a morphology similar to that of modern turtles, except that the skeleton was somewhat more flattened. The trend toward giantism was shown even in the turtles. One species which lived during the Cretaceous reached a length of 3.6 m (12 ft).

## Flying Reptiles

The vertebrates dominated the land and both fresh and salt waters at the beginning of the Mesozoic. However, they had not yet invaded the last major environment available to them: the air. This realm had been the exclusive domain of insects since the Carboniferous. The reptiles were the first of the vertebrates to fly. Flying reptiles are known as pterosaurs, and they evolved from the thecodonts. Before they could fly, pterosaurs had to develop the ability

to glide for short distances. One pterosaur had short, winglike structures as early as the Late Triassic, but the first flying vertebrate did not appear until the Jurassic.

The pterosaurs had a very long fourth finger from which a wide membraneous wing extended. This wing was somewhat like that of the modern bat. Pterosaur bones were rather thin and often hollow like those of modern birds. The hollow bones had openings in them which perhaps could be filled with air (**101**). The ability to fly allowed the pterosaurs to exploit new supplies of food, such as insects in flight. Furthermore, it provided some degree of protection against carnivorous dinosaurs.

Jurassic pterosaurs were small—no larger than robins (**103**). However, by the Cretaceous some varieties reached impressive sizes. *Pteranodon* had a wingspread of 9 m (27 ft) (Fig. 12.47), and a recently discovered pterosaur found in Texas had a wingspan of 15 m (50 ft) (**104**). This is larger than the wingspan of many jet fighters.

The small body size in comparison to wingspan and considerations of the degree of muscle development suggest that pterosaurs spent much of their time in the air soaring, perhaps like the modern albatross. Pterosaurs had relatively large brains and large eyes. Their large eyes helped them locate prey while soaring over land or water. They probably caught most of their prey while in flight, because it could not have been very easy for many of them to take off after landing. The teeth of one small pterosaur, *Pterodactylus,* pointed outward and presumably could be used for spearing small fish which swam just below the surface of the water.

There are two indications that the pterosaurs were warm-blooded. First, they must have been able to sustain physical activity over considerable periods of time while flying. Such a feat is next to impossible for a cold-blooded reptile (**103**). Second, some fossil remains show evidence that the pterosaur was covered with hair (**97**). Hair is a good insulator and it is very

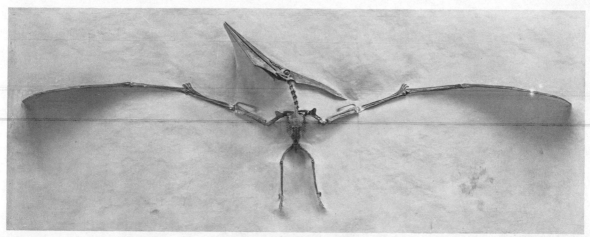

FIGURE 12.47 Restored skeleton of the flying reptile *Pteranodon ingens* from the Upper Cretaceous chalk deposits of western Kansas. (Courtesy of the Smithsonian Institution.)

useful to warm-blooded animals to help retain body heat. However, it would be a hindrance to cold-blooded animals because they must rely on absorption of heat from the sun's rays to raise their body temperatures. Hair, being an insulator, would slow down heat absorption. Pterosaurs reached their peak in the Late Jurassic and became extinct close to the end of the Cretaceous, not long before dinosaurs became extinct (**97**).

## Birds

Birds are warm-blooded animals covered with feathers. Their skeletal structures are similar to those of the reptiles, especially the small dinosaurs, from which they may have evolved (**105**). Remains of the earliest known bird, *Archaeopteryx,* have been found in a very fine-grained (lithographic) limestone of Late Jurassic age near Solenhofen, Germany. *Archaeopteryx* had a skeleton almost identical to that of a dinosaur, but the presence of impressions of feathers associated with the skeleton shows that it was a bird (Fig. 12.48). A backwards facing toe on each foot would have allowed this bird to grasp branches. However, the lack of a well-developed keel or breastbone in *Archaeop-*

*teryx* suggests that this bird was not able to fly, though it may have been able to glide for short distances. *Archaeopteryx* was, therefore, a transitional form between reptiles and modern birds.

Two thigh bones belonging to birds have recently been found in deposits of Late Jurassic age in eastern Colorado (**106**). These deposits are about the same age as the deposits in which *Archaeopteryx* are found, but the development of the thigh bones of the birds is much more like that of modern birds than *Archaeopteryx* (**106**). It is quite possible that this bird, which has not yet been named, was a true flyer.

By Cretaceous time, the breastbone of birds had been considerably enlarged. This adaptation permitted the attachment of powerful muscles for the working of wings. One Cretaceous bird, *Hesperornis,* had only vestiges of wings and had webbed feet. This bird was a strong swimmer able to dive below the surface of the water to catch fish.

## Mammals

The record of the early evolution of the mammals is fragmentary due to the scarcity of fossil remains, which are mostly comprised of teeth

FIGURE 12.48  Restoration of a Jurassic landscape showing the small dinosaur *Ornitholestes* in the foreground at the left, several birds of the genus *Archaeopteryx,* and a number of pterosaurs in the air and on the trees. (From a painting by Charles R. Knight, courtesy of the Field Museum of Natural History, Chicago.)

and fragments of jaws. The remains that have been found suggest that the Mesozoic mammals were relatively small and few in number. However, the brain cavities of the mammals were large when compared with those of the reptiles, and presumably they were more intelligent. It was probably this factor which enabled them to survive under the domination of the dinosaurs.

The mammals probably evolved from the therapsid reptiles during the Triassic (**107**). The earliest fossil mammals have been found in deposits of latest Triassic age, and by the Jurassic six orders had evolved (**107**) (Fig. 12.49). Each order is characterized by the shape of its teeth. Of these, the pantotheres were the basic stock from which the important Cenozoic mammals evolved. The teeth of the pantotheres were sharp and included well-developed canines.

Their diet probably consisted of insects, worms, grubs, and very small reptiles.

### Plants

The record of Mesozoic flora has been reconstructed from macrofossils and assemblages of spores and pollen. However, since it is not always possible to relate the spores and pollen to the plants that produced them, such reconstructions are subject to a degree of uncertainty. This is a less serious problem in upper Cretaceous deposits because many Late Cretaceous plants closely resembled modern forms.

Mesozoic floras differ markedly from those of the Paleozoic. Cycadeoides and cycads achieved prominence during the Jurassic and Early Cretaceous. These plants looked like

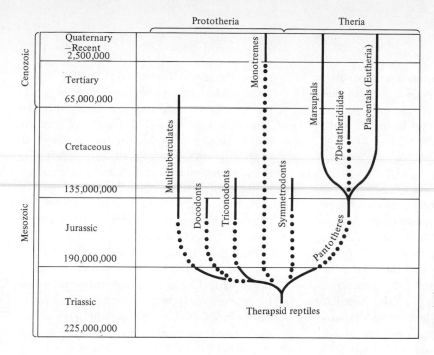

FIGURE 12.49 Evolution of the various orders of mammals from the therapsid reptiles. (After Z. Kieland-Jaworowska, Ref. 107.)

palms and grew more than 13 m (40 ft) high (Fig. 12.49). In the southern continents, the flora included ginkgoes and seed ferns in addition to cycadeoides. Seed ferns such as *Sphenopteris* and *Glossopteris* declined in importance through the Mesozoic.

Among the gymnosperms, conifers became increasingly important during the Mesozoic. This is especially true of the cedar-, pine-, and firlike forms. Ginkgoes, which were also widespread, occurred in the Arctic and central regions of North America, in Europe, and in Asia.

The most significant development in plant life in the Mesozoic was the appearance of flowering plants—the angiosperms. The unique adaptations of the angiosperms are in their reproductive organs (**108**):

1. The ovules of angiosperms are protected.
2. A stigma and style, structures in the female flower, allow fertilization soon after pollination.
3. Flowers which attract pollinating insects and birds are present in angiosperms. There are two distinct advantages of this type of pollination over wind pollination (such as occurs in the gymnosperms). First, winds are often unidirectional, and second, large amounts of pollen must be produced for successful wind pollination.
4. Seeds of the angiosperms may be carried by birds and mammals, both by passing through the intestinal tract and being stuck on the fur, feet, or feathers of these animals.

Angiosperms are divided into two subclasses: the monocotyledons, which are characterized by an uncovered seed or single seedleaf, a lack of bark and pith, and simple leaf veinations; and the more advanced dicotyledons, which have a covered or double seedleaf, bark, pith, and complex leaf veinations. Pollen from the monocotyledons appear in rocks as old as Jurassic and definite palm stems have been found in Middle Jurassic deposits (**109**). The explosive development of the angiosperms occurred during the Late Cretaceous. Only about 5 percent of the Early Cretaceous vascular flora was comprised of angiosperms, but by the end of the Cretaceous, approximately 95 percent of the vascular flora consisted of angiosperms. The dicotyledons, which first appeared near the boundary between the Early and Late Cretaceous, had become the dominant angiosperm form by the end of the period. Cretaceous monocotyledons include lilies and palms; among the Cretaceous dicotyledons are oak, birch, fig, willow, and a variety of shrubs.

## Extinctions

A large number of important animal groups became extinct at or near the end of the Mesozoic. Among the vertebrates were the dinosaurs, pterosaurs, plesiosaurs, ichthyosaurs, and mosasaurs; among the marine invertebrates were the ammonoids, belemnites, rudistids, exogyras, graphaeas, and inoceramids. Of the reptiles, only lizards, snakes, turtles, and crocodiles survived. Furthermore, numerous genera of foraminifera (marine protozoa) and coccoliths (marine algae) suddenly became extinct at the end of the Mesozoic. Among the land plants, cycadeoides became extinct and the number of gymnosperms were greatly reduced at the end of the Mesozoic. Numerous theories have been proposed to explain the mass extinctions:

1. Disease has been suggested as a cause of the extinction of the dinosaurs and related reptiles. However, there is no evidence of such disease in the fossil record. Furthermore, diseases generally affect one or at most only a few species, not a group as large and diverse as the dinosaurs.
2. Some sort of worldwide catastrophy at the end of the Mesozoic has been suggested. One such postulated catastrophy is a nearby supernova (**101**). However, there is no indication in the fossil record for this or any other catastrophy at the end of Mesozoic time. In the sequences that span the Cretaceous-Tertiary boundary, there is only a steady decline in certain groups.
3. A possible cause for the extinction of the dinosaurs and related reptiles is the rise of the mammals. Mesozoic mammals were not large enough to compete directly with the reptiles for food and living space, but they

may have eaten reptile eggs and/or reptile young. Modern reptiles leave their eggs virtually unprotected, and it is likely that the Mesozoic reptiles did the same. There are two problems with this suggestion. First, all of the mammals were small, and, judging by the number of Mesozoic fossil mammal remains which have been discovered so far, they were not very numerous. Second, the eggs and young of the large dinosaurs were quite large, perhaps too large for the small mammals to attack. On the other hand, birds may have played a role in reptilian extinctions (**110**). During the late Mesozoic, some birds were fairly large in size. Certain varieties of birds, such as the sea gull, have developed a taste for freshly hatched reptiles, such as marine turtles. It is more likely that birds, rather than mammals, played a greater role in the extinction of the dinosaurs and related groups.

4. It is possible that the change in terrestrial floras from dominantly gymnosperms and other lower plants to dominantly angiosperms during the Late Cretaceous played a role in the extinction of the herbivorous dinosaurs. Presumably, the early Mesozoic herbivorous dinosaurs ate gymnosperms and other lower plants, and it is possible that they could not adjust to the changing vegetation. One might wonder just what role the herbivorous dinosaurs played in the change in vegetation. Perhaps the herbivorous dinosaurs caused their own extinction by eating so selectively. Modern elephants in Africa sometimes destroy so many trees that they have difficulty surviving. The extinction of the herbivorous dinosaurs would almost certainly have caused the extinction of the large carnivorous dinosaurs who preyed on them.
5. A gradual withdrawal of the seas from the continents began during the Late Cretaceous (Campanian or Maestrichtian) (**111**). This regression of the seas many have been an important factor in the extinction of the ich-

thyosaurs, mosasaurs, plesiosaurs, and ammonites. The zones of shallow waters would have been greatly reduced, and if these animals lived primarily in shallow waters, this reduction would have significantly increased the competition for food among these groups.

6. The withdrawal of the seas at the end of the Cretaceous was apparently related to a general uplift of the continents at that time. This uplift would have helped drain swamps in which many herbivorous dinosaurs lived, and it would have resulted in a gradual cooling of the climate (see Chapter 14). Evidence of a late Mesozoic cooling of the climate comes from the study of changes in terrestrial vegetation and oxygen isotope studies of deep-sea cores (see Chapter 14) (**112**). A marked cooling of the climate could cause extinctions of the dinosaurs. The skin of dinosaurs is occasionally preserved, and there is no sign of any hair or feathers on it. If the dinosaurs were warm-blooded, they did not have any insulation to help keep in their body heat. This may not have been important if the climate was warm over most of the world, as it apparently was until the end of the Cretaceous. However, if the climate became significantly cooler, the lack of insulation could have been critical for both large and small dinosaurs (**101**). Even if the dinosaurs were cold-blooded, the decrease in temperatures may have been enough to cause the extinction of the dinosaurs. At the present time, large, cold-blooded reptiles are restricted to tropical and semitropical climates.

7. Continental drift may have played a role in the extinction of the dinosaurs. By the end of the Cretaceous Pangaea had become fragmented into six major continents. This would have prevented migration of many of the dinosaurs into a more suitable climate when the climate became cooler.

8. Overspecialization probably played a role in the extinction of many groups, especially the dinosaurs. The dinosaurs had become adapted to very specific environments, and when those enviroments changed late in the Cretaceous, they may not have been able to adapt to the new conditions.

From the foregoing discussion, it can be seen that no single cause can be identified for the mass extinctions at the end of the Cretaceous. Rather, the extinctions were probably caused by a combination of factors, the most important of which were uplift of the continents, withdrawal of the shallow epicontinental seas, cooling of the climate, fragmentation of the continents, competition, and overspecialization.

## MESOZOIC CLIMATE

### Triassic

It may be inferred from the pattern of paleoclimatic indicators that the climate during the Triassic was relatively warm and dry (Fig. 12.1a). No glacial deposits of this age have been confirmed and dune sandstones and red beds are common at low as well as high paleolatitudes. Coals are more restricted and are less important than coals of other periods.

### Jurassic

The distribution of Jurassic paleoclimatic indicators suggests that the Jurassic climate was warm, but not as dry as that of the Triassic (Fig. 12.1b). Reefs, evaporites, and red beds are generally confined to a zone within 30° of the paleoequator. Temperate or subtropical coals are widespread from 30° to 70° paleolatitude, and no Jurassic glacial deposits have been reported.

### Cretaceous

The Cretaceous climate was initially warm and rather moist (Fig. 12.1c). The flora was remarkably uniform throughout the world during the Early Cretaceous, with subtropical forms in latitudes up to 70° from the paleo-equator. Ferns and cycads from high latitudes are similar to those growing today in the sub-tropical South American rain forests. Creta-ceous coals are found on almost every continent at paleolatitudes ranging from 0° to 90°. Cool-ing of the climate began during the Late Creta-ceous and continued into the Tertiary. The major reason for this cooling was uplift of the continents.

## SUMMARY

At the beginning of the Mesozoic, all the conti-nents were part of one landmass, Pangaea. The breakup of the continents probably began dur-ing the Late Triassic with the separation of North America and Europe from Gondwana-land and the separation of Africa from South America, Australia, and Antarctica. During the Late Jurassic, Africa began to separate from India, Malagasy began to separate from Africa, Greenland began to separate from North America, Spain and Portugal began to rotate away from Europe, and Alaska began to rotate away from northern Canada. During the Late Cretaceous, Greenland began to separate from Europe and India began to separate from the Seychelles Islands.

The separation of North America and Eu-rope from Gondwanaland resulted in the for-mation of a series of fault basins (possible aulacogens) near the line of separation. As these continents continued to separate, geosyn-clines developed along the margins of the newly created Atlantic Ocean and Gulf of Mexico. The separation also resulted in the folding, faulting, uplift, and intrusion of the deposits of the Cordilleran Geosyncline during the Late Triassic and again during the Late Ju-rassic. The mountain range that formed during the latter deformation furnished sediments for the Coast Range Geosyncline west of the Cor-dilleran Geosyncline.

Geosynclinal deposition occurred on the Pa-cific and Tethyan margins of the continents as oceanic plates moved under these margins. As the Eurasian and African plates moved to-gether, a series of geanticlinal ridges was pro-duced within the northern Tethyan Geosyn-cline. Thick sequences of clastics resulted from the erosion of these uplifts.

Seafloor spreading occurred throughout the Mesozoic in the Pacific Ocean and began in the Atlantic and Indian oceans during the Late Tri-assic. Changes in the direction of seafloor spreading occurred during the Late Jurassic, Middle Cretaceous, and Late Cretaceous.

The Mesozoic saw the rise of the reptiles, which dominated almost all the available ter-restrial environments and the oceans and air as well. Some of the dinosaurs reached gigantic proportions, and the plesiosaurs and mosasaurs in the marine realm were almost as impressive. The mammals originated in the Late Triassic, but evolved very slowly. The extinctions at the end of the Mesozoic include the dinosaurs, pterosaurs, plesiosaurs, ichthyosaurs, mosa-saurs, and ammonoids. These extinctions may have been related to an uplift of the land, with-drawal of the shallow seas, cooling of the cli-mate, fragmentation of the continents, overspe-cialization, and competition for food and living space.

Horsetails, seed ferns, and conifers were the important floras of the Early and Middle Tri-assic, but from Late Triassic through Early Cretaceous, floras were dominated by ferns, cycadeoides, cycads, ginkgoes, and conifers. Angiosperms, which first appeared during the Jurassic, expanded rapidly into most terrestrial environments during the Late Cretaceous.

Mesozoic climates were generally warmer than those of either the Paleozoic or the Ceno-zoic. The Triassic was unusually arid, but the climate was more moist during the Jurassic and Cretaceous. However, beginning during the Late Cretaceous, worldwide climates began a long decline, which culminated in the development of continental glaciers during the late Tertiary and Quaternary.

## REFERENCES CITED

1. J. D. Phillips and D. Forsyth, 1972, *Geol. Soc. Amer. Bull.,* v. 82, p. 1579.
2. B. Cornet, A. Traverse, and N. G. McDonald, 1973, *Science,* v. 182, p. 1243; H. L. Cousminer and W. Manspeizer, 1976, *Science,* v. 191, p. 943.
3. W. H. Kanes and J. P. Conolly, 1971, *Geol. Soc. Amer. Abstr. Programs,* v. 3, p. 617.
4. R. D. Dallmeyer, 1975, *Geology,* v. 3, p. 243.
5. P. Toulmin, 1961, *Geol. Soc. Amer. Bull.,* v. 72, p. 775.
6. DSDP Scientific Staff, 1970, *Geotimes,* v. 17, no. 7, p. 14.
7. D. E. Hayes and P. D. Rabinowitz, 1975, *Earth Planet. Sci. Lett.,* v. 28, p. 105; DSDP Scientific Staff, 1975, *Geotimes,* v. 20, no. 12, p. 18.
8. C. A. Burk *et al.,* 1969, *Amer. Assoc. Petrol. Geol. Bull.,* v. 53, p. 1338.
9. W. C. Pitman III and M. Talwani, 1972, *Geol. Soc. Amer. Bull.,* v. 83, p. 619.
10. N. L. McIver, 1972, *Can. J. Earth Sci.,* v. 9, p. 54.
11. O. Renz, R. Imlay, Y. Lancelot, and W. B. F. Ryan, 1975, *Eclogae Geol. Helv.,* v. 68, p. 431; P. A. Rona, 1973, *Geol. Soc. Amer. Bull.,* v. 84, p. 2851.
12. U. Jux, 1961, *Louisiana Geol. Surv. Bull. 38.*
13. J. F. Dewey, 1977, *Geol. Soc. Amer. Abstr. Programs,* v. 9, p. 949.
14. E. D. Schneider and G. L. Johnson, 1971, *Rep. Inst. Geol. Sci.,* v. 70/13, ICSU/SCOR Symp., Cambridge.
15. G. Siedner and J. A. Miller, 1968, *Earth Planet. Sci. Lett.,* v. 4, p. 451; R. V. Dingle and R. A. Scrutton, 1974, *Geol. Soc. Amer. Bull.,* v. 85, p. 1467.
16. DSDP Scientific Staff, 1973, *Geotimes,* v. 18, no. 4, p. 16.
17. R. G. Markl, 1974, *Nature,* v. 251, p. 196.
18. DSDP Scientific Staff, 1974, *Geotimes,* v. 19, no. 11, p. 16.
19. I. McDougall and N. R. Ruegg, 1966, *Geochim. Cosmochim. Acta,* v. 30, p. 191; G. Amarl, V. G. Cordani, K. Kawashita, and J. H. Reynolds, 1966, *Geochim. Cosmochim. Acta,* v. 30, p. 159; K. M. Creer, J. A. Miller, and A. G. Smith, 1965, *Nature,* v. 207, p. 282.
20. N. Herz, 1977, *Geol. Soc. Amer. Bull.,* v. 88, p. 101; J. S. Marsh, 1973, *Earth Planet. Sci. Lett.,* v. 18, p. 317.
21. DSDP Scientific Staff, 1975, *Geotimes,* v. 20, no. 6, p. 22.
22. K. O. Emery *et al.,* 1975, *Amer. Assoc. Petrol. Geol. Bull.,* v. 59, p. 3.
23. R. L. Larson and J. W. Ladd, 1973, *Nature,* v. 246, p. 209.
24. P. J. Fox, B. C. Heezen, and G. L. Johnson, 1970, *Science,* v. 170, p. 1402; R. V. Dingle and H. C. Klinger, 1971, *Nature Phys. Sic.,* v. 232, p. 37; G. O. Allard and V. J. Hurst, 1969, *Science,* v. 163, p. 528.
25. D. A. Valencio and J. F. Vilas, 1970, *Nature,* v. 225, p. 262.
26. J. R. Heirtzler *et al.,* 1968, *J. Geophys. Res.,* v. 73, p. 2119.
27. DSDP Scientific Staff, 1970, *Geotimes,* v. 15, no. 11, p. 10.
28. Y. Kristoffersen and M. Talwani, 1977, *Geol. Soc. Amer. Bull.,* v. 88, p. 1037.
29. J. R. Heirtzler and D. E. Hayes, 1967, *Science,* v. 157, p. 185.
30. D. White and C. Schuchert, 1898, *Geol. Soc. Amer. Bull.,* v. 9, p. 343.
31. J. D. A. Piper, 1975, *Phys. Earth Planet. Int.,* v. 11, p. 36.
32. X. LePichon *et al.,* 1971, *J. Geophys. Res.,* v. 76, p. 4724.
33. S. W. Carey, 1958, *Continental Drift: A Symposium:* Geology Department, University of Tasmania, Hobart.
34. J. F. Sweeney, 1977, *Geol. Soc. Amer. Bull.,* v. 88, p. 41.
35. D. L. Clark, 1974, *Geology,* v. 2, p. 41.
36. M. Churkin, Jr., 1973, *Amer. Assoc. Petrol. Geol., Mem. 19,* p. 485.
37. R. Van der Voo and J. D. A. Zijderveld, 1969, *Konikl. Nederlandse Geol. Mijnbouw Verh.,* v. 26, p. 121; R. Van der Voo and A. Boessenkool, 1973, *J. Geophys. Res.,* v. 78, p. 5118.
38. E. J. W. Jones and J. I. Ewing, 1969, *Science,* v. 166, p. 102.
39. DSDP Scientific Staff, 1976, *Geotimes,* v. 21, no. 12, p. 19.

40. M. W. McElhinny, 1973, *Paleomagnetism and Plate Tectonics:* Cambridge University Press, Cambridge.

41. A. Hallam and B. W. Sellwood, 1976, *J. Geol.,* v. 84, p. 301; P. R. Vogt and O. E. Avery, 1974, *J. Geophys. Res.,* v. 79, p. 363.

42. K. Burke and J. F. Dewey, 1973, *J. Geol.,* v. 81, p. 406.

43. S. Moorbath and H. Welke, 1969, *Earth Planet. Sci. Lett.,* v. 5, p. 217; M. Brooks, 1973, *J. Geol.,* v. 81, p. 81; C. K. Brooks and A. J. W. Gleadow, 1977, *Geology,* v. 5, p. 539.

44. DSDP Scientific Staff, 1975, *Geotimes,* v. 20, no. 2, p. 24.

45. M. Talwani and O. Eldholm, 1977, *Geol. Soc. Amer. Bull.,* v. 88, p. 969.

46. S. H. Knight, 1953, *Wyo. Geol. Assoc. Guidebook, Eighth Ann. Field Conf.,* p. 65.

47. I. McDougall, 1961, *Nature,* v. 190, p. 1184; W. Compston, I. McDougall, and K. S. Heier, 1968, *Geochim. Cosmochim. Acta,* v. 32, p. 129; I. McDougall, 1963, *J. Geophys. Res.,* v. 68, p. 1535.

48. M. H. Johnstone, D. C. Lowry, and P. G. Quilty, 1973, *Roy. Soc. W. Aust. J.,* v. 56, p. 5.

49. R. L. Larson, 1977, *Geology,* v. 5, p. 57.

50. M. W. McElhinny, B. J. J. Embleton, L. Daly, and J. Pozzi, 1976, *Geology,* v. 4, p. 455.

51. W. T. C. Sowerbutts, 1972, *Nature,* v. 235, p. 435.

52. P. Molnar and J. Francheteau, 1975, *Nature,* v. 255, p. 128; D. Davies, 1968, *Nature,* v. 220, p. 1225.

53. DSDP Scientific Staff, 1972, *Geotimes,* v. 17, no. 7, p. 22.

54. J. G. Murtaugh, 1972, *Proc. 24th Int. Geol. Cong.,* sec. 15, p. 133.

55. R. K. Wanless, R. D. Stevens, G. R. Lachance, and C. M. Edmonds, 1968, *Geol. Survey Can., Pap. 67-2, Rep. 8,* p. 116; S. H. Wolfe, 1971, *J. Geophys. Res.,* v. 76, p. 5424; A. Larochelle and K. L. Currie, 1967, *J. Geophys. Res.,* v. 72, p. 4163.

56. C. K. Seyfert and J. G. Murtaugh, 1977, *Geol. Soc. Amer. Abstr. Programs,* v. 9, p. 1168.

57. P. R. May, 1971, *Geol. Soc. Amer. Bull.,* v. 82, p. 1285.

58. D. C. Krause, 1963, *Nature,* v. 200, p. 1280.

59. A. A. Abdel–Monem and J. L. Kulp, 1968, *Geol. Soc. Amer. Bull.,* v. 79, p. 1231; G. de V. Klein, 1969, *Geol. Soc. Amer. Bull.,* v. 80, p. 1825.

60. C. L. Drake, M. Ewing, and G. H. Sutton, 1959, "Continental Margins and Geosynclines: The East Coast of North America North of Cape Hatteras," *in Physics and Chemistry of the Earth,* v. 3: Pergamon Press, New York, p. 110.

61. N. D. Newell, 1955, *in* A. Poldervaart, ed., *Crust of the Earth:* Geol. Soc. Amer. Spec. Pap. 62, p. 303.

62. H. Gill, L. A. Sirkin, and J. A. Doyle, 1969, *Geol. Soc. Amer. Abstr. Programs,* part 7, p. 79; J. P. Owens, J. P. Minard, and N. F. Sohl, 1968, *in* R. Finks, ed., *Guidebook to Field Excursions:* NYS Geol. Assoc. 10th Ann. Mtg., Queens N.Y.

63. P. H. Mattson, 1960, *Geol. Soc. Amer. Bull.,* v. 71, p. 319.

64. G. Draper *et al.,* 1976, *Geol. Soc. Amer. Bull.,* v. 87, p. 1283.

65. A. A. Meyerhoff, K. M. Khudoley, and C. W. Hatten, 1969, *Amer. Assoc. Petrol. Geol. Bull.,* v. 53, p. 2494.

66. J. F. Sweeney, 1977, *Geol. Soc. Amer. Bull.,* v. 88, p. 41.

67. C. K. Seyfert, 1968, *Trans. Amer. Geophys. Union,* v. 49, p. 326; W. Hamilton, 1969, *Geol. Soc. Amer. Bull.,* v. 80, p. 2409.

68. C. H. Stephens, 1969, *Geol. Soc. Amer. Abstr. Programs,* part 5, p. 78.

69. M. D. Crittenden, Jr., 1969, *Geol. Soc. Amer. Abstr. Programs,* part 5, p. 18.

70. R. L. Armstrong and J. Suppe, 1973, *Geol. Soc. Amer. Bull.,* v. 84, p. 1375.

71. E. H. McKee and D. B. Nash, 1967, *Geol. Soc. Amer. Bull.,* v. 78, p. 669; F. W. McDowell and J. L. Kulp, 1969, *Geol. Soc. Amer. Bull.,* v. 80, p. 2379; G. H. Curtis, J. F. Everden, and J. I. Lipson, 1958, *Calif. Div. Mines Spec. Rep. 54;* M. A. Lanphere and B. L. Reed, 1973, *Geol. Soc. Amer. Bull.,* v. 84, p. 3773; P. Petö, 1974, *Geol. Soc. Amer. Bull.,* v. 85, p. 1269.

72. E. H. Bailey, M. C. Blake, Jr., and D. L. Jones, 1970, *U.S. Geol. Survey Prof. Pap. 700-C,* p. C-70.

73. D. L. Jones, 1975, *Geol. Soc. Amer. Abstr. Programs,* v. 7, p. 330; D. L. Jones, N. J. Silberling, J. Hillhouse, 1977, *Can. J. Earth Sci.,* v. 14, p. 2565–2577.

**74.** E. H. Bailey and M. C. Blake, 1970, *Geotectonics,* no. 3, p. 148; no. 4, p. 225.

**75.** W. R. Evitt and S. T. Pierce, 1975, *Geology,* v. 3, p. 433.

**76.** J. Suppe, 1969, *Geol. Soc. Amer. Bull.,* v. 80, p. 135.

**77.** A. J. Eardley, 1962, *Structural Geology of North America:* Harper and Row, New York.

**78.** R. Trumpy, 1960, *Geol. Soc. Amer. Bull.,* v. 71, p. 843.

**79.** J. F. Dewey, W. C. Pitman III, W. B. F. Ryan, and J. Bonnin, 1973, *Geol. Soc. Amer. Bull.,* v. 84, p. 3137.

**80.** V. J. Dietrich, 1976, *Geology,* v. 4, p. 145.

**81.** J. Aubouin, 1965, *Geosynclines:* Elsevier, New York; A. Bosellini and E. L. Winterer, 1975, *Geology,* v. 3, p. 279; M. D. Dimitrijević and M. N. Dimitrijević, 1973, *J. Geol.,* v. 81, p. 328.

**82.** R. Hesse, 1974, *Geol. Soc. Amer. Bull.,* v. 85, p. 859; W. G. Ernst, 1973, *Geol. Soc. Amer. Bull.,* v. 84, p. 2053.

**83.** R. W. Van Bemmelen, 1954, *Mountain Building:* Martinus Nijhoff, The Hague, Holland.

**84.** Y. Kawano and Y. Ueda, 1967, *in* T. Matsumoto, ed., *Age and Nature of the Circum-Pacific Orogenesis: Tectonophysics,* v. 4, p. 429.

**85.** C. Cradock, 1970, *in Geologic Maps of Antarctica,* Amer. Geog. Soc. Antarct. Map Folio Ser., 12, Pl. 23.

**86.** C. A. Landis and D. S. Coombs, 1967, *in* T. Matsumoto, ed., *Age and Nature of the Circum-Pacific Orogenesis: Tectonophysics,* v. 4, p. 501.

**87.** M. G. Audley-Charles, 1972, *Nature Phys. Sci.,* v. 240, p. 137.

**88.** A. Hallam, 1976, *Geol. Soc. Amer. Bull.,* v. 87, p. 47; M. S. Kashfi, 1976, *Geol. Soc. Amer. Bull.,* v. 87, p. 1486.

**89.** I. W. D. Dalziel *et al.,* 1974, *Nature,* v. 250, p. 291; M. Suárez, 1976, *Geology,* v. 4, p. 211.

**90.** H. Burgl, 1967, *in* T. Matsumoto, ed., *Age and Nature of the Circum-Pacific Orogenesis: Tectonophysics,* v. 4, p. 429.

**91.** R. E. Clemons and L. E. Long, 1971, *Geol. Soc. Amer. Bull.,* v. 82, p. 2729.

**92.** M. Halpern, 1967, *J. Geophys. Res.,* v. 72, p. 5133.

**93.** W. M. Neill, 1976, *Geology,* v. 4, p. 495.

**94.** DSDP Scientific Staff, 1971, *Geotimes,* v. 16, no. 4, p. 12.

**95.** J. W. Ladd, 1976, *Geol. Soc. Amer. Bull.,* v. 87, p. 969.

**96.** R. L. Larson, 1975, *Geology,* v. 3, p. 69.

**97.** A. S. Romer, 1966, *Vertebrate Paleontology:* University of Chicago Press, Chicago.

**98.** A. de Ricalés, 1969, *Comptes Rendus Acad. Sci.,* Paris, 268D, p. 782; K. A. Kermack, 1951, *Ann. Mag. Nat. Hist.,* v. 12, p. 830; R. T. Bakker, 1971, *Nature,* v. 229, p. 172.

**99.** R. T. Bakker, 1975, *Sci. Amer.,* v. 232, no. 4, p. 58.

**100.** A. J. Desmond, 1976, *The Hot-Blooded Dinosaurs: A Revolution in Paleontology:* Dial Press, New York.

**101.** D. A. Russell, 1972, *Can. J. Earth Sci.,* v. 9, p. 375.

**102.** J. O. Farlow, C. V. Thompson, and D. E. Rosner, 1976, *Science,* v. 192, p. 1123.

**103.** E. H. Colbert, 1951, *The Dinosaur Book: The Ruling Reptiles and Their Relatives,* 2nd ed.: McGraw-Hill, New York.

**104.** D. A. Lawson, 1975, *Science,* v. 187, p. 947.

**105.** J. H. Osmond, *Nature,* v. 242, p. 136.

**106.** J. L. M., 1978, *Science,* v. 199, p. 284.

**107.** Z. Keilan-Jaworowska, 1975, *Amer. Scientist,* v. 63, p. 152.

**108.** P. J. Regal, 1977, *Science,* v. 196, p. 622.

**109.** W. D. Tidwell, S. R. Rushforth, and J. L. Reveal, 1970, *Science,* v. 168, p. 835.

**110.** M. Cavanaugh, 1977, personal communication.

**111.** C. Schuchert, 1955, *Atlas of Paleogeographic Maps of North America:* Wiley, New York.

**112.** E. Dorf, 1964, *in* A. E. M. Nairn, ed., *Problems of Paleoclimatology:* Wiley, London, p. 13; A. Holmes, 1965, *Principles of Physical Geology:* Ronald Press, New York, p. 719; S. M. Savin *et al.,* 1975, *Geol. Soc. Amer. Bull.,* v. 86, p. 1499.

# The Tertiary Period

T HE TERTIARY was originally subdivided into five epochs, the Paleocene, Eocene, Oligocene, Miocene, and Pliocene, based on the percentages of modern species of invertebrates (Table 13.1). While the names have persisted, the establishment of the boundaries between the epochs has become more precise.

The Tertiary is characterized by an increase in the widths of the Atlantic and Indian ocean basins, while the Pacific Ocean basin decreased in width (Fig. 13.1). The Tethyan Geosyncline was transformed into a lofty mountain range as Africa and India collided with Europe and Asia. Erosion of these mountains resulted in the deposition of great floods of clastic sediments in the lowlands adjacent to the mountains. Deposition of thick sequences of sedimentary and volcanic rocks continued on the Pacific margins of the continents as oceanic plates moved under continental plates.

Mammals evolved rapidly during the Tertiary to occupy the various ecological niches once dominated by the dinosaurs. Primates appeared before the beginning of the Tertiary, and traces of the oldest hominids (two-legged primates) are found in rocks of middle Tertiary age. By the end of the Tertiary, the first member of the genus *Homo* had appeared.

Gradual cooling of the climate during the Tertiary may have been related to such factors as mountain building, epirogenic uplift of the continents, and migration of the continents into polar regions. Glaciation in Antarctica occurred as far back as the Eocene, and by late Miocene time, Antarctica had developed an ice cap.

## TERTIARY GEOGRAPHY

Australia and Antarctica probably began to separate during the early Tertiary about 54 million years ago (2), and India probably began to separate from the Seychelles Islands at about the same time. The Gulf of California and the Red Sea probably began to open during the Miocene, about 15 million years ago. This opening stopped for several million years and was resumed during the Pliocene 2 to 4 million years ago.

### Separation of Australia from Antarctica

The separation of Australia from Antarctica began sometime between the latest Paleocene and the Late Eocene, 37 to 54 million years ago.

1. Flood basalts of Paleocene and Eocene age are found in the southeastern part of Australia (3).
2. The oldest known sediments recovered from the deep ocean floor between Australia and Antarctica are of Late Eocene age, 37 to 45 million years old (4). Although these sediments are located very close to the continental margin of Australia, it is possible that a considerable amount of time elapsed be-

TABLE 13.1  **Subdivision of Cenozoic Based on Percentage of Modern Species**

| PERIOD | EPOCH | | MODERN SPECIES (PERCENT) | AGE OF THE BEGINNING (MILLIONS OF YEARS) |
|---|---|---|---|---|
| Quaternary | Recent | | 100 | .01 |
| | Pleistocene | Neogene | 90–100 | 1.8 |
| | Pliocene | | 50–90 | 5 |
| | Miocene | | 20–40 | 26 |
| Tertiary | Oligocene | | 10–15 | 37.5 |
| | Eocene | Paleogene | 1–5 | 54 |
| | Paleocene | | 0 | 65 (?) |

tween the separation of Australia and Antarctica and the deposition of the first sediments in this particular area.

3. The oldest magnetic anomaly between Australia and Antarctica is Anomaly M-22 which has been dated at about 53 million years old (earliest Eocene). This anomaly is located very close to the continental margin of Australia.

Based on the above information, it is believed that the separation of Australia and Antarctica began during the latest Paleocene, about 54 million years ago. This age agrees well with that of J. G. Sclater and R. L. Fischer for the beginning of separation of these two continents (**5**).

## Separation of Saudi Arabia from Africa

The Red Sea and the Gulf of Aden were formed sometime during the Tertiary by the separation of Saudi Arabia from Africa. The spreading ridge in the Red Sea makes a 60° bend where it is connected to the spreading ridge in the Gulf of Aden. The East African Rift System, a major series of rift zones (grabens), intersects the spreading ridge system at a triple junction where the ridge makes the 60° bend. The East African Rift System is probably an aulacogen and the triple junction is probably underlain by a mantle plume. The East African Rift System

may have formed at the same time as the spreading ridges in the Red Sea and the Gulf of Aden.

1. The oldest known sediments deposited on oceanic crust in the Red Sea are of late Miocene age, 5 to 12 million years old (**6, 7**).
2. Magnetic anomalies in the Red Sea have been correlated with anomalies of Oligocene age (**8**), but other correlations are possible. Therefore, magnetic anomalies can not be used to determine the time of opening of the Red Sea.
3. Middle Miocene sediments, which are thought to have been eroded from fault scarps bordering the Red Sea, may date from the early stages of opening of the Red Sea (**9**).
4. Volcanism which began in the East African Rift System in Kenya about 13.5 million years ago (**10**) may date the beginning of the formation of the East African Rift System.

From the evidence presented here, it seems likely that the Red Sea and the Gulf of Aden began to open during the Middle Miocene, about 15 million years ago.

## Separation of Baja California from Mexico

The Gulf of California began to open during the Tertiary (early Miocene to early Pliocene)

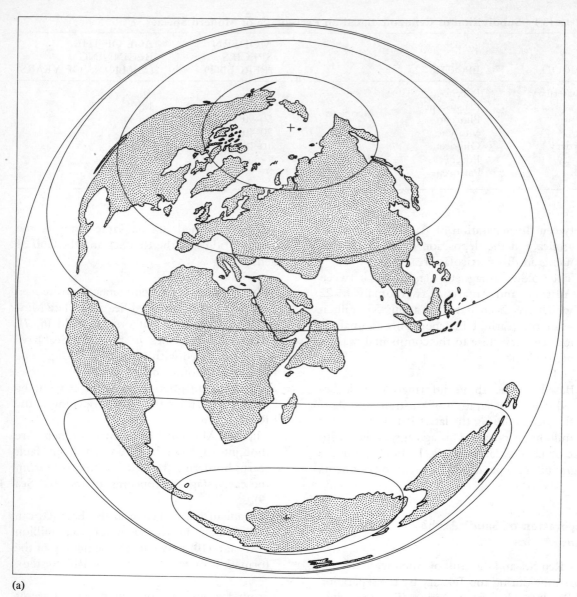

(a)

**FIGURE 13.1**  Reconstructions of the continents: **(a)** Earliest Tertiary, about 60 million years ago (Paleocene). **(b)** Middle Tertiary, about 30 million years ago (Late Oligocene). **(c)** Present. (Modified after Phillips and Forsyth, Ref. 1.)

by the separation of Baja California from Mexico.

1. The oldest known sediments deposited on ocean crust between Baja California and Mexico are of Miocene or Early Pliocene age (**10**).

2. Early Miocene to Pliocene marine strata on coastal plains bordering the Gulf of Southern California may have been deposited following the beginning of opening of the gulf (**7**).

3. The oldest identified magnetic anomaly in the Gulf of California has an age of about 3 million years (**11**).

(b)

D. G. Moore favors a middle (or perhaps early) Miocene age for the beginning of opening of the Gulf of California, followed by an interval during which seafloor spreading stopped and then resumed during the late Cenozoic (7). R. L. Larson suggests that the Gulf of California began to open during the Pliocene, about 4 million years ago. On the basis of the evidence presented above, a middle Miocene age is favored for the initial opening, followed by a period of quiescence lasting until 2 to 4 million years ago at which time spreading resumed.

The opening of the Gulf of California was accompanied by strike-slip movement along the San Andreas Fault north of the Gulf. Ap-

(c)

proximately 240 km of movement has occurred along the San Andreas since the Early Pliocene, about 5 million years ago (**12**).

## NORTH AMERICA

The coastal plains and continental shelves of North America received thick sequences of sediments throughout the Tertiary, particularly the Gulf and Atlantic coastal region, and in what are now the coast ranges of western North America (Table 13.2). Epicontinental seas were restricted during the Tertiary, and consequently, marine Tertiary deposits of the continental interior are limited to a small area bordering the Atlantic and Gulf coastal states. The Pacific coast continued as a leading plate margin riding over the subducting Pacific

TABLE 13.2  **Selected Tertiary Marine and Related Rock Units, Eastern and Western United States**

| STAGE | GULF COAST | MID-ATLANTIC COAST | S. CALIFORNIA |
|---|---|---|---|
| Pliocene | Citronelle Fm. | Walston Silt[a]<br>Beaverdam Sd.[a]<br>Yorktown Fm.[a]<br><br>Kirkwood Fm.<br>Cohansey[a]<br>(nonmarine) | Pico Fm.<br>Repetto Fm.<br>Bouse Fm.<br>Imperial Fm.<br>(Deeper Water) |
| Miocene | Pascagoula Clay<br>Hattiesburg Clay<br>Catahoula Ss.<br>(Tampa Ls. in Fla.) | Pensauken Fm.[a]<br>(nonmarine)<br>St. Mary's Fm.<br>Choptank Fm.<br>Calvert Fm.<br>Trent marl | Modelo or<br>Monterey Sh. |
| Oligocene | Vicksburg Ls.<br>(Byram Ls. in Fla.)<br><br>Jackson Fm.<br>(Ocala Ls. in Fla.) | Castle Hayne marl | Sespe Fm. |
| Eocene | Claiborne Gp.<br><br>Wilcox | Nanjemoy Fm.<br><br>Vincentown Sd.<br>Hornerstown marl | Llajas Fm. |
| Paleocene | Midway Gp. | Aquia Fm. | Martinez Fm. |

Compiled from published sources.
[a] Owens and Denny, Ref. **7**)

Ocean plate. The east coast was a trailing continental margin.

## Coast Ranges

Marine and nonmarine sediments were deposited during the Tertiary in a series of basins along the Pacific margin of North America in California, Oregon, Washington, British Columbia, and Alaska.

CENTRAL AND SOUTHERN CALIFORNIA  The seas began a gradual transgression in the Paleocene following the regression at the end of the Cretaceous. They reached their maximum extent during the Eocene and again in the Miocene. Marine deposits are largely clastics. The thickness of these deposits often exceeds 5000 m (15,000 ft). Subduction in this region was nearly over by the middle Oligocene as North America overrode a section of the East Pacific Rise. Late Miocene folding caused a partial withdrawal of the sea, and by the end of the Pliocene, the sea was restricted to the outer continental margin.

WASHINGTON, OREGON, AND NORTHERN CALIFORNIA.  Eugeosynclinal volcanic and sedimentary rocks of Tertiary age occur in western Washington, western Oregon, and northwestern California. Some basalts have pillow structures typical of the subaqueous extrusions. In the Olympic Peninsula of Washington, basalts have a composition identical to that of oceanic volcanics and were probably extruded on an oceanic ridge (**13**).

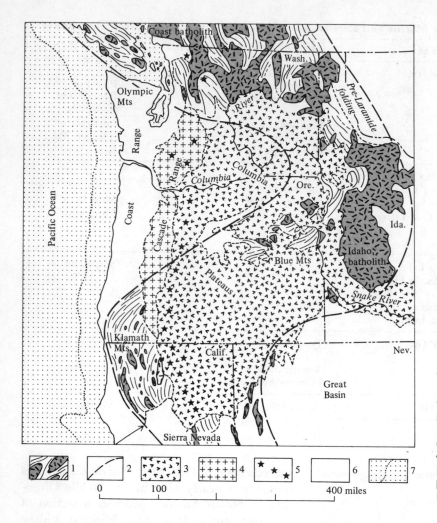

**FIGURE 13.2** Map of the Pacific Northwest showing the distribution of plateau basalts and andesitic volcanoes of Tertiary and Quaternary age 1 = Granites, 2 = Trend of foliation, 3 = Volcanics of Columbia Plateau and Snake River Plain, 4 = Volcanics of Cascades, 6 = Quaternary volcanoes, 7 = Ocean. (After P. B. King, Ref. 15, Reprinted by permission of Princeton University Press.)

Other volcanic rocks in the Coast Ranges were extruded subaerially, probably in an island arc environment (**14**).

Late Tertiary andesites occur in the Cascade Range in a narrow belt extending from northern California through central Oregon to central Washington (Fig. 13.2). These rocks may have been formed as a result of the underflow of the Pacific plate along a zone of subduction. This mechanism may account for folding and faulting of many Tertiary sequences in this region.

ALASKA.  Tertiary deposits of eugeosynclinal character are widely exposed in southern Alaska. Eugeosynclinal deposits extend from the mainland of Alaska westward through the Aleutian Islands. The western Aleutians are bordered on both the north and south by deep ocean and are underlain by oceanic crust. Early Tertiary land plant fossils have been found in sedimentary rocks of Amchitka Island, which is located near the center of the Aleutian chain. Therefore, the Aleutian Arc must have originated during or before the early Tertiary.

**FIGURE 13.3** Tertiary volcanics of the Snake River Plain in Idaho. Several different lava flows can be seen in the walls of the Snake River Canyon near Idaho Falls.

Metamorphism occurred in the western Aleutians during the Oligocene, about 30 million years ago (**16**), and granites were intruded in Alaska during the Oligocene between 38 and 26 million years ago (**17**). A major deformation which occurred in the Aleutians during the middle Miocene was accompanied by folding, faulting, and intrusion of granites (**18**).

### Columbia Plateau

About 340,000 square km (200,000 sq mi) of middle Tertiary basalt flows are exposed on the Columbia Plateau and Snake River Plain in the Pacific Northwest (Fig. 13.2). This area contains numerous flows which total more than 3000 m (10,000 ft) thick (Fig. 13.3). The thickness of individual flows averages a few tens of feet and the lavas travelled up to 170 km (100 mi) from the site of their eruption. Radiometric dating indicates that nearly 99 percent of

the lavas in central Oregon were erupted during a two-million-year period during the middle Miocene, about 15 million years ago (**19**).

### Basin and Range Province

The Basin and Range Province is an area of parallel mountains and valleys in Nevada, eastern California, western Utah, and western Arizona. Extrusion of rhyolitic and andesitic lavas and intrusion of granitic rocks occurred over most of the Basin and Range Province beginning during the latest Eocene or earliest Oligocene (**20**) (Fig. 13.4). This igneous activity continued throughout the Oligocene and into the lower Miocene, but there was a hiatus in volcanic activity during the middle Miocene from about 17 to 20 million years ago (**20**). Igneous activity resumed about 16 million years ago and continued to the end of the Tertiary with peaks at 3.5, 9.5, and 13 million years ago (**20**). Mio-

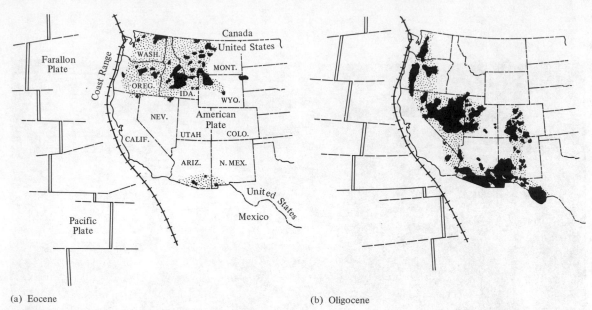

(a) Eocene

(b) Oligocene

FIGURE 13.4  Inferred Eocene (50 million years ago) and Oligocene (30 million years ago) plate boundaries adjacent to western North America. Distribution of continental igneous rocks in the western United States are shown by solid black areas (present distribution) and stippled pattern (inferred original extent). Double lines are midocean ridge crests, and crosshatched lines are trenches. (From Lipman, Prostka, and Christiansen, Ref. 20. Copyright © 1976 by the American Association for the Advancement of Science.)

cene and Pliocene activity was generally confined to the margins of the province.

Metamorphism occurred about 38 million years ago (Eocene–Oligocene boundary) and 26 million years ago (Oligocene–Miocene boundary) in northwestern Utah accompanied by folding and low-angle faulting (21). Low-angle faulting also occurred in this area up to about 12 million years ago (middle Miocene) and in southeastern California and west-central Nevada during the middle Miocene (21). It is somewhat of a mystery why deformation and metamorphism would occur so far from the edge of the continent without the intervening area being significantly affected. Part of the explanation may be that, during the Oligocene, the subduction zone under the Great Basin was dipping at a very low angle (20), and possibly was very close to the base of the continent under this part of Utah.

Beginning during the middle Miocene, the Basin and Range Province was uplifted and faulted (22). This faulting was responsible for the present topography of the province. At the time of the beginning of faulting, there was an increase in basaltic volcanism, although rhyolitic volcanism continued to dominate. The faults are normal faults, which are an indication of crustal tension (stretching). It has been suggested that the faulting was associated with a regional uplift caused by the overriding of a portion of the East Pacific Rise by North America (23). Prior to the middle Oligocene, subduction occurred along the entire west coast of North America (20). Beginning approximately during the middle Oligocene, the crest of a section of the East Pacific Rise was subducted beneath the continent. By middle Miocene time, the crest of the East Pacific Rise was under the Basin and Range Province. The pat-

## Central and Southern Rockies

During the early Tertiary, a series of uplifts developed in Montana, Wyoming, South Dakota, Colorado, Utah, and New Mexico. These uplifts are generally anticlinal and are bordered by steeply dipping reverse faults on one or two sides (Fig. 13.6). Erosion of these uplifts has supplied up to 300 m (10,000 ft) of shale, sandstone, and conglomerate to the adjacent basins, called *intermontane basins* (Fig. 13.7). Freshwater lakes formed in some of the basins, and these lakes record a history of cyclic deep- and shallow-water deposition. Some of the lakes are considered to have been playas (intermittent desert lakes) (**25**). Shales deposited in the lakes contain a rich fauna of fossil fish and insects.

In the Green River Basin of southwestern Wyoming, northwestern Colorado, and eastern Utah, lacustrine shales contain organic matter which can be converted into oil. These "oil shales" constitute a very large potential reserve of gas and oil. The Green River Formation in this basin was deposited about 49 million years ago during the later part of the early Eocene and early part of the middle Eocene (**26**).

In the Bighorn Basin of northwestern Wyoming, older Paleozoic sedimentary rocks overlie Tertiary deposits (Fig. 13.8). These Paleozoic rocks are not overturned, and therefore the strange sequence must be a result of faulting rather than folding. Apparently, blocks slid down a very gentle slope under the influence of gravity and with fluid pressure bouying up the blocks. The source of the slide blocks was probably an uplifted area in the Beartooth Mountains 28 to 48 miles to the west (**27**). Geologists do not know how quickly or why the blocks moved. However, the movement may have been associated with volcanism that occurred in the Beartooth Mountains shortly thereafter.

In the vicinity of Yellowstone National Park, Wyoming, volcanic conglomerates, tuffs, and breccias of middle and late Tertiary age overlie

**FIGURE 13.5** Fault scarp near Fairview Peak, Nevada. The photo was taken shortly after the event. (Photograph by Karl V. Steinbrugge.)

tern of magnetic anomalies in the floor of the Pacific Ocean indicates that the crest of the East Pacific Rise lies under eastern Nevada at the present time (**23**). Heating and expansion of the mantle under the Basin and Range Province may have caused the uplift and normal faulting.

The vertical displacements on the faults often exceed 3000 m (10,000 ft), and the movement is still going on today (Fig. 13.5). As the fault blocks were uplifted, the valleys between the ridges filled with thousands of feet of sediments. Renewed uplift in the Basin and Range Province occurred about 4.5 million years ago (**24**) at about the same time as the beginning of the second phase of opening of the Gulf of Southern California.

(a)

(b)

FIGURE 13.6  Folds produced during the Larimide orogeny: **(a)** Large anticline in southern Utah. (U.S. Bureau of Reclamation.) **(b)** East flank of a large anticline in the Bighorn Mountains, Wyoming.

FIGURE 13.7 Eocene lake deposits of Bryce Canyon National Park have been eroded into pillars and spires by the action of the wind and water. (Union Pacific Railroad Photo.)

FIGURE 13.8 Heart Mountain north of Cody, Wyoming, is capped by Paleozoic limestones. These are separated from the underlying Tertiary sediments by a gravity thrust fault.

FIGURE 13.9 Flat-lying late Tertiary volcanic rocks exposed on Carter Mountain along the South Fork of the Shoshone River west of Cody, Wyoming.

FIGURE 13.10 Folded Paleozoic limestones in Banff National Park, Canada. The folding occurred during the early Tertiary.

early Tertiary deposits (Fig. 13.9). Pollen analysis of these deposits shows that the middle and late Tertiary forests in the Yellowstone area were dominated by gymnosperms, mainly spruce, pine, hemlock, and cedar, and had lesser amounts of angiosperms, such as birch and hickory (**28**). As is typical of Tertiary floras, they contain species present in the region today as well as exotic species that are either extinct or are not now found in North America.

## Northern Rocky Mountains

Deposits in the northern Cordilleran Geosyncline were intensely folded, faulted, and intruded during the early Tertiary (Fig. 13.10). Seismic surveys and deep drilling indicate that the dip of thrust faults decreases with depth and that the basement is not involved in the deformation (Fig. 13.11). Intrusion of granites occurred in eastern Washington and Idaho during the Eocene. This activity began at approximately the Eocene-Paleocene boundary about 53 million years ago (**29**).

## The Great Plains

Continental clastics were deposited over a large area east of the Rocky Mountains during the early and middle Tertiary. In the Badlands of South Dakota, numerous fossilized remains of early mammals have been recovered from one such deposit, the Brule Clay of Oligocene age (Fig. 13.12). Tuffs interbedded with these sediments were derived from volcanoes lying to the west. A sea covered much of the southeastern part of the United States in early Tertiary time, but receded southward when the Great Plains were uplifted during the later part of the middle Tertiary.

## East Coast and Gulf Coast Geocline

Thick sequences of Tertiary sediments are exposed on the Atlantic and Gulf Coastal Plains (Fig. 6.16). Seismic surveys and drilling indicate that deposits of this age also underlie the continental shelf of the Atlantic Ocean and the Gulf of Mexico. The sediments are mostly sands, clays, and muds that reach a maximum thickness of 6000 m (20,000 ft) in the Mississippi River delta. Most of these deposits are marine in origin, but, in some, continental sediments occur on the landward side of the geosyncline. Limestones containing reef-building corals occur in Florida and the Bahamas. More than 2500 m (8000 ft) of Tertiary carbonates have been penetrated in drilling operations on the Bahama Banks.

In Paleocene time, an arm of the sea flooded the Mississippi River Valley as far up as southern Illinois. This area is called the Mississippi Embayment (Fig. 12.21). In the Gulf and Atlantic coastal regions, the position of the shoreline fluctuated considerably throughout the Tertiary as the seas slowly withdrew from the continent. During periods of regression, deltaic sands overlapped marine shales. The resulting interfingering of sands and shales created stratigraphic traps in which important reserves of oil and gas are found. By the end of the Tertiary, the seas were limited to the outer margin of the continent.

## The Greater Antilles

Volcanism and plutonism were widespread during the early and middle Eocene in Cuba, Hispaniola, and Puerto Rico (**31**). In Puerto Rico, volcanic and sedimentary rocks of late Paleocene or early Eocene through middle Eocene age unconformably overlie rocks of Cretaceous (Maestrichtian) age which were folded, faulted, and intruded before deposition of the underlying beds (**31**). Intrusion and deformation, therefore, occurred between 70 and 53 million years ago. The Tertiary deposits are intruded by granite batholiths that have been radiometrically dated at 41 million years (late Eocene) (**32**). Volcanic activity in the Greater Antilles ceased in the late Eocene (**33**). It is

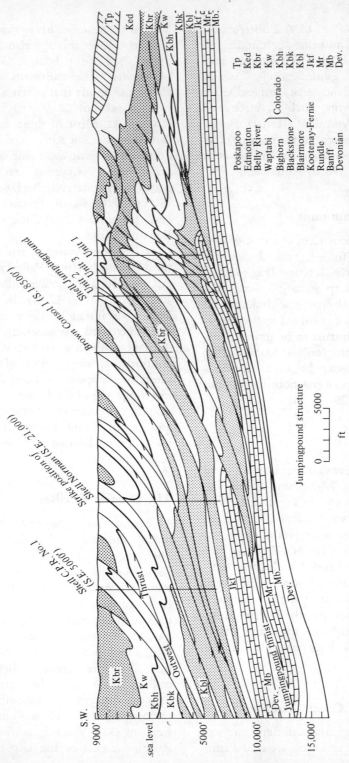

**FIGURE 13.11** Structural cross section across the Jumpingpond gas field in western Alberta, Canada. Note that the thrust faults flatten with depth. (After Fox, Ref. 30. Reproduced with the permission of The American Association of Petroleum Geologists.)

Tp    Poskapoo
Ked   Edmonton
Kbr   Belly River
Kw    Waptabi  ⎤
Kbh   Bighorn    ⎥ Colorado
Kbk   Blackstone  ⎦
Kbl   Blairmore
Jkf    Kootenay-Fernie
Mr    Rundle
Mb    Banff
Dev.   Devonian

**FIGURE 13.12** Oligocene sediments of continental origin in the Badlands National Monument of South Dakota.

interesting to note, however, that volcanism in the Lesser Antilles began about the same time (**34**). Possibly the direction of seafloor spreading changed during the late Eocene. Oligocene and younger deposits generally rest with marked angular unconformity on Eocene and earlier deposits (**33**). They are relatively undeformed in the Greater Antilles and consist of limestone and clastics.

### The Lesser Antilles (Leeward Islands)

The Lesser Antilles are an island arc extending southward from the eastern end of the Greater Antilles. This island arc is located on the eastern border of the Caribbean plate. The Caribbean plate at the present time appears to be stationary

with respect to the underlying deep mantle (**35**), but the American plate is moving under its eastern border. The oldest rocks in the Lesser Antilles are pillow lavas, cherts, and sodium-rich granitic rocks which have been radiometrically dated at late Jurassic-early Cretaceous (**36**). These rocks are found on the island of La Desirade. It is likely that they represent uplifted oceanic basement on which the later sedimentary and volcanic rocks were deposited. Volcanic activity has been continuous in the Lesser Antilles from the middle Eocene to the present time. Coral reefs that once fringed subaerial volcanoes have contributed to the formation of limestones associated with Tertiary volcanics. Clastic sediments were derived largely from volcanics, but some on Barbados (for example, the New Scotland Formation) may have been derived from continental rocks on the northern margin of South America. These clastics contain granitic and metamorphic rocks (including marble), and they may have been carried by submarine currents northward from South America. After having been deposited east of Barbados, the clastics were then been folded and uplifted as the American plate moved under the Caribbean plate. The clastics of the New Scotland Formation are quite similar to sediments cored from the Atlantic Ocean basin east of the island, which is consistent with the sequence of events described here.

## EUROPE

Tertiary geosynclinal deposits are widely exposed in the northern and southern Tethyan geosynclines. Marine sediments are also present near the outer borders of the continental interior, and continental clastics occur in regions adjacent to mountain belts.

### Northern and Southern Tethyan Geosynclines

The Northern Tethyan Geosyncline (on the southern border of Europe) and the Southern Tethyan Geosyncline (on the northern border of the African plate) were separated in the early Tertiary, but were joined together during the middle Tertiary as the African and European plates converged. The zone of joining of these two geosynclines is the Alpine-Himalayan Mountain Belt.

ALPINE REGION. During the early Tertiary, a series of geanticlinal ridges in the Northern Tethyan Geosyncline contributed a thick deposit of marine clastics (flysch) to the geosyncline. Late in the Oligocene, the deposits of the Alpine region were intensely folded and uplifted (**38**). The very large recumbent folds produced at this time are called *nappes* (Fig. 13.13). The mountains that resulted from this deformation were rapidly eroded, and a thick

layer of *molasse* (continental clastics) was deposited in the Peri-Alpine Depression to the north. The deposits range in age from late Oligocene to late Miocene. By late Miocene time, the zone of deformation had spread northward and the deposits of the Peri-Alpine Depression were folded. The Jura Mountains and the Carpathians were formed at this time. Vertical uplift without major faulting or folding continued throughout the remainder of the Tertiary.

In western Italy, recumbently folded Eocene and older rocks are unconformably overlain by Oligocene beds. Folding probably occurred near the end of the Eocene. The folding resulted in the formation of a series of geanticlinal ridges, which furnished thick sequences of clastics during the Oligocene in western Italy. In Miocene time, new ridges were uplifted to the west of the older ridges, and large blocks slid from these ridges into soft muds to the east. This resulted in a chaotic mass of exotic blocks in a shaly matrix, known as the *argille scagliose*. This unit, some of which is similar in many respects to the wildflysch of the Alps, contains blocks measuring thousands of feet across.

The cause of the Tertiary deformation within the Northern Tethyan Geosyncline was compression between the Eurasia plate and the Austroalpine plate (Fig. 13.14). Intensely metamorphosed rocks thought to be a part of the lower crust of the Italian plate are found on top of the nappes (**38**). Initial contact of Italy with the Eurasian plate probably occurred in the late Oligocene.

Paleomagnetic data indicate that the Italian Peninsula was rotated by about 43° after the early Paleocene, and then 25° counterclockwise since the middle Eocene (**39**). This rotation may have been associated with relative movements between Africa and Europe during the Tertiary. It is possible that Italy was part of the African plate during part of the Tertiary. Alternatively, the rotation may have occurred at the same time as the rotation of Sandinia and Corsica (early to middle Miocene).

Relative movements between *microplates*

**FIGURE 13.13** Folded sedimentary rocks in the Alps near Brienz, Switzerland. (Courtesy of CIBA-Geigy.)

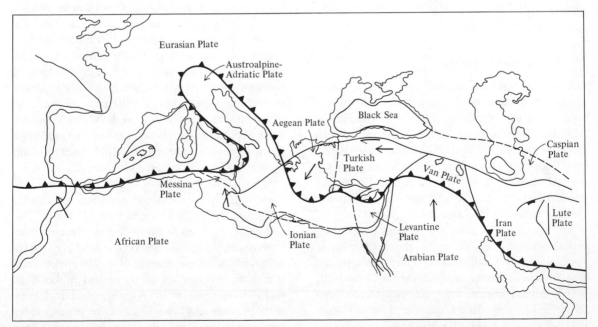

**FIGURE 13.14** Small plates (microplates) at the border between the Eurasian and African plates. Arrows indicate slip directions of plates with respect to the Eurasian plate. (Modified from Dewey, Pitman, Ryan, and Bonnin, Ref. 37.)

(very small plates) in the Alpine and Mediterranean region between the Eurasian and African plates resulted in further deformation of the margins of the Alps, Atlas, and Carpathian mountains and the development of several marginal seas that separated the microplates during times that crust was extended (Fig. 13.14).

MIDDLE EAST.   During the Tertiary, erosion of geanticlines on the northern border of the Southern Tethyan Geosyncline in the Middle East supplied thick sequences of clastic sediments to the geosyncline (**40**). These deposits grade southward into carbonates. The continental clastics of Miocene age are very thick. The principal folding in the central regions of the geosyncline also probably occurred in the Miocene (Fig. 13.15).

Folding of the Zagros Mountains occurred in late Tertiary time due to compression between the Arabian plate (then part of the African plate) and Eurasia (the Iran plate). The initial contact of Arabia with Eurasia probably occurred during the Miocene (**40**).

### North Atlantic Margin

Lavas were erupted during the early Tertiary along the borders of the North Atlantic in Ireland, Scotland, the Inner Hebrides, the Faeroe Islands, Spitsbergen, and in southeastern Greenland (Fig. 13.16). These lavas are generally flood basalts, and in some areas they are several thousands of meters thick. Columnar jointed basalts of early Tertiary age form the Giant's Causeway in Ireland (Fig. 13.17).

In Scotland, sedimentary rocks, including coals, are interbedded with the lavas. These rocks contain fossil plants which indicate an Eocene age for some of the lavas. In Greenland, fossil pollen indicates an age of latest Paleocene to earliest Eocene for a sequence of flood basalts up to 9 km thick (**41**). Radiometric dating indicates that the volcanism in Britain occurred primarily during the Paleocene and early Eo-

cene between about 50 and 60 million years ago (**42**). A peak in igneous activity occurred about 53 million years ago at approximately the boundary between the Paleocene and Eocene. The igneous activity in the North Atlantic region appears to be a continuation of volcanism that began with the separation of Greenland and Europe. It probably was associated with the mantle plume located beneath Iceland because the activity appears to lie at the end of an aseismic ridge which connects Iceland to both Europe and Greenland (Fig. 13.16).

The oldest known volcanic rocks on Iceland have been radiometrically dated at about 13 million years (middle Miocene) (**43**). The middle Miocene appears to have been a time of increase in volcanism associated with the Icelandic plume (**44**).

### Davis Straight

A sequence of flood basalts more than 8 km (5 mi) thick was erupted during the early Paleocene (late Danian), about 60 million years ago, on Disko Island in western Greenland (**45**) and on the southwest coast of Baffin Island in Canada at Cape Dyer (Fig. 13.16). These two areas lie at the end of an aseismic ridge which is thought to have been caused by a mantle plume underlying a hot spot. At the time of eruption of the lavas, Greenland and North America were separated by about 500 km (300 mi) as a result of seafloor spreading which began during the early Cretaceous (see Chapter 12).

The basalts at Disko Island contain nickel-iron ranging from tiny particles to masses weighing up to 24.5 metric tons (25 tons). The nickel-iron occurs in dikes cutting Precambrian basement rocks and therefore it is likely that its origin is within the mantle (**45**). The nickel-iron is rich in carbon and contains less nickel than nickel-iron meteorites. As suggested in Chapter 6, plumes and hot spots may be produced by meteorite impacts. It is possible that the nickel-iron at Disko Island is associated with such an impact. The high carbon content sug-

**FIGURE 13.15** The sedimentary rocks of the Zagros Mountains in southwestern Iran were folded during the Tertiary. (Aerofilms Limited, courtesy of the British Petroleum Co.)

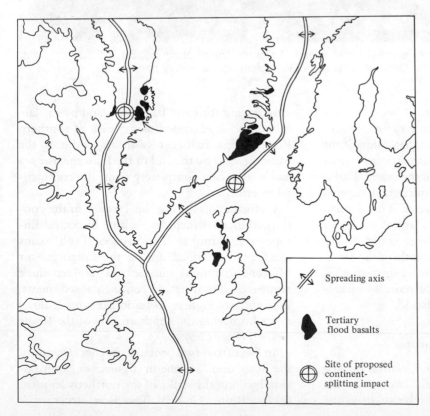

Spreading axis

Tertiary flood basalts

Site of proposed continent-splitting impact

**FIGURE 13.16** Areas of Tertiary flood basalts in the North Atlantic region.

**FIGURE 13.17** Columnar basalts of the Giant's Causeway, County Antrim, North Ireland. (NERC Copyright. Reproduced by permission of the Director, Institute of Geological Sciences, London Sw 7.)

gests that the meteorite may have been a carbonaceous chondrite. On impact, large meteorites will vaporize, but it is possible that a significant amount of the meteoritic material may condense in fractured rock within the crater. Disko Island is surrounded by a semicircular depression 210 km (126 mi) in diameter. This depression may mark the outer boundary of an impact crater. The base of such a large crater would probably be well within the mantle of the earth. Partial melting of mantle containing meteoritic material may explain the occurrence of nickel-iron in the basalts at Disko Island.

## Continental Interior of Europe

At the beginning of the Tertiary, central Europe was entirely above sea level. The sea transgressed onto the continental interior from the west during the early Paleocene, and by the late Oligocene it extended as far east as southern Russia. A gradual regression began during the Miocene, and by the end of the Pliocene, the sea had withdrawn completely from the continental interior.

Deformation during the Tertiary in the continental interior includes faulting in central Europe and folding in the Pyrenees. Fault basins (grabens) developed during the Oligocene or Miocene in central Europe and received thick sequences of marine and continental sediments. In southern France, volcanoes erupted along faults bordering the southern part of the Rhine Graben (Fig. 12.5).

In Germany, two probable impact structures, the Reis and Steinheim basins, lie approximately along the strike of the northern segment of the Rhine Graben. The Reis structure is

24 km (15 mi) in diameter and Steinheim is 3.5 km (2 mi) in diameter. Radiometric dating of glass from these structures indicates that they were formed 15 million years ago during the middle Miocene (**46**). This age is also supported by paleontological dating of lake beds filling the structures (**46**). These structures are about the same age as the major faulting in the Rhine Graben. The Houghton structure, a 17-km (10-mi) diameter, probable impact structure in northern Canada is of a similar age (**47**). An increase in folding and volcanism in many parts of the world and a time of change in the direction and rate of seafloor spreading occurred at the same time.

Stratigraphic, structural, and paleomagnetic evidence indicates that the islands of Sardinia and Corsica in the Mediterranean Sea were connected to Europe prior to the Miocene, but rotated about 30° in a counterclockwise direction sometime between the early and late Miocene (**48**). It is interesting to note that there is a large semicircular bay (the Gulf of Lion) located at the place where Sardinia and Corsica would have been connected to Europe prior to rotation (Fig. 12.5). It is possible that this bay was the site of a large meteorite impact during the middle Miocene. An impact-related plume may have caused the separation of the small part of Europe that became Sardinia and Corsica.

The Pyrenees Mountains separating France and Spain were folded at the close of Eocene time. This deformation was probably caused by movement of the Iberian Peninsula toward Europe by about 50 km (30 mi) (**49**).

## ASIA

### Northern and Southern Tethyan Geosynclines

The sea between the Northern and Southern Tethyanian geosynclines closed as India approached Eurasia during the early Tertiary. Early Tertiary deposits of the Southern Tethyan Geosyncline bordering India are dominantly limestones and marine clastics. The first major phase of deformation in the geosyncline occurred during the Oligocene and included intense folding and intrusion of granitic rocks. The cause of this deformation is probably the collision of India with Asia, which began about 40 million years ago (late Eocene) (**50**). A second major deformational episode occurred during the middle Miocene, when the deposits of the Southern Tethyan Geosyncline were again folded and faulted and great thrust sheets or nappes moved southward from the core of the Himalayan Orogenic Belt (Fig. 13.18). During and after both of these deformations, thick sequences of clastic sediments were deposited south of the Himalaya Mountains. As in the Northern Tethyan Geosyncline, the Tertiary deformations in the Southern Tethyan Geosyncline were probably the result of a collision between the India and Eurasia plates.

### India

Volcanism was widespread in the western continental interior of India during the early Tertiary. Numerous large dikes and massive intrusions occur along the western border of India, particularly in the area north of Bombay. Because the lavas of the Deccan Basalts here reach a thickness of about 3000 m (10,000 ft), this region was probably the center from which the basalts erupted. Flood basalts of the Deccan Basalts cover more than 550,000 square km (200,000 sq mi). These basalts overlie Paleocene beds and are themselves overlain by Eocene strata (**51**). Potassium–argon dates on the basalts range from 42 to 65 million years (**52**). Because of the possibilities of loss of argon it is likely that the true age is closest to 65 million years (earliest Paleocene). This age agrees well with the age of the Deccan Basalts determined from paleomagnetic studies (**53**). Some, and perhaps all, of the Deccan Traps may have been associated with separation of India from the Seychelles Islands, which probably began about 70 million years ago (**54**) (Fig. 13.19). Mafic

**FIGURE 13.18** Reverse fault offsetting Carboniferous sedimentary rocks, Lebung Pass, central Himalayas. (From Geological Survey of India, Memoir 23.)

intrusives on the Seychelles Islands are about the same age as those in the Deccan Basalts (**55**).

Marine Tertiary sediments are restricted to the outer margins of the continental interior of India. Very thick continental deposits from middle Miocene to Pleistocene in age, eroded from the Himalayas occur just south of the mountains.

### Indonesian Island Arc

Thick deltaic, estuarine, and shallow marine sequences were deposited in a zone bordering the Indonesian Island Arc in Paleocene and Eocene time. In the late Miocene, intense folding affected the Lesser Sunda Islands, southern Java, and southern Sumatra. Deep-sea drilling south of the Indonesian Island Arc has revealed that an increase in volcanism occurred in the area during the Miocene and again in late Pliocene time. Northwest Borneo contains thick sequences of folded and faulted Paleocene to Oligocene sediments of deep-water origin

overlain by shallow water, Miocene sandstone (**56**). The Indian, Phillipine, Eurasian, and Pacific plates have been moving toward each other since the beginning of separation of Australia and Antarctica. Presumably this is the reason for the Tertiary volcanism and deformation in the Indonesian region.

### Western Pacific Island Arcs

During the Tertiary, island arcs extended along the western margin of the Pacific from Kamchitka in northwestern Siberia to the Phillipines. The Tertiary history of the Western Pacific Island Arcs includes several deformational episodes, repeated transgression and regression of the sea, and deposition of thick marine and continental strata. Tertiary volcanic rocks are abundant in the island arcs and deformation during the Tertiary may have been more or less continuous as the seafloor was subducted beneath them. Episodes of especially intense folding occurred at the end of the Eo-

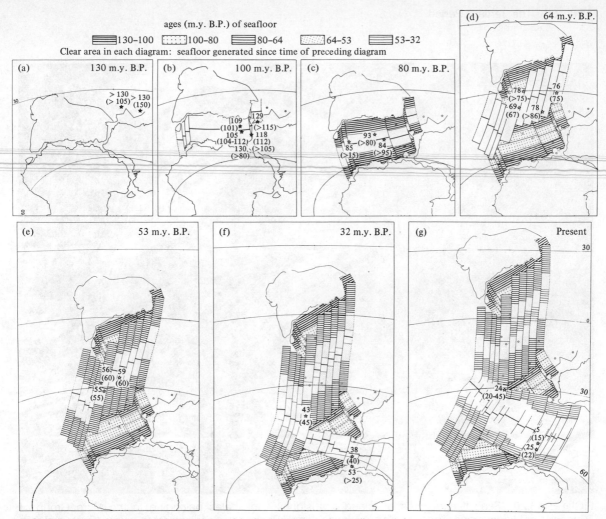

**FIGURE 13.19** History of movement of India relative to Australia and Antarctica. (Modified after Johnson, Powell, and Veevers, Ref. 54.)

cene, between the late Oligocene and early Miocene (beginning of Mizuko orogeny), between the middle and late Miocene, and between the late Pliocene and early Pleistocene (Fig. 13.20). Some, and perhaps all, of these deformations correspond to changes in the direction of seafloor spreading in the Pacific Ocean (Table 12.1). The Mizuko orogeny has been related to the opening of the Japan Sea behind the Japanese Islands (**57**), but marine magnetics do not support this suggestion. No symmetrical magnetic anomalies have been found on the Japan Sea floor (**58**). Intense volcanism is recorded for Miocene and late Pliocene time in

the Western Pacific Island Arcs. The continuing volcanism and tectonism suggest that throughout the Tertiary, the western Pacific was bordered by active island arcs adjacent to zones of subduction, as it is today. Sediments derived from the continents accumulated in marginal basins (i.e., Okhotsk, Japan, Phillipine, and China Seas) situated behind the island arcs. These basins overlie subduction zones.

## Continental Interior of Asia

The northern and central regions of the continental interior of Asia were generally above sea

**FIGURE 13.20**  Seacliff exposures of folded sandstones and shales belonging to the Muro Group of Oligocene or Miocene age. These rocks are exposed on Kii Peninsula in southwestern Japan. (Photo by W. R. Dickinson. Copyright © by American Geophysical Union.)

level during the Tertiary, but southern Asia was covered by a shallow sea during the Paleocene and Eocene. Molasse of Miocene and Pliocene age forms a clastic wedge that was derived from the Himalayan highlands to the south. From the Himalayan frontal thrust northward, east-west trending strike-slip faults cause a series of east-west trending valleys (Fig. 13.21). The north trending Baikal Rift is a graben which may be a midcontinent spreading center marking the incipient breakup of Asia (Fig. 13.21). It is interesting to note that a 100-km (60-mi) diameter meteorite impact structure lies along the strike

of the Baikal Rift. This structure, Popigay, formed about 42 million years ago (**60**). It is possible that the impact at Popigay is related to the rifting.

## AUSTRALIA

Marine Tertiary deposits are restricted to basins near the outer margins of Australia. These deposits are dominantly limestone, but minor clastics are also present. The sea transgressed into the marginal basins during the Paleocene and reached its maximum extent during the late

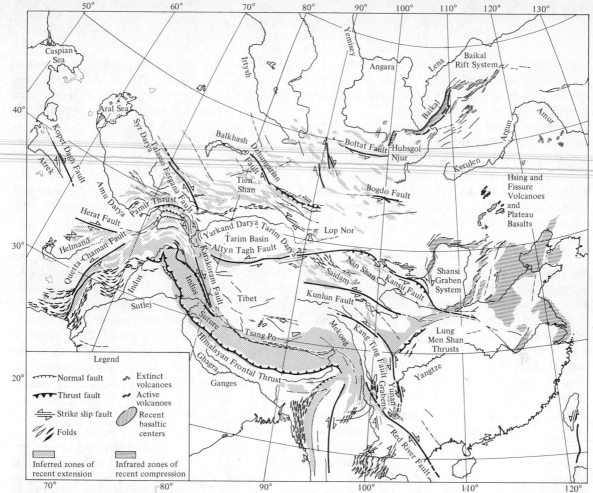

**FIGURE 13.21** Preliminary tectonic map of Asia. Bold lines represent faults of major importance—usually seismic and with very sharp morphology. Arrows indicate sense of motion on faults. For Tertiary folding, bold symbols indicate more prominent, more recent folds. The dotted areas indicate regions of inferred recent vertical motion associated with thrust faulting and compressional tectonics. Areas shaded by broken lines are covered by thick recent alluvial deposits and are dominated by horizontal extension and subsidence. (From Molnar and Tapponnier, Ref. 59. Copyright © 1975 by American Association for the Advancement of Science.)

Eocene and early Miocene. The seas regressed slowly from the continent throughout the remainder of the Tertiary. Coal-bearing terrestrial clastics of Tertiary age occur in basins near the center of the continent.

Tertiary volcanic rocks crop out in a broad, discontinuous belt along the eastern and southern border of Australia. The volcanics in Victoria range in age from Paleocene to Eocene (**61**), and may have been erupted when Australia and Antarctica began to drift apart. Based on the

age of magnetic anomalies, the separation of these two continents began about 53 million years ago (between the Paleocene and the Eocene) (**54**).

## NEW ZEALAND

At the beginning of the Tertiary, New Zealand was emergent, but it had been reduced to low relief. Transgression of the sea began during the Eocene and reached a maximum in middle or late Oligocene time. The sea began to recede before the end of the Oligocene, and there were no subsequent incursions for the duration of the Tertiary.

Tertiary sedimentary rocks are dominantly marine clastics that are interbedded with limestone and coal. The clastic deposits were eroded from geanticlinal ridges that bordered New Zealand. Andesitic volcanics are abundant and it is likely that New Zealand was an active island arc throughout much, if not all, of the Tertiary.

New Zealand was subjected to several strong orogenic episodes during the Tertiary. These deformations culminated during the late Pliocene and early Pleistocene with the Kaikoura orogeny (**61**). The Alpine Fault extends across New Zealand from the southwestern end of the South Island to the northeastern end of the North Island. Apparently, it is a transform fault that connects a west-dipping subduction zone north of New Zealand to an east-dipping subduction zone south of New Zealand. It has been proposed that there has been about 480 km (300 mi) of horizontal displacement on this fault and that movement may have begun during the Pliocene (**61**).

## AFRICA

Limestones and minor clastics were deposited during the Paleocene and early Eocene in northwestern Africa. However, the uplift of a geanticlinal ridge in the geosyncline resulted in the deposition of a thick sequence of clastic sediments during the late Eocene. Deformation also occurred at the end of the Oligocene and again at the end of the Miocene, a time when volcanism was widespread and granitic batholiths were intruded. The extensive deformation, volcanism, and plutonism indicates that this area was an active island arc during the Tertiary. The Canary Islands off the west coast of Africa may, in part, result from Tertiary volcanism. The eastern islands in the chain may be fragments of a microplate derived from Africa.

Marine Tertiary deposits are widespread in northern and central Africa. In Egypt and Libya, the Tertiary sequence is more than 3000 m (10,000 ft) thick. Marine deposits are dominantly limestone with minor quartz sandstone and shale (Fig. 13.22). These deposits are essentially undeformed in contrast to the intensely folded and faulted Tertiary strata to the north, east, and west. Apparently, this area was protected from compressional forces that developed during the joining of the African and Eurasian plates.

A series of rift zones cutting through the eastern half of the continent (the East African Rift System) are among the most conspicuous geologic features in Africa (**62**) (Fig. 13.23). The rift valleys often contain elongate lakes, such as Lake Nyasa and Lake Tanganyika, and they are commonly bordered by volcanoes such as Mt. Kilimanjaro and Mt. Kenya. As discussed earlier in the chapter, the East African Rift System may be an aulacogen formed 15 million years ago when the Red Sea and the Gulf of Aden began to open (Fig. 13.24). The rift system does not seem to be an active spreading center because the blocks on either side of the rift do not appear to be moving apart nearly as fast as seafloor moves away from midocean ridges.

## SOUTH AMERICA

Continued uplift in the Andean Geosyncline during the Tertiary caused the seas to become even more restricted than they had been at the end of the Cretaceous. Marine deposition in Paleocene time was limited to a rather narrow coastal strip along the westernmost margin of South America. At this time, South America was moving away from North America. In Eocene time, however, the seas transgressed through a gap in the Andean highlands and formed a shallow epicontinental sea east of the highlands.

Folding and volcanism were widespread at the end of the middle Eocene and at the end of the early Oligocene. Radiometric dating indi-

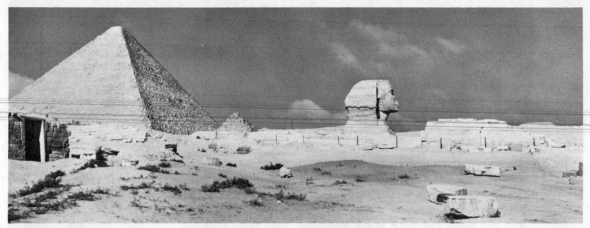

FIGURE 13.22 The Sphinx and the Great Pyramid, near Cairo, Egypt. These structures are built largely of nummulite-bearing Eocene limestone. (Photo by Eleanor Catena.)

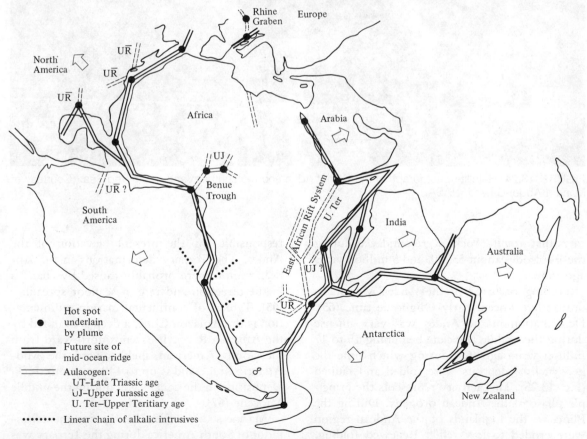

FIGURE 13.23 Rift systems in western Europe and Africa.

**FIGURE 13.24** Satellite photograph of the Gulf of Aden and the southern end of the Red Sea. (Courtesy of Monam Abdel-Gawad, NASA.)

cates that granitic rocks were emplaced during these episodes about 55, 30, and 5 million years ago (**64**).

Folding occurred in the Andes Mountains during late Eocene-early Oligocene time (**65**). Deformation in the Andes was very intense during the middle Miocene beginning 15 to 17 million years ago (**66**) during which time the geosynclinal deposits were folded and faulted (Fig. 13.25). This activity represents the principle phase of the Andean orogeny. During the Pliocene, the highlands of the Andean region were eroded to low relief. Renewed folding, faulting, and uplift during the late Pliocene are responsible for the present elevation of the Andes. The Tertiary deformations in the Andean region were probably caused by changes in the direction and rate of seafloor spreading (**65**). Typical of continental margin sedimentation is the Amazon Cone, a deep sea fan fed by the Amazon River. It extends northward from the South American shelf toward the Mid-Atlantic Ridge and is up to 14 km (8 mi) thick. Deposition in the cone began during the middle Miocene (**67**).

Marine sedimentation in the continental interior of South America during the Tertiary was confined to the margins of the continent.

**FIGURE 13.25** Folded sedimentary rocks in the Andes Mountains of Peru. These rocks are of Middle Cretaceous age and they were folded during the Tertiary.

Clastics in this region were derived both from the craton and from the eastern slope of the Andes.

## ANTARCTICA

The Tertiary history of Antarctica is very similar to that of South America. Sedimentary and volcanic rocks of geosynclinal character were deposited in a relatively narrow belt extending along the Pacific border from south Victoria Land to the Palmer Peninsula. As in South America, orogenic and volcanic activity during the Tertiary were probably associated with the eastward movement of an oceanic plate under the continent.

The Scotia Island Arc, located between Antarctica and South America, is similar in configuration to the Lesser Antilles Island Arc in the Caribbean region. Both are convex toward the east and both are underlain by a westward-dipping Benioff zone. It is likely that the Tertiary volcanic rocks of the Scotia Arc resulted from a westward movement of an oceanic plate beneath the arc.

## OCEAN BASINS

The Atlantic and Indian oceans increased in width as the continents bordering them continued to separate and as new seafloor was created at the crests of the ridges in the middle of the Atlantic and Indian oceans. New seafloor was also created in the Pacific Ocean basin, but this basin decreased in area as the continents moved toward the center of the basin.

Deep drilling by the *Glomar Challenger* and piston coring have resulted in the recovery of Tertiary sediments from all of the oceans of the world. The sediments include siliceous and calcareous oozes, red clays, and turbidites. The siliceous oozes consist of the remains of radiolarians, diatoms, silicoflagellates, and sponge spicules (Fig. 13.26). Calcareous oozes contain the skeletons of foraminifera, coccoliths, and discoasters. The so-called red clays, found in areas of low organic productivity, are generally chocolate brown and are only rarely red. They consist of windblown dust, volcanic ash, organic debris, and meteoritic dust. Turbidites of Tertiary age are common in the deep ocean adjacent to the continents, but some are found thousands of miles from land (**68**).

The age of the oldest sediments in the ocean basins generally increases with increasing distance from midocean (spreading) ridges. The age of volcanic rocks also generally increases with distance from midocean ridges, but there are some areas where young volcanics occur far from such ridges. The Hawaiian Islands, for example, are composed of late Tertiary to Quaternary volcanics, but they are thousands of miles from the crest of the East Pacific Rise.

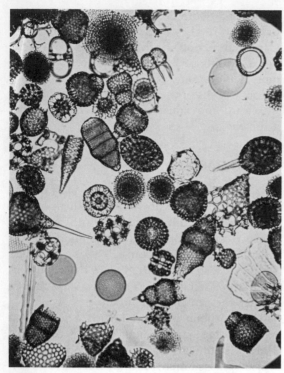

**FIGURE 13.26** Lower and Middle Miocene radiolarian assemblages from the western tropical Pacific recovered by the drilling ship *Glomar Challenger*. (Photos courtesy of Annika Sanfilippo.)

## Magnetic Anomalies and Seafloor Spreading

The trend of magnetic anomalies and transform faults may be used to determine the direction of seafloor spreading relative to the seafloor on the other side of a spreading ridge. Magnetic anomalies are generally oriented perpendicular to the direction of seafloor spreading at the time of their formation. Transform faults are almost always parallel to the direction of seafloor spreading.

There is an excellent correlation of the times of change in direction of seafloor spreading (summarized in Table 12.1, p. 369) with times of orogenic activity (mountain building, folding, metamorphism, and intrusion) on the continents (Table 12.2, p. 384). This suggests that changes in the pattern of seafloor spreading may be related to orogenies.

PACIFIC OCEAN.   Based on the orientation of magnetic anomalies and transform faults in the North Pacific, changes in the direction of seafloor spreading relative to the seafloor on the other side of the East Pacific Rise occurred 53 million years ago (the Paleocene-Eocene boundary) between anomalies 21 and 22) and between two and four million years ago (Pliocene) between anomalies 2A and 3A (**69**) (Fig. 13.27 and Table 12.1). A change in the direction of seafloor spreading relative to the underlying mantle occurred at about 42 million years ago (late Eocene) based on the date of the bend in the Hawaiian-Emperor Chain (**71**) (Fig. 13.28). This chain of volcanic islands, seamounts, and guyots is thought to have been produced by the movement of the lithosphere over a deep mantle plume (see Chapter 6). For the past 42 million years, the seafloor has been moving in a west-northwesterly direction, but between about 70 and 42 million years ago the seafloor moved in a north-northwesterly direction relative to the underlying deep mantle. Deep-sea drilling by the *Glomar Challenger* indicates a movement of the seafloor of 9 cm (3.5 in.) per year relative to the underlying plume (**72**) between 70 and 42 million years ago.

About 25 million years ago (at approximately the Oligocene-Miocene boundary) seafloor spreading began on the Galapagos Ridge, and the rate of seafloor spreading changed in the western Pacific (**73**). Moreover, seafloor spreading on the Chile Ridge (east of the East Pacific Rise) ended during the middle Miocene about 15 million years ago (**74**).

Magnetic anomalies on the crest of the Juan de Fuca ridge in the Pacific Ocean off the coast of Oregon trend north-northwest, but those more than 100 km from the ridge crest trend due north-south (**69**). Evidently the direction of seafloor spreading has changed in this area within the last several million years. The change in trend occurs between anomalies 2A

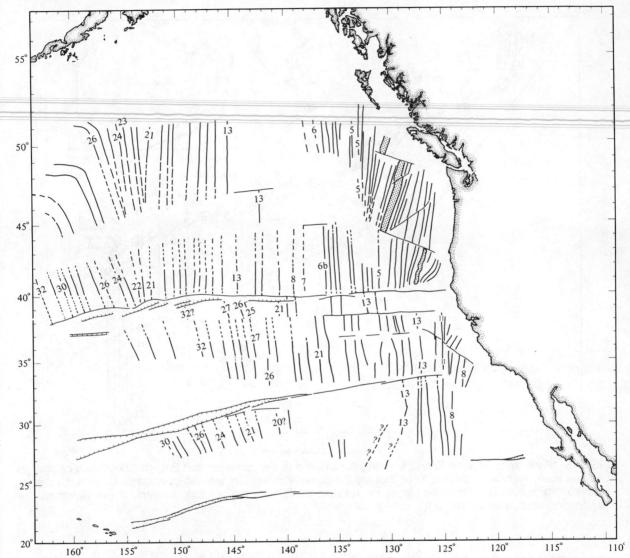

**FIGURE 13.27** Fracture zones and magnetic anomalies (with their numbers) in the northeastern Pacific Ocean. (From Menard and Atwater, Ref. 70.)

and 3A (**69**) which would place the time of change between 2 and 4 million years ago.

**ATLANTIC OCEAN.** The presence of Eocene sediments in the Norwegian Basin between Greenland and Norway indicates that seafloor spreading began there during or before the Eocene. There are two extinct spreading ridges in the Norwegian Basin (**75**) (Fig. 13.29). The distance from the 1000-fathom depth contour to the easternmost extinct ridge is approximately equal to the distance from the 1000-fathom depth contour to anomaly 16 in the Loften Basin to the northeast. The age of this anomaly is about 42 million years and this may be the time when active spreading shifted to a

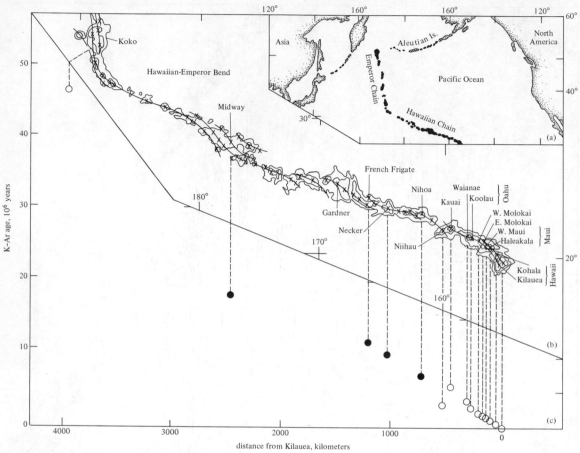

**FIGURE 13.28** Map of north Pacific showing the location of the Hawaiian and Emperor Seamount chains. Crosses mark major topographic highs that are presumably coincident with major volcanic centers. Circles indicate radiometrically determined ages on volcanic rocks from the volcanic centers. (From Dalrymple, Landphere, and Jackson, Ref. 71.)

new extinct spreading axis to the west (**76**). Spreading on the new extinct western axis lasted until about 15 million years ago, at which time the spreading activity shifted westward again, this time to the Iceland–Jan Mayen Ridge (Fig. 13.29). At about the same time, around 15 million years ago, the direction of seafloor spreading changed on the Mid–Atlantic Ridge south of Iceland (**77**).

INDIAN OCEAN. During the Tertiary, seafloor spreading occurred in the Indian Ocean

on: (1) the Mid–Indian Ocean Ridge and its continuation to the northwest, the Carlsburg Ridge; (2) the Southwest Indian Ocean Ridge and its continuation to the southwest, the Atlantic–Indian Ridge; and (3) a now extinct ridge between India and Australia. Seafloor spreading began on the Mid–Indian Ocean and Carlsburg ridges during the Late Cretaceous, about 70 million years ago, at which time India and the Seychelles Islands, began to separate. Prior to the beginning of seafloor spreading in the Red Sea (possibly during the middle Miocene), the

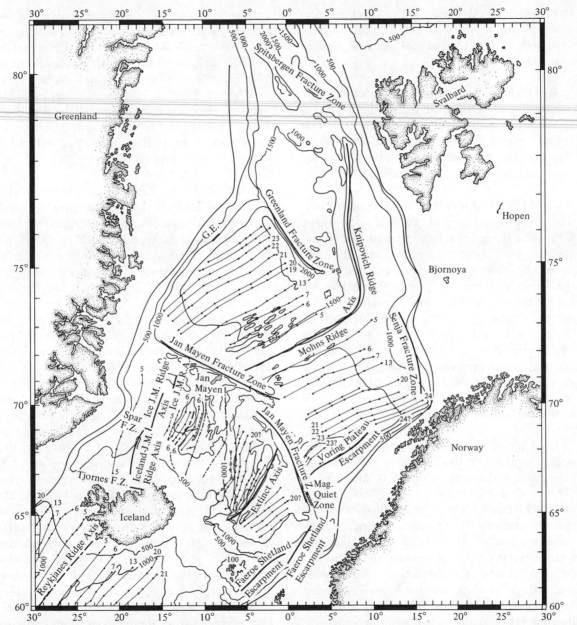

**FIGURE 13.29** Magnetic anomalies in the Norwegian-Greenland Sea. (From Talwani and Eldholm, Ref. 75.)

Carlsburg Ridge ended in a major fault zone near the continental margin of northeast Africa and Arabia. Seafloor spreading on the ridge between India and Australia began during the Late Jurassic and ended during the Oligocene about 32 million years ago (**54**) (Fig. 13.19). Spreading on the southwest Indian and Atlantic-Indian Ridges began during the Cretaceous and is still going on today, but at a rather slow half-rate of 0.8 cm per year. There appear to have been intervals during which this ridge was inactive, one between 13 and 27 million years ago, and a second between 42 and 53 million years ago (**78**).

RED SEA. Following opening of the Red Sea during the Oligocene or Middle Miocene, spreading ceased in the Red Sea (**6**). The spreading ended during the latest Miocene or Pliocene and it resumed less than 2 million years ago during the Pleistocene (**6**).

GULF OF SOUTHERN CALIFORNIA. The spreading history in the Gulf of Southern California is remarkably similar to that in the Red Sea. Following an opening during the early or middle Miocene (**7**), spreading ceased during the latest Miocene or Pliocene. Spreading resumed between 2 and 4 million years ago (during the Pliocene) (**3**). The amount of opening during the second stage was 200 to 400 km (120 to 240 mi). The opening of the Gulf of Southern California resulted in strike-slip faulting along the San Andreas Fault. There is evidence for 240 km (134 mi) of movement on this fault in the last 5 million years (**12**).

LABRADOR SEA. The youngest magnetic anomaly in the Labrador Sea between Greenland and North America is anomaly 20 (**79**) dated at 48.5 million years (**80**). The distance between this anomaly and the crest of the extinct spreading ridge in the center of the Labrador Sea is about 25 km (15 mi), and assuming a rate of seafloor spreading equal to that which occurred during the formation of the older magnetic anomalies, seafloor spreading ended in the Labrador Sea about 42 million years ago. It is interesting to note that seafloor spreading ended in the Loften Basin at about the same time.

## Relation Between Seafloor Spreading and Orogenies

The changes in the pattern of seafloor spreading at 53, 42, 25, 15, and 2–4 million years ago all correspond to times of orogenic activity in the circum-Pacific region. A similar relationship was found for Mesozoic time (see Chapter 12). Furthermore, meteorite impact epochs occur at 42, 15, and 2–4 million years ago. As discussed in Chapter 12, this relationship suggests that meteorite impacts may trigger changes in convection pattern and these may cause changes in the direction of seafloor spreading which in turn may cause orogenic activity.

MEDITERRANEAN SEA. At the beginning of the Tertiary, the Mediterranean Sea was part of the Tethys Sea. It was open to the east and probably to the west as well (Fig. 13.1a). As Africa and Europe converged during the Tertiary, the ocean between them narrowed (Fig. 13.1b). During the Oligocene, Italy and Europe collided, forming the Alps, and during the middle Miocene (or perhaps a bit earlier) the Red Sea began to open. Collision of Saudi Arabia and Europe occurred sometime during the Miocene (**40**), and this sealed off the eastern end of the Mediterranean. The western end of the Mediterranean may also have been sealed off during the late Miocene, as there is paleontological evidence for a land bridge of this age between Spain and North Africa (**81**). At the end of the Miocene (Messinian age, about 6 million years ago), there was widespread deposition of salt (about 1,000,000 cubic km; 200,000 cubic mi) throughout the Mediterranean region for a period of about one million years (**81**). Deposition of salt occurred not only on the margins of the Mediterranean (several thousands of feet of salt was deposited in Sicily), but in the deeper parts of the sea as well. The *Glomar Challenger* has recovered salt, gypsum, and anhydrite from the western Mediterranean basin. Salt deposition can occur in a deep marine basin, but it has been suggested that this salt was formed by the complete evaporation of the sea. This would have produced a basin whose bottom was more than 3000 m (10,000 ft) below present sea level. Evidence for complete evaporation of the Mediterranean includes the presence of (**82**):

1. Canyons up to 1000 m (3000 ft) below sea level cut into the continental margins bordering the Mediterranean
2. Anhydrite and chalcedony similar to that produced in sabkha (arid desert flats just above high-tide level) now 3000 m (10,000 ft) below sea level
3. Structures resembling stromatolites (stromatolites can form only in waters less than about 50 m (150 ft) deep) now in very deep water
4. Rapid variation in the bromine content of the salt, which is thought to be indicative of deposition in rather shallow water
5. Red beds associated with the salt

The evidence, while not conclusive, strongly suggests complete evaporation. However, more deep drilling is needed to test the validity of this hypothesis. If the Mediterranean did dry up completely, the subsequent filling of the basin would have been a very dramatic event. Seawater would probably have entered at the Straits of Gibraltar creating a gigantic waterfall delivering perhaps 60 million liters (30 million gallons) of seawater per second, which is 1000 times the discharge at Niagara Falls (**82**).

## TERTIARY LIFE

The Tertiary Period marks the beginning of the time of modern life. The mass extinctions of the dinosaurs, the swimming reptiles, and the flying reptiles at the end of the Mesozoic left numerous ecological niches empty at the beginning of the Tertiary. This resulted in adaptive radiation among many groups, and a large number of new species developed. The bird population increased, and for a time, giant birds became the most important predators. As the carnivorous mammals increased in size and numbers, they became the dominant predators. Herbivorous mammals also increased during the Tertiary, and consequently, the Cenozoic is known as the Age of Mammals. The mammals moved into virtually every environment including the air and sea, as well as every terres-

trial environment.

Other important changes took place in the sea. The bony fish increased in numbers and variety following the extinction of the marine reptiles. Significant changes in marine invertebrates took place as well. Lyell's subdivision of the Tertiary (Table 13.1) was based on the gradual modernization of invertebrate faunas.

### Protozoans

A major expansion of the Foraminifera, particularly the larger varieties, occurred during the Tertiary. The large, disk-shaped nummulites and orbitoids are extremely abundant in many Eocene and Oligocene limestones of the Tethyan, Caribbean, and Indo-Pacific regions. The small planktonic Foraminifera are important in correlating Tertiary sequences because of their widespread distribution and rapid evolution. Their ecological zonation has been useful in determining changes in water depth within depositional basins.

### Invertebrates

Invertebrate faunas of the Tertiary were dominated by the pelecypods and the gastropods. The pelecypods *Venericardia* and *Ostrea* are important worldwide index fossils in Eocene and younger sequences; *Pecten* and *Venus* are important in Miocene and younger deposits. Gastropods, such as *Turritella,* were abundant in Tertiary shelf faunas. Other important invertebrates include corals, echinoids, crinoids, starfish, and bryozoans.

### Fish, Amphibians, and Reptiles

During the Tertiary, the teleosts (bony fish) achieved a position of dominance among the marine and freshwater vertebrates. Numerous well-preserved fossil fish have been recovered from shales deposited in intermontane lakes in the Rocky Mountains (Fig. 13.30). Sharks were prominent in Tertiary oceans, and shark teeth are common fossils, especially in deposits of

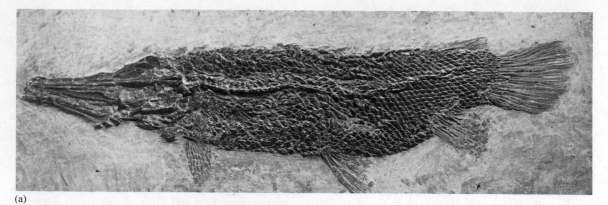

(a)

(b)

**FIGURE 13.30**   Fossil fish from the Eocene Green River Shale of southwestern Wyoming. [(a) Courtesy of the Buffalo Museum of Science; (b) courtesy of the Field Museum of Natural History, Chicago.]

Miocene age. Some of these teeth are more than 15 cm (6 in.) long (Fig. 13.31). The shark that possessed these teeth may have been 23 m (70 ft) long.

The scarcity of fossil amphibians and reptiles suggests that these creatures were greatly reduced in numbers during the Tertiary. Tertiary amphibians include frogs, salamanders, and toads. Turtles, crocodiles, lizards, and snakes were the only reptiles to survive the extinctions at the end of the Mesozoic.

## Birds

One of the interesting evolutionary trends during the Tertiary was the development of giant flightless birds. In the absence of competition from large carnivores, many birds lost the power of flight. Their great size enabled them to prey on numerous Tertiary mammals.

*Diatryma,* which lived in Wyoming during the Eocene, occupied an ecological niche left by the carnivorous dinosaurs. It was a stocky bird nearly 2.1 m (7 ft) tall and had a massive head (Fig. 13.32). These giant birds became extinct when large, carnivorous placental mammals appeared. Other flightless giants evolved in South America, Australia, New Zealand, and Madagascar before the arrival of large carnivores. However, only the rheas of South America, the ostriches of Africa, and the emus and cassowaries (the only large, flightless birds) near Australia are alive today. Perhaps these were able to survive because they are fast runners.

Most orders of modern birds had appeared by the Eocene, and the majority of fossil birds of the Tertiary are quite modern in aspect. However, at least one variety of toothed birds lived during the Cenozoic. The partial remains of a giant, toothed bird were recently found in

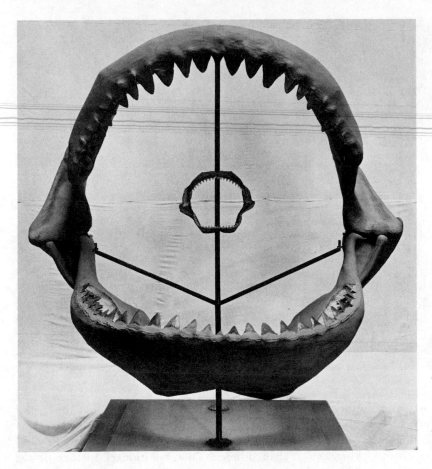

**FIGURE 13.31** Reconstruction of the jaws of a very large Miocene shark. The jaws of a large modern shark appear small in comparison. (Courtesy of the Field Museum of Natural History, Chicago.)

beds of Pliocene age in California. The wing span of this bird has been estimated at 4.9 m (16 ft).

## Mammals

Tertiary mammals are dominantly marsupials and placentals, both of which stem from the pantotheres (see Chapter 12). Born alive, marsupials are tiny and extremely immature at birth, and they must be transferred immediately to the mother's pouch for continued nourishment and development. The young of placental mammals are born in a more advanced state. Their delayed birth is made possible by the development of a highly efficient nutrient connection, the placenta, between the mother and

the fetus. As a result of this delayed birth, the young have a much better chance of survival. Many placental mammals are relatively independent only a few hours after birth.

In Eurasia, Africa, and North America, placental mammals replaced the marsupials during the early Tertiary. However, South America and Australia became isolated from the other continents before the evolution of the placental mammals. Consequently, marsupials thrived on these continents during most of the Tertiary.

Paleocene mammals were of small to medium size and had five-toed feet. The increase in the size of many mammals in Eocene time is reminiscent of the trend toward giantism in the dinosaurs. This is a common trend, since metabolic efficiency generally increases with the size

**FIGURE 13.32** The large flightless bird *Diatryma* lived during the Eocene in Wyoming. It grew up to 7 ft tall and had a skull as large as the skull of a small horse. (Courtesy of the Field Museum of Natural History, Chicago.)

of an animal. Another important evolutionary development among the mammals was the increase in brain size. The brain of the lower mammals is rather small and smooth, whereas the more advanced mammals have a larger brain with a highly involuted surface.

ARCHAIC HOOFED MAMMALS. Primitive mammals ranged from very small carnivores to large browsing forms. The condylarths and amblypods were the most important of the Paleocene herbivores, but both became extinct during the Eocene. The small and relatively agile condylarths were characterized by five toes and a long tail (Fig. 13.33). The amblypods were stockier and had sharp canine teeth. Uintatheres were strange-looking amblypods with three pairs of horns (Fig. 13.34).

ODD-TOED HOOFED MAMMALS. The odd-toed hoofed mammals have a large central toe which bears most or all of the weight. This group includes the horse, the rhinoceros, and the extinct titanotheres and chalicotheres. The evolutionary history of the horse is documented by an extensive fossil record (Fig. 13.35). The remains of the oldest known horse, *Hyracotherium* (formerly *Eohippus*), have been found in late Paleocene and Eocene strata (**83**) (Fig. 13.35). This horse was the size of a small dog and had four toes on each forefoot and three toes on each hind foot. Its teeth exhibit the characteristics of browsing animals. During the middle Tertiary, horses increased in size and their toes decreased in number. In the Miocene, most horses changed their feeding habits from browsing to grazing as meadows and plains became more extensive. One-toed horses similar to the modern horse appeared in the Pliocene.

Titanotheres were large mammals similar to the rhinoceros (Fig. 13.36). They were typified by the brontotheres, which were as much as

FIGURE 13.33 Reconstruction of an Early Tertiary landscape, Wyoming, showing two condylarths (*Phenacodus*). (Courtesy of the American Museum of Natural History.)

FIGURE 13.34 Reconstruction of an Eocene landscape in Wyoming. Two uintatheres watch a group of tiny, primitive horses (*Hyracotherium*). (Courtesy of the Field Museum of Natural History, Chicago.)

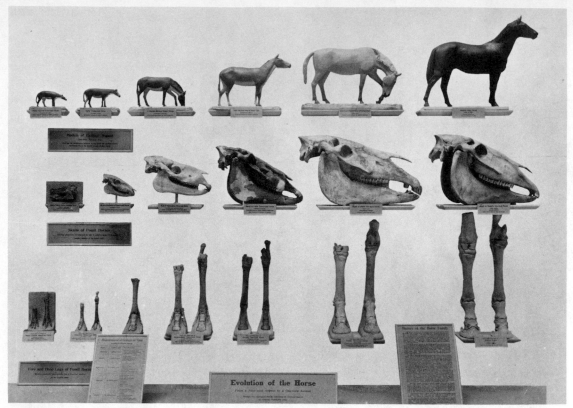

**FIGURE 13.35** Stages in the evolution of the horse from the Eocene to the present: Top row: Changes in the relative size of the body of the horse. Middle row: Changes in the size of the skull. Bottom row: Changes in the size and configuration of the fore and hind legs of the horse. (Courtesy of the Field Museum of Natural History, Chicago.)

2.4 m (8 ft) high at the shoulder. Titanotheres had four forefoot digits and three hindfoot digits, stout limbs, and bony-knob horns. They were prevalent in western and North America during the early Tertiary but became extinct in the Oligocene.

Chalicotheres had clawlike feet and horselike bodies (Fig. 13.37). These strange-looking mammals probably used their claws for digging up roots and tubers. Chalicotheres appeared during the Eocene and became extinct during the Pleistocene.

Several groups of rhinoceroses appeared during the Eocene. The early forms were small and hornless. Some were adapted for running; others were semiaquatic. Very large rhinoceroses developed in late Oligocene time. One of these giants, *Baluchitherium,* was 3.5 m (18 ft)

high at the shoulder and was probably the largest land mammal that ever lived.

**EVEN-TOED HOOFED MAMMALS.** The oreodonts thrived between late Eocene and the early Pleistocene and were common during the Oligocene in western North America (Fig. 13.37). They are related to the modern camel. The first true camels appeared in the Eocene in North America as small four-toed forms. As in the evolution of the horse, the camels increased in size as their toes decreased in number. During the Pliocene, camels migrated from North America to Asia, Africa, and South America.

Pigs and peccaries appeared in the Eocene, and giant pigs evolved during the Oligocene and Miocene (Fig. 13.37). The skull of one specimen is nearly 1 m (3 ft) long.

**FIGURE 13.36** Reconstruction of a group of titanotheres (*Brontops robustus*) from Early Oligocene deposits of North America. (Courtesy of the Field Museum of Natural History, Chicago.)

ELEPHANTS, DEER, AND CATTLE. The proboscideans first appeared during the Eocene. Typical of the early elephants was *Paleomastodon,* a small, short-tusked form which lived in northern Africa and southern Asia. In Miocene time, elephants migrated to Europe and North America. Miocene elephants included semiaquatic shovel-tusked forms. Elephants with large upper tusks appeared during the Pliocene. Deer and cattle first appeared in Eurasia in Oligocene time. These forms did not migrate to North America until the Pleistocene.

PRIMATES.  Primates are the most highly developed group of mammals, and they include hominids, apes, monkeys, lemurs, tree shrews, tarsiers, and so on. With the exception of hominids, apes, and baboons, all primates are tree dwellers, as most probably were their ancestors. The anatomy of the primates represents an adaptation to this environment. Tree dwellers must be agile and able to grasp branches as they move from tree to tree. Stereoscopic vision, which is a benefit of the forward eye orientation of the primates, provides depth perception.

Primates evolved from the insectivores during the late Cretaceous. The main primate groups of the early Tertiary were the lemurs, lorises, and tarsioids (Fig. 13.38). The anthropoids, which include monkeys, apes, and humans, did not appear until the Oligocene. The remains of *Parapithecus* (a monkeylike form) and *Aegyptopithecus* (an apelike form) have been found in the Fayum Depression of Egypt (**85**) (Fig. 13.39). *Propliopithecus,* whose remains have been retrieved from Oligocene deposits of Africa, may be an ancestor of modern hominids (**77**). The similarity of its dentition to that of modern humans provides the basis for this suggestion.

Fossil remains of *Dryopithecus* have been found in late Oligocene and early Miocene deposits of Europe, Asia, and Africa (**86**) (Fig. 13.40). *Dryopithecus* has a low, rounded fore-

**FIGURE 13.37** Restoration of a Miocene landscape in western Nebraska with wild hogs (in the foreground) and a chalicothere browsing on the lower branches of a tree. At the far left is a herd of oreodonts. (From a painting by Charles R. Knight, courtesy of the Field Museum of Natural History, Chicago.)

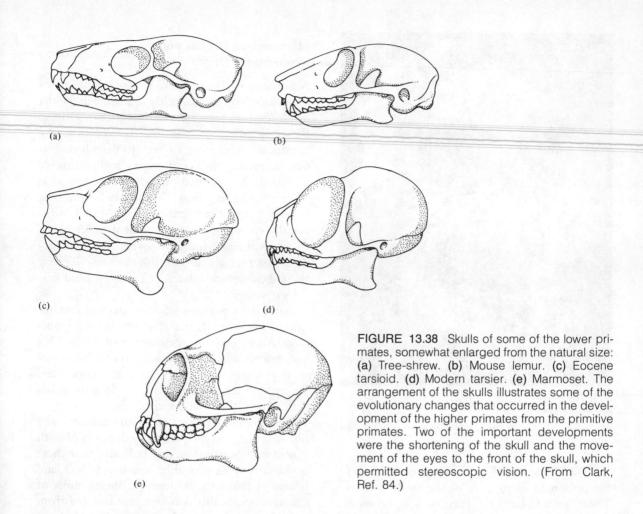

(a)

(b)

(c)

(d)

(e)

**FIGURE 13.38** Skulls of some of the lower primates, somewhat enlarged from the natural size: **(a)** Tree-shrew. **(b)** Mouse lemur. **(c)** Eocene tarsioid. **(d)** Modern tarsier. **(e)** Marmoset. The arrangement of the skulls illustrates some of the evolutionary changes that occurred in the development of the higher primates from the primitive primates. Two of the important developments were the shortening of the skull and the movement of the eyes to the front of the skull, which permitted stereoscopic vision. (From Clark, Ref. 84.)

head, moderate brow ridges, and large canine teeth and is considered to be a direct ancestor of modern humans (**86**). Between 10 and 15 million years ago, *Dryopithecus* gave rise to *Ramapithecus, Sivapithecus,* and *Gigantopithecus.*

Hominids are members of the family Hominidae. This group includes living humans and their direct ancestors. Members of Hominidae share certain ecological and physical characteristics. For example, they are bipedal, use their hands for manipulation of objects, and have similar dental characteristics. *Ramapithecus* is the earliest known hominid. Remains of *Ramapithecus* have been found in Europe, India, Africa, and China in strata of middle and late Miocene and early Pliocene age. *Ramapithecus* was probably an erect biped about one meter (3 ft) tall. The small size of the incisors and canines implies a dependence on the use of the hands for grasping and tearing vegetation. *Ramapithecus* may have used simple tools such as sticks and rocks, but there is no evidence of this in the fossil record. The similarity of the teeth of *Ramapithecus* to those of *Homo sapiens* indicates that this hominid could be a direct ancestor of the Australopithicines and hence of *Homo sapiens.* Australopithicines appeared during the Pliocene, but since they continued into the Pleistocene, they will be discussed in Chapter 14.

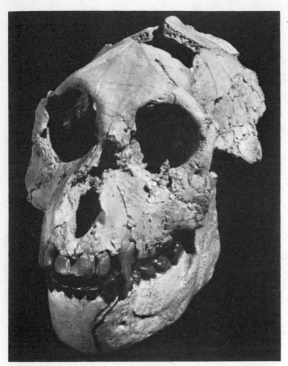

FIGURE 13.39 Skull of *Aegyptopithecus,* an Oligocene primate, probably a close relative to the ancestors of man. (Photo courtesy of E. L. Simons.)

OTHER MAMMALS.   The creodonts were the dominant carnivores of the early Tertiary. These include catlike forms, such as *Oxyaena,* and doglike forms. Creodonts expanded greatly in size and numbers in Miocene time, but all had become extinct by the end of the Pliocene. The aquatic cetaceans, which appeared in the middle Eocene, include carnivores such as porpoises and herbivores such as whales. Seals, sea lions, and walruses did not appear until Miocene time.

The rodents, including squirrels, rats, mice, and porcupines, emerged first in the Paleocene, and the lagomorphs (rabbits) appeared during the Eocene. The edentates, which include anteaters, sloths, glyptodonts, and armadillos, arose in Paleocene time and were an important part of the late Pliocene fauna of South America.

## Mammalian Faunas and Continental Drift

According to Björn Kürten, the evolution of mammals was significantly influenced by the displacements between continents (**87**). During the early stages of mammalian evolution, the continents were close enough to permit migration between the continents, and primitive mammals had gained worldwide distribution by the early Cretaceous. In the late Cretaceous, however, the continents had separated sufficiently to prevent many faunal migrations. Each continent produced several different orders of mammals, and during the early Tertiary there were seven isolated or partly isolated faunal provinces.

After the separation of Australia and Antarctica in the early Tertiary, Australia was completely isolated from the other continents. This isolation resulted in the adaptive radiation of such primitive mammals as the monotremes (spiny anteater and platypus) and the marsupials (kangaroos, wombats, and bandicoots).

The presence of certain fossil edentates and notoungulates in early Tertiary deposits of both North and South America indicates that these continents were probably connected by a land bridge at that time. However, during most of the Tertiary, South America was isolated from the other continents. About six orders of mammals evolved in South America during the Tertiary. These include the archaic placental mammals and several orders of marsupials. Adaptive radiation resulted in the development of a wide variety of mammals which occupied a number of different environments. Many of the marsupials of South America and Australia resemble the more advanced placental mammals. Carnivorous marsupials of South America included foxes, wolves, and cats. *Thylacosmilus* had very large canine teeth which were remarkably similar to those of the sabertooth feline, *Smilodon,* a placental mammal which lived during the Pleistocene. The rodents migrated to South America during the middle Tertiary. They

**FIGURE 13.40**   Restoration of the Miocene primate *Dryopithecus* (*Proconsul*). (Courtesy of the British Museum of Natural History.)

probably migrated from Eurasia by way of North and Central America across a narrow body of water to South America. Since monkeys originated in Africa and are not found in North America, it is presumed that they somehow were able to cross the Atlantic Ocean to South America.

Africa was isolated from the other continents during the early Tertiary. The Tethys Sea was still an open ocean and a shallow sea separated Spain and Africa. Between four and six orders of mammals evolved in Africa during this episode of isolation. These mammals include the elephants, conies, and aardvarks. Approximately 16 orders originated in Eurasia and North America. Among them are the insecti-

vores, bats, primates, cats, dogs, bears, hoofed mammals, rodents, and rabbits.

Several continents that were isolated during the early Tertiary were rejoined in the middle and late Tertiary. Africa was connected to Europe in the Oligocene as a result of its northward drift. Faunal migrations between Africa and Europe resulted in the dispersal of many forms, including elephants and mastodons. South America remained isolated until the late Pliocene, when a land bridge was established between North and South America. Ground sloths, porcupines, and armadillos migrated from South America northward, and horses, camels, deer, dogs, cats, and bears migrated southward. The archaic placental mammals and marsupials of South America, which could not compete with the more advanced placental mammals, became extinct. Of the mammals native to South America, only a few edentates survived. In the late Tertiary, the westward movement of North America resulted in the formation of a land bridge between Alaska and Siberia. This route permitted an intermixing of North American and Eurasian faunas.

By the end of the Tertiary, the number of faunal provinces had been reduced from 7 to 4 and the number of mammalian orders had declined from 30 to 18. Only Australia retained its primitive fauna until the introduction of advanced mammals by European settlers.

## Plants

Many families of modern plants have existed since the early Tertiary. Paleocene and Eocene temperate forests included poplar, birch, cedar, and alder. By the middle Tertiary, angiosperms had become adapted to a wide variety of different environments ranging from the hot, humid tropics to the colder polar regions. A number of modern genera may be traced to the middle Tertiary. Temperate forests of the Oligocene included oak, beech, chestnut, and conifers in the lowland regions. In the Miocene, forests diminished somewhat in importance, and grasslands reached the height of their development. As the climates became cooler in the middle and late Tertiary, a variety of different plant assemblages developed. Temperate and cool-temperate forests expanded, and the ranges of subtropical plants became more restricted. Boreal forests were dominated by conifers such as spruce, pine, and fir, shrubs and herbaceous plants.

## CLIMATE

The worldwide cooling that began during the late Cretaceous was interrupted by a warming trend between the Paleocene and the Eocene. During this warm interval, tropical and subtropical climates were widespread. Palm trees grew as far north as Germany and the states of Washington and Alaska; and corals lived 10° to 20° north of their present ranges (**46**). A climatic cooling which began during the early Oligocene resulted in a northward shifting of climatic belts (**88**) (Fig. 13.41). The fossil plant assemblage from a lignite in Vermont indicates that the climate in that area during the late Oligocene was still significantly warmer than that of today. The depositional environment was probably similar to that of the swamps in the southeastern United States (**89**).

Plant fossils from the Vienna Basin indicate that the climate of central Europe changed from subtropical-humid during the early Miocene, to subtropical-dry in the middle Miocene, and then to warm-dry in the late Miocene. Studies of mollusks in the circum-Pacific region indicate that there was a general warming trend in this area from the early to the middle Miocene (**90**). However, the climate of the Pliocene was in general cooler and drier than the Miocene.

## Glaciation

A study of cores taken in the deep ocean near Antarctica indicates that glacial ice existed there as early as Eocene time (**91**). Rafted debris has been found in cores from the north Pacific and Bering Sea in deposits as old as late Miocene (**92**). Glaciation of late Miocene and Pliocene age has been reported from the St. Elias Mountains in southern Alaska, but this may have been alpine glaciation (**93**). Glacial deposits in the Jones Mountains of Antarctica are overlain by lava flows which are probably between 7 and 10 million years old (late Miocene) (**94**). The extent of these deposits suggests that they were the result of continental glaciation, that is, the Antarctic Ice Cap.

Glacial deposits of late Pliocene age are widespread. Basaltic scoria overlying glacial sediments in the Taylor Valley in Antarctica has been dated at approximately 2.6 million years. Since these deposits are above the level of the

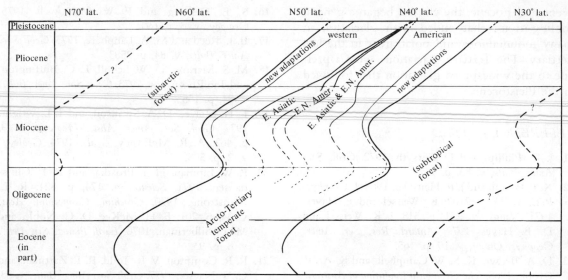

**FIGURE 13.41** Generalized indication of the migration of forests during the Tertiary in western North America. (After Dorf, Ref. 88.)

present glaciers, glaciation must have been more extensive than today. Since there is presently a continental ice sheet on Antarctica, continental glaciation presumably began more than 2.6 million years ago (**95**).

In the Sierra Nevada Mountains of California, till is underlain and overlain by volcanic rocks that have been dated at 3.0 and 2.7 million years old respectively (**96**). In Iceland, glaciation began about three million years ago. Icelandic glaciation occurred about 3.06, 2.90, 2.80, 2.69, 2.53, 2.47, and 2.33 million years ago (**97**). Glacial deposits more than two million years old have been reported from the Soviet Union and South America (**98**).

## SUMMARY

With the continued separation of the continents during the Tertiary, the Atlantic and Indian oceans widened while the Pacific Ocean decreased in size. Extensive tectonic activity occurred in the geosynclines bordering the Pacific Ocean and the Tethys Sea as oceanic plates moved under the continental margins. In the

Alpine and Himalayan regions, there was intense deformation during the Oligocene and Miocene as a result of compression between major continental plates. Changes in the direction of seafloor spreading, which occurred at 53, 42, 25, 15, and 2–4 million years ago, correspond to times of orogenic activity and also to times of increase in the numbers of large meteorite (astroid) impacts.

The seas transgressed into the interiors of the continents during the Paleocene and Eocene but were less extensive than they had been in the Mesozoic. An irregular withdrawal of the seas began in the Miocene, and by the end of the Tertiary, the seas were confined to the outer margins of the continents.

Mammals expanded rapidly in size, number, and variety during the Tertiary, coming to inhabit most of the available environments by the end of Eocene time. The Miocene primates *Dryopithecus* and *Ramapithecus* are probably ancestors of modern humans, and as the Tertiary ended, *Australopithecus* was making primitive tools.

Following a warming trend during the Paleocene and Eocene, the climate began a general cooling trend that ended with glaciation in many mountainous and polar areas in the late Tertiary. The Tertiary glaciations were a prelude to the widespread glaciation that followed in the Pleistocene.

## REFERENCES CITED

1. J. D. Phillips and D. Forsyth, 1972, *Geol. Soc. Amer. Bull.,* v. 83, p. 1579.
2. X. LePichon and J. R. Heirtzler, 1968, *J. Geophys. Res.,* v. 73, p. 2102; K. Weissel and E. Hayes, 1971, *Nature,* v. 231, p. 518; J. K. Weissel and D. E. Hayes, 1972, *Antarct. Res. Ser. Amer. Geophys. Union,* v. 19, p. 165.
3. D. A. Brown, K. S. W. Campbell, and K. A. W. Crook, 1968, *The Geological Evolution of Australia and New Zealand:* Pergamon Press, New York.
4. DSDP Scientific Staff, 1973, *Geotimes,* v. 18, no. 7, p. 14.
5. J. G. Sclater and R. L. Fisher, 1974, *Geol. Soc. Amer. Bull.,* v. 85, p. 683.
6. D. A. Ross *et al.,* 1973, *Science,* v. 179, p. 377; DSDP Scientific Staff, 1972, *Geotimes,* v. 17, no. 7, p. 24.
7. D. G. Moore, 1973, *Geol. Soc. Amer. Bull.,* v. 84, p. 1883; J. P. Owens and C. S. Denny, 1974, *Geol. Soc. Amer. Abstr. Programs,* v. 5, p. 61–62.
8. R. W. Girdler and P. Styles, 1974, *Nature,* v. 247, p. 7.
9. I. Evans, 1977, *Geol. Soc. Amer. Abstr. Programs,* v. 9, p. 970; M. A. Johnson, D. R. Lawrence, W. H. Kanes, and C. S. Bartlett, *Geol. Soc. Amer. Abstr. Programs,* v. 9, p. 1039.
10. J. D. Fairhead, J. G. Mitchell, and L. A. J. Williams, 1972, *Nature Phys. Sci.,* v. 238, p. 66.
11. R. Larson, 1972, *Geol. Soc. Amer. Bull.,* v. 83, p. 3345.
12. K. W. Ehlert and P. L. Ehlig, 1977, *Geol. Soc. Amer. Abstr. Programs,* v. 9, p. 415; S. E. Joseph and T. E. Davis, 1977, *Geol. Soc. Amer. Abstr. Programs,* v. 9, p. 443.
13. W. Glassley, 1974, *Geol. Soc. Amer. Bull.,* v. 85, p. 785.
14. P. D. Snaveley, 1968, *Amer. J. Sci.,* v. 266, p. 454.
15. P. B. King, 1959, *The Evolution of North America:* Princeton University Press, Princeton, N.J.
16. S. E. DeLong and F. W. McDowell, 1975, *Geology,* v. 3, p. 691.
17. B. L. Reed and M. A. Lanphere, 1973, *Geol. Soc. Amer. Bull.,* v. 84, p. 2583.
18. M. S. Marlow, D. W. Scholl, E. C. Buffington, and Tau Rho Alpha, 1973, *Geol. Soc. Amer. Bull.,* v. 84, p. 1555.
19. E. H. McKee, D. A. Swanson, and T. L. Wright, 1977, *Geol. Soc. Amer. Abstr. Programs,* v. 9, p. 463; A. R. McBirney, *et al.,* 1974, *Geology,* v. 2, p. 585.
20. P. W. Lipman, H. J. Prostka, and R. L. Christiansen, 1971, *Science,* v. 174, p. 821; R. L. Armstrong, 1970, *Geochim. Cosmochim. Acta,* v. 34, p. 201; E. H. McKee, D. C. Noble, and M. L. Silberman, 1970, *Earth Planet. Sci. Lett.,* v. 8, p. 93.
21. R. R. Compton, V. R. Todd, R. E. Zartman, and C. W. Naeser, 1977, *Geol. Soc. Amer. Bull.,* v. 88, p. 1237; G. A. Davis, K. V. Evans, E. G. Frost, S. H. Lingrey, and T. J. Shackelford, 1977, *Geol. Soc. Amer. Abstr. Programs,* v. 9, p. 943.
22. D. C. Noble, 1972, *Earth Planet. Sci. Lett.,* v. 17, p. 472.
23. H. W. Menard, 1961, *Science,* v. 132, p. 1737.
24. E. A. Hay, 1976, *Geology,* v. 4, p. 763.
25. R. C. Surdam and C. A. Wolfbauer, 1975, *Geol. Soc. Amer. Bull.,* v. 86, p. 335.
26. H. P. Eugster and L. A. Hardie, 1975, *Geol. Soc. Amer. Bull.,* v. 86, p. 319.
27. W. G. Pierce, 1957, *Amer. Assoc. Petrol. Geol. Bull.,* v. 41, p. 591.
28. L. H. Fisk and P. DeBord, 1974, *Geol. Soc. Amer. Abstr. Programs,* v. 6, p. 441.
29. F. K. Miller and J. C. Engels, 1975, *Geol. Soc. Amer. Bull.,* v. 86, p. 517.
30. F. G. Fox, 1959, *Amer. Assoc. Petrol. Geol. Bull.,* v. 43, p. 992.
31. P. H. Mattson and E. A. Pessagno, Jr., 1971, *Science,* v. 174, p. 138; P. H. Mattson, 1966, *Proc. 3rd Caribbean Geol. Conf.,* p. 49.
32. A. H. Barabas, 1971, *Geol. Soc. Amer. Abstr. Programs,* v. 3, p. 498.
33. J. W. Ladd, 1976, *Geol. Soc. Amer. Bull.,* v. 87, p. 969.
34. C. Schuchert, 1968, *Historical Geology of the Antillean-Caribbean Region:* Hafner, New York.
35. T. H. Jordan, 1975, *J. Geophys. Res.,* v. 80, p. 4433.

36. L. K. Fink, Jr. *et al.*, 1971, *Proc. 6th Caribbean Geol. Conf.*, Abstr.

37. J. F. Dewey, W. C. Pittman III, W. B. F. Ryan, and J. Bonnin, 1973, *Geol. Soc. Amer. Bull.*, v. 84, p. 3139.

38. J. F. Dewey, 1977, *Geol. Soc. Amer. Abstr. Programs*, v. 9, p. 949; W. G. Ernst, *Geol. Soc. Amer. Bull.*, v. 84, p. 2053.

39. W. Lowrie and W. Alvarez, 1974, *Nature*, v. 251, p. 285; W. Lowrie and W. Alvarez, 1975, *J. Geophys. Res.*, v. 80, p. 1579.

40. S. J. Haynes and H. McQuillan, 1974, *Geol. Soc. Amer. Bull.*, v. 85, p. 739.

41. N. J. Soper, C. Downie, A. C. Higgins, and L. I. Costa, 1976, *Earth Planet. Sci. Lett.*, v. 32, p. 149.

42. S. Moorbath and H. Welke, 1969, *Earth Planet. Sci. Lett.*, v. 5, p. 217; M. Brooks, 1973, *J. Geol.*, v. 81, p. 81; C. K. Brooks and A. J. W. Gleadow, 1977, *Geology*, v. 5, p. 539.

43. S. Moorbath, H. Sigurdsson, and R. Goodwin, 1968, *Earth Planet. Sci. Lett.*, v. 4, p. 197.

44. P. R. Vogt, 1972, *Nature*, v. 240, p. 338.

45. J. M. Bird and M. S. Weathers, 1977, *J. Geol.*, v. 85, p. 359.

46. J. G. Dennis, 1971, *J. Geophys. Res.*, v. 76, p. 5394.

47. T. Frisch and R. Thorsteinsson, 1977, *Geol. Soc. Amer. Abstr. Programs*, v. 9, p. 1244.

48. W. Alvarez, 1976, *Geol. Soc. Amer. Bull.*, v. 87, p. 891; G. Chabrier and G. H. Mascle, 1977, *Geol. Soc. Amer. Bl.*, v. 77, p. 1694; E. Argand, 1922, *Proc. 13th Int. Geol. Cong.*, fasc. 1, p. i71; K. A. De Jong, M. Manzoni, T. Stavenga, F. Van Dijk, R. Van der Voo, and J. D. A. Zijkerveld, 1973, *Nature*, v. 243, p. 281.

49. X. LePichon, 1972, *J. Geophys. Res.*, v. 77, p. 391.

50. P. Molnar, W. P. Chen, T. J. Fitch, P. Tapponnier, W. E. K. Warsi, and F. T. Wu, 1977, *Proc. Symp. Himalayan Geol.*

51. D. N. Wadia, 1966, *Geology of India:* St. Martin's Press, New York.

52. P. Wellman and M. W. McElhinny, 1970, *Nature*, v. 227, p. 595.

53. P. Molnar and J. Francheteau, 1975, *Nature*, v. 255, p. 128.

54. B. D. Johnson, C. McA. Powell, and J. J. Veevers, 1976, *Geol. Soc. Amer. Bull.*, v. 87, p. 1560; J. G. Sclater and R. L. Fisher, 1974, *Geol. Soc. Amer. Bull.*, v. 85, p. 683.

55. D. Davies, 1968, *Nature*, v. 220, p. 1225.

56. K. O. Emery and Z. Ben Arraham, 1972, *Amer. Assoc. Petrol. Geol. Bull.*, v. 56, p. 839.

57. S. Uyeda and A. Miyashiro, 1974, *Geol. Soc. Amer. Bull.*, v. 85, p. 1159.

58. N. Isezaki and S. Uyeda, 1973, *Marine Geophys. Res.*, v. 2, p. 51.

59. P. Molnar and P. Tapponnier, 1975, *Science*, v. 189, p. 419.

60. V. L. Masaytis, M. V.Mikhaylov, and T. V. Selivanovskaya, 1975, *The Popigay Meteorite Crater:* Nauka Press, Moscow.

61. D. A. Brown, K. S. W. Campbell, K. A. W. Crook, *The Geological Evolution of Australia and New Zealand:* Pergamon Press, New York.

62. J. H. Illies, 1970, *in* J. H. Illies and St. Mueller, eds., *Graben Problems:* E. Schweizerbartsche Verlagsbuchhandlung, Stuttgart, p. 4; W. T. C. Sowerbutts, 1972, *Nature*, v. 235, p. 435.

63. K. Burke and J. F. Dewey, 1973, *J. Geol.*, v. 81, p. 406.

64. S. Quirt *et al.*, 1971, *Geol. Soc. Amer. Abstr. Programs*, v. 3, p. 676; J. W. Stewart, J. F. Evernden, and N. J. Snelling, 1974, *Geol. Soc. Amer. Bull.*, v. 85, p. 1107.

65. R. Charrier, 1973, *Earth Planet. Sci. Lett.*, v. 20, p. 242.

66. E. Farrar and D. C. Noble, 1976, *Geol. Soc. Amer. Bull.*, v. 87, p. 1247.

67. J. E. Damuth and N. Kumar, 1975, *Geol. Soc. Amer. Bull.*, v. 86, p. 863.

68. M. Ewing, R. Houta, and J. Ewing, 1969, *J. Geophys. Res.*, v. 74, p. 2477.

69. P. R. Vogt and G. R. Byerly, 1976, *Earth Planet. Sci. Lett.*, v. 33, p. 185.

70. H. W. Menard and T. Atwater, 1968, *Nature*, v. 219, p. 463.

71. G. B. Dalrymple, M. Lanphere, and E. D. Jackson, 1974, *Geol. Soc. Amer. Bull.*, v. 85, p. 727.

72. A. Kaneps, 1978, *Geotimes*, v. 28, no. 1, p. 21.

73. D. A. Clague and R. D. Jarrard, 1973, *Geol. Soc. Amer. Bull.*, v. 84, p. 1135; R. Hey, 1977, *Geol. Soc. Amer. Bull.*, v. 88, p. 1404.

74. W. J. Morgan, P. R. Vogt, and D. F. Falls, 1969, *Nature*, v. 222, p. 137.

75. M. Talwani and O. Eldholm, 1977, *Geol. Soc. Amer. Bull.*, v. 88, p. 969.

76. P. R. Vogt and O. E. Avery, 1974, *J. Geophys. Res.*, v. 79, p. 363.

77. P. J. Fox, W. C. Pitman, III, and F. Shepard, 1969, *Science,* v. 165, p. 487.

78. H. W. Bergh and I. O. Norton, 1976, *J. Geophys. Res.,* v. 81, p. 5221.

79. Y. Kristofferson and M. Talwani, 1977, *Geol. Soc. Amer. Bull.,* v. 88, p. 1037.

80. J. R. Heirtzler *et al.,* 1968, *J. Geophys. Res.,* v. 73, p. 2119.

81. L. A. Smith, 1977, *Geotimes,* v. 22, no. 3, p. 20.

82. R. Kühn and K. J. Hsü, 1974, *Geology,* v. 2, p. 213; K. J. Hsü, 1972, *Sci. Amer.,* v. 227, no. 6, p. 27.

83. G. L. Jepsen and M. O. Woodbine, 1969, *Science,* v. 164, p. 543.

84. W. E. LeGros Clark, 1965, *History of the Primates:* University of Chicago Press, Chicago.

85. E. L. Simons, 1964, *Sci. Amer.,* v. 211, no. 1, p. 28.

86. E. L. Simons, 1967, *Sci. Amer.,* v. 217, no. 6, p. 28.

87. B. Kürten, 1969, *Sci. Amer.,* v. 220, no. 14, p. 54.

88. E. Dorf, 1964, *in* A. E. M. Nairn, ed., *Problems in Paleoclimatology:* Wiley, New York, p. 13.

89. J. A. Wolfe and E. S. Barghoorn, 1960, *Amer. J. Sci.,* v. 258-A, p. 388.

90. W. O. Addicott, 1966, *Science,* v. 155, p. 583.

91. S. V. Margolis and J. P. Kennett, 1970, *Science,* v. 170, p. 1085.

92. DSDP Scientific Staff, 1971, *Geotimes,* v. 16, no. 11, p. 12.

93. G. H. Denton and R. L. Armstrong, 1969, *Amer. J. Sci.,* v. 267, p. 1121.

94. R. H. Rutherford *et al.,* in press, *Tertiary Glaciation in the Jones Mountains:* SCAR Symposium on Antarctic Geology, 1970.

95. R. L. Armstrong, W. Hamilton, and G. H. Denton, 1968, *Science,* v. 159, p. 187.

96. R. R. Curry, 1966, *Science,* v. 154, p. 770.

97. I. McDougall *et al.,* 1977, *Geol. Soc. Amer. Bull.,* v. 88, p. 1.

98. J. H. Mercer, 1969, *Science,* v. 164, p. 823.

# The Quaternary Period 14

I N THE LATE 1800s geologists were puzzled by the "drift" that covers much of northern and central Europe. They observed that the random mounds and ridges were commonly composed of unsorted sediments quite unlike ordinary waterlain deposits. Some cobbles and boulders in the drift differed considerably from the underlying bedrock but could be matched with distant outcrops. It was suggested that these deposits had been carried by icebergs or by the waters of the biblical flood, but these explanations failed to account for many of the characteristic features of the drift.

Those who believed that the flood or icebergs were not adequate depositional mechanisms sought an alternative explanation. James Hutton and Ventez-Sitten were among the first scientists to propose that glaciers were the source of the drift in the vicinity of the Alps. Ventez-Sitten, a Swiss civil engineer, demonstrated that sediment deposited by modern glaciers was identical to drift found well beyond the present extent of those glaciers. In 1832, A. Bernhardi proposed that continental glaciers from the north had at one time extended into central Europe and were responsible for the lowland drift (1). The great naturalist Louis Agassiz later demonstrated that continental glaciers had at one time also covered much of North America. He further suggested that glaciation occurred in both Europe and North America during a "great ice period." By the middle of the nineteenth century, Agassiz had become a leading proponent of the glacial theory.

## QUATERNARY CHRONOLOGY

The Quaternary period is divided into two epochs, the Pleistocene, which includes the "ice ages," and the Holocene or Recent, which comprises much of postglacial time. An early attempt to date the beginning of the Pleistocene was based on the assumption that approximately 250,000 years were required for continental ice sheets to form, advance, and recede. The existence of four distinct glacial episodes provided an age of one million years for the beginning of the Pleistocene and the onset of the ice ages. Although the currently accepted date for the beginning of the epoch is about 1.8 million years ago, the original figure was a surprisingly accurate estimate.

## RECONSTRUCTION OF QUATERNARY CLIMATES

The shell chemistry of fossil Foraminifera (forams) depends on the temperature of the ocean in which they lived and constructed their calcium carbonate tests. Harold Urey found that when seawater evaporates, the water vapor is slightly enriched in the lighter oxygen isotope, oxygen 16, and a slight concentration of

**497**

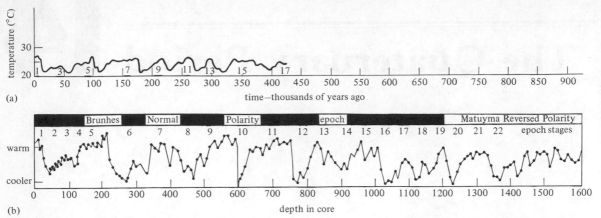

(a)

(b)

**FIGURE 14.1** Variations in paleotemperatures as determined by oxygen isotope studies: **(a)** Caribbean Sea. (After Emiliani, Ref. 3. Copyright © 1966 by the University of Chicago.). **(b)** Equatorial Pacific. (After Shackleton and Opdyke, Ref. 4.)

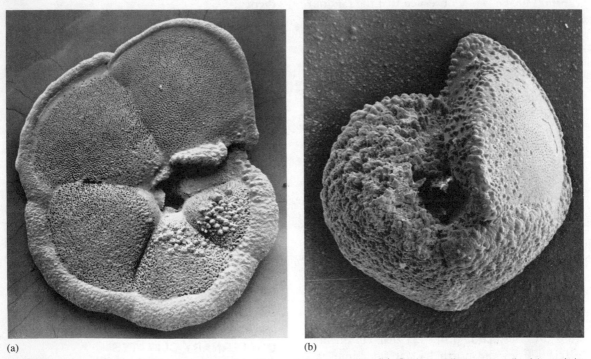

(a)                                                                  (b)

**FIGURE 14.2** Pleistocene foraminifera: **(a)** *Globorotalia menardii.* **(b)** *Globorotalia truncatulinoides,* right coiling. (Photo courtesy of J. P. Kennett.)

the heavier isotope, oxygen 18, remains in the seawater (2). Since the rate of evaporation generally increases with temperature, the amount of oxygen 18 enrichment depends on the temperature of the seawater. When compared with modern shell chemistry and seawater temperatures, the ratio of oxygen 18 to oxygen 16 in the tests of fossil planktonic forams provides

the temperature of the surface water of the ocean in which they lived (Fig. 14.1).

The relative abundance of certain forams is also a function of the temperature of the seawater in which they lived. *Globigerina pachyderma,* for example, is restricted to cold climates, whereas *Globorotalia menardii* is found principally in warm climates (Fig. 14.2a). *Globorotalia truncatulinoides* (Fig. 14.2b) has a unique characteristic of coiling to the left in cold water and to the right in warm water. Glacial conditions will tend to lower oceanic temperatures. An abundance of left over right coiled tests in midlatitude ocean sediments might indicate that glacial conditions had once occurred there (Fig. 14.3).

Terrestrial climates may be inferred from the study of the remains of fossil plants, such as leaves, pollen, and spores, and fossil insects. Pollen and spores in ancient lake sediments provide an indication of the sequences of vegetation that once grew near the lake. For example, a change in the dominant vegetation from tundra to spruce, to pine, and then to oak would indicate a progressive warming of the climate. Insect remains also give an approximation of temperature at the time of deposition. Fossil insects, pollen and spores are considered to be indispensable in climate reconstructions (Figs. 14.4 and 14.5).

## PLIOCENE-PLEISTOCENE BOUNDARY

The boundary between the Pliocene and the Pleistocene is generally placed at the beginning of the widespread glaciation. However, since continental glaciation may have begun as early as 60 million years ago (see Chapter 13), it is not possible to define the Pleistocene as the time of the ice ages.

The type section for the basal Pleistocene is located in southern Italy. Here the boundary between the Pliocene and Pleistocene has been placed at the first appearance of coldwater invertebrates, such as the pelecypod *Arctica islan-*

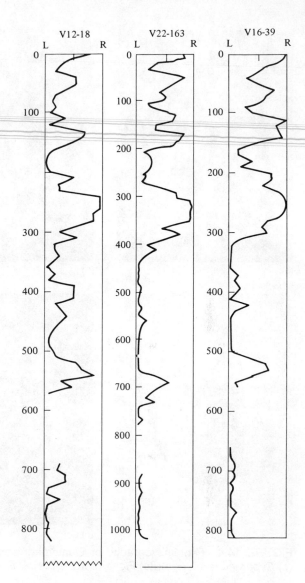

**FIGURE 14.3** Correlation of three cores from the South Atlantic as defined by changes in the direction of coiling of *Globorotalia truncatulinoides.* The scale runs from 100 percent left coiling (left) to 100 percent right coiling (right). Numbers to the left of the columns are depths in the cores, in centimeters. The percentage of right-coiling forms increases with increasing temperature of the surface waters. Warmer surface waters occurred during interglacial and interstadial periods. (From Ericson and Wollin, Ref. 5. Copyright © 1968 by the American Association for the Advancement of Science.)

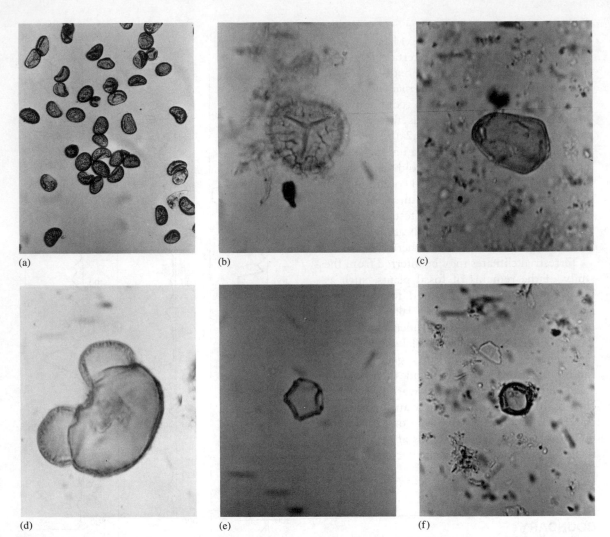

**FIGURE 14.4** Photomicrographs of Quaternary spores and pollen: **(a)** Fern spores. **(b)** *Lycopodium.* **(c)** *Cyperaceae* (sedge). **(d)** *Picea* (spruce). **(e)** *Alunus* (alder). **(f)** Rosaceae.

*dica* and the foram *Hyalinea baltica*. This boundary also marks the extinction of most discoasters (a type of microalgae) and the first appearance of large numbers of *Globorotalia truncatulinoides* (**7**). From studies of cores taken in the south Atlantic, David Ericson and Goesta Wollin found that these floral and faunal changes occurred at the beginning of the Olduvai (Gilsa) event, approximately 1.8 million years ago (**8**). The basal Pleistocene marine beds in southern Italy have been assigned to the

Calabrian stage (Fig. 14.6). Sedimentary units that have been classically assigned to the basal or Villafranchian stage of the terrestrial Pleistocene are found in northern Italy along the Po River. The fauna of the Upper Villafranchian includes one-toed horses, the first true elephants, and distinctive species of dogs, carnivores, rodents, deer, and other vertebrates. Marine fossils found interbedded with terrestrial deposits of the Upper Villafranchian stage are quite similar to those found in deposits of the

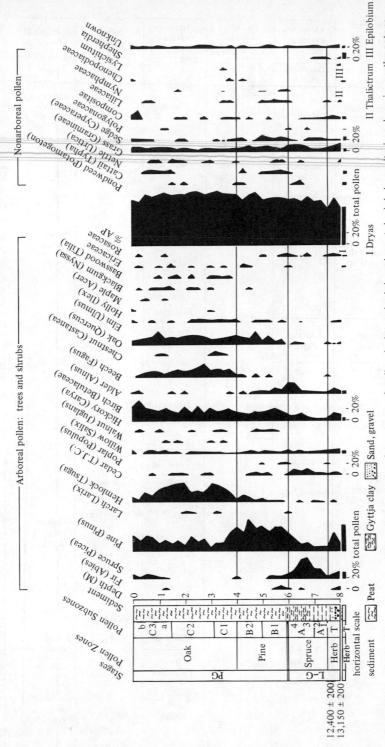

**FIGURE 14.5** Pollen diagram showing the relative frequency of pollen in late glacial and postglacial deposits in a bog in northeastern New York. Pollen zones are based on the relative frequency of pollen at each level sampled. Radiocarbon dates (12,400 ± 200 and 13,150 ± 200 years B.P.) provide a minimum age for the herb (tundra) pollen zone. The decrease in boreal (northern) vegetation and increase in vegetation preferring a warmer climate is an indication of the gradual warming of the climate. The late rise in spruce corresponds to a minor reversal in the general warming trend in this area. (From Connally and Sirkin, Ref. 6.)

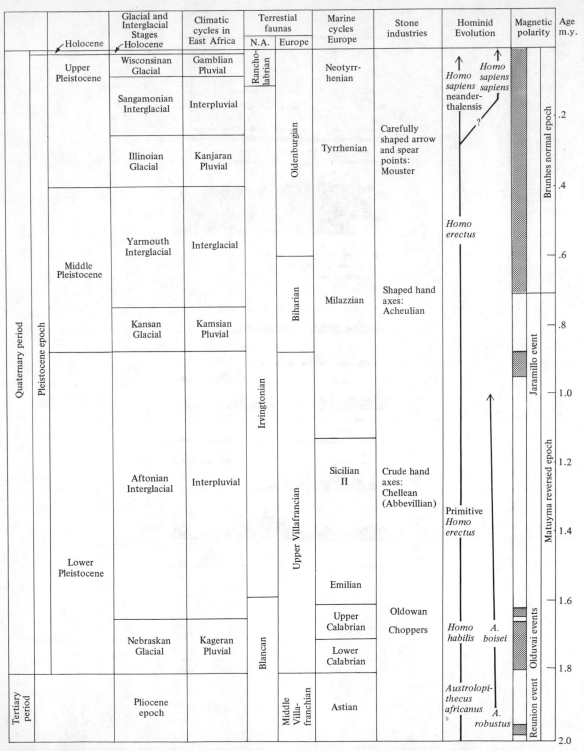

**FIGURE 14.6** A provisional stratigraphic chart comparing climatic cycles, marine cycles, stone industries, and the ranges of the ancestors of man.

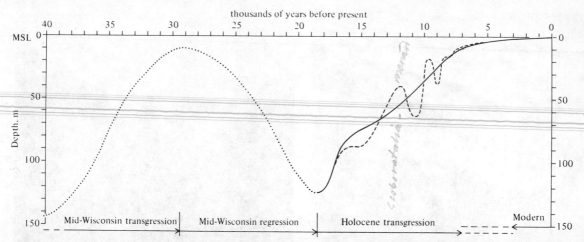

**FIGURE 14.7** Variation in sea level throughout the past 40,000 years based on radiocarbon dates of fresh and salt marsh peats and shallow marine fossils plotted against depth below sea level where they were found. (From Curray, Ref. 13. Copyright © 1965 by Princeton University Press, reprinted with permission.)

Calabrian stage. Therefore, the Upper Villa-franchian stage is probably the same age as the Calabrian, which is basal Pleistocene. In the Olduvai Gorge in Tanzania, volcanic rocks interbedded with sediments containing an Upper Villafranchian fauna have been dated at approximately 1.75 million years old (**9**). This date supports the conclusion that the Pleistocene began 1.8 million years ago if the Villafranchian stage correlates with the Calabrian stage. However, there are conflicting dates for this boundary. Radiometric dating of the Calabrian (and, therefore, basal Pleistocene) deposits indicates that they may be 2.0 to 2.2 million years old (**10**). Others place the Pliocene-Pleistocene boundary near the top of the Olduvai Event, which would place the boundary at about 1.6 million years old (**11**). Obviously all three of these dates can not be correct. We favor a date of 1.8 million years for the Pliocene-Pleistocene boundary.

## PLEISTOCENE-HOLOCENE BOUNDARY

Although no universal criteria for the boundary between the Pleistocene and the Holocene have

been established, the dissipation of the continental ice sheets provides acceptable time planes. The ice sheets disappeared from North America about 5000 years ago and from northern Europe 8000 years ago (**12**). Stratigraphers variously place the Pleistocene-Holocene boundary at either the midpoint of worldwide warming of the oceans or at the midpoint of the rise of sea level. The midpoint in the warming of the oceans occurred between 11,000 and 12,000 years ago. At this time there was an increase in the abundance of *Globorotalia menardii,* several species of planktonic radiolaria, and warm-water coccoliths in the midlatitudes. The midpoint in the rise of sea level occurred about 13,500 years ago, when ice still covered much of North America (Fig. 14.7).

The International Quaternary Association's Holocene Commission recommends placing the Pleistocene-Holocene Boundary at 10,000 years *B.P.* (before present).

## DISTRIBUTION OF PLEISTOCENE GLACIERS

Modern glaciers cover only about 10 percent of the earth's land surface. Most of these are lo-

**FIGURE 14.8** Active glaciers. **(a)** Southern Victoria Land, Antarctica. (U.S. Navy Photograph, courtesy of the U.S. Geological Survey.) **(b)** The Alps. (Courtesy of Francois Arbey.) **(c)** Bylot Island, just north of Baffin Island. (Courtesy of the National Air Photo Library, Department of Energy, Mines and Resources, Canada.) **(d)** The Bagley Ice Field, South Alaska. (Courtesy of British Petroleum Co.)

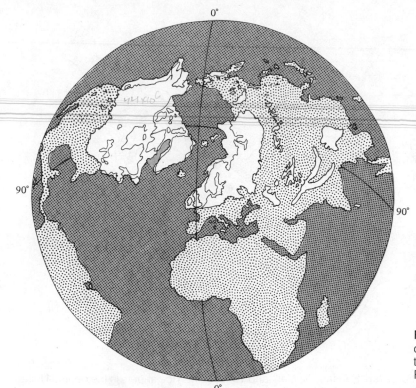

**FIGURE 14.9** Extent of Pleistocene glaciation (shaded area) in the northern hemisphere. (After Flint, Ref. 1. Copyright © 1970 by John Wiley, New York.)

cated in Antarctica, Greenland, Iceland, and mountainous terrains (Fig. 14.8). Based on the outer limits of glacial deposits, Pleistocene glaciers covered more than three times that area, or about 44 million square km (17 million square mi) (Fig. 14.9 and Table 14.1). Seismic studies and test drilling in Greenland and Antarctica indicate that the ice reaches thicknesses of 3000 to 4000 m (10,000 to 13,000 ft), respectively (**13**). The North American and Scandinavian ice sheets were probably of comparable thickness. It has been estimated that the Pleistocene ice sheets had a volume of 76.7 million cubic km (18.4 million cubic mi). A corresponding vol-

TABLE 14.1  **Approximate Land Areas Covered by Pleistocene Glaciers**[a]

| CONTINENT | SQUARE MILES | SQUARE KILOMETERS |
|---|---|---|
| North America, Greenland, Hawaii | 7,128,227 | 18,512,391 |
| Europe and Iceland | 2,779,404 | 7,208,145 |
| Asia | 1,523,505 | 3,951,000 |
| South America | 335,472 | 870,000 |
| Africa | 732 | 1,900 |
| Australia, New Zealand, and New Guinea | 11,580 | 30,000 |
| Antarctica | 5,325,136 | 13,810,000 |
| Total | 17,105,056 | 44,383,436 |

[a] From Flint (**1**).

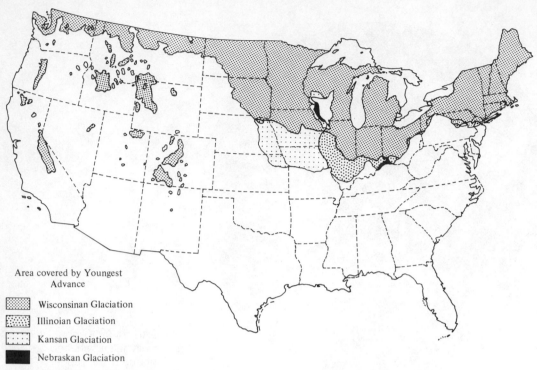

Area covered by Youngest
Advance

▨ Wisconsinan Glaciation
▦ Illinoian Glaciation
▤ Kansan Glaciation
■ Nebraskan Glaciation

**FIGURE 14.10** Borders of the Nebraskan, Kansan, Illinoian, and Wisconsinan ice sheets in the central United States based on the extent of the glacial drift of different ages. Where older drift is buried beneath younger drift, the borders are not shown. (From Denny, Ref. 14.)

ume of seawater converted to glacial ice would require a drop in sea level of about 197 m (644 ft) (**1**). However, loading of the crust would result in a flow of mantle from below the ice that would tend to cause a rise in sea level. The actual lowering of sea level would probably be only about 133 m (431 ft). At the present time, about 26.25 million cubic km (6.3 million cubic mi) of ice occupies the continents. If it all melted, this volume of ice would raise the sea level about 65 m (210 ft) (**1**).

## MULTIPLE GLACIATIONS

Evidence of multiple glaciations has been found in North America, South America, Iceland, Europe, and Asia. The most convincing evi-

dence is the presence of till sheets or outwash gravels separated by paleosols, gumbotils (weathered tills), or marine sediments. In the Alps, at least 12 distinct outwash deposits, each separated from its neighbors by paleosols, indicate that 12 or more separate episodes of glaciation occurred in that area (Table 14.2). At least eight glacial episodes have been identified in northwestern Europe and 18 glacial advances occurred in central Europe. Interglacials are represented by paleosols, unconformities, or interbedded marine and terrestrial deposits.

T. C. Chamberlain and F. Leverett were among the first to record evidence of multiple glaciations in North America. In the Great Plains region, they found four major glacial stages and lithologically distinct tills separated by gumbotils. At least 13 separate glacial ad-

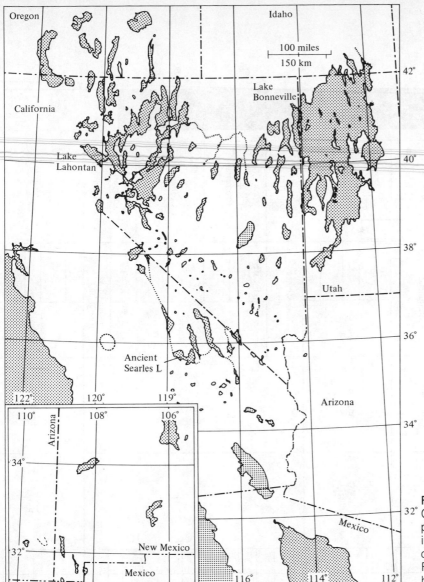

**FIGURE 14.11** Lakes of the Great Basin in the western part of the United States during the Pleistocene at the time of maximum glaciation. (From Flint, Ref. 1. Copyright © 1971 by John Wiley, New York.)

vances have since been documented in central North America (Fig. 14.10 and Table 14.2).

## Pluvial and Interpluvial Features

Striking fluctuations in climate occurred during the Pleistocene in areas distant from the glaciers. Many regions that today have warm-dry and cool-dry climates experienced cooler pluvial (rainy) climates. Large, deep lakes formed in the western United States, Europe, Asia, and Australia. Parts of the Sahara and the Gobi deserts were also covered by lakes (Fig. 14.11). These lakes left wave-cut cliffs and beach deposits far up on the modern hillsides. Wave-cut terraces occur as much as 300 ft above Utah's Great Salt Lake, a remnant of Lake Bonneville (Fig. 14.12). Cores taken from the Great Salt Lake reveal several cycles of lacustrine silts, shoreline sands, and salt (**14**). The presence of

**TABLE 14.2  Suggested correlations of North American and European glacial and interglacial stages. Stippled boxes in column 2 indicate cold episodes.**

Geomagnetic time scale — Million years: .10 .20 .30 .40 .50 .60 .70 .80 .90

**N.W. Europe**
- Weichsel ●
- Eem ●
- Warthe ●
- Saalian — Treene ●, Drente ●
- Holstein ●
- ●
- Elsterian ● ★
- ● ★
- Cromerian III ●
- Glacial B Cromerian ●
- Cromerian II ●
- Glacial A Cromerian ◐
- Cromerian I ○
- Menapian ○
- ○

**European Alps**
- Würm Glacial
- Riss-Würm Interglacial — II, I
- Riss Glacial
- Mindel-Riss Interglacial — II, I
- Mindel Glacial ★
- Günz-Mindel Interglacial — III, II, I
- Günz Glacial
- Donau-Gunz Interglacial ◐ — III, I
- Donau Glacial

**British Isles**
- Flandarian
- Drensian (Weichsel)
- Ipswichian
- Gippian
- Hoxnian
- Anglian
- Cromerian
- Baventian

**Rocky Mountains North America (Pacific NW)**
- Pinedale (Vashon)
- (Olympia)
- Late Bull Lake (Salmon Springs)
- Early Bull Lake
- Sacajawa Ridge
- Cedar Ridge

**Temperature variations in ocean cores**
- A.: 1, 3, 4, 5, 6, 7, 8, 9, 10, 11, 12, 13, 14, 15, 16, 17
- B.: 1, 2, 3, 4, 5, 6, 7, 8, 9, 10, 11, 12, 13, 14, 15, 16, 17, 18, 19, 20, 21, 22

**Mid Continent North America**
- Recent
- Wisconsinan Glacial — Late, Early
- Sangamonian Interglacial
- Illinoian Glacial — Late, Middle, Early
- Yarmouthian Interglacial ● ★
- Kansan Glacial — Late, Middle, Early

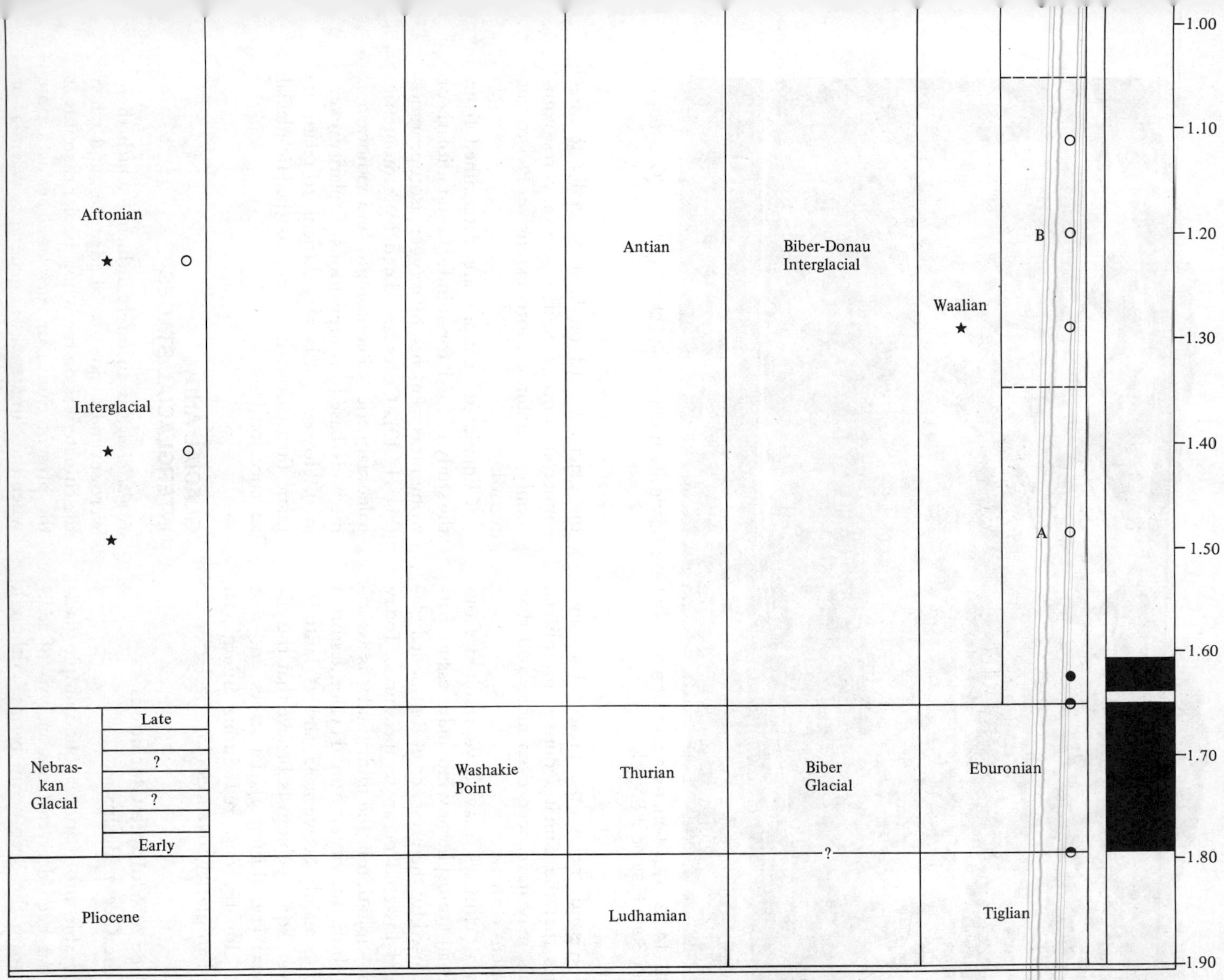


★ Radiometric Date    Paleomagnetic Control: ● Normal ○ Reversed ◒ Boundary between Normal and Reversed

**FIGURE 14.12** Wave-cut terraces of the ancient Lake Bonneville north of Salt Lake City, Utah. (Courtesy of the Utah State Historical Society.)

shoreline sands below the present lake level indicates that the climates during interpluvial intervals may have been even more arid than the present climate.

Pluvial climates may have resulted from changes in atmospheric wind and pressure patterns caused by the presence of continental glaciers and lowered ocean temperatures. Today most precipitation in the midlatitudes is associated with the activity of frontal cyclonic storms occurring mainly between 35° and 50° latitude. However, some geologists believe that the climate was arid during glacial stages and that pluvial climates occurred during interglacial rather than glacial advances (**15**).

### Evidence of Multiple Glaciations from the Ocean Basins

Considerable information about multiple glaciations has been derived from analysis of sediments from the deep-ocean basins. Although the glacial record on land generally features numerous unconformities, ocean cores commonly contain a more extensive Pleistocene record.

Climatic fluctuations are determined from the study of fossil foraminifera and radiolaria or from measurements of oxygen isotope ratios (Fig. 14.13). Cores are dated by radiometric, paleomagnetic, paleontologic, and radiocarbon methods (based on the positions of polarity reversals). However, it is very difficult to relate climatic fluctuations in a core to specific glacial advances on land.

### GLACIAL AND INTERGLACIAL STAGES

A *glacial age* is an interval during which there were several major glacial advances, and when the climate was on the whole cooler than it is at the present time. An *interglacial age* is a time when the climate was in general as warm as or

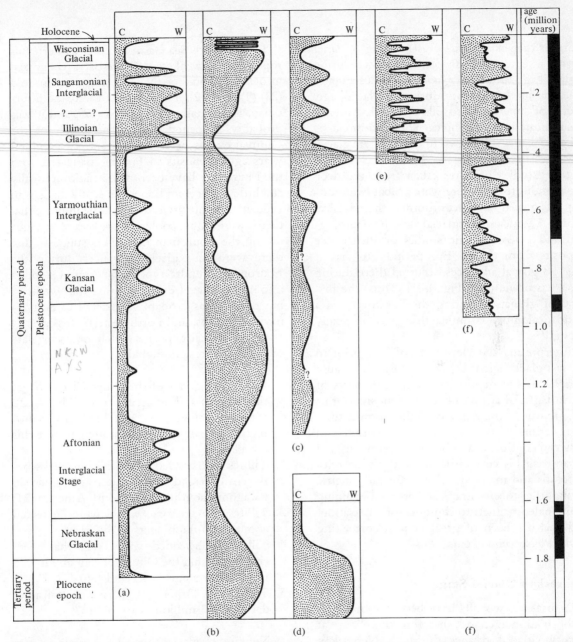

**FIGURE 14.13** Generalized climate curves for the Quaternary period based on: **(a)** Variation in the frequency of *Globorotalia menardii*. (After Ericson and Wollin, Ref. 5. Copyright © 1968 for the American Association for the Advancement of Science.) **(b)** Planktonic foraminifera from the northern Gulf of Mexico. (After J. H. Beard, Esso Production Research Co., Ref. 17.) **(c)** Planktonic foraminifera from the southern oceans. (After Kennett, Ref. 18.) **(d)** Planktonic foraminifera from the Mediterannean, the Caribbean, and the Gulf of Mexico. (After Lamb, Ref. 19.) **(e)** Paleotemperatures of the Caribbean as determined by oxygen isotope studies. (After Emiliani, Ref. 3; Copyright © 1966 by the University of Chicago.) **(f)** Paleotemperatures in the equatorial Pacific as determined by oxygen isotope studies. (After Shackleton and Opdyke, Ref. 4.) The time scale at the left has been dated by various radiometric and paleomagnetic techniques discussed in the text. C = Cold; W = Warm.

warmer than it is at the present time. A *glacial stage* is a sequence of deposits laid down during a glacial age and an *interglacial stage* is a sequence of deposits laid down during a glacial age. Study of superimposed tills in North America has indicated that during each of the four classical glacial ages, from two to five separate glacial advances have occurred. Moreover, oxygen isotope studies of cores taken from the deep ocean indicate that there were probably a number of glacial advances during the classical North American interglacial ages. Paleontological and paleomagnetic studies of Pleistocene deposits from central Europe indicate that at least 18 glacial advances occurred there during the Pleistocene (30) (Fig. 14.1). Thus, the history of glaciation during the Pleistocene is a good deal more complex than was originally thought.

Interpreting the Pleistocene history prior to the last glacial age is especially difficult because of erosion and deposition by the ice sheets of the last glacial age and because deposits of this age are out of the range of radiocarbon dating. Correlation of Pleistocene events within and between continents and between continents and ocean basins is very difficult due to the scarcity of fossils and materials suitable for radiometric dating. Pleistocene stratigraphers are beginning to use paleomagnetic techniques for correlation, and this has been of great help in correlating Early Pleistocene events.

## Nebraskan Glacial Stage

In Nebraska, two tills have been designated as Nebraskan in age, and, therefore, there were at least two glacial advances during the Nebraskan (1). In Kansas, deposits thought to correlate with those of the Nebraskan glacial stage are underlain by a volcanic ash that has been dated at between 1.9 and 1.97 million years by the fission track method (21). Therefore, the Nebraskan probably began less than 1.9 million years ago.

In the North Sea Basin of northwestern Europe, seven glacial stages have been recognized in deposits younger than 1.8 million years old (22) (Table 14.2). The oldest of these, the Eburonian, is normally magnetized (22), and it was probably deposited during the oldest of the Olduvai events between 1.80 and 1.64 million years ago. Deposits in France correlated with the Eburonian have been dated radiometrically at a little under 2 million years (23). Changes in vegetation in northwestern Europe indicate that there were four cooler-than-average periods during the Eburonian (22). This suggests that there were four glacial advances during the Eburonian. Tundra-type vegetation extended into northwestern Europe during the advances, but the ice was probably restricted to Scandanavia, the Alps, and northwestern Russia.

Other glacial deposits which may correlate with the Nebraskan include:

1. Outwash gravels of the Biber Glacial Stage in the Alps and adjacent areas (Table 14.2). However, there have not been any radiometric or paleomagnetic studies to test this correlation.
2. Tillites interbedded with volcanic rocks between 1.86 and 1.67 million years old in Patagonia, southernmost South America (24)
3. Tillites in Argentina, which were deposited about 1.7 million years ago (25)
4. Tillites interbedded with lavas that were erupted during the Olduvai event about 1.75 million years ago in Iceland (26)
5. Tillites in Antarctica between volcanics dated at 1.7 million years old (27)

Nebraskan glacial deposits contain a late Blancan mammalian fauna, and the end of the Blancan is thought to be approximately at the end of the Nebraskan (28). The end of the Blancan has been dated at between 1.5 and 1.6 million years (1), and therefore, the end of the Nebraskan occurred between 1.5 and 1.6 million years ago.

## Aftonian Interglacial Stage

A deep soil developed on Nebraskan glacial deposits during the Aftonian interglacial age. Rainfall during this interval was probably comparable to that of today, but warmer temperatures during parts of the Aftonian Age resulted in a dryer climate (1). In Nebraska, reversely magnetized volcanic ash interbedded with deposits thought to have been laid down during the Aftonian have been dated at 1.2 million years old (21). Waalian deposits in northwestern Europe may be the equivalent of deposits of the Aftonian stage in the central United States (Table 14.2). Deposits correlated with the Waalian stage are overlain by a basalt which has been dated at 1.3 million years by the potassium-argon method (29). This date supports the correlation of deposits of the Waalian stage with those of the Aftonian stage.

Studies of foraminifera from deep-sea cores indicate that there were as many as five cooler-than-average periods during the Aftonian (Fig. 14.13a). Furthermore, changes in vegetation in northwestern Europe indicate that at least one cooler-than-average period occurred during the Waalian, during which time pine forests migrated into central Europe. Studies of glacial *loess* (windblown silt carried from glacial outwash plains) indicate that there were four glacial advances during the Waalian (Aftonian) time (30). Thus, it is evident that the term "interglacial" is somewhat inappropriate because glacial advances occurred during these times.

## Kansan Glacial Stage

There are three tills of Kansan age in Nebraska (1). Evidently there were at least three separate glacial advances during the Kansan. Till of probable Kansan age is overlain by a normally magnetized volcanic ash dated at 600,000 years old (21). Thus, the Kansan ended more than 600,000 years ago.

Glaciation occurred during Menapian time in the North Sea Basin in northwestern Europe (Table 14.2). Paleomagnetic study indicates that the Menapian sediments were deposited during the Matuyma reversed geomagnetic epoch, just after the Jarimillo event (22). This would date the Menapian deposits at between 890,000 and 690,000 years. The Menapian may correlate with the Kansan, and if so, it is likely that the Kansan began about 890,000 years ago. It has been suggested that the Elster deposits rather than the Menapian deposits correlate with deposits of the Kansan glacial stage (1). However, deposits correlated with the Elster have been radiometrically dated at between 220,000 and 400,000 years (31).

Vegetation changes in northwestern Europe indicate that there were three cooler-than-average periods within Menapian time (22). These three cool periods may correspond to the three Kansan glacial advances. During the Menapian, tundra-type vegetation migrated into northwestern Europe, but the glaciers were probably confined to Scandinavia, the Alps, and northwestern Russia during this interval.

The Mindel glacial stage in the Alps is commonly correlated with the Kansan glacial stage of North America. However, potassium-argon dating of volcanic rocks within deposits correlated with the Mindel glacial stage have been radiometrically dated at between 220,000 and 300,000 years old (32), and the Kansan stage is more than 600,000 years old.

## Yarmouthian Interglacial Stage

A deep soil developed in Kansan drift during the Yarmouthian interglacial stage. The climate during at least part of the Yarmouthian was probably much like that of the present, but perhaps slightly warmer and drier (1). The Yarmouthian stage probably correlates with the Cromerian deposits in northwestern Europe (Table 14.2). There are three or four glacial substages within the Cromerian, two of which are designated *Glacial A* and *Glacial B* (22).

Glacial A deposits were laid down during the last reversal of the earth's magnetic field 690,000 years ago (**22**).

Oxygen isotope studies of cores from the equatorial Pacific indicate that there were five times during the Yarmouthian when the oceans were significantly cooler than they are today (**4**) (Table 14.2). This indicates that there were probably five glacial advances during the Yarmouthian.

The Günz glacial stage in the Alps contains three substages (**33**). It has commonly been correlated with the Nebraskan glacial stage in North America. However, this may not be correct. Paleomagnetic studies indicate that deposits immediately overlying outwash gravels correlated with the Gunz contain a reversal that is thought to be the reversal at the beginning of the Brunhes normal geomagnetic epoch (**34**). If this is the case, the Gunz deposits are probably about 690,000 years old, whereas the Nebraskan probably began more than 1.5 million years ago. At least part of the Gunz may correlate with the Cromerian glacial age. Potassium-argon dating of volcanic rocks in river terrace deposits thought to correlate with the Gunz glacial stage are between 340,000 and 400,000 years old (**32**). These dates appear to conflict with the paleomagnetic data, but perhaps they were determined on a younger Gunz substage. Even so, the dates still seem to be too young, suggesting that the Gunz ended about 450,000 years ago (Table 14.2).

## Illinoian Glacial Stage

There are three different Illinoian tills in Illinois and Indiana (**1**). Since they differ little in their degree of weathering, it is likely that they were all deposited over a relatively short time interval. Several paleoclimatic indicators indicate that a marked cold period began about 400,000 years ago (Fig. 14.13). This may be the time of the beginning of the Illinoian glacial stage.

The Elsterian deposits of northwestern Europe may correlate with deposits of the Illinoian

glacial stage (Table 14.2). As discussed earlier, radiometric dates on deposits correlated with the Elsterian range from 220,000 to 400,000 years (**31**). However, the Elsterian probably ended 250,000 years ago (Table 14.2), which would suggest that the younger dates may be in error.

The Mindel glacial stage in the Alps may also correlate with the Illinoian. Potassium-argon dates on volcanic rocks in terrace deposits of the Rhine River indicate that the outwash gravels correlated with the Mindel are between 220,000 and 300,000 years old (**32**). Again, we would suggest that the younger dates are in error. Glacial deposits in Hawaii found above volcanics dated at 270,000 years and below volcanics dated at 190,000 years (**35**) may also correlate with the Illinoian.

## Sangamonian Interglacial Stage

The Sangamonian interglacial stage is widely represented by both sediments and a deep soil developed in Illinoian glacial deposits. This soil is much thicker and more mature than that developed since the retreat of the last ice sheet (**1**), and therefore, it may have taken a considerable time to form.

260,000 TO 150,000 YEARS AGO. The interval between 260,000 and 150,000 years ago was, for the most part, nearly as warm as today. However, oxygen-isotope studies of deep-sea cores indicate that there were at least two episodes during this interval that were significantly cooler than average (Fig. 14.1). These may have been times of glacial advance.

Elevated coral reefs on the island of Tonga in the South Pacific are dated at 260,000 years old (**36**). These reefs formed during a high stand of sea level, presumably during an interglacial or interstadial age. They may have formed at the beginning of the Sangamonian interglacial age.

Holstein deposits in northwestern Europe and the Mindel-Riss interglacial deposits in the

Alps may correlate with deposits in the lower part of the Sangamonian stage (Table 14.2). Rhine River terrace gravels which have been correlated with the Mindel-Riss have been dated at 140,000 to 150,000 years (**32**).

150,000 TO 125,000 YEARS AGO. Oxygen-isotope studies of deep-sea cores indicate that the oceans were unusually cold between about 150,000 and 125,000 years ago (Fig. 14.1). Studies of foraminifera indicate a cold interval at about the same time or perhaps a little later (Fig. 14.14). Several lines of evidence indicate that there was a major glacial advance at this time:

1. Subaerial deposits have been found off the Bahamas at a depth of 32 m (100 ft) below sea level (**40**). These deposits have been dated at between 125,000 and 105,000 years old, and they are thought to have formed during a time of lowered sea level associated with a major glacial advance.
2. Glaciation occurred in Hawaii about 135,000 years ago (**35**).
3. In Yellowstone National Park, rhyolite flows having a potassium–argon date of 120,000 to 130,000 years are directly overlain by deposits of the Bull Lake glaciation (**41**).
4. Early Saalian (Drente) deposits in northwestern Europe probably correlate with Riss I glacial deposits, which have been dated at about 116,000 years (**32**).

The cold interval between 150,000 and 120,000 years ago may correlate with one of the Illinoian glacial advances, or it may have occurred during the Sangamonian. As discussed earlier, glacial advances can occur during an "interglacial" period. It is possible that this cold interval took place within the Sangamonian. However, it is not possible to be certain of where this cold interval belongs until radiometric dates are obtained on the Illinoian tills. Unfortunately, this is probably unlikely due to the lack of material suitable for dating.

125,000 TO 75,000 YEARS AGO. Temperatures for the interval between 125,000 and 75,000 years ago are indicated by oxygen-isotope studies of deep-sea cores and glacial ice from Greenland and Antarctica, from the study of foraminifera from deep-sea cores, and from radiometric dating of elevated coral reefs (Fig. 14.14). Temperatures were higher than those at the present time for the intervals between 81,000 and 89,000 years, between 90,000 and 108,000 years, and between 109,000 and 125,000 years. Minor glacial advances probably occurred in the intervals between these warm periods.

Radiometric dating of elevated coral reefs and raised marine terraces in New Guinea, Barbados, Egypt (bordering the Red Sea), Morocco (bordering the Mediterranean Sea), Hawaii, and Bermuda indicate that sea level was 7 to 12 m (23 to 39 ft) higher than that of the present time about 120,000 years ago (**4, 42,** and **43**). An interglacial is a time when glaciers would have been less extensive than they are today, and, consequently, sea level would be as high or higher than it is today. Therefore, the interval between 125,000 and 75,000 years ago might represent the latter part of the Sangamonian interglacial stage.

### Wisconsinan Glacial Stage

Deposits of the Wisconsinan glacial stage have been extensively studied throughout the world. There were four or five major advances of the glaciers during the Wisconsinan Age. Oxygen-isotope studies of deep-sea cores and radiometric dating of elevated coral reefs indicates that only partial retreat of the glaciers occurred between these advances. The intervals between glacial advances would be therefore designated as *interstadials* rather than interglacials. Glacial advances occurred during *stadials*.

EARLY WISCONSINAN. As defined by Richard Foster Flint, the Early Wisconsinan lasted from about 75,000 to 55,000 years B.P.

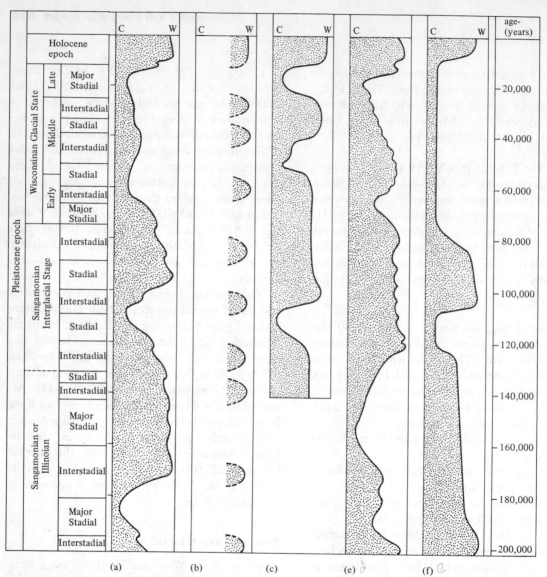

**FIGURE 14.14** Generalized climatic curves for the Late Pleistocene and Holocene epochs based on: **(a)** Paleotemperatures of the Caribbean as determined by oxygen isotope studies. (After Emiliani, Ref. 3. Copyright © 1966 by the University of Chicago.) **(b)** The study of raised coral reefs from the Barbados Islands. (After Mesolella, Ref. 37. Copyright © 1969 by the University of Chicago.) **(c)** The relative abundance of *Globigerina rubescens* from the northern Indian Ocean. (After Frerichs, Ref. 38, copyright © 1968 by the American Association for the Advancement of Science.) **(d)** Paleotemperatures in the equatorial Pacific as determined by oxygen isotope studies. (After Shackleton and Opdyke, Ref. 4.) **(e)** Variations in the frequency of *Globorotalia menardii.* (After Ericson and Wollin, Ref. 5. Copyright © 1968 by the American Association for the Advancement of Science.)

(Fig. 14.15). Two glacial advances have been proposed for the midcontinent region of North America during that interval; one ending about 73,000 years B.P. and a second ending about 58,000 years B.P. (Fig. 14.15). Oxygen-isotope studies of the equatorial Pacific reveal a cold period centered at about 70,000 years, which would correspond to the first Wisconsinan gla-

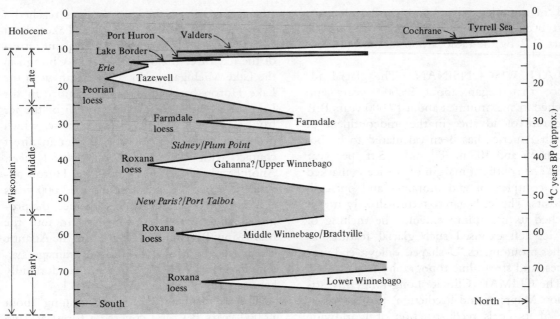

**FIGURE 14.15** Diagrammatic sketch showing the advance and retreat of the ice sheet during Middle and Late Wisconsinan time in central North America. Nonglacial intervals are in italics. (Revised from Flint, Ref. 1. Copyright © 1971 by John Wiley, New York.)

cial advance. Radiometric dating of elevated coral reefs in Barbados and New Guinea indicate a relatively warm period occurred about 62,000 years ago. This period may correspond to a time between the first and second glacial advances. Glaciation in Hawaii dated at 55,000 years B.P. (**35**) may correspond to the second glacial advance in the Early Wisconsinan.

In northern Europe, glacial till assigned to the early Würm rests on Warthe till. Since the Warthe deposits are not significantly weathered, these two glacial deposits are probably relatively close in age and are separated by an interstadial period rather than an interglacial. The Warthe glacial deposits were probably deposited during the Early Wisconsinan (Table 14.2). Late Bull Lake glacial deposits in the Rocky Mountains were probably also deposited at this time.

MIDDLE WISCONSINAN. The Middle Wisconsinan lasted from about 55,000 to 25,000

years B.P. (Fig. 14.15). The glacial retreat which began during the latter part of the Early Wisconsinan ended about 50,000 years B.P. This retreat was followed by an advance, retreat, advance, and then retreat of the ice (Fig. 14.15). Elevated coral reefs in Barbados dated at about 40,000 years B.P. (**43**) formed the first of these retreats. During the last retreat, about 30,000 years B.P., sea level was significantly higher than at the present time and warm-temperate forests migrated into areas which were formerly covered by the southern margin of the ice sheet. This warm interval includes the Farmdalian substage in North America and the Paudorf interstadial in Europe. A relatively high stand of sea level, which occurred in the mid-Atlantic region about 30,000 years B.P., corresponded with a warm interval in the United States (**44**). Deposits of this age in southern England include the "Arctic Plant Bed" that contains fossil mosses and dwarf birches along with the bones of rhinoceroses and mammoths. These and similar deposits have been dated at between 29,000 and 32,000 years B.P. by the radiocarbon method (**1**).

LATE WISCONSINAN. The glacial advance that began about 25,000 years ago reached its maximum at about 19,000 years B.P. The rate of advance in the midcontinent of North America has been calculated to be between 35 and 105 m (80 and 125 ft) per year (**44**). The southern margin of the ice is marked by a complex of end moraines and outwash deposits. The ice began to retreat shortly after it reached its maximum extent. The melting of the ice left exposed such glacial features as roches moutonnées, U-shaped valleys, end moraines, and streamline topography (Fig. 14.16).

The CLIMAP (Climate Long-range Investigation Mapping and Prediction) project has as one of its goals reconstruction of worldwide climates for the past 18,000 years, based on surface temperatures of the sea, extent of glacial ice, and sea level changes (**45**). This data bank now assists scientists studying the Quaternary to understand the last major glaciation and subsequent climatic changes. It will also increase our ability to answer the question, "Will we have another ice age?"

One of the results of the CLIMAP project is the recognition of a worldwide cooling of the climate between about 12,500 and 13,000 years B.P., which interrupted the general warming trend following the last maximum glacial advance. This cold interval is called the Oldest Dryas and it was followed by a general warming trend lasting until about 5000 years B.P.

THE GREAT LAKES REGION. The Wisconsinan ice sheet reached its maximum extent in the Great Lakes region about 18,000 years ago, at which time the southern margin of the ice sheet extended from southern New York State across southern Ohio, central Indiana, and central Illinois to central Iowa (Fig. 14.10). The retreat of the ice was not uniform, but took place as a series of retreats separated by minor

advances. One of these readvances reached a maximum just south of the Great Lakes about 14,500 years ago. During the subsequent retreat of the ice, three major lakes—Lake Chicago in the Lake Michigan basin, Lake Saginaw in the Lake Huron basin, and Lake Maumee in the Lake Erie basin—formed just south of the ice front (Fig. 14.17a). Lakes such as these, which occupy a basin in front of a glacier in direct contact with ice, are called *proglacial lakes*. Another glacial advance, the Port Huron advance, reached a maximum about 13,000 years ago (Fig. 14.17b). During this advance, the proglacial lakes drained southwestward into the Mississippi River and eastward into the Atlantic Ocean via the Mohawk-Hudson drainage system in New York. The ice retreated rapidly between 12,750 and 12,000 years B.P.

The ice readvanced again beginning about 12,000 years B.P. and covered a forest at Two Creeks, Wisconsin. Wood from beneath till formed during the advance has been dated at about 11,850 years B.P. (**1**). This advance is called the Great Lakean (formerly Valderan) advance. A final advance, the Algonquin (Cochrane) occurred in Ontario about 10,000 years ago.

Glacial Lake Agassiz formed west of the Great Lakes as the ice receded from that region. This lake covered more than 200,000 sq km (73,000 sq mi) and lasted from about 12,500 to 7500 years B.P. (**1**). Drainage from this lake, which was first southward into the Gulf of Mexico, shifted eastward into the Great Lakes and finally northward into Hudson Bay.

NORTHEASTERN UNITED STATES. During its maximum extent, the southern margin of the Late Wisconsinan ice sheet extended from northern Pennsylvania and northern New Jersey across southern New York State to Long Island and Cape Cod (Fig. 14.10). The retreat of the Late Wisconsinan ice sheet from the northeastern United States probably began about 18,000 years ago. The time of the retreat has been determined in the Hudson Valley from

**FIGURE 14.16** Erosional and depositional features produced by Pleistocene glaciers: **(a)** Small tarn in a cirque in the Klamath Mountains of northern California—notice the striated *roche moutonnée* in the foreground. **(b)** U-shaped valley along the "Million Dollar Highway" in western Colorado. **(c)** Recessional moraines east of Hudson Bay, Canada. **(d)** Deeply scoured topography in the Northwest Territories, Canada. [**(c)** and **(d)** Courtesy of the National Air Photo Library, Department of Energy, Mines and Resources, Canada.]

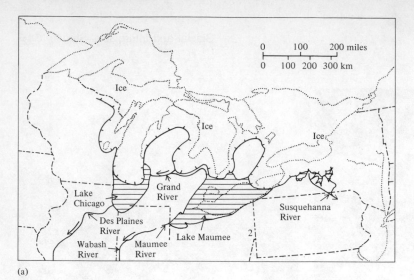

(a)

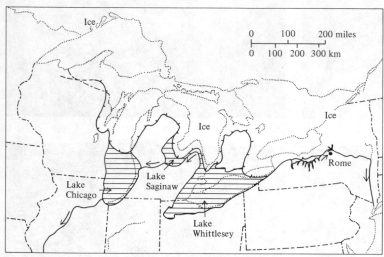

(b)

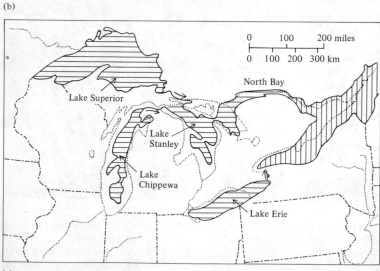

(c)

**FIGURE 14.17** Evolution of the Great Lakes: **(a)** At the time of Lake Maumee (ca. 13,750 B.P.). **(b)** At the time of the Port Huron readvance (ca. 13,000 years B.P.). **(c)** At the time of Lake Chippewa. (After Flint, Ref. 1; copyright © 1971 by Wiley, New York.)

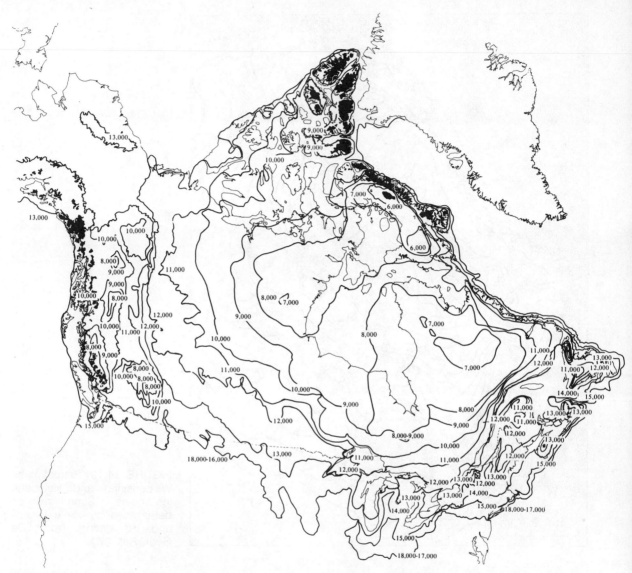

**FIGURE 14.18** Retreat of the Wisconsinan and Holocene ice sheets in North America. Numbers refer to the age of the ice margin along the adjacent line. (After a map published by the Geological Survey of Canada.)

radiocarbon dates and pollen stratigraphy in lake and bog sediments (**46**). The age of the tundra or herb pollen zone, which indicates the presence of tundra vegetation and therefore very cold climate, in a given bog on a mapable glacial surface, provides an approximate age for the beginning of deglaciation in a specific region.

The retreat of the ice was slow at first, but by 15,000 years B.P., the ice front had retreated from northern Pennsylvania into central New York and southern New England (Fig. 14.18). The ice margin reached the northern Hudson region by about 13,000 years B.P. and the St. Lawrence lowlands by about 12,600 years B.P. Extensive proglacial lakes occupied river val-

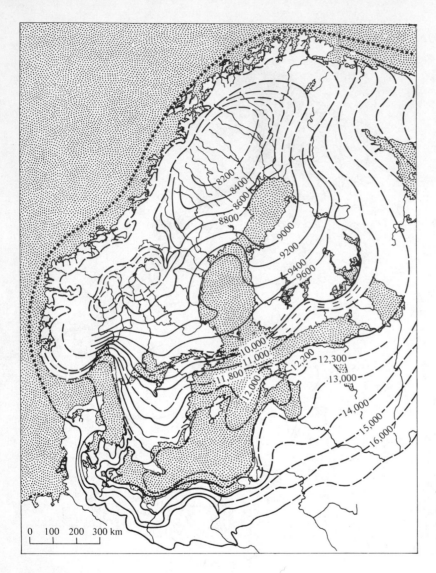

**FIGURE 14.19** Retreat of the Wisconsinan and Holocene ice sheets in Scandinavia as determined by varve counts and $^{14}C$ dating. (After De Geer, Ref. 47.)

leys between the ice margin and the end morains during the glacial recession. Glacial lakes Albany in the Hudson Valley and Hitchcock in the Connecticut Valley are examples of such lakes.

About 12,000 years B.P., before extensive crustal rebound occurred along the area which had been the southern margin of the ice sheet, an arm of the sea (Champlain Sea) invaded the St. Lawrence basin (Fig. 14.17c). At about the same time, the sea invaded the lower Hudson Valley in New York State. The marine invasion ended about 10,000 years B.P., when rebound elevated the crust along the St. Lawrence River.

EUROPE. The maximum advance of the Scandinavian ice sheet during the Late Wisconsinan occurred between about 20,000 and 17,000 years B.P. (**1**). At this time, the southern margin of the ice sheet extended from the continental shelf surrounding the British Isles,

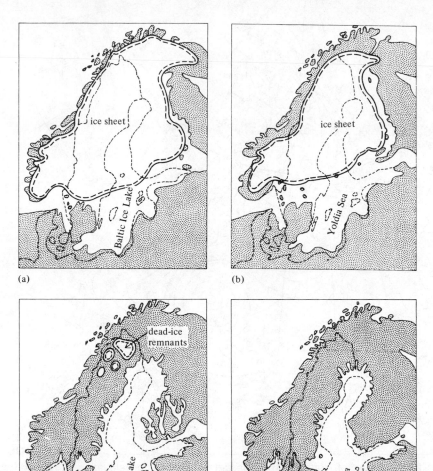

**FIGURE 14.20** Four stages in the evolution of the Baltic region during the Holocene: **(a)** Baltic Ice Lake (ca. 8500 years B.P.). **(b)** Yoldia Sea (ca. 7900 years B.P.). **(c)** Ancylus Lake (ca. 7200 years B.P.). **(d)** Littorina Sea (ca. 5200 years B.P.). (After Lundquist, Ref. 48. Copyright © 1965 by Wiley, New York.)

across central Europe to northwestern Siberia (Fig. 14.9). It is uncertain whether or not the ice sheet in the British Isles coalesced with the main ice sheet at this time, but these two ice sheets were separate soon after the onset of deglaciation. By 13,000 years B.P., the ice still covered most of the Scandinavian Peninsula (Fig. 14.19). Glacial readvances still occurred as late as the Loch Lomond readvance in the Scottish Highlands during the younger Dryas about 10,800 to 10,400 years ago. As the ice continued to re-

treat, a proglacial lake, the Baltic Ice Lake, developed south of the ice margin (Fig. 14.20). Approximately 9000 years B.P. sea level rose sufficiently to form an inland sea, the Yoldia Sea. Postglacial rebound isolated the Baltic basin from the sea and created the Ancylus Lake, which persisted through the melting of the last remnants of the Scandinavian ice sheet. About 6000 years B.P. the sea again occupied the basin. The Littorina Sea decreased in size as the Baltic Shield continued to be uplifted.

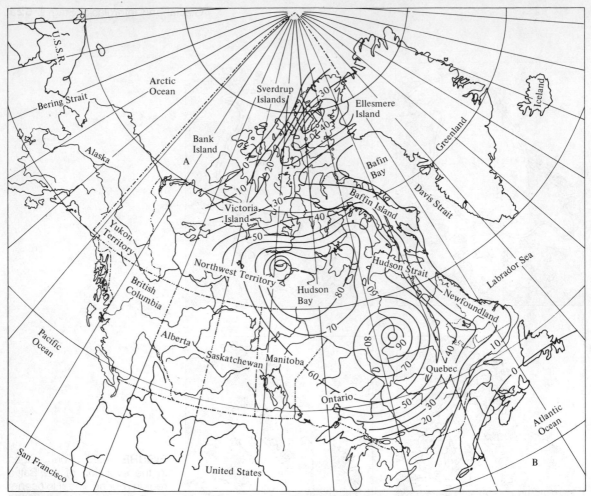

**FIGURE 14.21** Postglacial uplift (in meters) of North America as determined by the elevation of marine deposits 6000 years old. (After Andrews, Ref. 52.)

## THE HOLOCENE

If we define the Holocene as beginning about 10,000 years ago, sea level has risen 33 m (110 ft) during the intervening time (Fig. 14.7). As a result, much of the continental shelves has been flooded. Holocene sediments have accumulated in river deltas, flood plains, and beaches. In most areas these deposits are relatively thin owing to the short interval of deposition. However, in many estuaries Holocene deposits may exceed 30 m (100 ft).

Palynological studies of Holocene sediments have indicated that 7000 to 5000 years ago climates were generally warmer than those of today. This interval has been referred to as the climatic optimum or Hypsithermal interval. In western Australia, there is evidence for a 1.2 to 2.5 m (3.5 to 7 ft) fall in sea level over the past 4000 to 5000 years (**48**). This suggests that glaciers have been expanding during the last 4000 to 5000 years. Presumably the climate has on the whole been cooling slightly during this time. Within the last 5000 years, several episodes of cooling and glacial expansion have occurred. Based on data provided by a study of growth rings of bristlecone pines in southwestern United States, the midpoints of the cooler periods are at 3000 B.C., 800 B.C., 1025 A.D., and 1600 A.D. (**50**). Some geologists and climatologists believe that the midpoint in the postglacial interstadial (or interglacial) stage has

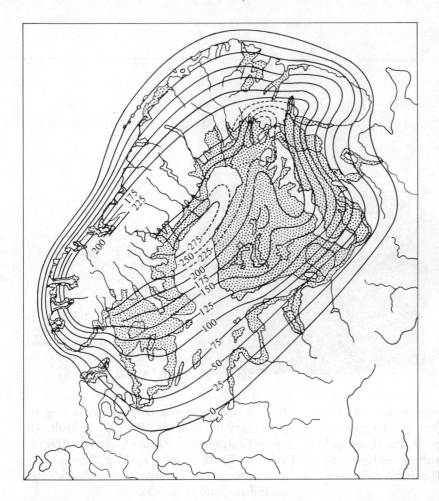

**FIGURE 14.22** Postglacial uplift of Fennoscandia as determined by the highest strand mark of the ocean during the late Quaternary. (After Wright, Ref. 53.)

been reached and that the northern hemisphere is in a cooling trend which will result in a new glacial stage. However, there appears to be a warming trend in the southern hemisphere (**51**).

## POSTGLACIAL REBOUND

The weight of glacial ice depressed the earth's crust and caused a slow flow of the mantle from beneath the ice-loaded crust. The amount of crustal depression during Pleistocene glaciations was approximately 1 m for each 3.6 m of ice. When the ice melted, the crust slowly rose as the mantle material flowed back under the deglaciated crust. This uplift is still going on in

northern North America, Scandinavia, and Asia, where the rate of uplift may exceed 30 cm (1 ft) per century (Figs. 14.21 and 14.22).

## TECTONISM AND VOLCANISM

The tectonic and volcanic history of the Quaternary is a result of movements of crustal and oceanic plates which began in Tertiary time. Many mountains and plateaus were uplifted without appreciable folding. The Colorado Plateau, for example, was uplifted nearly 1500 m (5000 ft) and the Appalachian Mountains were uplifted almost 1000 m (several thousand feet) during the late Tertiary and early

**FIGURE 14.23** Trace of the San Andreas Fault showing a stream offset in a right-lateral sense. (Photo courtesy of the U.S. Geological Survey.)

Quaternary. Uplift of the continental interior was in part responsible for the lowering of sea level during the Quaternary. In the Basin and Range Province, normal faulting and volcanism, which began in the Tertiary, continued into the Quaternary.

Folding occurred along the Pacific margins of the continents as crustal plates continued to move away from the Mid-Atlantic and Mid-Indian Ocean ridges. Strike-slip faulting was particularly prevalent in these regions. The San Andreas Fault System in California (Fig. 14.23) forms the boundary between sections of the American plate and the Pacific plate (Fig. 14.24). In the region of the San Andreas Fault, the American plate is moving southwestward and the Pacific plate is moving northwestward relative to Europe. The combination of these motions produces a right-lateral displacement along the fault (Fig. 14.24). The Queen Charlotte and Merriweather faults to the north may be of the same origin (Fig. 14.24).

Other major strike-slip faults active during the Quaternary include the Philippine Fault, the Alpine Fault in New Zealand, and the Atacama Fault in South America. The Alpine Fault marks the junction between the Pacific plate and Indian plate (Fig. 6.3).

Widening of the North Atlantic has occurred at a rate of about 2.5 cm (one inch) per year, with a total widening of about 48 km (39 mi) since the beginning of the Pleistocene. The South Atlantic has widened approximately 80 km (48 mi), and the southern Indian Ocean has widened about 120 km (72 mi) in that time.

The prevalence of earthquakes and active volcanoes in the "ring of fire" around the Pacific Ocean basin indicates that oceanic plates are moving under the Pacific margins of the continents at the present time (Fig. 14.25). However, volcanism in the western United States appears to have been less intense during the Holocene epoch than during the late Tertiary and Pleistocene. Currently, the only active

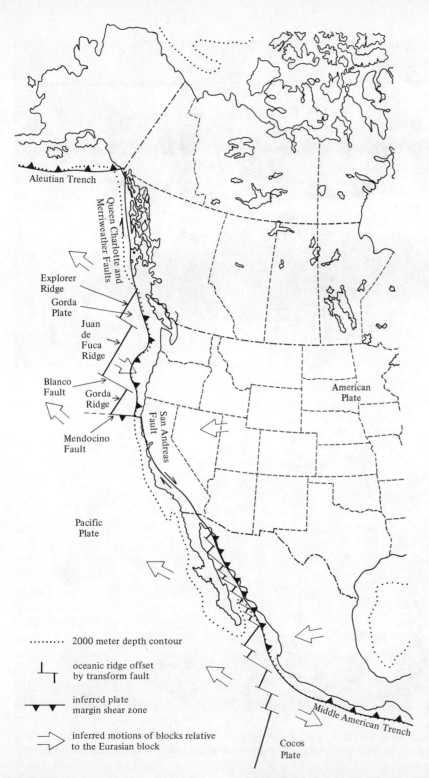

Aleutian Trench

Queen Charlotte and Merriweather Faults

Explorer Ridge

Gorda Plate

Juan de Fuca Ridge

Blanco Fault

Gorda Ridge

Mendocino Fault

San Andreas Fault

American Plate

Pacific Plate

Middle American Trench

Cocos Plate

·········  2000 meter depth contour

⊥⊤  oceanic ridge offset by transform fault

▼▼  inferred plate margin shear zone

⟹  inferred motions of blocks relative to the Eurasian block

**FIGURE 14.24** Interpretation of movements along oceanic ridges and major faults in and adjacent to western North America.

(a)

(b)

FIGURE 14.25 Quaternary volcanoes: **(a)** Mt. Rainier in the western part of the state of Washington. (Union Pacific Railroad photo.) **(b)** Mt. Fuji, an active volcano. (Courtesy of the National Park Association of Japan.)

volcano in the contiguous United States is Mt. Lassen, although Mt. Rainier and Mt. Shasta have hot regions at depth. Volcanism in the African Rift System produced such peaks as Mt. Kenya and the extensive ash deposits in the Olduvai Gorge. In the Atlantic Ocean basin, continued volcanism produced new volcanic islands such as Tristan da Cunha and, most recently, Surtsey, on the crest of the Mid-Atlantic Ridge.

Volcanic activity occurred throughout the Pleistocene in Yellowstone National Park. Explosive volcanism which produced widespread tuffs occurred at 1,920,000, 1,220,000, and 600,000 years ago (**54**). Eruptions come from a large caldera which is close to the center of the park and covers about one-quarter of the park (Fig. 14.26). Heat flow is very high in the park (**46**), as one would expect from the presence of numerous hot springs and geysers (Fig. 14.27). It is likely that there is an active magma chamber at depth which is part of a still molten batholith (**56**). Volcanic activity increases in age in a westerly direction (Fig. 14.26), and therefore it is likely that the Yellowstone region is a hot spot underlain by a mantle plume (**57**).

The volcanism in Yellowstone includes both felsic (rhyolitic) and mafic (basaltic) activity. To the west on the Snake River plain, recent volcanic activity is largely mafic in character. Some of this activity is quite recent, and volcanic features are well displayed at craters of the Moon National Monument (Fig. 14.28).

## CAUSES OF THE ICE AGES

Numerous theories have been proposed to explain why the earth has experienced repeated glaciations during much of the late Cenozoic. Any explanation of Pleistocene ice ages should not only account for the initial onset of cold climates, but also explain why the glaciations have been cyclical. Many geologists believe that any acceptable theory should also apply to prior ice ages.

### Initiation of the Ice Ages

Studies of changes in vegetation and oxygen-isotope studies of marine fossils indicate that there has been a general decrease in temperatures throughout the world which began during the Late Cretaceous (Fig. 14.29). To be sure, there have been reversals in this trend, but on the whole temperatures have been decreasing. Presumably the glaciations which have occurred during the late Tertiary and Quaternary are the result of this long term temperature decline. This decline may have been caused by one or more of the following factors: uplift of the continents, polar wandering, continental drift, closing of the Isthmus of Panama, volcanism, meteorite impact, mountain building, and galactic rotation.

UPLIFT OF THE CONTINENTS. During the Late Cretaceous, shallow seas were widespread on the continents throughout the world. Beginning sometime between the Campanian and the Maestrichtian, the seas began a general withdrawal. By Miocene time, even before continental glaciers began to form, the seas had almost completely withdrawn from the continents. Presumably this withdrawal of the seas is related to a general uplift of the continents. The beginning of the uplift coincides almost exactly with the beginning of the temperature decline, and it is very likely that the uplift played a major role in the decline.

Because air temperature normally decreases with altitude (at a lapse rate of about 6.5°C per 1000 m or 3.6°F per 1000 ft), the continents would have cooled through increased elevation. Furthermore, an increase in land area would also increase the rate of wind erosion, and increase the amount of dust in the atmosphere and the opacity of the atmosphere. Dust would also have provided condensation nuclei for water vapor, leading to increased cloud cover and precipitation. Widespread volcanism in the Tertiary would have contributed dust to the

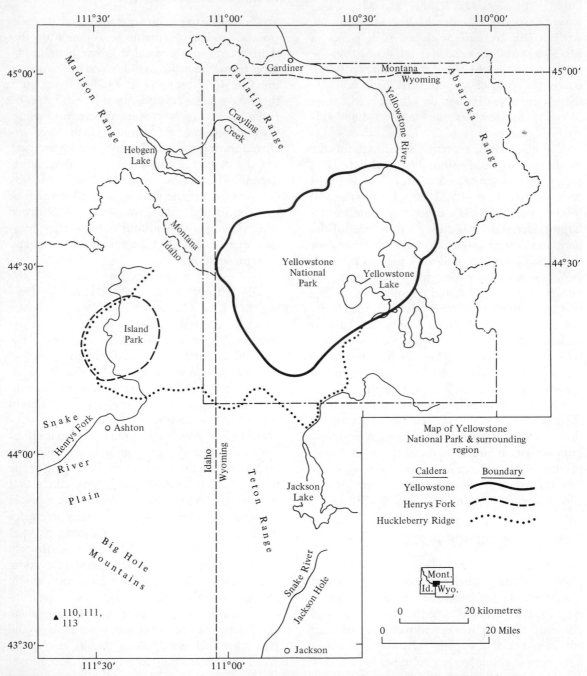

**FIGURE 14.26** Map of Yellowstone National Park surrounding areas showing the outlines of major calderas. The calderas were recognized and mapped by R. L. Christiansen. (After Reynolds, Ref. 54.)

FIGURE 14.27 Old Faithful geyser in Yellowstone National Park.

FIGURE 14.28 Small spatter cones in Craters of the Moon National Monument.

| Late Cretaceous | Paleocene | Eocene | Olig. | Miocene | Pliocene |
|---|---|---|---|---|---|
| tropical | | | | | |
| subtropical | | | | | |
| warm temperate | | | | | |
| cool temperate | | | | | |
| subarctic | | | | | |

— · — · — Western Europe ———— Western United States

**FIGURE 14.29** Curves illustrating climatic changes in western Europe and western United States during the Tertiary. (From Dorf, Ref. 58. Copyright © 1964 Wiley, London.)

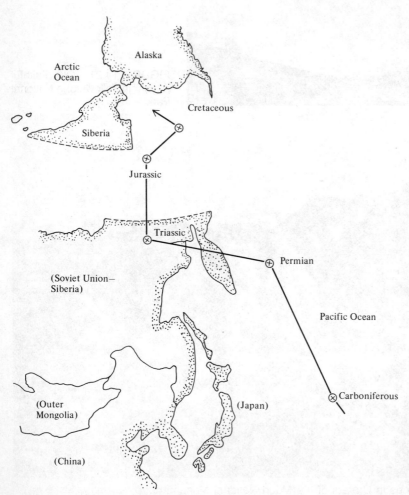

Arctic Ocean

Alaska

Cretaceous

Siberia

Jurassic

Triassic

(Soviet Union—Siberia)

Permian

Pacific Ocean

(Outer Mongolia)

(Japan)

Carboniferous

(China)

(a)

**FIGURE 14.30 (a)** Polar wandering curve for the Late Pleistocene and Mesozoic relative to Asia. **(b)** Polar wandering curve for the Late Pleistocene, Mesozoic, and Cenozoic relative to Antarctica.

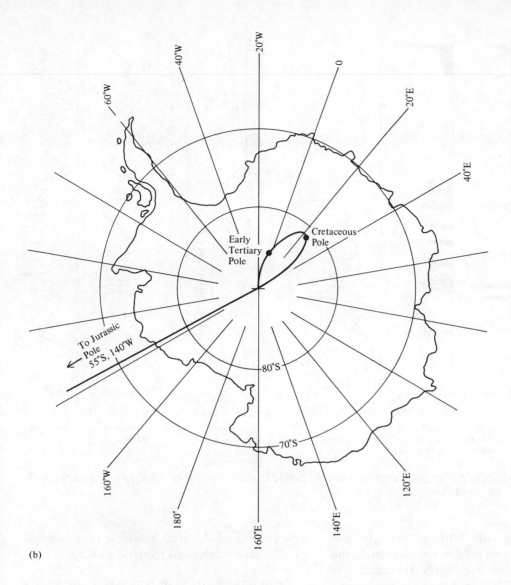

(b)

atmosphere and may have supplemented this process.

POLAR WANDERING AND CONTINENTAL DRIFT. It is likely that polar wandering and continental drift played a significant role in the cooling of the earth during the Tertiary. Throughout the Paleozoic and during most of the Mesozoic, the North Pole was located in the Pacific Ocean region (Fig. 14.30). In the Tertiary, however, movements of the conti-

nents were such that the pole migrated into the Arctic Ocean. If, as proposed in our reconstruction of the continents, the northeastern tip of Siberia was separated from the rest of Siberia prior to the separation of Europe and Greenland, there would have been a free interchange between the waters of the Arctic and Pacific oceans during the Paleozoic, Mesozoic, and early Tertiary (Fig. 14.30). The gap between the northeastern tip of Siberia and the rest of Siberia narrowed as Greenland drifted away from

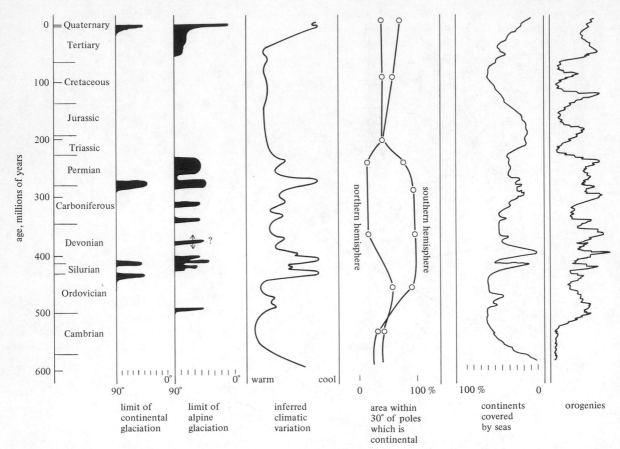

age, millions of years

limit of
continental
glaciation

limit of
alpine
glaciation

inferred
climatic
variation

area within
30° of poles
which is
continental

continents
covered
by seas

orogenies

**FIGURE 14.31** Comparison of glacial episodes with periods when continents were near the poles and periods of deformation and uplift.

Europe. By the late Tertiary, the gap had closed; the pole was then in approximately the same position relative to North America that it occupies today. The location of the pole over a body of water with limited circulation would have helped to cool the Arctic region because of the lack of interchange with warmer equatorial waters. This may have contributed to the initiation of glaciation in that region.

In the southern hemisphere the South Pole was located in the Pacific Ocean throughout most of the Mesozoic. Both the pole and Antarctica migrated toward their present positions, and the South Pole reached the margin of Antarctica during the Cretaceous (Fig. 14.31). The

movement of Antarctica into a polar region could have accelerated climatic cooling.

CLOSING OF THE ISTHMUS OF PANAMA. The closing of the Isthmus of Panama during the Pliocene also may have been a factor influencing the onset of a glacial climate. This closing would have strengthened the Gulf Stream and increased precipitation in the northern hemisphere, thereby contributing to the accumulation of glacial ice.

VOLCANISM. Volcanism may also have played a role in the initiation of the ice ages. Studies of deep-sea cores have indicated that the last two million years has been a time of greatly increased volcanic activity (**59**).

Volcanic ash from recent eruptions has had a profound effect on atmospheric opacity and, consequently, on worldwide temperatures. For example, the eruption of Krakatoa in 1883 resulted in a dispersal of volcanic dust in the atmosphere throughout the whole world for a period of several years. This dust caused an increased amount of sunlight to be reflected into space and as a result solar radiation was significantly decreased for a period of three years after the eruption (**60**). Similar decreases in solar radiation have been observed following other major explosive volcanic eruptions (**60**). The worldwide warming trend which began about 1912 has been tied to the lack of major explosive volcanic eruptions since that time (**60**). A prolonged period of increased volcanic activity could therefore account for the initiation of glaciation.

METEORITE IMPACT. It is possible that meteorite impacts played some role in initiation of Pleistocene glaciation. It is interesting to note that there are at least eight meteorite impact craters more than 1 km (0.6 mi) in diameter that were formed within the last 2.5 million years (Table 12.1). Three of these are more than 10 km (6 mi) in diameter, and one may be 250 km (150 mi) in diameter. The impact which produced the 250 km diameter crater may have been the source of a huge swarm of tektites dated at 690,000 years (**61**). This impact occurred at the time of the last reversal of the earth's magnetic field and, in fact, the impact may have caused the reversal (**62**).

The explosion produced by a large meteorite impact would eject a very large volume of dust into the atmosphere. This dust, like that produced by a large volcanic eruption, would produce a short-term cooling of the earth. A very large impact might affect convection within the earth (**63**) and this might increase the amount of volcanic activity over a period of tens or hundreds of thousands of years. This could produce

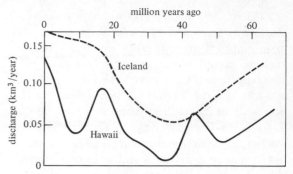

**FIGURE 14.32** Variations in the rate of extrusion of volcanic rocks on the Hawaiian-Emperor Seamount Chain and on the Greenland-Iceland-Faeroe aseismic ridge. (After Vogt, Ref. 64.)

a long-term cooling of the earth sufficient to initiate glaciation.

Peter Vogt has observed that volcanic eruptions from hot spots appear to be synchronous worldwide and that peaks of activity occurred in the Late Eocene (42 million years ago), Middle Miocene (15 million years ago), and Pleistocene (less than 2 million years ago) (**64**) (Fig. 14.32). This relationship appears to hold true for island arcs as well (**64, 65**). These are not only times of climatic cooling (Fig. 14.29), but also times of major meteorite impacts, all of which are associated with major tektite swarms (**63**) (Table 12.1).

MOUNTAIN BUILDING. Known periods of continental glaciation began about 2290, 850, 700, 610, 435, 290 and 1 million years ago. All of the Phanerozoic glacial periods (the last three) follow rather closely times of major mountain building. The Taconic orogeny began 442 million years ago, the Hercynian orogeny climaxed about 315 million years ago, and the main phase of the Andean orogeny began 15 million years ago. Mountains tend to accumulate snow which often lasts through the summer, and snow reflects 90 percent of the incoming solar radiation. Therefore, an increase in mountains may tend to produce a general decrease in temperatures.

GALACTIC ROTATION. The solar system orbits the center of the Milky Way Galaxy in an elliptical orbit. It takes about 274 million years to complete one period of rotation (**67**). The length of time between the present glacial episode and the Permo-Carboniferous glaciation is about 290 million years, very close to the 274-million-year galactic year. This relationship has led to the suggestion that there is a connection between glaciation and galactic rotation (**67**). As the solar system orbits the galactic center, the central force per unit mass from the galaxy varies periodically due to the elliptical orbit of the solar system. Glaciation tends to occur when the galactic force is at a minimum. However, it is not at all clear how this might cause glaciation, and furthermore, this relationship does not appear to hold true for Late Proterozoic glacial episodes.

CONCLUSION. Episodes of widespread deformation, volcanism, and mountain building, maximum withdrawal of the seas, and location of a geographic pole on or near a continent are also associated with most Precambrian and Phanerozoic continental glaciations (Fig. 14.28). Therefore, it is likely that these factors are very important in the formation of continental ice sheets.

## Causes of Multiple Glaciation

Cyclical glaciation requires controlling factors which are repetitive, such as:

1. Astronomical control, which involves variations in the amount of solar radiation which the northern hemisphere receives during a year
2. Atmospheric control, which depends on variations in the carbon dioxide, water vapor, and particulate content of the atmosphere. Excess volcanic ash from major eruptions may also have an effect
3. Oceanic control, which involves freezing and thawing of the Arctic Ocean
4. Magnetic field control, which involves fluctuations in the strength and polarity of the earth's magnetic field

ASTRONOMICAL CONTROL. Glaciations may have been triggered by changes in the amount of solar radiation (insolation) received at various latitudes at different times of the year. Cyclic changes in the shape of the earth's orbit, in the inclination of the earth's axis, and eccentricity of the earth's orbit might bring about such changes. This mechanism was proposed by the Yugoslavian meteorologist M. Milankovitch. Milankovitch postulated that cool summers in the northern hemisphere would favor the accumulation of glacial ice. The earth's axis is now inclined at approximately 23.5° to the plane of the earth's orbit around the sun. It is this inclination that produces the seasons. However, the angle of inclination of the axis varies between 21.5° and 24.5°. This angle fluctuates in a 40,000-year cycle. A greater inclination would produce warmer summers and colder winters; a lesser inclination would produce warmer winters and cooler summers (Fig. 14.33).

The earth's orbit around the sun is an ellipse of slightly varying eccentricity. At minimum eccentricity, the orbit is almost circular; at maximum eccentricity, the ratio of the semiminor axis to the semimajor axis is about 0.998. The eccentricity of the earth's orbit varies in a 92,000-year cycle.

The earth wobbles like a top, and thus the rotational poles "point" in various directions in space (Fig. 14.34). For example, the North Pole now points approximately at the North Star, but this has not always been the case. The precession of the equinoxes—the shift in the seasons through time due to this wobble—varies in a 21,000-year cycle. At the present time, the earth's orbit is near maximum eccentricity, and summer occurs when the earth is farthest from the sun. These conditions are favorable for the accumulation of glacial ice (Fig. 14.34).

Based on these cycles, the variation in sum-

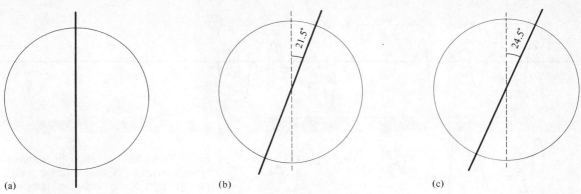

**FIGURE 14.33** Effect of variations in angle of inclination of earth's axis on climate: **(a)** Hypothetical, inclination of axis vertical, no seasons. **(b)** Cooler-than-average summers, warmer-than-average winters—favorable for glacial ice. **(c)** Warmer-than-average summers, cooler-than-average winters—unfavorable for such accumulation.

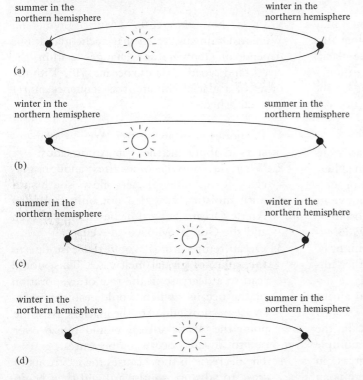

**FIGURE 14.34** Effect of precession and changes in the ellipticity of the earth's orbit on climate: **(a)** Warmer-than-average summers, cooler-than-average winters—unfavorable for accumulation of glacial ice. **(b)** Cooler-than-average summers, warmer-than-average winters—favorable for accumulation of glacial ice. **(c)** and **(d)** Average summers and winters—unfavorable for accumulation of glacial ice. Note that orbits **(a)** and **(b)** are more elliptical than average; **(c)** and **(d)** are almost circular.

mertime insolation for northern hemisphere latitudes may be calculated and plotted against time (Fig. 14.35). In general, there is fairly good correlation between this curve and paleotemperatures based on oxygen–isotope studies (**68**) (Fig. 14.35). For example, the maximum of Late

Wisconsinan glaciation occurred about 19,000 years ago, and there is a minimum in the insolation curve at 22,000 years B.P.

It has been widely assumed that the amount of radiation emitted by the sun (the *solar "constant"*) does not vary. However, recent studies

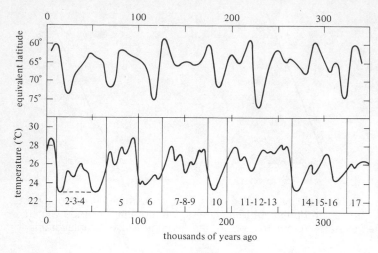

**FIGURE 14.35** Variation of summertime insolation in the northern hemisphere compared with the temperatures of surface water of the Atlantic Ocean. (From Broecker, Ref. 67. Copyright © 1966 by the American Association for the Advancement of Science.)

have indicated that the solar "constant" may show small variations (**69**). It is possible that climatic fluctuations leading to alternate glacial and interglacial climates could be due to long-term fluctuations in the energy output by the sun. Unfortunately, however, there is no way to test this possibility.

ATMOSPHERIC CONTROL. Variations in the carbon dioxide content of the atmosphere have been suggested as a possible cause of multiple glaciations, since carbon dioxide above its present level of 0.03 percent of the atmospheric gases might result in warmer worldwide climates. The increase in carbon dioxide might be caused by forest fires, volcanism, or other imbalance in the carbon cycle. Conversely, a decrease in carbon dioxide might lead to an increase in glacial ice.

Periodic increases in volcanic dust in the atmosphere may play a role in causing multiple glaciations. As discussed earlier, an increase in dust tends to lower temperatures. There is an excellent correlation between the rate of volcanic glass accumulation and the beginning of glaciation in a core taken in the South Pacific Ocean between New Zealand and the southern tip of South America (**65**) (Fig. 14.36).

OCEAN CONTROL. Maurice Ewing and William Donn suggested that cyclic glaciations depend on temperature changes within the Arctic and North Atlantic oceans (**70**). With the onset of a glacial climate, this sequence might be established:

1. Evaporation of an ice-free Arctic Ocean results in abundant snow in Arctic lands.
2. When the snow becomes thick enough, glaciers form. The glaciers flow southward until moisture is added not only from the Arctic Ocean, but from the Atlantic Ocean and the Gulf of Mexico as well.
3. Ocean temperatures would decrease due to the influx of glacial meltwater. This would lead to a decrease in the rate of evaporation of the oceans, which would result in a decrease in snow added to the glaciers. At some point the Arctic Ocean would freeze over due to lowered ocean temperatures.
4. The decrease in snow causes the ice fronts to cease to advance southward and then begin to retreat northward.
5. The glaciers largely disappear from North America and Asia as the climate warms. The Arctic ice pack would then begin to melt and the cycle would begin again.

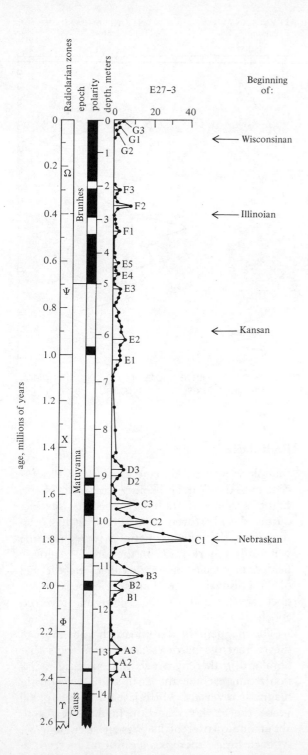

### MAGNETIC FIELD CONTROL.

There appears to be a relationship between temperature and the strength of the earth's magnetic field. Direct measurements of these two parameters for the past 60 years show that temperature generally decreases as the magnetic field decreases (**61**) (Fig. 14.37). A similar relationship for the past 470,000 years has been shown by comparison of temperatures determined from oxygen-isotope studies of foraminifera from ocean cores with intensity of the earth's magnetic field as determined by paleomagnetic studies of the intensity of magnetization of sediments from the same core (Fig. 14.38). It may be that the magnetic field has a direct effect on climate or that a third factor influences both climate and the magnetic field simultaneously (**61**).

There is an interesting correlation of the beginning of glacial stages with reversals in the earth's magnetic field (Table 14.2): the beginning of the Nebraskan corresponds in time to the beginning of the Olduvai (Gilsa) event, the Kansan began at the end of the Jarimillo event, and the Glacial A division of the Cromerian began at the boundary between the Matuyma and the Brunhes polarity epochs. The beginning of the Illinoian correlates approximately with the end of a proposed short event within the Brunhes dated at about 420,000 years (**65**). A similar correlation of the beginning of glaciation with magnetic reversals has been found in Iceland (**22**). Here, tills are found just above reversals at: (1) the base of the Mammoth event (3.06 million years ago), (2) the base of the Kaena event (2.9 million years ago), and (3) at

**FIGURE 14.36** Variation in the rate of accumulation of volcanic glass (in milligrams per 1000 years per square centimeter) during the past 2.5 million years in a core taken from the South Pacific between Victoria Land, Antarctica, and New Zealand. Notice how the peaks of volcanism correspond closely to the beginning of glacial stages. (After Huang and Watkins, Ref. 65. Copyright © 1976 by American Association for the Advancement of Science.)

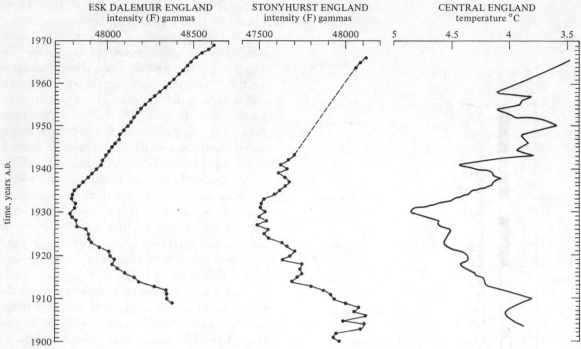

**FIGURE 14.37** Correlation of the intensity of the earth's magnetic field with temperatures in central England for the period from 1909 to 1969. Both magnetic intensity and temperatures were measured directly. (After Wollin, Ericson, and Wollin, Ref. 71.)

the top of the Kaena event (2.80 million years ago). It is possible that there is a relationship between reversals in the earth's magnetic field and temperature decreases which result in glaciation. The rate of volcanic activity has been correlated with magnetic reversals (**63**), and it may be that increases in volcanism rather than the magnetic reversals cause climatic change.

## PLEISTOCENE LIFE

A striking feature of Pleistocene life is the distinction between cold and warm faunas and floras. Extensive migrations of animal and plant assemblages occurred in response to alternating glacial and interglacial climates and to the establishment of new connections between continents and islands. Many new forms appeared and many extinctions occurred, especially in formerly isolated groups.

### Radiolaria

Radiolaria were abundant during the Pleistocene in both warm and cold waters (Fig. 14.39). Eight species of radiolaria became extinct during the Pleistocene (Fig. 14.40). It has been suggested that the extinctions were in some way caused by reversals in the earth's magnetic field. One species became extinct near the end of the Olduvai event. Four species became extinct at the end of the Matuyma reversed epoch.

The magnetic field of the earth traps most of the radioactive particles and cosmic rays that bombard it, thereby preventing harmful radiation from reaching the earth's surface. During magnetic reversals, which last for about 10,000 years, the earth's magnetic field decreases to about one-quarter of its present strength (**75**). At such times, there is a significant increase in

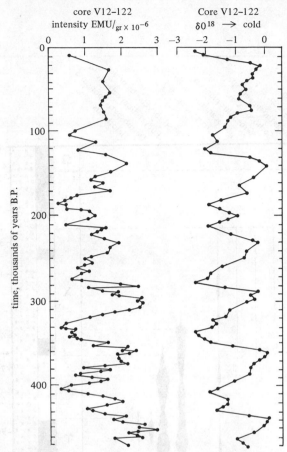

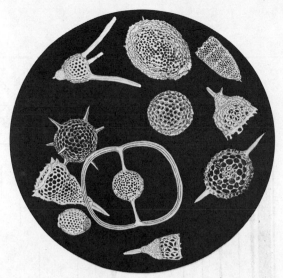

**FIGURE 14.39** Antarctic radiolaria. (From Hays and Opdyke, Ref. 74. Copyright © 1967 by the American Association for the Advancement of Science.)

**FIGURE 14.38** Correlation of the intensity of the earth's magnetic field (as determined from paleomagnetic measurements on deep sea cores) with temperatures (as determined by oxygen isotope studies) for the past 450,000 years. (After Wollin, Ericson, and Wollin, Ref. 71.)

may be due to temperature declines since volcanism and glaciation often occur at magnetic reversals. Or it may be that chemical changes in seawater produced by volcanism caused extinctions.

## Terrestrial Faunas

The Lower Pleistocene faunas of southern Europe are distinctly warm-temperature assemblages which included true elephants, rhinoceroses, cattle, and one-toed horses. The contemporaneous cold fauna was comprised mainly of mastodons, woolly mammoths, and woolly rhinoceroses (Figs. 14.42 and 14.43).

The Middle Pleistocene faunas of Europe included warm-climate forms such as straight-tusked elephants, rhinoceroses, hippopotamuses, pigs, cattle, horses, bison, elk, deer, bear, a variety of rodents, and human beings. Cold-climate forms are similar to those of the Early Pleistocene.

The fauna of Europe during the Wurm glacial stage was comprised of woolly mammoths,

radiation reaching the earth's surface. The radiation would be especially intense if a large solar flare or nearby supernova (**76**) occurred during intervals of reduced magnetic field (Fig. 14.41). Whether the increases in radiation would produce an appreciable effect on evolution or extinctions has not been established (**77**). However, experimental results indicate that very low magnetic fields have harmful effects on a large variety of organisms (**78**). Alternatively, the correlation between extinctions and reversals

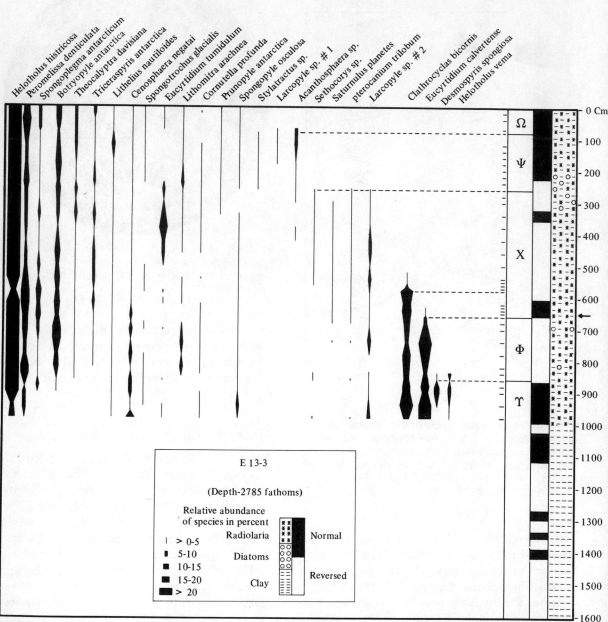

**FIGURE 14.40** Ranges of radiolarian species correlated with magnetic stratigraphy. The black parts of the column at the left are periods of normal polarity and the white parts are periods of reversed polarity. Faunal zones at the left, designated by Greek letters, are based on the extinction or near extinction of certain species of radiolaria. Notice how these boundaries often occur in the vicinity of magnetic reversals. (From Hays and Opdyke, Ref. 74. Copyright © 1967 by the American Association for the Advancement of Science.)

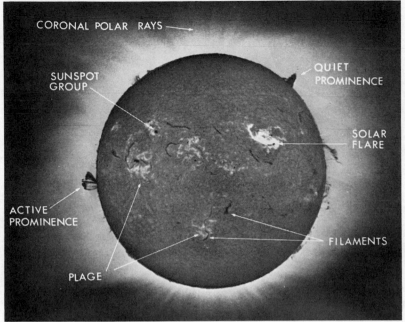

FIGURE 14.41 Photograph of the sun's surface taken in the light of H-alpha, showing a solar flare. Large quantities of radiation are given off by the sun during flares. (Courtesy of Lockheed Solar Observatory, Lockheed Missiles and Space Co.)

woolly rhinoceroses, large cave bears, deer, and human beings (Figs. 14.44 and 14.45). The cold-climate faunas of North America had many similarities to those of Eurasia because mammals were able to migrate between these continents by way of the Bering Strait during times of lowered sea level. The warm-climate faunas of Europe and Africa were also similar and included buffalo, gazelles, and sheep.

A remarkable record of Late Pleistocene warm-climate fauna is preserved in the La Brea tar pits in Los Angeles (Fig. 14.46). The tar pits have yielded very well-preserved bones of thousands of mammals, such as those of the imperial elephant, the sabertooth cat, wolves, lions, horses, bison, camels, and giant ground sloths (Fig. 14.47). Radiocarbon analysis indicates that the bed (6 m or 18 ft thick) in which these fossils occur ranges in age from 20,000 to 12,000 years B.P. (**79**).

The connection of South America and North America allowed northward and southward migration of two distinct faunas that resulted in mass extinctions in both groups. The South American fauna, which migrated northward, included glyptodonts, camel-like guanacos, the giant ground sloth, armadillos, opossums, monkeys, giant anteaters, tapirs, and rodents.

Of the large mammals of the Pleistocene, most became extinct between 10,000 and 5000 years B.P. The list includes mammoths, mastodons, sabertooth cats, Irish elk, and giant ground sloths. Horses and camels disappeared from North America at about this time. The bison is one of the few large mammals which persisted in North America. Changes in climate and increased competition due to migrations over land bridges may have caused some extinctions, but these factors were operative throughout the Pleistocene and it is difficult to understand why the extinctions should have been concentrated near the beginning of the Holocene. The increased skill of human hunters, as evidenced by the development of more efficient weapons and improved hunting techniques, may have been the principal cause of the

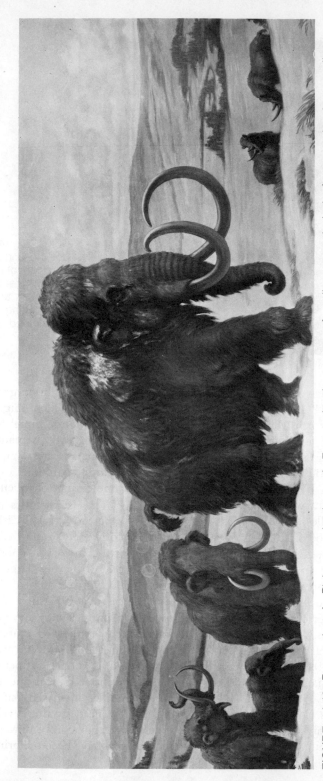

**FIGURE 14.42** Restoration of a Pleistocene landscape in Europe showing a group of woolly mammoths in the foreground and two woolly rhinoceroses behind the mammoths at the right. (From a painting by Charles R. Knight, courtesy of the Field Museum of Natural History, Chicago.)

FIGURE 14.43 Restoration of a mastodon. The imperial mastodon stood nearly 4.5 m (14 ft) high at the shoulders. (Courtesy of the Field Museum of Natural History, Chicago.)

FIGURE 14.44 Restoration of a Pleistocene landscape showing two cave bears. (Courtesy of the Field Museum of Natural History, Chicago.)

**FIGURE 14.45**  The "Irish Elk" pictured here was actually a very large deer with antlers measuring up to 4 m (12 ft) across. (From a painting by Charles R. Knight, courtesy of the Field Museum of Natural History, Chicago.)

early Holocene extinctions (**80**). However, the pattern and magnitude of extinctions of birds at the end of the Pleistocene is similar to that of the mammals, and therefore some factor other than hunters may have caused the extinction of both groups (**81**).

**Terrestrial Vegetation**

A distinct floristic change occurs between the Pliocene and Pleistocene. In the northern hemisphere, late Pliocene floras include species that generally do not occur in the early Pleistocene (**82**). The Pliocene floras that are preserved in

Pleistocene deposits are generally restricted in distribution. A greater variety of Pliocene floras persisted into the Pleistocene in the southern hemisphere where glaciation was much less extensive.

Floral assemblages also responded to the alternating glacial and interglacial climates. During glacial advances, the cold, temperate, and warm vegetation migrated toward the equator. When the ice receded, floral assemblages migrated back toward the poles. At the height of the glacial advance during the Late Wisconsinan (about 19,000 years B.P.) in eastern North America, oak forests extended southward to

FIGURE 14.46 Restoration of a Pleistocene landscape at Rancho La Brea showing the sabertooth cat, *Smilodon*, and other animals. (Courtesy of the Field Museum of Natural History, Chicago.)

FIGURE 14.47 Restoration of a Pleistocene landscape showing large ground sloths and armored armadillos. (From a painting by Charles R. Knight, courtesy of the Field Museum of Natural History, Chicago.)

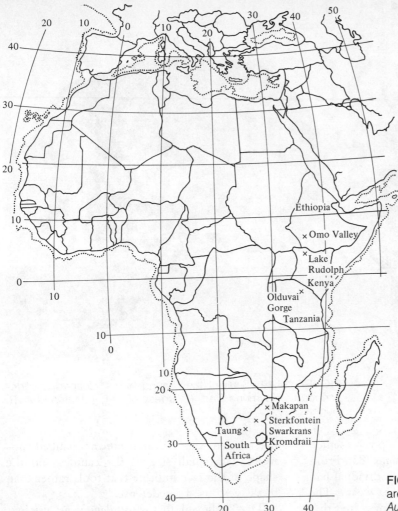

**FIGURE 14.48** Map indicating areas where fossil remains of *Australopithecus* have been found.

## Hominids

The record of hominid evolution is gradually being reconstructed through careful search of

Florida, pine and spruce forests grew on the exposed southern continental shelf, spruce forests occupied the eastern and central upland of the United States, and tundra vegetation bordered the ice margin. As the ice receded northward, the climate warmed, and these plant assemblages also migrated northward. During interglacials and the postglacial warming trend, plant associations now existing in the southeastern coastal plain of the United States flourished as far north as southern New York.

likely sites of habitation by interdisciplinary scientific teams. The record is far from complete, but some broad evolutionary trends are evident (Fig. 14.6).

EARLY PLEISTOCENE HOMINIDS. In 1924 a miner sent Raymond Dart a box of fossil-bearing breccia from a cave near the railroad station at Taung, South Africa (Fig. 14.48). Among the fossils was an almost complete skull imbedded in the breccia. For many months Dart worked to free the skull from the enclosing rock. The skull that emerged was that of a hominid unlike any that had yet been found. It was similar in some respects to the skull of a human, but it also resembled that of an ape

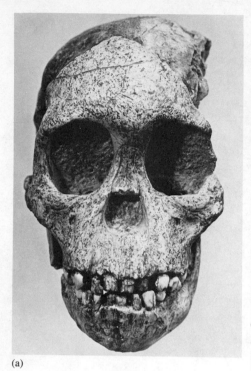

(a)

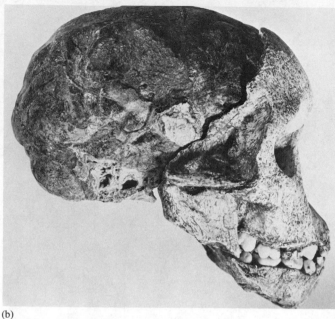

(b)

**FIGURE 14.49** Skull of the "Taung baby" cleaned and described by Raymond Dart. (Courtesy of P. T. Tobias and A. R. Hughes.)

(Fig. 14.49). Dart named the group represented by the skull *Australopithecus africanus* (**83**). From the teeth, he concluded that the individual had died around the age of five or six years. The unworn character of the molars suggested that this hominid was a carnivore. It is interesting to note that among the modern primates, only humans eat meat regularly.

Full grown, this specimen of *Australopithecus africanus* would have weighed 41 kg (90 lb) and would have been about 1.3 m (4 ft) tall (Fig. 14.50), although the average height of *Australopithecus africanus* was 1.45 m (4'9") (**84**). Its cranial capacity would have been about 600 cc as an adult, compared with the average capacity of 1350 cc for modern humans. However, since *Australopithecus* was relatively small, the ratio of brain weight to body weight was relatively large. This ratio has been estimated at 1 : 42, versus 1 : 47 in modern humans. From the position of the opening at the base of the skull and from the shape of the pelvis, it was

determined that *Australopithecus* walked upright. The small size of the canines and the shape of the face indicate that tools rather than teeth were used for defense.

Dart believed that *Australopithecus africanus* was a direct ancestor, or at least a very close relative, of modern *Homo sapiens*. Many of Dart's contemporaries did not accept this idea because they felt that Asia was the birthplace of the hominids. Until that time the oldest known remains of hominid ancestors had been found in Java and China. But with subsequent discoveries of australopithecines in Africa, the view that humans had originated in Asia was finally abandoned.

In 1936 Robert Broom found the remains of a hominid in a cave at Sterkfontein, South Africa. Although he assigned this specimen to a new genus, it is now believed to belong to the species *Australopithecus africanus*. However, this specimen and most of the other australopithecines from the South African caves have a

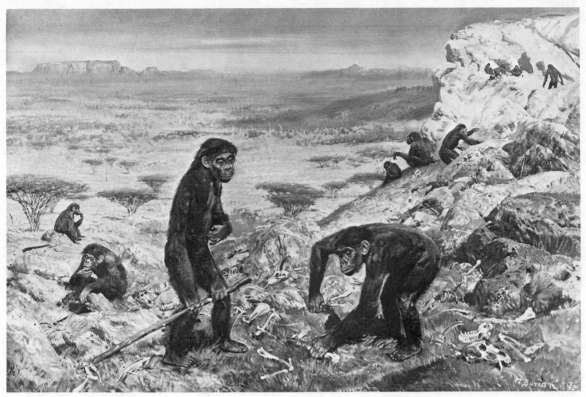

**FIGURE 14.50** Restoration of *Australopithecus* as they may have looked 3 or 4 million years ago. (Courtesy of Zdenek Burian. From *Prehistoric Man* by Augusta and Burian, Artia, Prague, Czechoslovakia.)

smaller brain capacity (averaging 442 cc) (**84**) than Dart's original find. The larger brain capacity may indicate that Dart's original find is a more advanced form, perhaps transitional to the genus *Homo*. Faunal evidence supports this suggestion (**84**). A recent study of the fauna at Sterkfontein indicates an age of 3.0 to 2.5 million years for the australopithecine–bearing strata (**85**).

In 1948 Broom found the remains of a distinctly different homonid in a cave at Swartkrans, less than a mile from Sterkfontein (**86**). This hominid was more robust than *Australopithecus africanus,* and Broom named it *Paranthropus robustus*. However, most investigators now believe that the specimen is similar enough to *Australopithecus africanus* to be called *Australopithecus robustus*. The robust form aver-

aged about 54.5 kg (120 lb) and 1.53 m (5′0″) tall (**84**). It has heavy bones, flattened molars, and a thick skull. Like the modern gorilla, *Australopithecus robustus* has a sagittal crest on the top of its skull. The sagittal crest was probably used for the attachment of powerful jaw muscles. The flattened molars and powerful jaws may indicate that the robust form was herbivorous.

No stone tools have been found associated with the remains of *Australopithecus africanus* (**87**), but bone tools may have been used. Dart noticed that many baboon skulls from the South African caves had a peculiar double depression. He attributed this to a blow from the humerus of an antelope—the distal end of this bone fits perfectly into the depression. The unusual abundance of such "weapons" among the

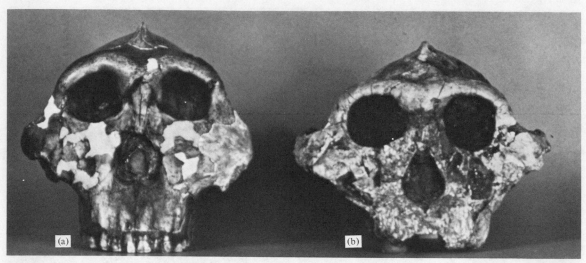

**FIGURE 14.51** Skulls of *Australopithecus* (*Zinjanthropus*) *boisei* from: **(a)** Tanzania. **(b)** Lake Rudolph, Kenya. (Courtesy of R. B. Leakey.)

fossils in the caves supports his suggestion. Other bones in the caves include slender, pointed bones that may have been produced during the extraction of bone marrow. However, some investigators question Dart's conclusion and believe all of the observed broken bones could be naturally caused (**87**).

Louis and Mary Leakey have made extensive studies of the fossiliferous deposits of Olduvai Gorge in Tanzania, Africa. In this 25-mile canyon they found many crude artifacts in the lowest exposed unit, Bed I (**88**). In 1959, a nearly complete skull (Fig. 14.51) was found by Mary Leakey. Subsequent excavations have resulted in the discovery of numerous bones and artifacts on a "living floor" (Fig. 14.52). The reconstructed skull resembles that of *Australopithecus robustus,* but it is somewhat more massive. Both skulls have a prominent sagittal crest. However, Louis Leakey assigned the skull to a new genus and species, *Zinjanthropus boisei.* Most authorities, however, feel that it is only a new species, *Australopithecus boisei.*

Bed I also yielded the remains of another hominid, which Leakey named *Homo habilis* (**89**) (Fig. 14.53). The cranial volume of this

hominid was 673 cc, and the average for similar specimens is 640 cc (**84**). Clearly this species is more advanced than *Australopithecus africanus,* and faunal and radiometric dating seems to indicate that it is younger (**84**, **91**). However, Richard Leakey, the son of Louis and Mary Leakey, believes that *Australopithecus africanus* and *Homo habilis* lived at the same time. It is likely that *Australopithecus africanus* gave rise to *Homo habilis* approximately 1.75 million years ago, the age of the earliest specimens which may be assigned to the genus *Homo* (**85**, **87**). A recent find was made at Sterkfontein of remains closely resembling *Homo habilis* in a bed containing a fauna dated at 2.0 to 1.5 million years (**85**). These remains are associated with stone tools and they lie above a bed containing remains belonging to *Australopithecus africanus.*

Tools found in Bed I at Olduvai include stone choppers and scrapers. While it is not certain, it is likely that they were made by *Homo habilis.* The tools have been assigned to the Oldowan Tool Industry. Bed I, which contains a fauna similar to that of the Upper Villafranchian of southern Europe, an interbedded volcanic tuff, has been dated at approximately 1.75

**FIGURE 14.52** Fossiliferous strata near Olduvai Gorge, Tanzania. (Courtesy of R. B. Leakey.)

million years by the potassium–argon method; a basalt flow near the base of Bed I has been dated at 1.8 million years (**92**). Thus, Bed I is dated as earliest Pleistocene.

Australopithecines have also been found near Lake Turkana (formerly Lake Rudolph) in Kenya and in the Omo Valley of Ethiopia (**93**). The strata in which the fossils occurred have been dated at 2.5 million years (**94**). Choppers associated with these remains are among the oldest known artifacts (**95**). In the Omo Basin, teeth and jaw fragments of hominids have been found in deposits as old as 3.1 million years (**96**). Teeth resembling those of *Australopithecus africanus* and either *Australopithecus* (*Zinjanthropus*) *boisei* or *Australopithecus* (*Paranthropus*) *robustus* have been found in four different horizons dated at 3.1, 2.4, 2.3 and 2.1 million years. However, tools have not yet been found associated with these hominid fossils. The oldest known australopithecine remains are two jaws which were found by Mary Leakey, and they have been dated at 3.35 to 3.75 million years.

Java is the only locality outside of Africa where the remains of australopithecines have been found (**97**). These remains resemble those of *Australopithecus* (*Paranthropus*) *robustus* and they are found in rocks dated at about 1.5 million years old (**97**).

A brain case and two fossil teeth were discovered in Bed II at Olduvai Gorge in association with stone tools of the Chellean (Abbevillian) culture (**89**). The skull, which is relatively large and has heavy brow ridges, may represent a more advanced hominid than *Homo habilis;* indeed, it may belong to a primitive variety of *Homo erectus.* The tools include hand axes, a significant advance over the choppers of the Oldowan culture. Bed II ranges in age from about 1.6 to 1.0 million years.

Richard Leakey has recently discovered the remains of a hominid strikingly similar to *Homo erectus* near Lake Turkana in beds dated at between 1.3 and 1.6 million years old (**97**). What is remarkable about this find is that it was made in

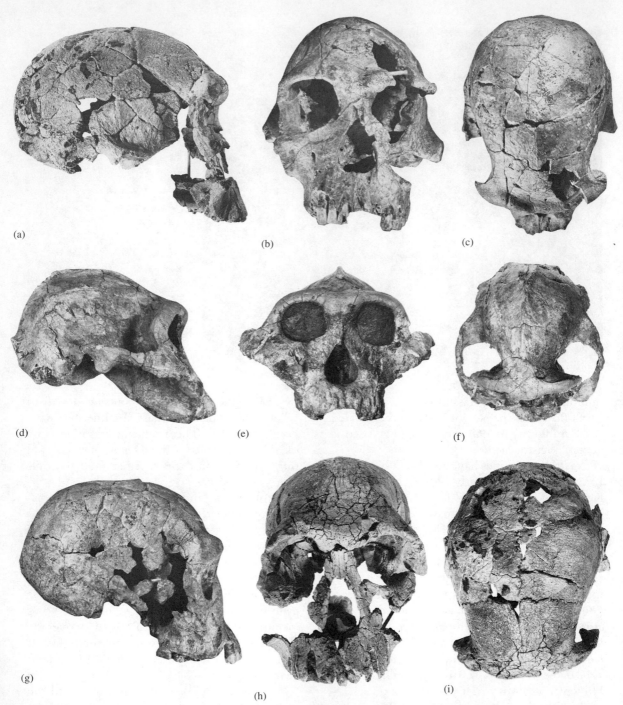

(a)

(b)

(c)

(d)

(e)

(f)

(g)

(h)

(i)

**FIGURE 14.53** Comparison of skulls from three Plio-Pleistocene hominids from Lake Turkana, Kenya: **(a)** to **(c)** *Australopithecus africanus.* **(d)** to **(f)** *Australopithecus boisei.* **(g)** to **(i)** *Homo habilis.* (From R. E. Leakey, Ref. 90. Reproduced with permission of National Museum of Kenya.

**FIGURE 14.54** Restoration of *Homo erectus*. (Courtesy of Zdenek Burian. From *Prehistoric Man* by Augusta and Burian, Artia, Prague, Czechoslovakia.)

beds which also contain remains of *Australopithecus boisei*.

MIDDLE PLEISTOCENE HOMINIDS. The Middle Pleistocene may be defined as extending from the end of the Jarimillo event (the beginning of the Kansan Glacial Age) about 890,000 years ago to the beginning of the Illinoian (Elsterian) glacial age about 400,000 years ago.

Eugene Dubois found the remains of a relatively advanced hominid in Middle Pleistocene deposits of Java. The average cranial volume of this hominid, known as the "Java man," is between 900 and 1000 cc; the skull is characterized by massive brow ridges (Fig. 14.54). Originally designated *Pithecanthropus erectus,* this post-Villafranchian hominid is now designated as *Homo erectus*. Basalt associated with the remains of *Homo erectus* has been dated at about 500,000 years (**1**).

Remains of *Homo erectus* were found by Davidson Black in a cave near Peking, China; nearby was charcoal, which indicates that *Homo*

*erectus* probably used fire. In addition, many of the skulls have been broken open, suggesting the possibility of cannibalism. A jaw fragment unearthed near Heidelberg, Germany, resembles that of *Homo erectus* and indicates that this hominid may have ranged into central Europe.

Pre-Neanderthal human remains have been found in Arago Cave in southeastern France, and these may belong to an advanced form of *Homo erectus* (**98**). These remains are probably between 500,000 and 250,000 years old.

LATE PLEISTOCENE HOMINIDS. *Homo sapiens* does not appear in the fossil record until about 250,000 years ago. Numerous finds in Europe document the skeletal morphology and habits of these early humans during the Riss–Wurm interglacial stage and the Wurm glacial stage. Skull remains of a primitive variety of *Homo sapiens* have been found near Steinheim, Germany, and Swanscombe, England, in deposits more than 200,000 years old.

Remains of a probable subspecies, "Neanderthal man" (*Homo sapiens neanderthalensis*), were first found in the Neander Valley in Germany. This early human lived from 110,000 to 35,000 years ago in a broad area covering much of Africa, Asia, and Europe. A typical specimen had a prominent brow ridge, large jaws, and a massive frame (Fig. 14.55). Neanderthals also had very large brain cases which averaged about 1500 cc. Tools associated with this hominid include carefully shaped points, hand axes, and needles. The Neanderthal subspecies disappeared about 35,000 years ago, perhaps having been killed off and/or assimilated by a more advanced form of *Homo sapiens*.

Modern human (*Homo sapiens sapiens*), also known as "Cro-Magnon man," almost certainly occupied much of Africa before 50,000 years B.P. and may have occupied this area

**FIGURE 14.55** Reconstruction of a male Neanderthal. (Courtesy of the Field Museum of Natural History, Chicago.)

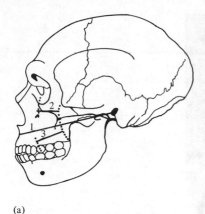 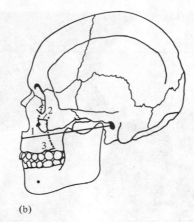

(a)                    (b)

**FIGURE 14.56** Comparison of (a) the Neanderthal skull with (b) a modern skull. (From Howells, Ref. 99.)

more than 100,000 years B.P. (**100**). Their remains have also been found in Java and Borneo in deposits more than 40,000 years old. By 30,000 years B.P., they were widely distributed over much of the world. They are characterized by much less prominent brow ridges, a more pointed chin, and a higher cranium than Neanderthals (Figs. 14.56 and 14.57). Their cranial capacity averaged about 1350 cc, slightly less than that of Neanderthals. Among their more advanced cultural skills was painting. Beautiful cave paintings in southern Europe were produced by "Cro-Magnon man" between 30,000 and 9000 years ago (**101**) (Fig. 14.58).

There is a marked similarity in the shape (but not the size) of the skulls of most representatives of *Australopithecus africanus* (all except the original "Taung baby"), *Homo habilis, Homo erectus,* and *Homo sapiens neanderthalensis.* All have a low, sloping forehead and heavy brow ridges (Fig. 14.57). It is not hard to imagine a simple evolutionary sequence from *Australopithecus africanus* to Neanderthals. However, it is not certain from what species *Homo sapiens sapiens* evolved. It is probable that modern human beings and Neanderthals evolved from a common ancestor (possibly an advanced form of *Homo erectus*). It is not known when *Homo sapiens sapiens* began to separate from the hominid that gave rise to *Homo sapiens neanderthalensis.* It must have been more than 50,000

years B.P. (the oldest well-established date on *Homo sapiens sapiens*), but it is not possible to say more than that at the present time. There are no known transitional forms between *Homo sapiens sapiens* and *Homo erectus.* Some sort of isolation must have been necessary for two different subspecies to evolve from a common stock.

The repeated glaciations resulted in the migration of humans through Eurasia. Eventually, they reached what is now the Bering Strait and crossed the exposed continental shelf, known as Beringia, into North America. Sparse evidence has come to light for human occupation of North America prior to 11,000 years B.P. Artifacts have also been reported in deposits which have radiocarbon ages between 24,000 and 13,000 years B.P. (**102**). In Mexico an obsidian blade was found beneath a log dated at about 23,000 years B.P. In Illinois an artifact has been found in deposits predating the Late Wisconsinan glacial advance and thus perhaps more than 23,000 years old (**103**). A skull found near Laguna Beach, California, has been dated by the radiocarbon method at between 18,620 and 15,680 years. Other evidence of occupation in North America prior to 11,000 years B.P. comes from Pennsylvania, the northern Yukon, southern California, Peru, Texas, Argentina, Columbia, northwestern Canada, and Utah (**104**).

**FIGURE 14.57** Restoration of *Homo sapiens sapiens* in central Europe. (From a painting by Maurice Wilson, courtesy of the Trustees of the British Museum of Natural History.)

**FIGURE 14.58** Paintings on the wall of a cave near Lascaux, France. (Courtesy of the French Embassy Press and Information Division.)

There is abundant radiocarbon data to indicate that large numbers of people migrated to North America about 11,000 years ago. Migration was slowed considerably when a rise in sea level resulted in the flooding of the Bering Strait about 9000 years ago.

## SUMMARY

The chronology of the Quaternary has been developed through the use of radiometric dating, magnetic polarity measurements, and paleoclimatic reconstructions. The Pliocene-Pleistocene boundary has been dated at about 1.8 million years at the type locality of the basal Pleistocene. At least four major glacial episodes occurred during the Pleistocene: 1.8 to 1.65, 0.90 to 0.75, 0.40 to 0.25, and 0.075 to 0.001 million years ago. During each of the four glacial ages, there were two or more separate glacial advances, separated by partial retreat of the ice sheet. During the three interglacial ages between the glacial ages, there were glacial advances, but these were not as extensive as during the glacial ages.

During the last cold interval, there were four separate glacial advances, reaching their maximum extent about 72,000, 55,000, 35,000, and 19,000 years ago. The last retreat of ice sheets began shortly after the maximum advance, and by 5000 years B.P. the ice sheets had disappeared from North America and Europe. Tundra vegetation and spruce, pine, and oak forests migrated successively behind the retreating ice.

Postglacial rebound followed the melting of the Late Wisconsinan ice sheets. Tectonism and volcanism continued throughout the Quaternary in the circum–Pacific region and in the boundary regions of the crustal plates. Strike-slip faulting was common in these regions as the Atlantic and Indian oceans widened at rates ranging from two to four centimeters per year.

The worldwide cooling that led to the formation of Pleistocene glaciers began during the late Cretaceous and may have been related to uplift of the continents, volcanism, polar wandering, and continental drift. Cyclic glaciation has been attributed to changes in insolation due to changes in the angle and direction of tilt of the earth's axis and changes in the ellipticity of the earth's orbit. Alternatively, cyclic glaciation may have been related to a freezing and thawing of the Arctic Ocean, volcanism, or magnetic field fluctuations.

Pleistocene marine life was marked by the development of cold- and warm-water species which migrated in response to changing oceanic temperatures. Terrestrial faunas included mammoths, mastodons, and woolly rhinoceroses in cold climates, and elephants, rhinoseroses, horses, and cattle in warmer climates. The australopithecines which emerged during the lastest Tertiary gave rise to *Homo habilis,* which in turn gave rise to *Homo erectus* in the Early Pleistocene. Emerging in the Late Pleistocene, *Homo sapiens* was first represented by Neanderthals who appeared about 100,000 years ago. Modern humans (*Homo sapiens sapiens*) appeared between 100,000 and 50,000 years ago.

## REFERENCES CITED

1. R. F. Flint, 1971, *Glacial and Quaternary Geology:* Wiley, New York.
2. C. Emiliani, 1958, *Sci. Amer.,* v. 198, no. 2, p. 54.
3. C. Emiliani, 1966, *J. Geol.,* v. 74, p. 109.
4. N. J. Shackleton and N. D. Opdyke, 1973, *Quaternary Res.,* v. 3, p. 39.
5. D. B. Ericson and G. Wollin, 1968, *Science,* v. 162, p. 1227.
6. G. G. Connally and L. A. Sirkin, 1971, *Geol. Soc. Amer. Bull.,* v. 82, p. 989.
7. F. T. Banner and W. H. Blow, 1967, *Micropaleont.,* v. 13, p. 133.
8. D. A. Emilia and D. F. Heinrichs, 1969, *Science,* v. 154, p. 349.
9. L. S. B. Leakey, R. Protsch, and R. Berger, 1968, *Science,* v. 162, p. 559.
10. C. Savelli and R. Mezzetti, 1977, *INQUA Abstr.,* p. 400.
11. B. U. Haq, W. A. Berggren, and J. A. VanCouvering, 1977, *X INQUA Abstr.,* p. 194.
12. J. C. Frye, H. B. Willman, M. Rubin, and R. F. Black, 1968, *U.S. Geol. Survey Bull. 1274-E.*
13. J. R. Curray, 1965, *in* H. E. Wright and D. G. Frey, eds., *The Quaternary of the United States:* Princeton University Press, Princeton, N.J. p. 723.
14. C. S. Denny, compiler, 1970, *The National Atlas of the United States of America:* U.S. Dept. of the Interior, Geol. Survey, p. 76.
15. G. J. Kukla, F. Brown, and R. Shuey, 1977, *X INQUA Abstr.,* p. 259; A. J. Eardley *et al.,* 1873, *Geol. Soc. Amer. Bull.,* v. 84, p. 211.
16. R. W. Fairbridge, 1964, *in* A. E. M. Nairn, ed., *Problems in Paleoclimatology:* Wiley, New York, p. 356; W. G. Deuser, E. H. Ross, and L. S. Waterman, 1976, *Science,* v. 191, p. 1168.
17. J. H. Beard, 1969, *Geol. Soc. Amer. Abstr. Programs,* Part 7, p. 256.
18. J. P. Kennett, 1969, *Antarct. J.,* v. 4, p. 178.
19. J. L. Lamb, 1969, *Geol. Soc. Amer. Abstr. Programs,* Part 7, p. 256.
20. O. L. Bandy and R. E. Casey, 1969, *Antarct. J.,* v. 4, p. 170.
21. J. Boellstorff, 1973, *Isochron/West,* no. 8, p. 39; J. Boellstorff, 1973, *IX INQUA Abstr.,* p. 31.
22. H. M. van Montfrans, 1971, *Earth Planet. Sci. Lett.,* v. 11, p. 226; H. M. van Montfrans and J. Hospers, 1969, *Geologic en Mijnbouw,* v. 48, p. 565; W. H. Zagwijn *et al.,* 1971, *Geologie en Mijnbouw,* v. 50, p. 41.
23. G. H. Curtis, 1967, *in* W. W. Bishop and J. C. Clark, eds., *Background to Evolution in Africa:* University of Chicago Press, Chicago.

24. J. H. Mercer, R. J. Fleck, E. A. Mankinen, and W. Sander, 1975, *in* R. P. Suggate and M. M. Cresswell, eds., *Quaternary Studies:* Royal Society of New Zealand, Wellington, p. 223.
25. J. H. Mercer, 1969, *Science,* v. 164, p. 823.
26. I. McDougall *et al.,* 1977, *Geol. Soc. Amer. Bull.,* v. 88, p. 1.
27. G. H. Denton *et al.,* 1970, *Antarct. J.,* v. 5, p. 15.
28. C. W. Hibbard, 1971, personal communication.
29. A. Azzaroli, 1970, *Giornale di Geologia,* v. 35, p. 111.
30. J. Fink and G. J. Kukla, 1977, *Quaternary Res.,* v. 7, p. 363; K. W. Butzer, 1971, *Environment and Archeology* (2nd ed): Aldine-Atherton, Chicago.
31. T. van der Hammen, T. A. Wijmstra, and W. H. Zagwijn, 1971, *in* K. K. Turekian, ed., *Late Cenozoic Glacial Ages:* Yale University Press, New Haven, p. 391; P. Evans, 1971, *in* W. B. Harland and E. H. Francis, eds., *The Phanerozoic Time-Scale:* Geol. Soc. London, special publication 5, p. 35.
32. V. J. Frechen and H. J. Lippolt, 1965, *Eiszeitalter und Gegenwart,* v. 16, p. 5.
33. A. Holmes, 1945, *Principles of Physical Geology:* Ronald Press, New York.
34. G. J. Kukla, 1975, *in* K. W. Butzer, ed., *After the Australopithecines:* Mouton, The Hague, p. 99.
35. S. C. Porter, M. Stuiver, and I. C. Yang, 1977, *Science,* v. 195, p. 61.
36. F. W. Taylor and A. L. Bloom, 1977, *INQUA Abstr.,* p. 462.
37. K. J. Mesolella, R. K. Matthews, W. S. Broecker, and D. L. Thurber, 1969, *J. Geol.,* v. 77, p. 250.
38. W. E. Frerich, 1968, *Science,* v. 159, p. 1456.
39. S. J. Johnsen, W. Dansgaard, H. B. Clausen, and C. C. Langway, Jr., 1972, *Nature,* v. 235, p. 429.
40. R. P. Steinen, R. S. Harrison, R. K. Matthews, 1973, *Geol. Soc. Amer. Bull.,* v. 84, p. 63.
41. G. M. Richmond, 1970, *Quaternary Res.,* v. 1, p. 3.
42. C. E. Stearns and D. L. Thurber, 1965, *Quaternaria,* v. 7, p. 29; L. S. Land *et al.,* 1967, *Geol. Soc. Amer. Bull.,* v. 78, p. 993; T-L. Ku, M. A. Kimmel, W. H. Easton, and T. J. O'Neil, 1974, *Science,* v. 183, p. 959; H. H. Veeh and R. Giegengack, 1970, *Nature,* v. 226, p. 155.
43. A. L. Bloom *et al.,* 1974, *Quaternary Res.,* v. 4, p. 185.
44. L. A. Sirkin, 1977, *Ann. N.Y. Acad. Sci.,* v. 288, p. 206; J. P. Kempton and D. L. Gross, 1971, *Geol. Soc. Amer. Bull.,* v. 82, p. 3245.
45. A. McIntyre *et al.,* 1976, *Science,* v. 191, p. 1131.
46. G. G. Connally and L. A. Sirkin, 1972, *Wisconsinan History of the Hudson-Champlain Lobe:* Geol. Soc. Amer. Mem. 136.
47. E. H. De Geer, 1954, *Geol. Foren. Stockholm Forh.,* v. 76, p. 303.
48. J. Lundquist, 1965, *in* K. Rankana, ed., *The Quaternary:* Wiley, New York, p. 139.
49. P. E. Playford and A. E. Cockbain, 1976, *in* M. R. Walter, ed., *Stromatolites:* Elsevier Scientific Publ. Co., New York, p. 389. D. J. J. Kinsman and R. K. Park, 1976, *in* M. R. Walter, ed., *Stromatolites:* Elsevier, New York, p. 421.
50. V. C. LaMarche, 1974, *Science,* v. 183, p. 1043.
51. P. E. Damon and S. M. Kunen, 1976, *Science,* v. 193, p. 447.
52. J. T. Andrews, 1969, *in Earth Science Symposium on Hudson Bay:* Can. Geol. Survey Pap. 65-53, Ottawa, p. 49.
53. W. B. Wright, 1937, *The Quaternary Ice Age:* Macmillan, London.
54. R. L. Reynolds, 1977, *J. Geophys. Res.,* v. 82, p. 3677.
55. P. Morgan, D. D. Blackwell, R. E. Spafford, and R. B. Smith, 1977, *J. Geophys. Res.,* v. 82, p. 3719.
56. G. P. Eaton, *et al.,* 1975, *Science,* v. 188, p. 787.
57. D. M. Hadley, G. S. Stewart, and J. E. Ebel, 1976, *Science,* v. 193, p. 1237.
58. E. Dorf, 1964, *in* A. E. M. Nairn, ed., *Problems in Paleoclimatology:* Wiley, London, p. 13.
59. J. P. Kennett and R. C. Thumell, 1975, *Science,* v. 187, p. 497.
60. H. Wexler, 1952, *Sci. Amer.,* v. 186, no. 4.
61. R. A. Schmidt, 1962, *Science,* v. 138, p. 443.
62. B. P. Glass and B. C. Heezen, 1967, *Sci. Amer.,* v. 217, no. 1, p. 32.
63. C. K. Seyfert and J. G. Murtaugh, 1977, *Geol. Soc. Amer. Abstr. Programs,* v. 9, p. 1168.
64. P. R. Vogt, 1972, *Nature,* v. 240, p. 338.
65. T. C. Huang and N. D. Watkins, 1976, *Science,* v. 193, p. 576.
66. D. Ninkovich and W. L. Donn, 1976, *Science,* v. 194, p. 899.
67. W. S. Broecker, 1966, *Science,* v. 151, p. 299.
68. W. L. Donn and M. Ewing, 1968, *Meteorol.*

Monogr., v. 8, p. 100; W. L. Donn and M. Ewing, 1966, *Science,* v. 152, p. 1706; J. D. Hays, J. Imbrie, N. J. Schackleton, 1976, *Science,* v. 194, p. 1121.

69. G. W. Lockwood, 1975, *Science,* v. 190, p. 560; J. A. Eddy, 1976, *Science,* v. 192, p. 1189.

70. J. Steiner and E. Grillmair, 1973, *Geol. Soc. Amer. Bull.,* v. 84, p. 1003.

71. G. Wollin, D. B. Ericson, and J. Wollin, 1974, *in Les méthodes quantitatives d'étude des variations du climat au cours du Pléistocène,* Colloques Internationaux du C.N.R.S., No. 219: Centre National de la Recherche Scientifique, Paris, p. 273.

72. R. A. Bryson, 1974, *Science,* v. 184, p. 753.

73. J. P. Kennett and N. D. Watkins, 1970, *Nature,* v. 227, p. 930.

74. J. D. Hays and N. D. Opdyke, 1967, *Science,* v. 158, p. 1001.

75. P. J. Smith, 1967, *Geophys. J. Roy. Ast. Soc.,* v. 12, p. 321.

76. M. A. Ruderman, 1974, *Science,* v. 184, p. 1079.

77. C. J. Waddington, 1967, *Science,* v. 158, p. 913.

78. I. K. Crain, 1971, *Geol. Soc. Amer. Bull.,* v. 82, p. 2603.

79. T. Y. Ho, L. F. Marcus, and R. Berger, 1969, *Science,* v. 164, p. 1051.

80. P. S. Martin and H. E. Wright, Jr., eds., 1967, *Pleistocene Extinctions:* Yale University Press, New Haven, Conn.

81. D. K. Grayson, 1977, *Science,* v. 195, p. 691.

82. L. A. Sirkin and J. P. Owens, in press, "Palynology of Tertiary Sediments in the Mid-Delmarva Peninsula."

83. R. A. Dart, 1925, *Nature,* v. 115, p. 195.

84. P. V. Tobias, 1970, *in* S. L. Washburn and P. C. Jay, eds., *Perspectives on Human Evolution:* Holt, Rinehart and Winston, New York, v. 3, p. 377.

85. A. R. Hughes and P. V. Tobias, 1977, *Nature,* v. 265, p. 310.

86. R. Broom and J. T. Robinson, 1952, *The Swartkrans Ape-men:* Memoir, Transvaal Museum.

87. R. G. Klein, 1977, *Science,* v. 197, p. 115.

88. L. S. B. Leakey, 1965, *A Preliminary Report on the Geology and Fauna: Olduvai Gorge 1951–1961:* Cambridge University Press, Cambridge, v. 1; L. S. B. Leakey, 1951, *A Report on the Evolution of the Hand-Axe Culture in Beds I–IV: Olduvai Gorge:* Cambridge University Press, Cambridge.

89. L. S. B. Leakey, 1961, *Nature,* v. 189, p. 649.

90. L. S. B. Leakey, 1976, *Amer. Scientist,* v. 64, p. 174.

91. T. D. White and J. M. Harris, 1977, *Science,* v. 198, p. 13.

92. L. S. B. Leakey, J. F. Evernden, and G. H. Curtis, 1961, *Nature,* v. 191, p. 478.

93. R. E. F. Leakey, 1970, *Nature,* v. 226, p. 223.

94. F. J. Fitch and J. A. Miller, 1970, *Nature,* v. 226, p. 226.

95. M. D. Leakey, 1970, *Nature,* v. 226, p. 228.

96. F. C. Howell, 1969, *Nature,* v. 223, p. 1234; F. C. Howell, 1968, *Nature,* v. 219, p. 567.

97. K. Butzer and G. L. Isaac, eds., 1975, *After the Australopithecines:* Mouton, the Hague, p. 889.

98. H. De Lumley and M. De Lumley, 1973, *Yearbook Phys. Anthrop.,* v. 17, p. 162.

99. W. W. Howells, 1975, *in* R. H. Tuttle, ed., *Paleoanthropology: Morphology and Paleoecology:* Mouton, the Hague, p. 389.

100. J. C. Vogel and P. B. Beaumont, 1972, *Nature,* v. 237, p. 50.

101. P. V. D. Stern, 1969, *Prehistoric Europe:* Norton, New York.

102. C. V. Haynes, 1969, *Science,* v. 166, p. 709.

103. P. J. Munson, 1965, *Science,* v. 150, p. 1722.

104. W. N. Irving and C. R. Harrington, 1973, *Science,* v. 179, p. 335; J. L. Bada, R. A. Schroeder, and G. F. Carter, 1974, *Science,* v. 184, p. 791; R. S. MacNeish, 1976, *Amer. Scientist,* v. 64, p. 316.

# The Phanerozoic
# Time Scale

A

| GEOLOGIC TIME UNITS[a] | | | | | | Age of the Beginning (Million Years) |
|---|---|---|---|---|---|---|
| Era | Period | Epoch | Age North America | Age Europe | | |
| Cenozoic | Quaternary | Pleistocene and Recent | | | | |
| Cenozoic | Tertiary | Pliocene | | Astian | | 1.8 |
| | | | | Pontian (Tabianian) | | 5 |
| | | Miocene | | Sarmatian (Messinian) | | 6.5 |
| | | | | Tortonian | | 18.5 |
| | | | | Helvetian | | |
| | | | | Burdigalian | | 26 |
| | | | | Aquitanian | | |
| | | Oligocene | | Chattian | | ? |
| | | | | Rupelian | | 31.5 |
| | | | | Tongrian | | 37.5 |
| | | Eocene | Jacksonian | Ludian | | 45 |
| | | | | Bartonian | | |
| | | | Claibornian | Auversian | | 49 |
| | | | | Lutetian | | |
| | | | Wilcoxian | Cuisian | | 53.5 |
| | | | | Ypresian | | |
| | | Paleocene | Midwayan | Thanetian | | 58.5 |
| | | | | Montian | | |
| | | | Laramian | Danian | | 65 |
| Mesozoic | Cretaceous | Upper Cretaceous | Montanan | Maestrichtian | | 70 |
| | | | | Senonian | Campanian | 76 |
| | | | | | Santonian | 82 |
| | | | Coloradoan | | Coniacian | 88 |
| | | | | Turonian | | 94 |
| | | | Dakotan | Cenomanian | | 100 |
| | | | Washitan | | | |
| | | Lower Cretaceous | Fredericksburgian | Albian | | 106 |
| | | | Trinitian | Aptian | | 112 |
| | | | | Neocomian | Barremian | 118 |
| | | | | | Hauterivian | 124 |
| | | | | | Valanginian | 130 |
| | | | | | Berriasian | 136 |
| | Jurassic | Upper Jurassic (Malm) | | Tithonian | Purbeckian | 141 |
| | | | | | Portlandian | 146 |
| | | | | | Kimmeridgian | 151 |
| | | | | Oxfordian | | 157 |
| | | | | Callovian | | 162 |
| | | Middle Jurassic (Dogger) | | Bathonian | | 167 |
| | | | | Bajocian | | 172 |
| | | Lower Jurassic (Lias) | | Toarcian | | 178 |
| | | | | Pliensbachian | | 183 |
| | | | | Sinemurian | | 188 |
| | | | | Hettangian | | 192.5 |
| | Triassic | Upper Triassic | | Rhaetian | | |
| | | | | Norian | | |
| | | | | Karnian | | 205 |
| | | Middle Triassic | | Ladinian | | |
| | | | | Anisian | | 215 |
| | | Lower Triassic | | Scythian | Olenekian | |
| | | | | | Indulan | 225 |

## GEOLOGIC TIME UNITS[a]

| Era | Period | Epoch | Age — North America | Age — Europe | | Age of the Beginning (Million Years) |
|---|---|---|---|---|---|---|
| Paleozoic | Permian | Upper Permian | Ochoan | Chideruan (Tartarian) | | 230 |
| | | | Guadalupian | Kazanian | | 240 |
| | | | | Kungurian | | 256.5 |
| | | Lower Permian | Leonardian | Artinskian | | 266.5 |
| | | | Wolfcampian | Sakmarian | | 280 |
| | Carboniferous — Pennsylvanian | Upper Pennsylvanian | Virgilian | Stephanian | Uralian | 292.5 |
| | | | Missourian | | Gshelian | |
| | | Middle Pennsylvanian | Desmoinesian | Moscovian (Westphalian) | | 312.5 |
| | | | Atokan | | | |
| | | Lower Pennsylvanian | Morrowan | | | |
| | | | Springeran | Namurian | | 325 |
| | Carboniferous — Mississippian | Upper Mississippian | Chesteran | | | |
| | | | Meramecian | Viséan | | 337.5 |
| | | Lower Mississippian | Osagian | Tournaisian | | 345 |
| | | | Kinderhookian | Etroeungtian | | |
| | Devonian | Upper Devonian | Conewangoan | Famennian | | 353 |
| | | | Cassadagan | | | |
| | | | Chemungian | Frasnian | | 359 |
| | | | Fingerlakesian | | | |
| | | Middle Devonian | Taghanican | Givetian | | |
| | | | Tioughniogan | | | |
| | | | Cazenovian | Eifelian | | 370 |
| | | Lower Devonian | Onesquethawan | Coblenzian | Emsian | 374 |
| | | | Deerparkian | | Siegenian | 390 |
| | | | Helderbergian | Gedinnian | | 395 |
| | Silurian | Upper Silurian (Cayugan) | Keyseran | Dowtonian (Pridoli) | | |
| | | | Tonolowayan | Ludlovian | | |
| | | | Salinan | | | |
| | | Middle Silurian (Niagaran) | Lockportian | | | |
| | | | Cliftonian | Wenlockian | | 410 |
| | | | Clintonian | Llandoverian | | |
| | | Lower Silurian | Alexandrian | | | 435 |
| | Ordovician | Upper Ordovician (Cincinnatian) | Richmondian | Ashgillian | | 438 |
| | | | Maysvillian | Caradocian | | 444 |
| | | | Edenian | | | |
| | | Middle Ordovician (Mohawkian) | Trentonian | | | |
| | | | Blackriveran | | | 456 |
| | | | Chazyan | Llandeilian | | 458 |
| | | | | Llanvirian | | 467 |
| | | Lower Ordovician | Canadian | Arenigian (Skiddavian) | | 495 |
| | | | | Tremadocian | | |
| | Cambrian | Upper Cambrian (Croixian) | Trempealeauan | (Lingula Flags) | | 515 |
| | | | Franconian | | | |
| | | | Dresbachian | | | |
| | | Middle Cambrian | Albertan | Menevian | | 540 |
| | | | | Solvan | | |
| | | Lower Cambrian | Waucoban | Comleyan | | 560 |
| | | | | Tortonian | | 570 |

[a] Dates from W.B. Harland, A.G. Smith, and B. Wilcock, Eds., 1964. The Phanerozoic Time-scale: A symposium: Geol. Soc. London, v. 120s.

# The Major Fossil Groups

MANY ORGANISMS are entirely soft-bodied and are not generally preserved as fossils. The following brief discussion is restricted to those groups which have left a significant fossil record, and it is meant to supplement information in the preceding sections of the text.

## THE PROTISTS

The single-celled protists are the oldest and most primitive of the fossil groups. Most of these organisms can only be seen with a microscope, but some are several centimeters (almost an inch) in size. Nearly all reproduce asexually by splitting in two.

Modern blue-green algae consist of simple cells joined into long filaments. The filaments are not generally preserved as fossils, but some varieties of blue-green algae precipitate calcium carbonate in concentrically laminated structures called stromatolites and occur as fossils in rocks as old as 3.0 billion years. Other cells resembling algae are found in rocks up to 3.4 billion years old.

Diatoms, a more advanced variety of algae, secrete minute siliceous skeletons known as frustules (Figs. B.1 and B.2). They appear much later in the fossil record than blue-green algae, and they are present today in both marine and freshwater environments. Locally, they are abundant enough to form a diatomaceous ooze. The consolidation of such an ooze produces a rock called diatomite. Charophytes (chara or

gyrogonites) are the calcareous reproductive bodies of freshwater algae known as stoneworts.

The skeletons of coccoliths are typically very small, discoid in shape, and composed of calcium carbonate (Fig. 12.34). Coccoliths are temperature sensitive and have been used as temperature indicators in oceanic sediments. Their skeletons are important constituents of some marine sediments, especially chalk (Fig. B.1c).

The dinoflagellates are solitary, planktonic marine organisms that may have a shell composed of silica or calcium carbonate (Fig. B.3). The shell is characterized by the presence of an equatorial band girdling it. Dinoflagellates are useful in making correlations of marine deposits. Silicaflagellates have silica skeletons formed into nets and rings.

The animallike protozoans include foraminifera and radiolaria (Figs. B.4 and B.5). Both are abundant in modern oceans, and judging from their fossil records they were abundant in Paleozoic, Mesozoic, and Cenozoic oceans as well. Most foraminiferal tests are composed of calcite, but some are composed of chitin, while others are made up of very fine sand or agglutinated and covered with the tests of much smaller organisms. They have very distinct morphologic characteristics (Figs. B.2 and B.4) and are widely used as guide fossils, especially in deposits that lack macrofossils. The radiolarian test is generally composed of silica and has a

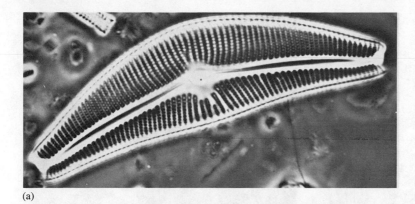
(a)

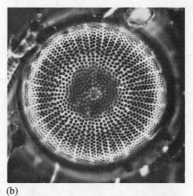
(b)

(c)

**FIGURE B.1** Diatoms and coccoliths. **(a)** *Cymbella mexicanum,* a diatom X1000 from Pickerel Lake, South Dakota. **(b)** *Stephanodiscus niagarae,* X1000, a diatom from Pickerel Lake, South Dakota. **(c)** Scanning electron micrograph of coccoliths and diatoms from a marine environment, X1000. [**(a)** and **(b)** From Haworth, Ref. 1. **(c)** Taken by V. G. Shah of Physical Research Laboratory, Ahmedabad, India, from Ref. 2; copyright ©, 1977 by American Association for the Advancement of Science.]

central capsule and outer rings or rays (Fig. B.5). The tests from both foraminifera and radiolaria commonly make up a significant percentage of a rock (Table B.1).

## MULTICELLULAR ANIMALS
### Archaeocyathans

The Archaeocyathans are an extinct group of invertebrates which had a cone-shaped, double-walled, porous skeleton (Fig. B.6). The affinity of this group is somewhat uncertain. These organisms have been included with the sponges and with the corals, but it is now generally agreed that they should be placed in a separate phylum. They were abundant enough in Early and Middle Cambrian seas to form reeflike deposits. Because of their limited time range, archaeocyathans are very useful as guide fossils.

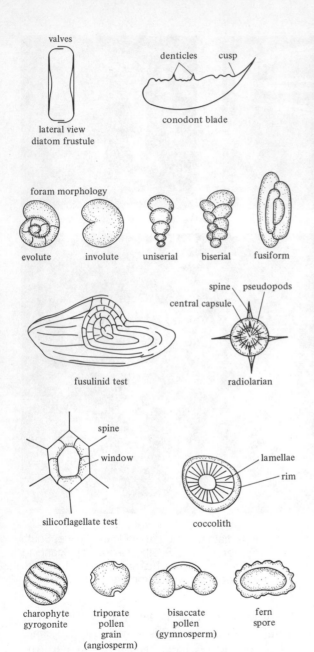

FIGURE B.2 Morphology of some microfossils.

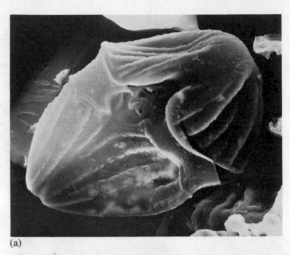

(a)

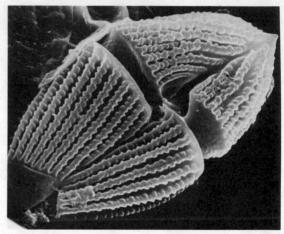

(b)

FIGURE B.3 *Dinogymnium,* a dinoflagellate from the Upper Cretaceous of the coastal plain of northern New Jersey. (Courtesy of F. S. May, from Ref. 3; copyright © 1976 by American Association for the Advancement of Science.)

## Sponges

Sponges are sessile (attached) organisms which live mainly in shallow, marine environments both as individuals and in colonies (Fig. B.7).

They are the most primitive of the multicellular animals and possess no nervous tissue, circulatory system, or digestive system. The body walls contain numerous small openings through which water flows into a series of canals in the wall of the sponge. After food particles are removed from the water in the canals, the water is forced into a central cavity and out of an opening in the top of the organism. A network of organic, calcareous, or sili-

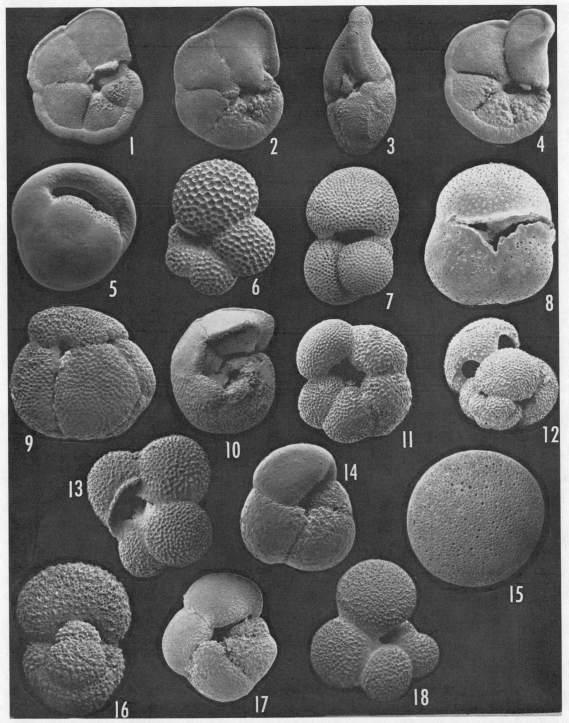

**FIGURE B.4** Foraminifera from the Recent of the Gulf of Mexico. (Courtesy of J. P. Kennett.)

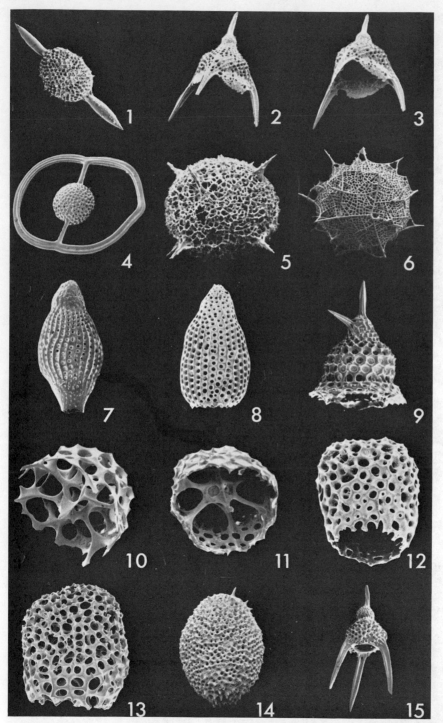

**FIGURE B.5** Radiolaria from a deep-sea core taken near Antarctica. (Courtesy of J. P. Kennett.)

TABLE B.1  **Important Microfossils**

| TYPE | | DESCRIPTION | GEOLOGIC RANGE | HABITAT | SIZE |
|---|---|---|---|---|---|
| Bacteria | | Cellulose<br>Single cells, chains | Precambrian—Recent | All | ca.1μ |
| Coccolithaceae | | Calcareous<br>Imperforate and perforate plates | Cretaceous—Recent | Marine | 2–3μ |
| Silicoflagellata | | Siliceous<br>Rings, nets, etc. | Cretaceous—Recent | Marine | 5–20μ |
| Dinoflagellata | | Cellulose, proteinaceous<br>Cell wall with furrows | Jurassic—Recent | Fresh, brackish, marine | 10–20μ |
| Desmidaceae | | Cellulose<br>Single cell constricted in middle | Tertiary—Recent | Fresh water | 0.1–0.5 mm. |
| Diatomaceae | | Siliceous<br>Single cell, discoidal, rhombic, elliptic, triangular | Cretaceous—Recent | Fresh, brackish, marine | 5–60μ |
| Charophyta | | Calcareous<br>Spheres, ovoids, spiral aspect | Devonian—Recent | Fresh, brackish | 1–3 mm. |
| Spores | | Cellulose<br>Single or groups, spheres, ovoids, etc. | Precambrian—Recent | Terrigenous, fresh, brackish, marine | 1–900μ |
| Pollen | Gymnospermae | Cellulose<br>Single cells, several cells, masses | Devonian—Recent | Terrigenous, fresh, brackish, marine | 10–130μ |
| | Angiospermae | | Jurassic—Recent | | |
| Discoasteridae | | Calcareous<br>Discs, asters | Up. Cretaceous—Tertiary | Marine | 20–35μ |
| Hystrichosphaerida | | Proteinaceous<br>Spheres, ovoids, spiny, smooth | Precambrian—Recent | Brackish, marine | 10–400μ |
| Chitinozoa | | Proteinaceous?<br>Tubes, flasks, urn-shaped | Cambrian–Mississippian | Marine | 50μ–1 mm. |
| Foraminifera | | Calcareous, arenaceous, chitinous<br>Simple to complex chambers, and chamber linings | Cambrian–Recent | Marine | 30 |
| Radiolaria | | Siliceous<br>Single cells, spherical or bell-shaped | Precambrian—Recent | Marine | 50–500μ |
| Ostracoda | | Calcareous, chitinous<br>Carapaces, appendages | Ordovician—Recent | Fresh, brackish, marine | 0.5–5 mm. |
| Scolecodonts | | Chitinous<br>Jaws, cusps | Cambrian–Recent | Brackish, marine | 50μ–5 mm. |
| Conedonts | | Phosphatic<br>Plates, cuspate jaws | Cambrian—Triassic | Marine | 100μ–3 mm. |

Adapted from Wilson, L. R., Ref. 1.

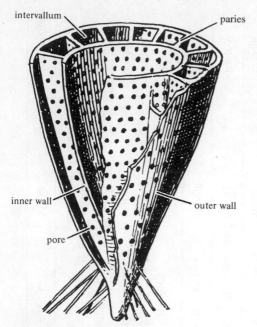

FIGURE B.6 Diagrammatic sketch of an archaeocyathan. These fossils have porous inner and outer walls composed of calcium carbonate. (From Moore *et al.,* Ref. 4; copyright © 1952 by McGraw-Hill, by permission of McGraw-Hill Book Co.)

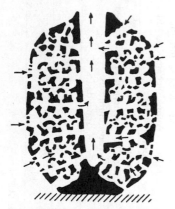

FIGURE B.7 Diagrammatic cross section of a sponge. Arrows indicate direction of water circulation. (After de Laubenfels, Ref. 5.)

(a)

(b)

FIGURE B.8 Modern hexacorals. (a) *Stylopora pistillate,* a collonial coral. (Courtesy of J. W. Wells.) (b) A solitary coral.

## Coelenterates

The coelenterates include hydras, jellyfish, corals, and stromatoporoids. This group is characterized by tissues that are not differentiated into organs. Most coelenterates have stinging tentacles that are used to seize and kill small organisms for food.

Jellyfish are free-swimming forms that lack a rigid skeleton. Although they lack hard parts, molds and carbon films of jellyfishlike forms have been preserved as fossils.

ceous spicules imbedded within the tissue walls acts as a skeletal framework. Sponges may be preserved as casts or molds, but commonly only the more resistant spicules are preserved.

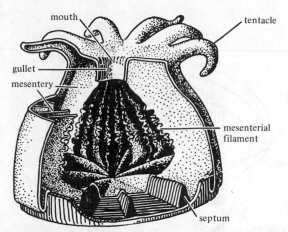

**FIGURE B.9** Diagrammatic sketch of a coral polyp. (From Moore *et al.,* Ref. 4; copyright © 1952 by McGraw-Hill, Inc. Used by Permission of McGraw-Hill Book Co.)

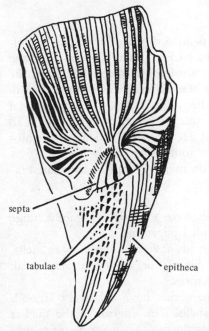

**FIGURE B.10** Diagrammatic sketch of the exoskeleton of a rugose coral. (From D. Hill, Ref. 5.)

Corals have been important reef-building organisms since the early Paleozoic. They occur as individuals (solitary corals) and in colonies that may be several feet across (Fig. B.8). The fleshy part of the coral, the *polyp,* possesses a number of armlike tentacles which surround a slitlike mouth (Fig. B.9). Most corals have a hard, calcareous exoskeleton that is the only part of the coral preserved in the fossil record. Sea anemones, which belong to the same class as the stony corals, do not have an exoskeleton.

In many corals, calcareous *septa* radiate inward from the outer walls or *theca* of the exoskeleton (Fig. B.10). There are four primary septa in the Paleozoic corals of the order Rugosa, whereas there are six primary septa in the Mesozoic and Cenozoic hexacorals of the order Scleractinia. Additional septa are added between the primary septa so that septa occur in multiples of four in the rugose corals and in multiples of six in the hexacorals. Tabulate corals have platforms or *tabulae* within the skeleton that are added during successive growth stages of the coral. Other corals have a series of bubble-like partitions called *dissepiments* within the skeleton.

## Bryozoans

Almost all bryozoans are colonial organisms. Some secrete a branchlike calcareous exoskeleton, while others occur as incrustations on other marine forms. The bryozoan polyps bear a superficial resemblance to coral polyps, but are generally smaller and more highly developed (Fig. B.11). Bryozoan polyps have more complex nervous and digestive systems than the corals.

## Brachiopods

The brachiopods are anatomically related to the bryozoans, but their skeletal morphology is totally different. The shells of brachiopods consist of two unequal valves. The valves are symmetrical about a plane perpendicular to the hinge line (Fig. B.12). Living brachiopods are attached to the seafloor by means of a fleshy stalk called the *pedicle* that protrudes through an opening in the pedicle valve of the shell

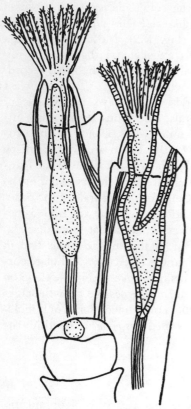

**FIGURE B.11** Diagrammatic sketch of a bryozoan polyp. (From R. S. Bassler, Ref. 7.)

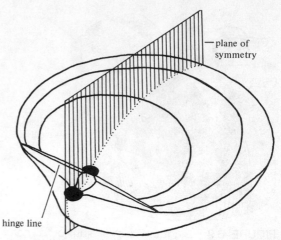

**FIGURE B.12** Diagrammatic sketch of a brachiopod shell, dorsolateral view. (From Williams and Rowell, Ref. 8.)

(Fig. B.13). The other valve, known as the brachial valve, contains the brachidium, an internal support for the animal's tissues. The external ornamentation of the shell may include concentric growth lines and *plications* (ridgelike features) radiating from the beak area.

The two classes of brachiopods, Inarticulata and Articulata, are differentiated on the basis of the manner in which their valves are connected. The valves of inarticulate brachiopods have no definite hinge structures, whereas those of the articulate brachiopods have a definite hinge line with teeth along it. Of the two classes of brachiopods, the inarticulate variety is the more primitive. Their valves are composed of chitin, whereas the valves of articulate brachiopods are generally composed of a double layer of calcite.

Teeth of the pedicle valves of articulate brachiopods fit into sockets on the brachial valves. The contraction of muscles attached from the pedicle to the brachial valve opens the valves in a leverlike fashion (Fig. B.13). When the animal dies and the muscles relax, the valves tend to remain closed. Hence, fossil articulate brachiopods are often found intact.

There are six major groups of articulate brachiopods:

1. The orthids have a relatively straight hinge line and radial plications.
2. Strophomenids have concavo-convex shells, a straight hinge line, and relatively weak radial structures.
3. Most pentamerids have relatively smooth biconvex shells. The interior of the shell is divided near the beak with a series of radiating partitions which are used for the attachment of muscles to the shell.
4. Spiriferids have a spiral coiled internal support for the brachia.
5. Rhynchonellids generally have pointed beaks and a biconvex shell with prominent plications.

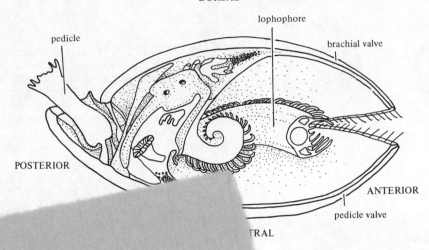

DORSAL

lophophore

pedicle

brachial valve

POSTERIOR

ANTERIOR

pedicle valve

TRAL

**FIGURE B.13** Diagrammatic sketch of the internal organs of a brachiopod. (From Williams and Rowell, Ref. 8.)

...hells

...aped

...ollusks

The varied classes of mollusks have provided many important index fossils. The most important are the pelecypods (clams, muscles, oysters, and scallops), gastropods (snails), and cephalopods (nautiloids, ammonoids, belemnoids, octopuses and squids). Mollusks possess well-developed nervous and circulatory systems, and they are more advanced than the brachiopods.

PELECYPODS.    Pelecypods superficially resemble brachiopods in that they have two valves (Fig. B.14). However, the pelecypod shells are nearly equal in size and shape, and they are commonly symmetrical about a plane passing between the valves. Furthermore, the internal structure is entirely different from that of brachiopods. The valves of the pelecypod close by muscle contractions and open when the muscle releases. When the organism dies, the two valves of the shells open and are commonly separated. Thus, fossil pelecypods generally consist of single valves. Pelecypod shells have three layers: an outer chitinous layer, a middle layer composed of prismatic calcium carbonate, and an inner layer of laminated aragonite. The inner layer often exhibits a mother-of-pearl luster. Pelecypods are found in both salt and fresh water.

GASTROPODS.    The gastropods are a diverse group both in morphology and in habitat. Most have coiled shells, although certain groups have reduced or eliminated their shells. They are found in marine, freshwater, and terrestrial environments. The terrestrial forms have air-breathing lungs and occupy nearly all terrestrial environments.

CEPHALOPODS.    The cephalopods are represented by both coiled and straight marine forms. They may have external shells, as in the nautiloids and ammonoids, an internal skeleton, as in belemnoids and squids, or only restricted skeletal structures, as in octopuses (Fig. B.15). Nautiloid and ammonoid shells have chambers which represent growth stages and which are separated by partitions or *septa*. A tubelike siphuncle connects the gas-filled chambers to the living animal. The cephalopod regulates its buoyancy by adjusting the amount of gas in the

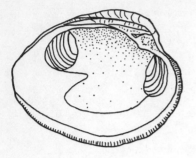

**FIGURE B.14** Diagrammatic sketch of the pelecypod *Venus mercenaria*. (From Shrock and Twenhofel, Ref. 9; copyright © 1953 by McGraw-Hill. Used by permission of McGraw-Hill Book Co.)

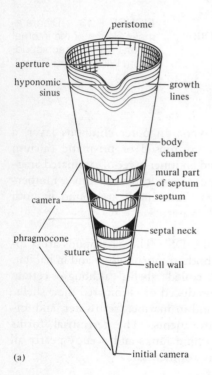

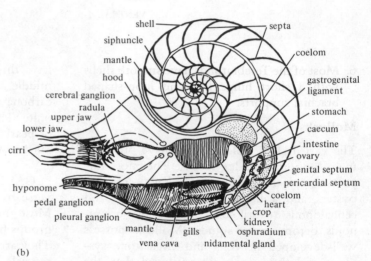

**FIGURE B.15** Diagrammatic sketches of nautiloid cephalopods. **(a)** Exoskeleton of a straight nautiloid. **(b)** Median longitudinal section of a coiled nautiloid. (From Teichert and Moore, Ref. 10.)

chambers. The line of intersection of the septa with the test is the *suture* (Fig. B.16). The increasing complexity of cephalopod sutures from late Paleozoic through Mesozoic time has provided a nearly continuous zonation for marine sequences.

Belemnoids are an extinct group of cephalopods which resembled the modern squid. They had a bullet-shaped internal skeleton consisting of a chambered phragmocone and a calcareous guard (Fig. 12.38). The guards are common fossils in Mesozoic rocks.

## Annelids and Similar Organisms

Annelids are soft-bodied, segmented worms. Their fossil remains include carbon films on black shale, trails and burrows, and very small toothlike jaws called scolecodonts. Conodonts are similar in size and appearance to scolecodonts, but their biological affinity is somewhat uncertain (Fig. B.17). A recent discovery of the carbonized remains of conodont animals (**12**) indicates that they were bilaterally symmetrical, free swimming, and probably fed

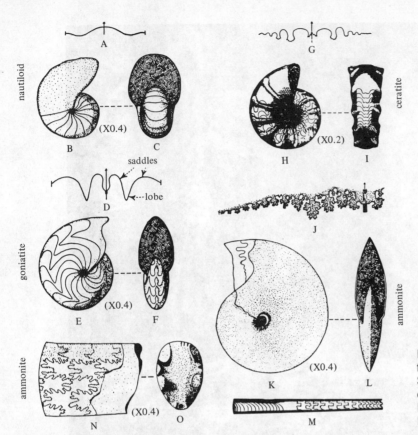

**FIGURE B.16** Sutures on four types of cephalopods. (From Shrock and Twenhofel, Ref. 9; copyright © 1953 by McGraw-Hill, Inc. Used by permission of McGraw-Hill Book Co.)

on phytoplankton at or near the surface of the water. The toothlike conodonts were located within the digestive tract and may have acted as a filtering system, retaining some food particles and rejecting others. The conodont animals that have been found are about 70 mm long and they do not seem to be directly comparable to any living or fossil animal (**12**) (Fig. B.18). They may belong to the protochordates, and may have been a direct ancestor to the vertebrates.

## Arthropods

The arthropods are a relatively advanced group of invertebrates that includes insects, spiders, scorpions, centipedes, crustaceans, and trilobites. They are characterized by the presence of a segmented exoskeleton composed of calcium carbonate or chitin.

For the purposes of correlation, trilobites are one of the most important fossil groups. The name refers to the three lobes extending the length of the body. The exoskeleton of the trilobite is divided into the head (cephalon), midsection (thorax), and tail (pygidium) (Fig. B.19). Like many other arthropods, trilobites molted their shells as they grew, and therefore one organism could produce many fossil remains. Frequently, however, the cephalon, thorax, pygidium became separated either after death or after molting and complete trilobites are somewhat uncommon. Trilobites are classified according to shell morphology, such as the relative size of the cephalon and pygidium, the number of body segments, and the presence or absence of eye structures.

**FIGURE B.17**   Ordovician conodonts from western Newfoundland (X28). (From L. E. Fahraeus, Ref. 11.)

FIGURE B.18 A carbon replica of a conodont animal about 6 cm (2 in) long from a limestone of Mississippian age in central Montana. Type specimen, University of Montana Museum of Paleontology. (Collected by Melton and Horner in 1969, photo from Ref. 12, courtesy of W. Melton.)

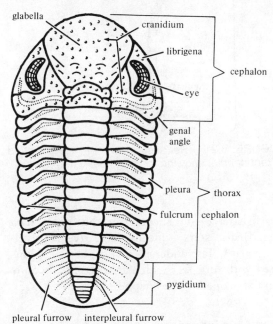

FIGURE B.19 Diagrammatic sketch of the exoskeleton of a trilobite. (From H. J. Harrington, Ref. 13.)

Other significant arthropods in the fossil record are the eurypterids, ostracods, and insects. Eurypterids were scorpionlike forms that had a segmented exoskeleton (Figs. 10.48 and 10.49). They were one of the largest of the Paleozoic invertebrates and reached a maximum length of 6 feet (2 meters). Ostracods are bivalved crustaceans whose tests range in size from a few hundredths of a millimeter to a centimeter or two (Fig. B.20). They are useful in correlations since most species have a short time span and they were widely distributed in fresh, brackish, and marine environments. Their tests are common enough and small enough to be retrieved from well cuttings and cores.

## Echinoderms

The echinoderms are advanced invertebrates that typically possess a bilateral symmetry or a fivefold radial symmetry. The major representative groups are the crinoids, cystoids, asteroids, and echinoids.

Most crinoids, cystoids, and blastoids attach themselves to the sea bottom by rootlike appendages. The rootlike structures and the presence of a "stem" and a tentacle-bearing calyx

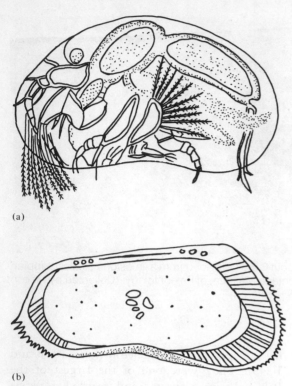

(a)

(b)

**FIGURE B.20** Diagrammatic sketch of the modern ostracod, *Cypris*. **(a)** Interior anatomy. **(b)** Interior of left valve. (From D. L. Jones, Ref. 14.)

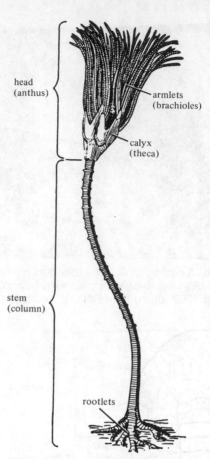

head (anthus)

armlets (brachioles)

calyx (theca)

stem (column)

rootlets

**FIGURE B.21** Reconstruction of the blastoid *Orophocrinus* showing the principal parts. (From O. Fay, Ref. 15.)

give these organisms an appearance that is somewhat like that of a plant (Figs. B.21 and B.22). The tentacles are used to trap food and are commonly arranged in multiples of five. Crinoid stems are composed of a series of disk-shaped segments with a round or star-shaped central canal. When the animal dies, the stem segments are generally separated. Fragments of the stem are abundant fossils in many limestones.

The echinoids are unattached and include sea urchins and sand dollars. These forms have circular or heart-shaped shells which may be covered with spines. Some spines exhibit distinctive morphological features and are useful as guide fossils.

The asteroids, which include starfish and brittle stars, are mobile benthonic forms. Most starfish have five short arms, but some have ten. Brittle stars typically have five relatively long arms. The undersides of the arms are covered with tube feet which are used for locomotion and opening pelecypod shells.

## The Hemichordates

Intermediate between the invertebrates and the vertebrates are the hemichordates, which possess a cartilaginous, rodlike supporting structure, the *notochord*. While the notochord is not a backbone, it is believed to represent an early stage in the development of a spinal column.

Graptolites are a group of extinct colonial organisms that were formerly classified with corals and bryozoans, but modern fossil extraction methods have shown that the graptolites are closely related to certain living colonial hemichordates which have a chitinous exoskeleton. Graptolite colonies consist of a stemlike structure, the nema, with a number of cups, *zoecia,* that house polyplike organisms. The individuals were apparently interconnected and shared biologic functions. The graptolite colonies consist of one or more branchlike structures called *stipes* that were often attached to a floating organ or to seaweed.

## Vertebrates

The vertebrates belong to the phylum Chordata. Members of this phylum are characterized by the possession at some time during their life histories of a notochord as the primary axial skeletal element and a dorsal, tubular nerve cord. Vertebrates are chordates that have a vertebral column. Vertebrates, like all life, evolved from simple, one-celled organisms. The earliest of these were probably similar to modern bacteria or algae. These evolved into ciliates or flagellates which are one-celled organisms with cilia or flagella used for locomotion. From these, complex protozoa or colonial protozoa evolved. In these forms, there was probably limited specialization of cells.

The first true multicellular relatives of the vertebrates were probably hydralike coelenterates related to the modern jellyfish and corals. From these, flatwormlike organisms similar to the platyhelminthids (such as planarians) evolved. These gave rise to an organism ancestral to the echinoderms. There are several important similarities between the vertebrates and the echinoderms: embryonic development, the presence of a dorsal nerve cord, and a similar biochemistry. From the echinoderm ancestor, an organism resembling the modern sea-squirt evolved. Sea-squirts have a notochord in their larval stage. These gave rise to a form which

**FIGURE B.22** The modern crinoid *Cenocrinus asterius* frp, Discovery Bay, Jamaica. This is an isocrinoid 1 m (3 ft) high with an arm span across the disk of 60 cm (24 in) across. (Photographed from the *Nekton Gamma* in 1972, courtesy of D. B. Macurda, Jr.)

probably resembled the modern *Amphioxus,* a small chordate which for the most part lives buried in the sand (though it can swim). *Amphioxus* has a well-developed notochord and is a filter feeder. Primitive fish probably evolved from an *Amphioxus*-like organism.

FISH. The first vertebrates were jawless fish known as ostracoderms. They are first found in Late Cambrian deposits, and they reached their peak during the Silurian. They are represented today by the lamprey and hagfish. Ostracoderms were covered by a bony armor plate that is generally all that is preserved of the organism. They were jawless and had a single nostril high on their head. They were small (up to one foot in length), nonpredatory, filter feeders. They lived by straining food through their gills. They were probably sluggish animals, spending much of their life on the sea bottom.

The ostracoderms gave rise to a group of jawed fish known as placoderms. This group included both predatory and herbivorous forms. They first appeared during the Silurian, reached their peak during the Devonian, and became extinct by the end of the Permian. The placoderms can be divided into two groups, the acanthodians, or spiny sharks, and the arthrodires, or jointed-neck fish. The acanthodians were sharklike in appearance, fully clad in bony scales, and had paired fins composed of spines with a small web of skin between them. The arthrodires were more common. Their head and gill regions were covered by a great bony shield. Another ring of bony armor sheathed much of the rest of the body.

Sharks and rays are thought to stem from the placoderms. Sharks have a cartilaginous skeleton and form an important group in the modern marine environment. The bony fish include the ray-finned fish, the lobe-finned fish, and the lungfish. Of these groups, the most prolific are the ray-finned fish which have been found in a wide variety of marine and freshwater environments. This group constitutes the majority of modern fish.

AMPHIBIANS.  Labyrinthodonts are a primitive group of extinct amphibians that have intricately infolded tooth enamel. Since lobe-finned fish possess similar teeth, it is thought that the amphibians evolved from lobe-finned fish (Fig. B.23). Amphibians are cold-blooded and spend part of their lives in water. Most amphibians are covered with scales, which shows their close relationship to fish. They lay their eggs in water and the young hatch and develop in water. Young amphibians breathe by means of gills, but later develop lungs.

REPTILES.  The reptiles emerged late in the Carboniferous through continued adaptation of the amphibian stock to a terrestrial existence. The evolutionary improvement in this case was the development of the fertilized, shelled egg (the amniote egg) from which the young hatch on dry land. The shell protects the egg from drying out and thus frees the reptiles from dependence on water. Furthermore, the reptilian young are placed out of the grasp of aquatic predators. Like amphibians, reptiles are cold-blooded. They have a bony skeleton and are generally covered by scales or bony plates. Reptiles are thought to have evolved from labyrinthodont amphibians.

BIRDS.  Birds are warm blooded animals that have such structures as wings and hollow bones that allow them to fly. They appear reptilian in that they lay eggs, but they resemble mammals in their feeding and care for their young.

MAMMALS.  Mammals are warm-blooded animals, most of which are covered with hair or fur. Almost all mammals give birth to live young, which they suckle with milk. However, there are two primitive mammalian forms that lay eggs, the duckbilled platypus and the echidna, an ant-eating mammal. The origin of the mammals is traced to mammal-like reptiles that gave rise to the monotremes, marsupials, and insectivores (Fig. B.23). The placental mammals, which give birth to live young, were derived from the insectivores, a group of primitive mammals. The insectivores were of small size, much like their modern form, the shrew.

## MULTICELLULAR PLANTS

The two major groups of multicellular plants are the bryophytes and the tracheophytes. The bryophytes include mosses and liverworts. While they are not a very important group in the fossil record, the mosses are useful as ecological indicators.

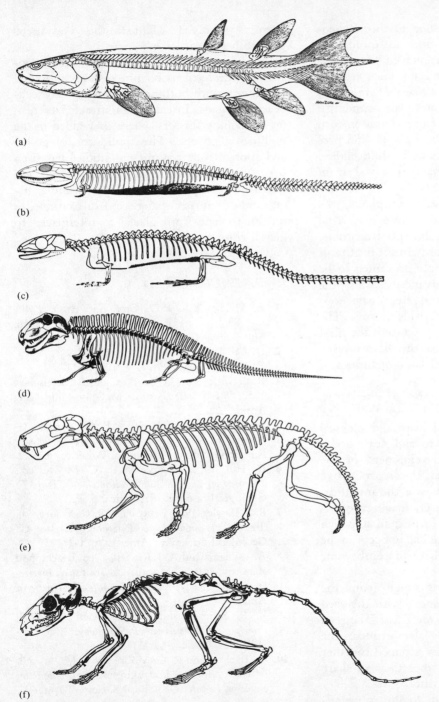

**FIGURE B.23** Series of skeletons illustrating the evolution of placental mammals from crossopterygians: **(a)** *Eusthenopteron,* a crossopterygian. **(b)** *Pholidogaster,* an early labyrinthodont. **(c)** *Hylonomus,* one of the earliest reptiles. **(d)** *Sphenacodon,* a pelycosaur. **(e)** *Lycaenops,* a primitive therapsid. **(f)** *Tupia,* a tree shrew. (From Romer, Ref. 16; copyright © 1967 by the American Association for the Advancement of Science.)

The vascular plants belong to the tracheophytes. These plants are more advanced than the bryophytes, and they evolved from one-celled, marine nonvascular plants such as blue-green algae. The one-celled marine forms gave rise to nonvascular freshwater plants similar to modern freshwater green algae. These in turn gave rise to nonvascular land plants, and perhaps to mosses and fungi as well. Multicellular, vascular land plants probably evolved from these nonvascular land plants.

The most primitive vascular plants, the psilopsids, are the oldest known terrestrial plants. These have stems, but no true roots, small leaves, if any, and spore cases at the tips of the stems. The psilopsids are presumed to be ancestral to the more advanced forms, the lycopsids (club moss and scale trees), sphenopsids (horse-tails) and pteropsids (ferns). The lycopsids and pteropsids produced the first treelike plants and formed the first forests. These plants also produced large quantities of spores in their reproductive cycles.

The sphenopsids gave rise to the ferns, which in turn gave rise to the seed ferns. The development of the seed was a biological breakthrough comparable to and nearly contemporaneous with the development of the shelled egg. Important in the reproductive process in the higher plants was the modification of the spore to an entirely male reproductive product, the pollen, formed in a pollen cone. Fertilization of the egg had to occur in the female cone or flower, where the embryonic seed then developed.

The seed-bearing plants range from the lower Carboniferous beginning with the seed ferns which evolved from the ferns. The seed ferns, conifers, and two later groups, the ginkgoes and the cycads, are grouped together with other gymnosperms since they all share the "naked seed" characteristic.

The seed ferns gave rise to the more advanced gymnosperms, which include the modern pines, spruce, fir, and cedar. The angiosperms evolved from the advanced gymnosperms.

In the most advanced plants, the angiosperms, the fertilization process was modified by flowers with both pollen and egg and seed-bearing organs. Insects were attracted to nectar-containing flowers where they aided in the pollination process. The application of pollen and spores to environmental studies has been particularly beneficial in tracing forest and grassland composition and determining sequences of climatic change, as, for example, in the Pleistocene from glacial to interglacial to glacial climates.

## REFERENCES CITED

1. E. Y. Haworth, 1972, *Geol. Soc. Amer. Bull.,* v. 83, p. 157; L. R. Wilson, 1956, *Micropaleontology,* v. z, pp. 1–6.
2. D. Lal, 1977, *Science,* v. 198, p. 997.
3. F. S. May, 1976, *Science,* v. 193, p. 1128.
4. R. C. Moore, C. G. Lalicker, and A. G. Fischer, 1952, *Invertebrate Fossils:* McGraw-Hill, New York.
5. N. D. de Laubenfels, 1955, Porifera, *in* R. C. Moore, ed., *Treatise on Invertebrate Paleontology, Part E:* Geological Society of America, p. E22.
6. D. Hill, 1956, Rugosa, *in* R. C. Moore, ed., *Treatise on Invertebrate Paleontology, Part F:* Geological Society of America, p. F233.
7. R. S. Bassler, 1953, Bryozoa, *in* R. C. Moore, ed., *Treatise on Invertebrate Paleontology, Part G:* Geological Society of America, p. G1.
8. A. Williams and A. J. Rowell, 1965, Brachiopod anatomy, *in* R. C. Moore, ed., *Treatise on Invertebrate Paleontology, Part H:* Geological Society of America, p. H6.
9. R. R. Schrock and W. H. Twenhofel, 1953, *Principles of Invertebrate Paleontology:* McGraw-Hill, New York.
10. C. Teichert and R. C. Moore, 1964, Mollusca 3, *in,* R. C. Moore, ed., *Treatise on Invertebrate Paleontology, Part K:* Geological Society of America, p. K2.
11. L. E. Fahraeus, 1970, *Geol. Soc. Amer. Bull.,* v. 81, p. 2061.

**12.** W. G. Melton and H. W. Scott, 1972, *Conodont animals from the Bear Gulch Limestone, Montana:* Geological Society of America Special Paper 141, p. 31.

**13.** H. J. Harrington, 1959, General description of Trilobita, *in* R. C. Moore, ed., *Treatise on Invertebrate Paleontology, Part O:* Geological Society of America, p. O38.

**14** D. J. Jones, 1956, *Introduction to Microfossils:* Harper & Row, New York.

**15.** O. Fay, 1967, Introduction, *in* R. C. Moore, ed., *Treatise on Invertebrate Paleontology, Part S:* Geological Society of America, p. S298.

**16.** A. S. Romer, 1967, *Science,* v. 158, p. 1629, Dec. 29.

# Glossary

THIS GLOSSARY contains only the definitions of common rocks and textural terms relating to them. All other geologic terms have been defined in the text itself. Such terms are listed in the index. The page on which the definition is given is in boldface and the word defined is set in italic type in the text.

*amphibolite*. A metamorphic rock formed by intense (high-grade) metamorphism of a basalt. Principal minerals present include hornblende and plagioclase in approximately equal percentages.

*andesite*. An extrusive (volcanic) rock which is intermediate in composition between a rhyolite and a basalt. It is compositionally the equivalent of a diorite.

*anorthosite*. A plutonic (intrusive) igneous rock made up almost entirely of the mineral plagioclase. It is similar in composition to an andesite, but is not quite as rich in iron and magnesium.

*arkose*. Sandstone containing more than 15 percent feldspar.

*basalt*. An extrusive (volcanic) rock relatively rich in iron and magnesium. It is the common volcanic rock on many volcanic islands, such as Hawaii. It is compositionally the equivalent of gabbro and diabase.

*breccia*. A rock composed of angular fragments formed by fracturing of preexisting rocks. There are four types of breccias: fault breccias, volcanic breccias, impact breccias, and sedimentary breccias. *Fault breccias* are formed during faulting. *Volcanic breccias* are formed by explosive igneous activity. *Impact breccias* are formed during the impact of a large meteorite. *Sedimentary breccias* are formed by collapse of the roof of a cave or other similar processes.

*calcareous*. Composed of calcite (calcium carbonate).

*clastic*. Composed of fragments of rocks, minerals, or organic materials (such as broken shells). Sedimentary rocks are either clastic or crystalline.

*clay*. An unconsolidated sediment composed almost entirely of grains less than 1/256 mm in size. Most clay is composed of clay minerals.

*conglomerate*. A clastic sedimentary rock consisting chiefly of grains more than 2 mm in size.

*crystalline*. A type of sedimentary rock formed by direct precipitation of minerals from seawater. Crystalline also refers to coarse-grained igneous and metamorphic rocks such as granite, gneiss, and schist.

*diabase*. An igneous rock relatively rich in iron and magnesium intruded at a shallow depth; it is equivalent in composition to a basalt or gabbro. The grain size is generally intermediate between these two rock types.

*diamictite*. A sediment or sedimentary rock

which is poorly sorted and unlayered. It may form from glacial processes (a till is an example of a diamictite), by mudflows, or submarine landslides.

*diorite.*  A plutonic igneous rock intermediate in composition between a granite and a gabbro. It is the intrusive equivalent of an andesite.

*dolerite.*  Another term for diabase.

*dolomite.*  A sedimentary rock composed mostly of the mineral dolomite (calcium magnesium carbonate). It is generally thought to form by the replacement of limestone or a lime mud.

*dunnite.*  An intrusive igneous rock composed almost entirely of the mineral olivine.

*extrusive.*  Igneous rocks formed from lava poured out on the surface of the earth.

*felsic.*  Rich in feldspar or in the chemical components in feldspar (sodium, potassium, silicon, aluminum, and calcium).

*flint.*  A sedimentary rock, often grey or black in color, which is composed of microscopic crystals of quartz. Flint is a type of chert and breaks with a conchoidal (smooth curving) fracture.

*foliation.*  A planar structure within a metamorphic rock often formed by the parallel orientation of platy minerals such as mica in a schist or by alternating light and dark layers as in a gneiss.

*gneiss.*  A layered metamorphic rock formed by intense (high-grade) metamorphism of an igneous or sedimentary rock. Two types of gneisses are orthogneisses and paragneisses.

*granite.*  A felsic plutonic igneous rock containing abundant quartz and feldspar (microcline and/or plagioclase) and with a relatively low percentage of dark minerals (biotite, hornblende, or pyroxene). Granite may also form by metasomatism of a high-grade metamorphic rock. Granite is the intrusive equivalent of a rhyolite.

*gravel.*  Unconsolidated sediment consisting chiefly of grains more than 2 mm in size. Conglomerate is consolidated gravel.

*graywacke.*  A sedimentary rock composed of sand-sized grains of quartz, feldspar, and/or rock fragments and having more than 15 percent matrix between the sand-sized grains.

*greenstone.*  A metamorphic rock formed by moderate (low-grade) metamorphism of a basalt or andesite. The minerals actinolite (a green amphibole), chlorite, and epidote give greenstone its green color.

*high-grade.*  Formed by metamorphism at high temperatures and pressures.

*igneous.*  Formed by the crystallization of a magma (molten rock).

*intrusive.*  Formed by the intrusion of a magma beneath the earth's surface.

*iron formation.*  A sedimentary rock consisting of one or more iron-rich minerals (such as hematite, magnetite, greenalite, or siderite) interbedded with chert (which is often red in color).

*jasper.*  A sedimentary rock consisting of red chert.

*lime mud.*  Mud consisting of clay- and silt-sized grains of calcite (calcium carbonate). Many limestones form from the lithification of lime mud.

*limestone.*  A sedimentary rock composed for the most part of calcite (calcium carbonate). Limestone may be formed by direct precipitation from seawater, but most probably it forms by the accumulation of clastic grains of calcite formed by fragmentation of calcarous organic skeletons.

*lithification.*  The process of consolidation of loose sediment into a coherent sedimentary rock.

*low-grade.*  Formed by metamorphism at moderate temperatures and pressures.

*mafic.*  A mafic rock is one rich in magnesium and iron. Basalt, diabase, and gabbro are mafic rocks.

*magma.*  A molten (melted) rock such as a lava.

*marble.*  A rock formed by the high-grade metamorphism of a limestone (calcite marble) or dolostone (dolomite marble).

*matrix*. The finer material between larger grains.

*metamorphic rocks.* Rocks formed as a result of the change of preexisting sedimentary or igneous rock under the influences of temperature and/or pressure. Changes during metamorphism may include reduction in water content, increase in grain size, formation of new minerals, and increase in the degree of bonding of the mineral grains.

*metasomatism*. Chemical alteration of a rock involving the addition of some substances and the removal of others. Granitization is a type of metasomatism in which a granite is produced by the addition of silicon, potassium, and sodium while iron and magnesium are removed.

*mud*. An unconsolidated sediment composed of silt- and clay-sized particles.

*orthogneiss*. A gneiss formed by the metamorphism of an igneous rock such as granite.

*paragneiss*. A gneiss formed by the metamorphism of a sedimentary rock such as a graywacke.

*peridotite*. An intrusive igneous rock composed dominantly of olivine, but also containing significant amounts of pyroxene (enstatite and diopside). Some peridotites also contain garnet, plagioclase, and hornblende, and most contain small amounts of chromium-rich spinel (picotite). The mantle of the earth is thought to consist mostly of peridotite.

*phyllite*. A metamorphic rock usually formed by the low-grade metamorphism of a shale. It is coarser in grain size than a slate, but finer than a schist. The grains are too small to be seen with the naked eye, but the rock has a strong sheen.

*plutonic*. Formed at great depth, as a granite.

*porphyritic*. A textural term applied to igneous rocks having large grains set in a fine-grained matrix.

*porphyry*. An igneous rock with a porphyritic texture.

*precipitation*. The process of crystallization of a mineral from an aqueous (water-rich) solution. Evaporation of seawater will result in the precipitation of salt crystals on the bottom of the sea or on the bottom of a container.

*quartzite*. A rock formed by the metamorphism of a quartz sandstone.

*quartz sandstone*. A sandstone composed almost entirely of quartz grains.

*rhyolite*. A felsic extrusive (volcanic) rock which is compositionally the equivalent of a granite.

*sand*. An unconsolidated sediment composed of sand-sized (2 mm to 1/16 mm) grains. The grains may be composed of quartz, feldspar, rock fragments, fragments of broken skeletons, or any other sand-sized material.

*sandstone*. A sedimentary rock composed of sand-sized grains. The grains may be quartz, feldspar, rock fragments, fragments of broken skeletons, etc.

*schist*. A rock formed by the intense (high-grade) metamorphism of a shale. It is generally rich in mica and shows a strong foliation. Grains of mica may be seen with the naked (unaided) eye.

*sediment*. Unconsolidated (uncemented, unlithified) materials such as sand, clay, gravel, etc.

*sedimentary rock*. A rock formed either by the accumulation of mineral, rock, or organic fragments, or by the chemical precipitation of minerals on the sea bottom.

*serpentinite*. A rock consisting chiefly of the mineral serpentine. Serpentinite is generally formed by the alteration of peridotite by hot aqueous solutions.

*shale*. A sedimentary rock formed by the lithification of a mud.

*silt*. An unconsolidated sediment consisting chiefly of grains 1/16 to 1/256 mm.

*slate*. A rock formed by the metamorphism of shale at moderate temperatures and pressures. This rock is finer in grain size than a phyllite, but coarser than a slate. It has a slight sheen, but the grains are too small to be seen with the naked eye.

*syenite*. An intrusive igneous rock rich in feld-spar (especially microcline), but poor (less than 10 percent) in quartz.

*till*. An unconsolidated sediment consisting of unsorted, unlayered debris deposited by a glacier. Till often contains striated, faceted pebbles and boulders.

*tillite*. Rock formed by the consolidation (lithification) of till.

*ultramafic*. Very rich in iron and magnesium. Peridotite and dunnite are ultramafic rocks.

*volcanic*. Formed by extrusive igneous activity as would occur near a volcano.

# Index

Page numbers in **boldface** identify the page where the term is defined. Definitions of common rocks and textural terms are given in the Glossary.

80  81  82  9  8  7  6  5  4  3  2